Post-Harvest Physiology and Crop Preservation

NATO ADVANCED STUDY INSTITUTES SERIES

A series of edited volumes comprising multifaceted studies of contemporary scientific issues by some of the best scientific minds in the world, assembled in cooperation with NATO Scientific Affairs Division.

Series A: Life Sciences

Recent Volumes in this Series

Volume 43 – Advanced Topics on Radiosensitizers of Hypoxic Cells
edited by A. Breccia, C. Rimondi, and G. E. Adams

Volume 44 – Cell Regulation by Intracellular Signals
edited by Stéphane Swillens and Jacques E. Dumont

Volume 45 – Structural Molecular Biology: Methods and Applications
edited by David B. Davies, Wolfram Saenger, and Steven S. Danyluk

Volume 46 – Post-Harvest Physiology and Crop Preservation
edited by Morris Lieberman

Volume 47 – Targeting of Drugs
edited by Gregory Gregoriadis, Judith Senior, and André Trouet

Volume 48 – Neurotransmitter Interaction and Compartmentation
edited by H. F. Bradford

Volume 49 – Biological Effects and Dosimetry of Nonionizing Radiation
edited by Martino Grandolfo, Sol M. Michaelson, and Alessandro Rindi

Volume 50 – Somatic Cell Genetics
edited by C. Thomas Caskey and D. Christopher Robbins

Volume 51 – Factors in Formation and Regression of the Atherosclerotic Plaque
edited by Gustav R. V. Born, Alberico L. Catapano, and Rodolfo Paoletti

This series is published by an international board of publishers in conjunction with NATO Scientific Affairs Division

A Life Sciences B Physics	Plenum Publishing Corporation London and New York
C Mathematical and Physical Sciences	D. Reidel Publishing Company Dordrecht, The Netherlands and Hingham, Massachusetts, USA
D Behavioral and Social Sciences E Applied Sciences	Martinus Nijhoff Publishers The Hague, The Netherlands

Post-Harvest Physiology and Crop Preservation

Edited by
Morris Lieberman
Late of Beltsville Agricultural Research Center
Beltsville, Maryland

PLENUM PRESS • NEW YORK AND LONDON
Published in cooperation with NATO Scientific Affairs Division

Library of Congress Cataloging in Publication Data

NATO Advanced Study Institute on Post-Harvest Physiology and Crop Preservation (1981: Sounion, Greece)
Post-harvest physiology and crop preservation.

(NATO advanced study institutes series. Series A, Life sciences; v. 46)
"Proceedings of a NATO Advanced Study Institute on Post-Harvest Physiology and Crop Preservation, held April 28-May 8, 1981, in Sounion, Greece"—T.p. verso.
Bibliogrpahy: p.
Includes index.
1. Food crops—Physiology—Congresses. 2. Food crops—Preservation—Congresses. I. Lieberman, Morris, 1919- . II. North Atlantic Treaty Organization. III. Title. IV. Series.
SB175.N37 1981 631.5′6 82-3645
AACR2

ISBN 978-1-4757-0096-1 ISBN 978-1-4757-0094-7 (eBook)
DOI 10.1007/978-1-4757-0094-7

Proceedings of a NATO Advanced Study Institute on Post-Harvest Physiology and Crop Preservation, held April 28-May 8, 1981, in Sounion, Greece

Softcover reprint of the hardcover 1st edition 1983
A Division of Plenum Publishing Corporation
233 Spring Street, New York, N.Y. 10013

To
all those who labored
and labor to preserve crops after
harvest for the benefit of mankind

MORRIS LIEBERMAN

Morris Lieberman, the Director of the NATO Advanced Study Institute on Post-Harvest Physiology and Crop Preservation, passed away after completing the editing of the manuscripts of the present book. The loss of this great scientist and fine personality is deeply regretted by his colleagues and friends.

Lieberman contributed considerably to the knowledge of post-harvest physiology, particularly to the understanding of the bio-synthetic pathway leading to the hormone that regulates plant organ senescence, ethylene. In the sixties, while working with Mapson at Cambridge, he discovered methionine as a substrate for this synthesis. It was at the Post-Harvest Plant Physiology Laboratory at Beltsville, which he directed, that rhizobitoxine analogs were found to inhibit the main regulatory enzyme of the pathway, aminocyclopropanecarboxy-lic-acid synthetase. Morris Lieberman died deeply involved in the further analysis of the enzyme systems of this pathway.

Next to his enthusiasm for science, Morris had a vivid interest in many aspects of life and was a very kind and friendly person with a beautiful sense of humor. Nothwithstanding his sharp insight his judgment was mild. It was a privilege to be among his friends, only few of whom knew that he was so ill that he nearly had to stay away from the Institute that was his very conception and which he had been fostering with such zeal. It gave him deep satisfaction to direct the Institute and to edit this volume. We may accept the book with gratitude and respect as a memorial to a great man.

Johan Bruinsma
Co-Director

PREFACE

Emphasis in agricultural research for many years has concentrated on crop production. This emphasis has become more important in recent years with the realization that the population worldwide is outstripping the food supply. There is, however, another side to increasing the availability of the food supply. This simply involves preservation of the harvested crop for human consumption. The losses incurred in harvesting, handling, transportation, storage and marketing crops have become a greater problem as the distance from the farm to the ultimate consumer increases. In the Western world where modern transportation, storage facilities, and marketing technology are widely used, post-harvest technology requires a large input of energy which increases costs considerably. Therefore, losses are more significant and the ability to provide fresh fruits and vegetables, out of season, at reasonable costs will depend on reduced post-harvest losses throughout the marketing chain from the farm gate to the ultimate consumer.

The reduction in post-harvest losses depends on proper use of current technology and further developments derived from a broad spectrum of scientific disciplines. Biochemistry, plant physiology, plant pathology, horticulture, agronomy, physics, engineering and agricultural economics, all provide knowledge which has been useful and will be useful in the future for improving post-harvest technology and crop preservation. This volume records the Proceedings of the NATO Advanced Study Institute on Post-Harvest Physiology and Crop Preservation, held at Sounion, Greece, April 28 - May 8, 1981. Lecturers at the Institute represented all of these scientific areas, providing a broad spectrum view of current basic and practical information which impact on post harvest technology and crop preservation.

Notice was also taken, at the Institute, of the special problems of crop preservation in developing countries wherein conditions are quite different from those in the Western developed world. Quite a different technology is applicable to developing world post-harvest problems. Nevertheless, basic scientific information should still be useful in establishing improvements which prevent losses within the context of the primitive systems used. The economics of some of these problems were also discussed in the course of the Institute. Hopefully, the information in this volume will provide basic knowledge of current and emerging technology which will be useful in advancing post-harvest technology and crop preservation worldwide.

The Institute was brought into being through the efforts of many dedicated individuals. I am especially indebted to my Co-Director, Prof. Bruinsma, and the Organizing Committee members: Prof. Bangerth, Prof. Côme, Prof. Dekazos, Prof. Dilley, Dr. Rhodes, and Prof. Vlitos. Their invaluable help in arranging the program, obtaining financial support in their own countries, and selecting students made our Institute a success. A special thanks is due Prof. Dekazos and Dr. Karaoulanis, members of the local committee, for selecting the site and making all the local arrangements.

In addition to the generous support and sponsorship of NATO, the Institute was co-sponsored by the Science and Education Administration of the U.S. Department of Agriculture and the Greek Ministry of Agriculture. Financial support was also provided by the following organizations and companies: Air Liquide; BASF AG; Bayer AG; Centre Technique Interprofessionnel des Fruits et Legumes; Compagnie des Entreports et Gares Frigorifiques; Duphar; E.I. Dupont de Nemours and Company; Froilabo; Greek Ministry of Culture and Science; Hellenic Organization of Tourism; Hoechst AG; Plenum Press; Societé Francaise de Transports et Entreports Frigorifiques; Tate and Lyle Ltd; and Velsicol Chemical Corporation. Generous contributions by these organizations provided support for additional students and lecturers.

Finally I wish to acknowledge my sincere appreciation to Delores Sessions who carried the burden of extensive correspondence and other matters related to the Institute and retyped virtually all the manuscripts in the required format. I accept full responsibility for all editorial errors, which, despite my efforts, probably exist.

Morris Lieberman
Beltsville, Maryland
October 1981

CONTENTS

THE GENERAL BIOLOGY OF PLANT SENESCENCE AND THE ROLE OF NUCLEIC ACID IN PROTEIN TURNOVER IN THE CONTROL OF SENESCENCE PROCESSES WHICH ARE GENETICALLY PROGRAMMED

Harold W. Woolhouse

John Innes Institute
Colney Lane
Norwich NR4 7UH, U.K.

PART I. SOME GENERAL PROBLEMS CONCERNING THE BIOLOGY OF SENESCENCE

Definitions

There are two terms "ageing" and "senescence" which are widely used in reference to changes which impair the structure or functioning of living organisms. Medawar (1) defined ageing as referring to all those changes which occur in time, without reference to death as a consequence, indeed its use need not be confined to living organisms. This is a convenient definition in that it allows of a clear distinction of senescence as describing those changes which lead sooner or later to the death of an organism or some part of it. As Medewar puts it "It is a curious thing that there is no word in the English language that stands for the mere increase in years; that is for ageing silenced of its overtones of increasing deterioration and decay. At present we are obliged to say that Dorian Gray did not exactly 'age' though to admit that he certainly grew older. We obviously need a word for mere ageing, and I propose to use 'ageing' itself for just that purpose. 'Ageing' hereafter stands for mere ageing, and has no other innuendo. I shall use the word 'senescence' to mean ageing accompanied by that decline of bodily faculties and sensibilities and energies which ageing colloquially entails. Dorian Gray aged, but only his portrait disclosed the changes of senescence. I hope that makes it clear."

Occurrence of Senescence

Figure 1 summarizes the main senescence processes which occur in seed plants arranged as they occur in successive stages of the life cycle (2). The diagram emphasizes that we are dealing with a wide range of phenomena, many but not all of which are integral parts of the programme of development of the plant. If this is so then to understand senescence one must understand development, and the immensity of the task becomes apparent. It is a small step from senescence as a part of the process of development and differentiation to the view that "Natural" death is an inevitable consequence of differentiation (3). Clearly there are certain forms of cellular differentiation in both animals and plants in which the events which lead to death are assured from the moment that the cell, or organ of which it is a part, embarks upon its special differentiation.

It has long been recognized, for example, that the lignified conducting elements of xylem tissues in plants and the formation of enucleate erythrocytes in mammals are processes which have led these cells to die or to a state where no resumption of synthetic activity is possible (4). The operational difficulty in the study of these forms of senescence is in deciding how far back in the developmental sequence one should go; in the case of a xylem element it could be argued that once the crucial division takes place which brings a particular daughter cell into the zone where the balance of hormones or other determinent factors is appropriate for its differentiation as a xylem element, then everything which follows could, on the definition of senescence adopted here, be regarded as part of the senescence process. In practice, however, one might choose to focus attention more particularly on the later events in the programme in which the disorganization of the cell was becoming evident. Similarly, in the case of the erythrocyte, the point of departure is the formation of the enucleate cell at erythropoeisis, but again operationally, students of the senescence of erythrocytes will often choose to concentrate on those particular changes in the cell surface which occur after about 120 days of circulation in the blood stream, whereby the red cell is no longer recognized as "self" by the organism and is removed by phagocytosis. The important point here is that when considering those forms of senescence which are integral to development, it is well to bear in mind that the moment of embarkation on a particular line of development, be it as certain type of cell, or organ, is the point of departure in the sense that the future of that cell or organ is now irretrievable, nonetheless, it will often be a matter of operational choice influenced by the experience of the investigator, which will lead him to single out more specifically terminal events in the differentiation process for inclusion in the study of senescence.

Senescence and the Developmental Programme

The synchronous senescence of a whole population of plants, as in the ripening of a field of corn, provides a spectacular example of senescence manifested in the developmental programme. It is important to recognize however that bland phrases such as "senescence in the developmental programme," tell us little and tend if anything

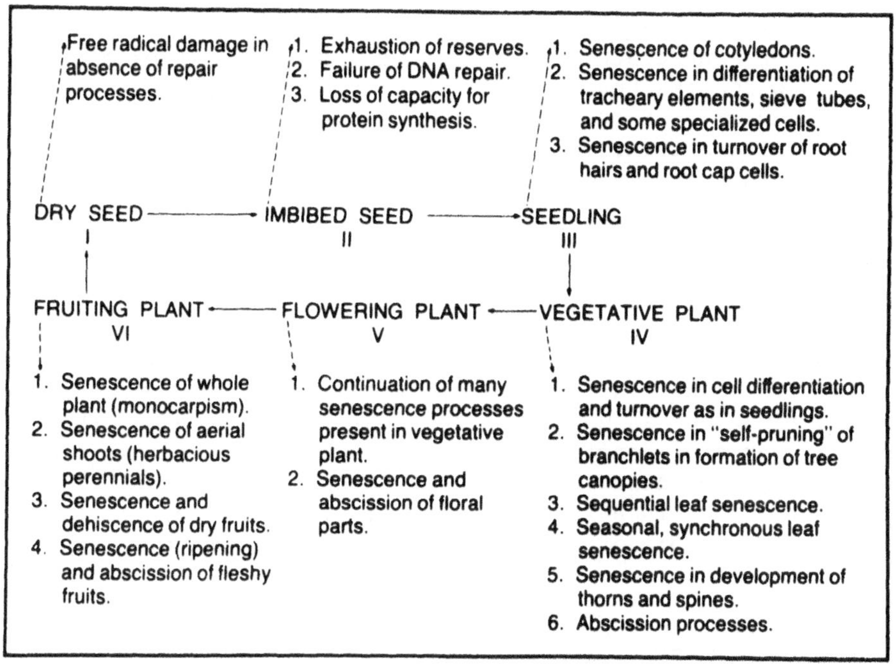

Fig. 1. A summary of the main senescence processes which are found in seed plants, arranged as they occur in succesive stages of the life cycle. Many but not all of the phenomena shown are integral parts of the developmental programme of the plant.

to obsure the multiplicity of mechanisms which may be involved. For convenience I shall distinguish three types of senescence arising from the programme of development.

(1) Direct or positively programmed senescence. I have in mind here those types of senescence in which there is released into the cell a specific set of hydrolytic enzymes which will break down its own macromolecules. If we consider some generalized form of sequentially arranged operator-operon model as controlling the production of the enzymic apparatus, then one could envisage a "suicide squad" or "death clock," a bank of genes coding for hydrolytic enzymes, whose activation would lead to the death of the cell. As we shall see there are several instances, as in the autolysis of the tail of tadpoles, the ripening of certain fruits and a number of other senescence phenomena, in which something along these lines seems to be happening.

(2) Indirect programming of senescence. There are a number of examples in the senescence of plant organs, such as leaves and minor branches, in which the senescence of the organ arises from unsuccessful competition with other organs, because it comes to occupy an unfavourable position relative to new organs which develop subsequently. The sequential senescence of the lower leaves when they come to be shaded by newer leaves developing on the stem above them, and the loss of minor branches within a tree canopy, are examples of this. Now the detailed terminal events in the senescence of such organs may involve the positive action of a hydrolytic enzyme "suicide squad" of the type postulated under category 1. Equally it could arise through starvation or some other manifestation of inter-organ competition; in which case the events in the senescence process may not be directly-coded as genes for degradative processes in the cells which die. They are, nonetheless, inimical to the development of that plant simply because it puts its subsequent leaves and branches at particular angles in particular places. In doing this it has sealed the fate of some of the leaves and branches which it used at an earlier stage.

(3) Allometric effects, mechanical and physiological imbalance. Allometric growth, that is the unequal growth of different parts of an organism, can lead to a state of mechanical or physiological instability (5). It is frequently observed in various species of trees that when a certain height is reached, the trunk may tend to snap (as in plantation-grown birches) or the plant becomes uprooted in the event of strong gales as commonly occurs in tall elms. It may be argued that in such cases the plant is contributing to its own demise as a result of its development to a state of mechanical instability. A more frequently canvassed view is that the senescence of trees may result from their developing to a state of physiological instability as a consequence of the gradual production of an increasing proportion of non-photosynthetic tissues; the branches, trunk and roots which comprise a respiratory burden which is greater than the canopy can sustain (6). The concept is attractive but there is, at the present time, no quantitative evidence to sustain this speculation. Maggs (7) suggested that the distance between the shoot apices and the roots may be important in the senescence of trees, possibly as a result of some constraint in the transport system of the plant. This hypothesis also lacks experimental evidence but the frequently observed "die-back" of distal branches in senescent trees, allied to the finding that terminal twigs from such plants will grow vigorously if rooted as cuttings, suggests that this hypothesis is worthy of further investigation.

Clonal Senescence

Many protozoa and unicellular plants can be maintained as clones, that is as specific cell lineages, for an apparently indefinite period, provided that an appropriate regime of sub-culturing is adopted.

Many perennial herbs, shrubs and trees can be propagated by tillers, suckers or cuttings so that a particular genotype is conserved over hundreds of years without an apparent decline in vigour. The apple cultivar Winter Pearmain was known in England in 1200 AD and was still in cultivation 600 years later (8). There are similar examples of clonally propagated varieties of cassava (*Manihot usitatissima*), date palms (*Phoenix dactylifera*) yams (*Ipomea batatas*), plums (*Prunus domestica*) and figs (*Ficus domestica*) some of which may have been grown from cuttings for 2,000 years (9). An early reference to the possibility that clones might undergo senescence is cited by the gardener Knight in 1795. In a letter to Sir Joseph Banks, he describes difficulties relating to the propagation of the older varieties of apples which make it clear that canker, now known to be caused by *Nectria galygena*, accounted for much of the problem, but continues "The wood of all the old fruits has long appeared to me to possess less elasticity and hardness, and to feel more soft and spongy under the knife, than that of the new varieties which I have obtained from seed. This defect may I think be the immediate cause of the canker and moss, though it is probably itself the effect of old age and incurable." He goes on to list similar effects in other species, the raspberry (*Rubus idaeus*) raised from seed is said to live for 20 years, but the "common elm" (*Ulmus sp.*) a species widely propagated vegetatively, is noted as an exception.

Among the many difficulties in most of the work from the time of Knight to the present day, is that of deciding whether the supposed decline in vigour of a given clone is due to cultural conditions, the accumulation of virus or other diseases or to genuinely endogenous factors. The supposed greater rejuvenation following sexual reproduction, referred to so frequently in the work of Krenke and his colleagues (10) could well be due to recombination of genes in the offspring. Whilst there is clearly room for more work on this subject, the evidence at the present time is strongly in favour of the potential immortality of meristemic cell lineages of many herbaceous and woody perennials, provided that methods of vegetative propagation exist. What is not disputed is that individual shrubs or trees will have a finite existence, even though cuttings taken from old and decrepit specimens can grow and develop to form normal healthy plants, which resume the growth rate of the parent tree when it was young. This situation has been variously summarized: "The tree is in fact not unlike a coral colony though a much more organized symbiotic community (W.E. Baker) and apart from accident, disease and some mechanical and nutritional considerations there seems no reason why it should not live indefinitely," (11). A similar view is that of Wynn-Edwards (12) "There is no universal life-span characteristic of the species as a whole, indeed in some cases the durability and resistance of the dead heartwood seem to be critical factors rather than the true viability of living cells."

Senescence at the Cellular Level

We can distinguish three types of cell from the standpoint of senescence and renewal, cells which are no longer capable of division, cells capable of undergoing mitosis and cells of the germ line where there exists the possibility of meiotic divisions.

(1) Post mitotic cells. The lifespan of post-mitotic cells varies greatly according to the species and the tissue in which they occur. Individual neurones in the brains of long-lived mammals may continue in a functional state for a hundred years or more (13). Similarly, cells in the xylem parenchyma of the wood of certain trees may remain active, carrying out a seasonal starch storage function for a hundred years or more.

Over the past forty years the so-called mutation hypothesis has held sway in much of the work on the origins of senescence at the cellular level, particularly amongst zoologists. Comfort (14) offers three reasons for this, the fact that mutative changes in cells would probably lead to eventual senescence if nothing else did; the prevalence of theories based upon somatic mutation in cancer and radiation research, and the readiness to which somatic mutation lends itself to speculative exercises in higher mathematics (15). The rich debate and often heated controversy which has surrounded this somatic mutation hypothesis have been admirably summarized by Comfort (16). It originated in the suggestion of King (17) that cosmic rays could cause the accumulation of mutagenic changes in the body. In successive refinements the possible nature of the hypothetical lesions was elaborated and, under the influence of physicists, radiation biologists and cancer research workers, the mathematical models gradually took leave of biological realities. The hypothesis was brought back into the ambit of experimental cell biology with the discovery that living cells possess highly specialized mechanisms for the repair of faults which arise in the genetic material (18,19).

If the accumulation of genetic damage is involved in cellular senescence it would be reasonable to expect that in organisms which depend upon populations of post mitotic cells for their functioning, there would be some relationship between the capacity for DNA repair and the lifespan and between the accumulation of damage in the DNA and other cells and the efficiency of the repair system in such cells.

Hart and Setlow (20) found a remarkable correlation between lifespan and the capacity for unscheduled DNA synthesis, a measure of the amount of repair (21), for a range of species (Fig. 2). In an attempt to standardize the comparison, equivalent cells of the primary fibroblasts were taken from each species and the same range of doses of U.V. radiation was used to induce the DNA damage in the cell samples for each species. Both the rate and the extent of unscheduled DNA synthesis after U.V. radiation were found to increase with the lifespan of the species (Fig. 3). The authors emphasize that such results must be interpreted with caution; the extent of ^{3}H-thymidine incorporation will be influenced by the amount of DNA per cell, the specific activity of the internal thy-

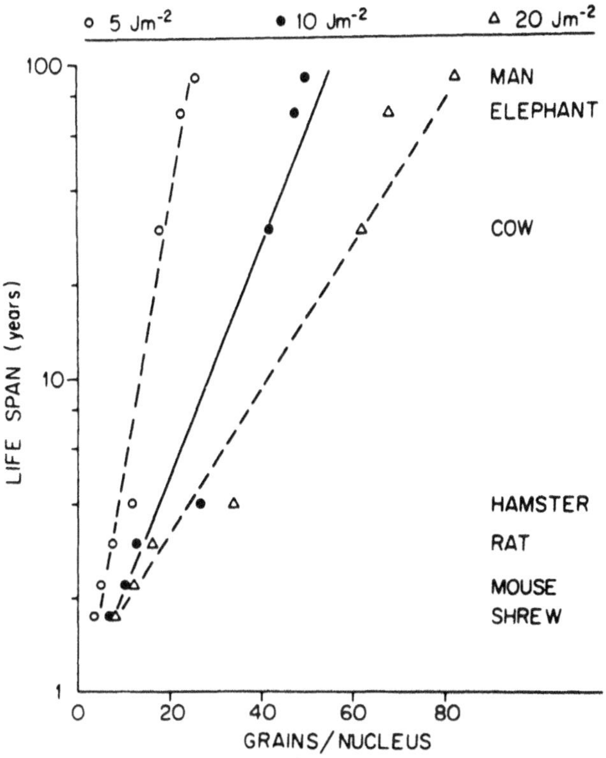

Fig. 2. The correlation between extent of DNA repair, measured in terms of unscheduled DNA synthesis at 13 hours after different UV fluences, and the estimated life-span of the species. (After Hart and Setlow (20)).

midilate pools, the average size of the repaired regions and the number of repaired sites per unit length of DNA. The authors are, however, confident that it is the last of these factors which is important for several reasons. (i) the fraction of the DNA dimerised in comparable U.V. exposures is the same for mouse, hamster, cow and man (22, 23, 24). (ii) Almost all mammalian cells contain about

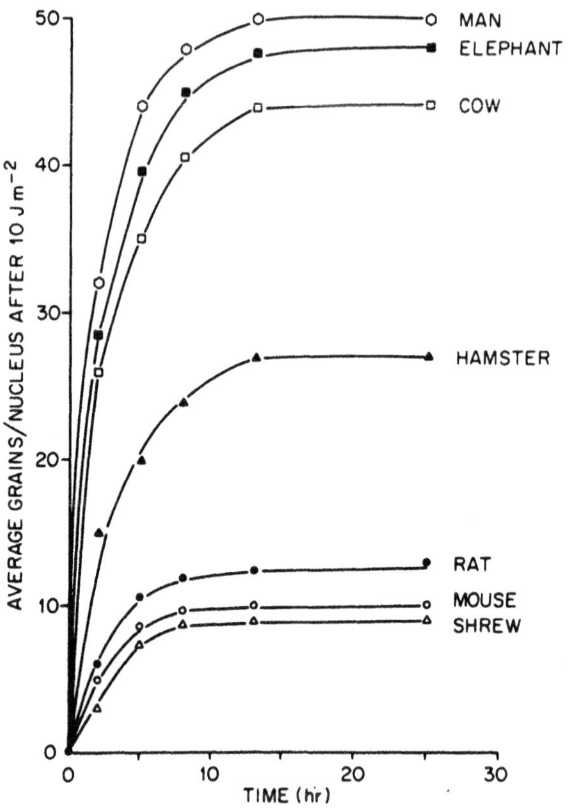

Fig. 3. The average amount of DNA repair, measured in terms of unscheduled DNA synthesis, as a function of time after exposure to a fluence of 10 Jm^{-2} of UV radiation, for seven different species. (After Hart and Setlow (20)).

the same amount of DNA, 6pg per nucleus, (25) so that the total DNA available for repair is constant. (iii) Small numbers of fibroblasts from the different species used in Setlow and Hart's experiments underwent scheduled DNA synthesis in the course of the experiment and all had similar grain counts per nucleus suggesting that the specific activities and the size of the thymidilate pools were not significantly different in the different species. (v) Measurements of the photolysis of bromodeoxyuridine incorporated into DNA in place of thymidine during excission repair in cells of mouse, hamster and man suggest that the average size of the repaired regions was similar in each species (26, 22).

If U.V. radiation is such a potent agent for inducing genetic damage in cells, it is reasonable to suppose that many plants, particularly those which grow at high altitude where the amounts of U.V. received are greatest must have very efficient DNA repair mechanisms. For example, in the Bristle Cone Pine, Pinus aristata, which grows at an altitude of 4-5,000m in the White Mountains of California, individual leaves can be retained in a functional state for up to 30 years (27) during which time they must sustain an enormous radiation load. There are numerous reports of visible light stimulating recovery from damaging doses of U. V. radiation in leaves of higher plants (28,29) but it is not known at the present time whether these effects are attributable to DNA repair. Initial attempts to detect excission repair of DNA in plant cells were unsuccessful (30,31), but with the development of more refined techniques such repair has now been detected in Chlamydomonas reinhardi (32) and in protoplasts of Daucus carota (33). In C.reinhardi, U.V.-sensitive strains were deficient in excission repair but there are, at the present time, no reports of comparative studies which might enable one to say whether or not there are correlations between the extent of the development of DNA repair systems and the lifespan of plant cells or the magnitude of the radiation load which they normally encounter.

(2) Cells having the potential for mitotic division. In some relatively unspecialized organisms, mitosis appears to prevent senescence. The head region of the flatworm Stenostomum consists of non-dividing cells; the rest of the body includes cells that contrive to divide, giving rise to daughter animals by renewed growth and differentiation of the posterior regions. If the head of the animal is repeatedly severed and allowed to regenerate a new posterior region, it soon becomes senescent and dies. If, on the other hand, the newly developed posterior region is taken and allowed to regenerate a new head then the organism can be clonally propagated for an indefinite period (34). The possibility cannot be excluded that selection may be taking place against cells which become senescent, so that these are simply lost without trace: it is interesting to speculate that an analogous selection against senescent cells may account for the apparent immortality of the meristems of perennial plants. An interesting parallel to the decline in regenerative capacity of the post-mitotic cells of Stenostomum is the case of frond development in the free-floating aquatic plant Lemna minor (35,36,37,38,39). This group studied the regeneration capacity of single parent fronds of L.minor when the successive daughter fronds were removed as soon as they were expanded. It was found that the successive daughter fronds became successively smaller as the parent frond became progressively more senescent before the final collapse of the parent frond. When small daughter fronds arising from a senescent parent were cultured further, they gave rise to fronds which were larger than themselves but took approximately six sub-culturings of this kind to regain the size of the original parent frond.

Another interesting example of loss of regenerative capacity in plants comes from work with cell suspension cultures (40). Clones of cells fully established from the roots of carrot (*Daucus carota*) can readily give rise to a new sporophytic plant when plated onto an appropriate medium. When the cells in suspension culture are sub-cultured through several generations and then plated, the capacity to regenerate a new plant (totipotency) is greatly reduced even though the growth rate of the cultured cells had not declined. However, in cases where a plant can be regenerated from an old suspension culture, other cells taken from the root formed on such a plant are found to have regained their totipotent potential.

Steward (41) suggested that changes in the cytoplasm of the cultured cells were the cause of the decline in capacity for regeneration and these are apparently corrected during the formation of a new plant.

Studies such as those on the senescence and rejuvenation of the fronds of *Lemna minor* suggest that, as in some flatworms and protozoa, there is an element of cell rejuvenation through the act of mitosis *per se*. It is well known that in most, if not all, plants the cells of the meristems are in symplastic contact with their neighbors by means of plasmodesmata, indeed it has been estimated that even the smallest meristemic cells have 1,000 to 10,000 plasmodesmatal connections with the adjacent cells (42) the frequencies, length and diameter of these connections shows wide variation according to species (43) and there is evidence that these may function as major channels for the transport of nutrients into the meristem (44). Figure 4 is a diagrammatic representation of current views of the structure and dimensions of a plasmodesma justaposed with a representation of the dimensions of a range of plant viruses drawn to the same scale (after Gibbs (45)). The evidence suggests that the genome of many viruses could pass readily through the plasmodesmata, particularly if they lack a desmosome; in some cases the viruses are small enough to pass directly through the plasmodesmata intact. Figure 5 shows the particles of strawberry latent ringspot virus passing through the plasmodesmata of *Chenopodium amaranticolor*. Notwithstanding this evidence, considerable success has been achieved in obtaining virus-free clonal stocks of a number of horticulturally important species by culturing meristems isolated from the shoot apices. This suggests that, if the viruses are able to enter the apices, they are unable to survive in the cytoplasm during mitosis. The detailed biochemical events in the mitotic cell are of enormous complexity and are very imperfectly known, for review see Prescott (46). The organelles and the ribosomes of mitotic cells remain intact throughout division but their associated biosynthetic activity shows large fluctuations. In the middle if interphase (G_2) there is a *de novo* synthesis and much increased activity of RNAase in cells of *Physarum*, (47) but with the onset of the M phase, the capacity of the cells for protein synthesis falls to a level which is only a

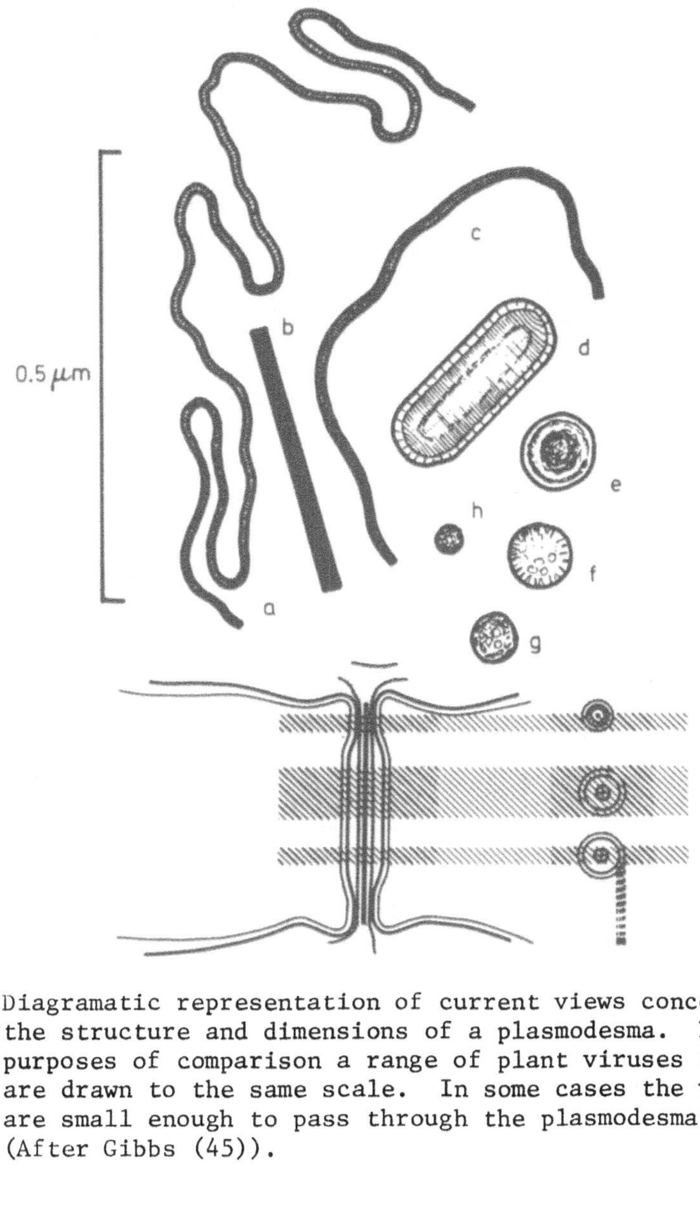

Fig. 4. Diagramatic representation of current views concerning the structure and dimensions of a plasmodesma. For purposes of comparison a range of plant viruses (a to h) are drawn to the same scale. In some cases the viruses are small enough to pass through the plasmodesmata intact. (After Gibbs (45)).

very small fraction of that during interphase. This decreased protein synthetic activity is associated with disaggregation of the polysomes (48,49,50) and is apparently caused by the development and persistence of an inhibitor of mRNA synthesis during this M phase due to a transient incompetence of the ribosomes (51).

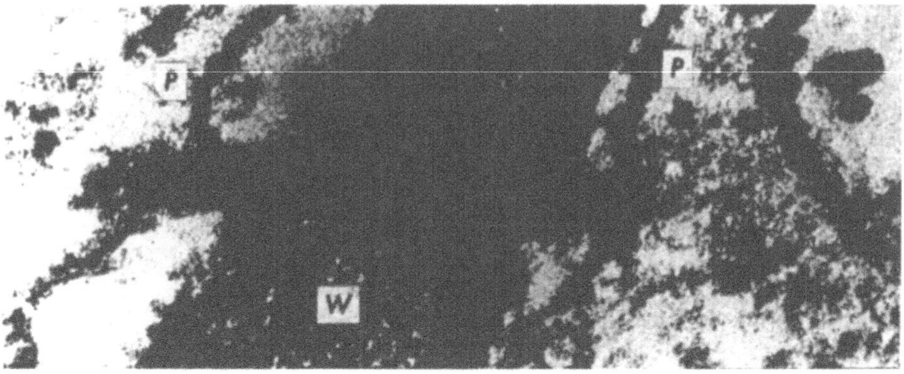

Fig. 5. Electron micrograph showing a chain of particles of strawberry latent ringspot virus passing through a plasmodesma of Chenopodium amaranticolor. W denotes the cell wall. (After I.M. Roberts and B.D. Harrison (1980) J.Gen.Virol. 7,47)

It is of interest that in dividing cells of pig kidney infected with vaccinia virus, the cells lose the ability to support the replication of the virus at this stage, (52) either because the virus is unable to recruit the inactivated cytoplasmic ribosomes needed for its replication at this stage or it may be degraded as a consequence of the increased ribonuclease activity in the cells. The relationship of this sequence of events to the cell cycle is shown in Figure 6. This scheme is of course conjectural but it serves to emphasize the possibility of some limited "scrubbing" or removal of unwanted components such as viruses, denatured nucleic acids or other damaged cytoplasmic components, at this stage of the cell cycle. In fairness it must be noted, however, that there is good evidence from work with HeLa cells (53) and synchronized Chinese hamster cells (49) that some long lived mRNAs synthesized during G_2 may persist through the M phase to be translated during resurgence of protein synthesis which begins in late telophase and proceeds into G_1. It may well be of course that such mRNA molecules are protected against any general scrubbing process by complexing with a protective protein or sequestration in a protected compartment of the cell in a manner not available to viral RNA or damaged cytoplasmic components.

Another mechanism for the renewal of cell components which might operate during mitosis is that certain genes could become active only during cell division. If such genes were to code for proteins which are required throughout the life of the cell, then the life of the cell would be determined by the half-lives of the

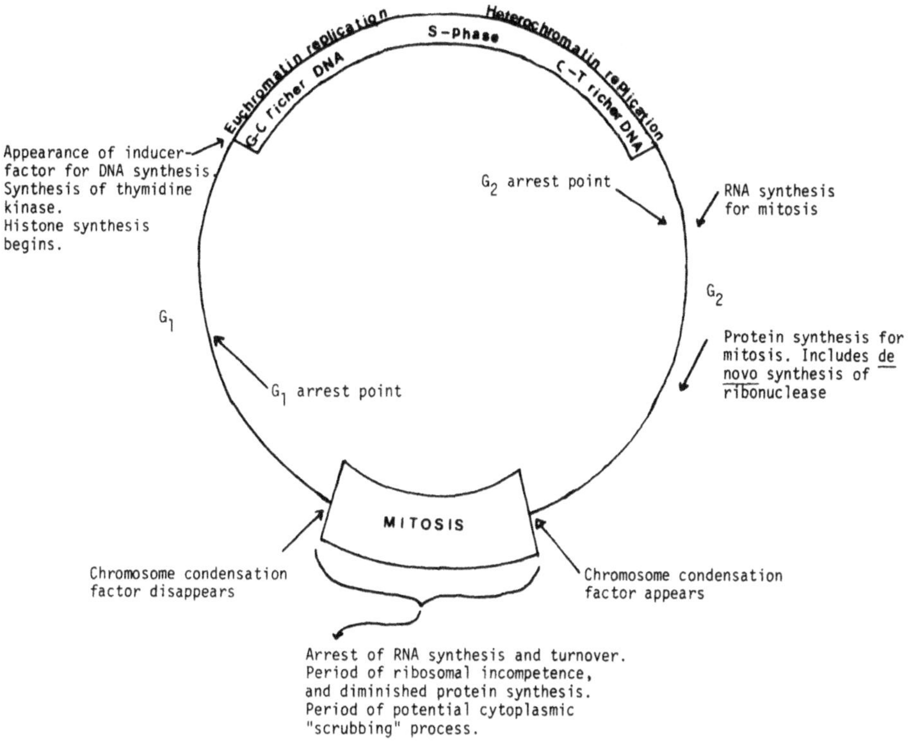

Fig. 6. Part-conjectural diagram describing events in the mitotic cell cycle showing potential stages at which "scrubbing" or removal of unwanted components such as viruses, denatured or faulty nucleic acids or other cytoplasmic components may occur. For discussion see text.

mRNA and protein molecules derived from them. There is no good evidence that this is so. Certainly the levels of particular enzymes varies greatly at different stages in the cell cycle and in relation to the stage of development of the tissue of which that cell is a part. It is also well established that different enzymes are turned over at very different rates, some having half-lives that may be measured in hours, others in days or even weeks (see Part II). As far as the author is aware, however, there are few if any cases from eukaryotic organisms of enzymes which are renewed only in the dividing cell.

(3) Meiosis in relation to cell rejuvenation. It is evident that whatever the cause of breakdown in the somatic cells of an organism there must be far reaching events in the cells of the germ-line which confer upon them the ability to sustain the continuity of

the species. There are, of course, the events of recombination in the genome of the cell which may bring together new groupings of genes more favorable to a particular environment, but this is not sufficient to explain the evident immortality which graces the germ-line, since there is no reason to suppose that the already existing arrangement of genes is not entirely adequate for the progress of the individual, but there is evidence of drastic changes in the cytoplasm of meiotic cells (54).

Autoradiographic studies in such widely different species as Lilium, Rhoeo discolor, Zea mays and Paeonia show that there is pronounced RNA synthesis in the pre-meiotic phase followed by a period of declining RNA synthesis throughout the meiotic prophase and a complete cessation of RNA synthesis from metaphase I until the end of meiosis (55,56,57,58,59) which could be due to the fact that during this period the pollen mother cells become sealed off by callose walls which may prevent the passage of precursors into the cells during this period. In Lilium and Trillium EM studies reveal a decline in the ribosome population from zygotene to pachytene (60). The ribosome population is restored during the meiotic mitoses. Concomitant with the prophase elimination of ribosomes, the endoplasmic reticulum changes from a plate-like appearance to circular profiles of paired membranes often arranged in the manner of concentric spheres, the membranes return to the normal form in the last stages of breakdown of the ribosomes. Correlated with these changes in the ribosomes and E.R. is a remarkable cycle of dedifferentiation and re-assembly of the mitochondria and plastids. In Tradescantia the mitocondria are elongate before meiosis and become spherical by pachytene and may lose the capacity to divide at this time (61). They subsequently lose internal structure until by meiosis I the christae are lost and they appear as simple rounded bodies 0.4nm in diameter. Redifferentiation of the mitochondria follows a similar time course to the return to normal of the ER.

The plastids undergo a similar cycle of drastic dedifferentiation followed by re-assembly (62). At pre-meiosis in Lilium longiflorum the proplastids of the pollen mother cells have a rudimentary lamellar system containing starch, and, insofar as can be judged from electron micrographs, are undergoing division. In leptotene the starch is lost, the membrane system breaks down and no more "apparently dividing" plastids are seen. In the transition from zygotene to pachytene the plastids attain their simplest form with scarcely any ribosomes present and a much reduced lamellar system. From the late tetrad stage onwards there is a re-appearance of starch, restoration of the plastid ribosomes and redevelopment of a lamellar system. There are similar drastic alterations in the cytoplasm of oocytes at meiosis and indeed recent work suggests that during this period the nucleus may become relatively isolated from the events occurring in the cytoplasm due to a transient closure of the nuclear pores.

The circumstantial evidence thus points very heavily towards a wholesale eradication of cytoplasmic macromolecules in meiosis which is at its height in the zygotene-pachytene transition. A basic reason for this "scrubbing out" may well be the need to remove pre-existing mRNAs in order to permit a re-programming of the cytoplasma by the post meiotic nuclei (54). It is of interest that very few plant viruses are seed born suggesting that if viruses were present in the pre-meiotic cells they would also succumb to this scrubbing process. It would therefore seem very probable that there is in fact the opportunity for a very wide ranging renewal of damaged components of the cytoplasm and its organelles at this stage, which may be of crucial importance in counteracting cytoplasmic changes which would otherwise give rise to senescence of the cell.

Concluding Remarks

We have seen that in plants senescence processes may be linked to the successive stages of the life cycle as the programme of development unfolds. We have considered at some length the various types of molecular damage which might be expected to cause senescence in plant cells and described the processes of DNA repair, cell selection and renewal at the time of cell division which prevent such processes from becoming dominent factors in plant senescence. It follows that in plants the majority of important senescence changes incurred in the course of the life cycle arise from programmed rather than environmental insults of a stochastic nature. In the second part of this paper I shall consider the turnover of nucleic acids and proteins as an example of the types of metabolic process through which the developmentally programmed forms of senescence may be initiated and controlled.

PART II.THE ROLE OF NUCLEIC ACID AND PROTEIN TURNOVER IN PLANT SENESCENCE WITH PARTICULAR REFERENCE TO LEAVES

It is now over a hundred years since the suggestion was first made that proteins in plant tissues were continually being broken down and resynthesized (63,64). This idea provided a framework for measurements of the synthesis and breakdown of proteins from tissues which gave rise to the concept of a protein cycle (65,66). This protein cycle involves continuous synthesis and breakdown of proteins, even in circumstances where the total protein content of the tissue may not be changing. Subsequent work has reinforced the concept of a protein cycle in many plant tissues; for simplicity we shall confine our attention to the case of leaves in this discussion. We are confronted by three problems. Firstly there is the question of how one shall measure the activity of this protein cycle - the problem of protein turnover as it has come to be known. The two further problems concern the mechanisms by which the synthetic and degradative aspects of this cycle are regulated. In the case of leaves it is well known that there is a net decline in protein content in the course of senescence (67,68,69,70). We shall therefore require to consider the regulation of protein synthesis and degradation and assess the extent to which changes in each of these processes (Figure 7) are responsible for the net loss of protein (71,72,73).

Protein Turnover in Leaves

(1) Enzymes. Measurement of protein turnover in plant tissues first became a practical proposition with the introduction of N^{15} and radioisotopes in the nineteen forties. There remain great practical obstacles to the accurate measurement of protein turnover in plants; I shall not dwell on the difficulties here, for a recent review see Davies (74). Published values of protein turnover in leaves are relatively few but suggest a half life of 3 - 5 days in several species (74). This is of course an average value for the leaf protein as whole and may derive from some proteins which are turning over rapidly and others which are turning over slowly or not at all. It therefore becomes necessary to purify to homogeneity specific proteins in order that their several rates of turnover may be measured. In the past it has proved very difficult to purify leaf proteins except for RuBP carboxylase which is present in large quantities. However, the recent introduction of immunological and affinity column techniques should make it possible to purify and measure the turnover of more leaf proteins in the near future.

If the significance of protein turnover in the events of senescence is to be understood it will be necessary to measure turnover of individual proteins since averaged values of turnover for unresolved protein mixtures can tell us relatively little. Clearly a protein carrying out an essential function, which is synthetized at one phase of development but not renewed thereafter, could become important in the declining functional capacity of a senescing cell. Such appears to be the case with the enzyme RuBP carboxylase which, in the leaves of some species, declines both in activity (75) and amount (76) in a manner which correlates approximately with the declining photosynthetic capacity of the leaf (77). Turnover studies on RuBP carboxylase have been carried out on leaves of *Perilla frutescens* (71) barley (78,79) and maize (80). When considering these studies caution must be exercised in respect of the methods used, but the general conclusion may be drawn that in *Perilla* and barley RuBP carboxylase is not turned over whereas in maize some turnover does occur. In *Perilla* and barley RuBP carboxylase is synthesized during the development of the leaves but formation of the enzyme ceases when expansion of the leaf lamina is completed, thereafter the amount of activity of the enzyme declines gradually in the mature leaf. In maize turnover of the RuBP carboxylase

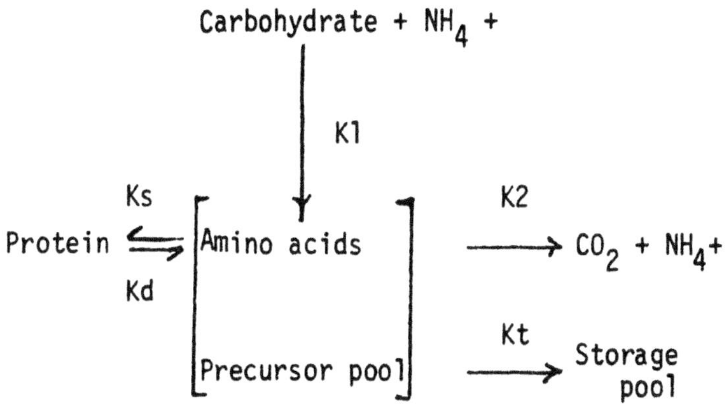

Fig. 7. A simple model for the turnover of proteins in the cell incorporating an element of recycling of amino acids. K denotes rate constants, Ks for synthesis of protein, Kd for degradation of protein, K1 and K2 for synthesis and degradation of amino acids and Kt for transfer to a storage pool. (After Davies (74)).

appears to continue in the mature leaf, albeit with degradation proceeding faster than synthesis in the later stages (80).

The measurement of turnover is perhaps of greater significance in the case of enzymes which are close to the threshold level at which they are required in the cell and those which occupy key positions in regulatory aspects of metabolism. Unfortunately the small quantities of enzyme present and the difficulties in purifying them have so far militated against measurements of turnover for most of these enzymes, there are however a number of observations which suggest that such measurements would be of interest. Enzymes of the reductive pentose phosphate pathway in chloroplasts do not all decline in activity at the same rate, which invites further study of the rates at which these proteins are synthesized and broken down (81). Nitrate reductase is an enzyme which shows a much more rapid turnover in leaf cells than the majority of the soluble proteins. In leaves of Perilla frutescens the half-life of the enzyme, based on activity measurements, was estimated to vary from about 240 minutes in mature leaves to 120 minutes in senescing leaves (Figure 8) (82). An analogous pattern of loss of activity was observed in cell free extracts of barley leaves (83). In the case of barley the nitrate reductase appeared to be converted to a cytochrome C reductase, a modification which may represent the first step in breakdown of the enzyme.

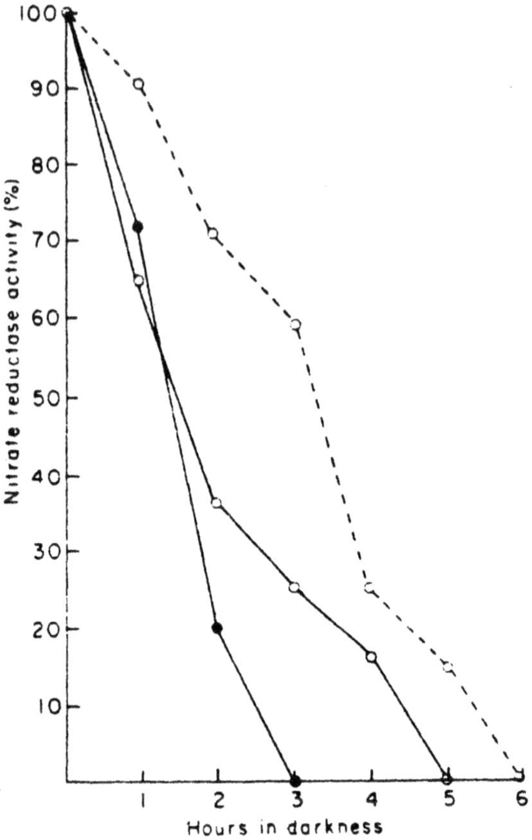

Fig. 8. Decrease in nitrate reductase activity following darkening in leaves of Perilla of different ages. 0——0, young leaves (15 days from emergence); 0---0, mature leaves (45 days from emergence); ●——●, senescent leaves (70 days from emergence. The data are expressed as percentages of the activity at the time of placing in darkness. Subsequent work has shown that loss of activity involves breakdown of the enzyme. (After Kannangara and Woolhouse (82)).

(2) Membrane and structural constituents. Changes in the structure and properties of membranes are a universal feature of senescence in plant cells: the evidence comes from electron microscopy (84), x-ray-diffraction studies on isolated membranes (85,86), and physiological measurements which show increasing permeability in cells of senescing tissues manifested as leakage, that is to say failure to retain accumulated solutes (87,88). Indirect supporting

evidence for involvement of membrane changes in senescence comes from a miscellany of sources, as for example the observation that Ca^{2+}, a well-known factor in the stabilization of plant cell membranes, will delay senescence in excised leaf discs of some species (89).

Evidence that the constituent proteins of membrane systems may turnover at very different rates may be exemplified by the case of the thylakoid membranes. One of the constituents of the thylakoid is a protein encoded and synthesized within the chloroplast, having a molecular weight of 32,000 (P-32,000) (90). The synthesis of P-32,000 is light-dependent and involves the formation of a 33,500 M.W. precursor which is processed by proteolytic cleavage to the P-32,000 form which is inserted into the thylakoid. Studies involving pulse-chase experiments with ^{3}H-leucine labelled fronds of Spirodela show P-32,000 to be turned over an order of magnitude more rapidly than the other major chloroplast polypeptides (91).

Although accurate measurements of turnover of individual leaf proteins are so few, the available evidence for turnover and age-dependent differential loss of particular proteins is sufficiently strong to suggest that we should look to the system regulating protein turnover as an important site at which the senescence process may be regulated. Underlying the turnover mechanism there are undoubtedly hormonal controls which will be dealt with by Professor Bruinsma elsewhere in this symposium but my present concern is the dynamic system which is actually the subject of these controls.

The Regulation of Protein Turnover in Leaves

(1) General considerations. It will be evident from the generalized scheme shown in Figure 9 that modulation of the level of a particular protein may be achieved by altering the balance of synthetic and degradative sides of the process. Moreover, since it is with relative rates of synthesis and degradation that we are concerned, the balance may be altered in favor of synthesis or degradation even when both processes are increasing or decreasing in absolute terms.

A further general problem in considering protein turnover in leaves arises from the fact that we have to consider three genomes as sites of transcription, the nucleus, chloroplasts and mitochondria and three sites of translation, the cytoplasm, chloroplasts and mitochondria. As we shall see these different sites of protein coding and synthesis may have important implications for the sequence of events involved in leaf senescence.

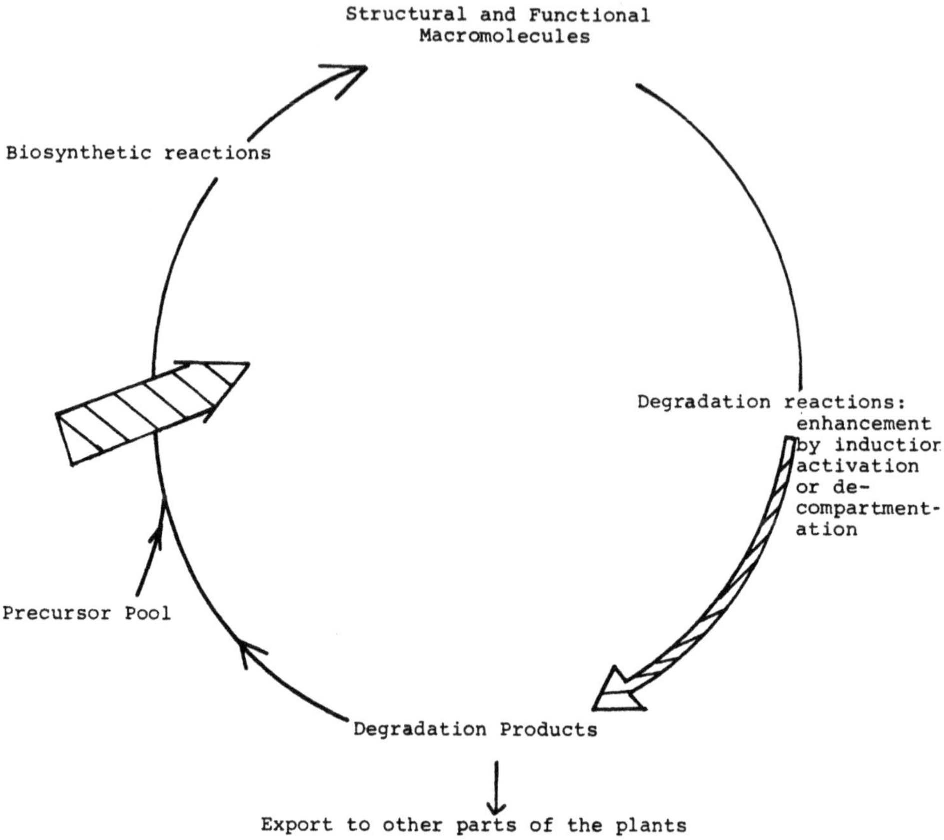

Fig. 9. Scheme for the turnover of proteins and other macromolecular constituents of the leaf in which a net loss of a component may arise from depressed activity amongst elements of the biosynthetic system or enhanced activity of enzymes on the catabolic side of the cycle. Some breakdown products are shown as being lost to the system by translocation whilst some are re-cycled through the biosynthetic pathways. For discussion of the implications of this process, see text.

(2) Control of Protein Synthesis. It seems to be a general rule that in prokaryotic cells and in the organelles of higher organisms the mRNAs are relatively shortlived and gene expression is controlled primarily at the level of transcription. The mRNAs of eukaryotes are longer-lived and control of protein synthesis occurs at the levels of both transcription and translation (92). This difference in extent of translational control in the 70s and 80s-ribosomal system of prokaryotes and eukaryotes respectively may be

what occasions the much greater complexity of the 80s-ribosomal based translation complex (Table 1).

a) Chloroplasts. The main evidence for changes in the biosynthetic side of protein turnover in relation to senescence comes from studies of the photosynthetic apparatus. A consistent feature associated with the declining photosynthetic capacity of the senescing leaf is loss of RuBP carboxylase, the large sub-unit of which is coded for and synthesized within the chloroplast. A detailed study of the electron transport chain in senescing leaves of *Phaseolus vulgaris* has revealed a declining activity associated with a lesion the PQ-Cyt-f section of the chain (93,73). Cyt-f is a component of the electron transport chain which is also encoded and synthesized within the chloroplast (94).

In leaves of *Perilla* there is a loss of capacity for chloroplast protein synthesis in the course of leaf senescence, although cytoplasmic protein synthesis is maintained (95). Similar preferential loss of capacity for synthesis of proteins in the chloroplasts has been demonstrated in leaves of cucumber and *P. vulgaris*. Declining chloroplastic protein synthesis associated with a loss of polysomes and the cessation of chloroplast RNA synthesis. In *P. vulgaris* arrest of chloroplast RNA synthesis coincides with the completion of leaf expansion and has been traced to the loss of chloroplast RNA polymerase activity. Chloroplast RNA polymerase is encoded in the nucleus and synthesized in cytoplasmic ribosomes (96) suggesting that the regulation of this enzyme may be a key to the shut-down of protein synthesis in the chloroplasts (97,98). There is as yet no evidence as to whether synthesis of the chloroplast RNA polymerase is under transcriptional control in the nucleus or translational control in the cytoplasm.

There is no evidence at present to suggest that the pattern of events involving an early shut-down of chloroplast protein synthesis is a feature of leaf maturation and early senescence in all species; thus in wheat for example, Brady and Scott (99) have shown that protein synthesis continues in the chloroplasts of senescing leaves; chloroplast rRNA synthesis is also maintained in senescing leaves of tobacco (134).

There is one set of circumstances in which arrest of chloroplast protein synthesis becomes a major factor in leaf senescence; this concerns the senescence which occurs when leaves are deprived of light. It frequently happens in species growing in dense swards that a particular value of the leaf area index is maintained because there is a sequential senescence of the lower leaves as they become shaded by younger leaves developing above, (100). There are probably a number of factors conspiring to promote this correlative senescence of the older leaves. For example, in low light the

Table 1. Factors involved in translational control in prokaryotes and eukaryotes

Soluble Protein Factors from Escherichia coli

	Molecular weight	Mole factor per mole ribosome	Function
I. Initiation			
IF-1	9,000	0.18	Promotes IF-2 and IF-3 functions
IF-2a	115,000	0.15	Binds Met-tRNA: GTPase
IF-2b	90,000		
IF-3	21,000	0.14	mRNA binding: dissociation
II. Elongation			
EF-Tu	44,000	5-7	Binds AA-tRNA: GTPase
ET-Ts	30,000	1	GTP-GDP exchange on EF-Tu
EF-G	80,000	1	Translocation: GTPase
III. Termination			
RF-1	44,000	0.02	Recognizes UAA and UAG
RF-2	47,000	0.02	Recognizes UAA and UGA
RF-3	46,000	--	Promotes RF-1 and RF-2 GTPase

Table 1. Continued

Soluble Protein Factors from Rabbit Reticulocytes

	Molecule weight	Number of polypeptides	Function
I. Initiation			
eIF-1	15,000	1	Promotes mRNA binding
eIF-2	150,000	3	Forms ternary complex with Met-tRNA
eIF-3	~700,000	~9	Promotes Met-tRNA and mRNA binding dissociation
eIF-4A	49,000	1	Promotes mRNA binding
eIF-4B	80,000	1	Promotes mRNA binding
eIF-4C	17,500	1	Promotes Met-tRNA binding
eIF-4D	16,500	1	Stimulates Met-puromycin synthesis
eIF-5	150,000	1	Required for 80 S complex formation
Co-EIF-2	22,000	1	Stimulates ternary complex formation
Cap binding	24,000	1	Binds to cap of mRNA
II. Elongation			
$EF\text{-}1_{\alpha}$	~55,000	1	Forms ternary complex with AA-tRNA: binds to Rb: GTPase
$EF\text{-}1_{\beta}$	30,000	1	Promotes exchange of GTP GDP on $EF\text{-}1_{\alpha}$
$EF\text{-}1_{\gamma}$	55,000	1	
EF-2	100,000	1	Translocation: GTPase
III. Termination			
RF	56,500	1	Promotes termination. GTPase

stomata will tend to close a common concomitant of leaf senescence, (101). The transpiration stream through these leaves will thus be much reduced leading to a depletion in the supply of cytokinins from the roots, which appear to be essential for the maintenance of protein synthesis in plant cells (102). There are other, more direct effects of light, in regulating chloroplast protein synthesis. Firstly light is necessary for photophosphorylation which provides the energy for protein synthesis; in the absence of light chloroplast protein synthesis ceases and apparently cannot be sustained by energy sources external to the plastid.

Evidence is accumulating to suggest that chloroplast protein synthesis may also be controlled at the transcriptional level involving modulation of the activity of chloroplastic RNA polymerase. In leaves of *Perilla* there is a loss of capacity for chloroplast protein synthesis in the course of leaf senescence, although cytoplasmic protein synthesis is maintained (95). Similar preferential loss of chloroplast activity has been noted in leaves of cucumber and *Phaseolus vulgaris*. The decline of protein synthesis in the chloroplasts is associated with a loss of polysomes and the cessation of chloroplast RNA synthesis (103). In *P. vulgaris* the arrest of chloroplast rRNA synthesis at the time of completion of leaf expansion has been traced to the loss of chloroplast RNA polymerase, an enzyme which appears to be coded for in the nucleus and synthesized on cytoplasmic ribosomes (96) suggesting that the regulation of this enzyme may be a key to the initiation of shut-down in the plastid metabolism,(97, 98). In this context it is of interest that when leaves are caused to re-green by appropriate manipulation of the plant, there is an early resumption of chloroplast RNA polymerase activity (98).

The chloroplast genome is made up of 45 μm double circles of DNA which codes for two proteins present in large amounts, the large sub-unit of RuPCase and the 32,000 Dalton thylakoid protein (104) and at least 100 other proteins whicn are made in much smaller quantities. Silverthorne and Ellis (105) have shown that the relative proportions of the two major proteins synthesized by the chloroplast, changes in the course of leaf development in spinach due to changes in the amounts of translatable RNA for each. Bogorad et al. (106) have isolated a protein of MW 26-29,000 which specifically stimulates transcription of maize chloroplast DNA by the maize chloroplast RNA polymerase and have proceeded to show that transcription of different parts of the chloroplast genome, isolated by restriction endonucleases, is differentially influenced by this "stimulator" protein. The chloroplast genome also codes for two sets of ribosomal genes per DNA molecule (107,108,109) and for about 30 tRNA species (110). Figure 10 provides a working hypothesis for a mechanism by which modulators like the stimulator protein described above, may be produced by nucleus and cytoplasm and passed into the plastid. The model is based on the envelope-

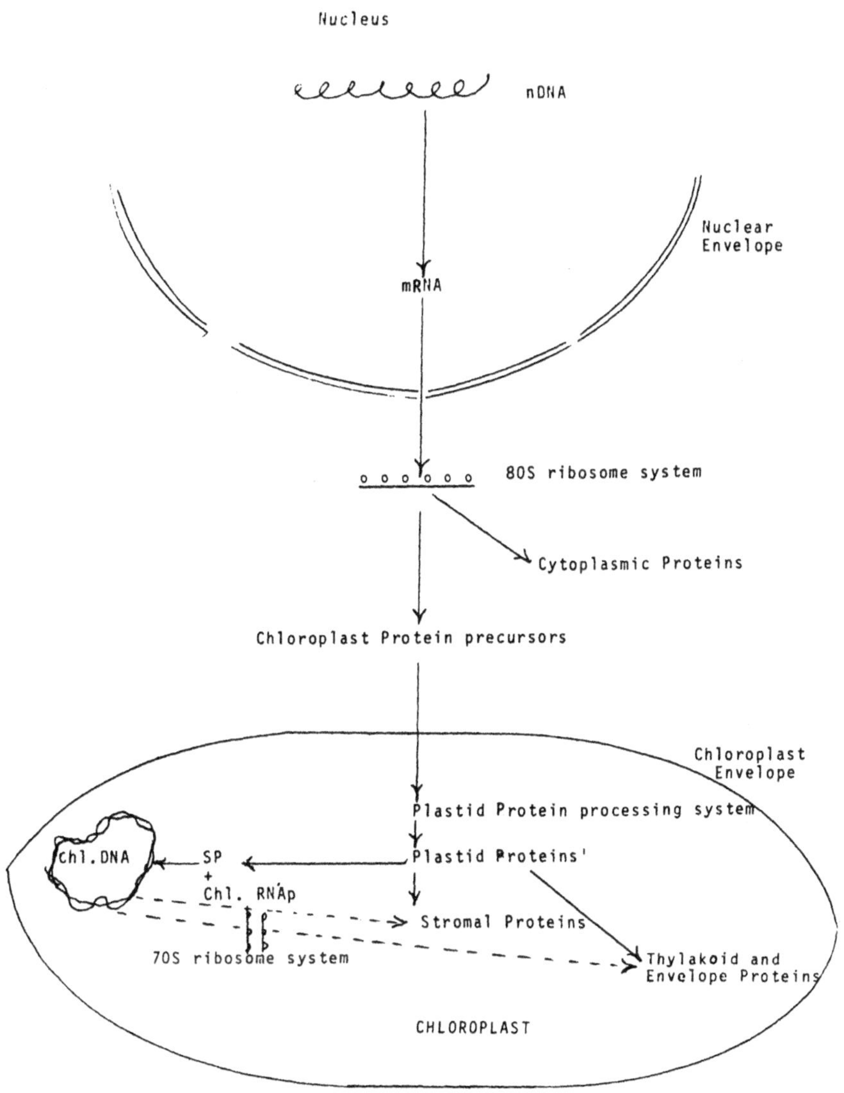

Fig. 10. Postulated mechanism for regulation of interactions between nucleus and chloroplast. Proteins synthesized within the cytoplasm which are destined for the plastid are identified by a precursor segment which is removed after entry into the plastid stroma. SP indicates a stimulator protein modifying the extent of transcription of particular genes in the chl.DNA. ChlRNAp indicates chloroplast RNA polymerase, an enzyme encoded in the nucleus.

carrier mechanism which appears to be involved in the transport of the small sub unit of RubPCase into chloroplasts and may well serve as a general mechanism. Through a mechanism of this type we may envisage a ready means whereby the chloroplast RNA polymerase activity could be modified (105, 106) or totally arrested (98).

It is frequently observed both in gas-exchange studies (111) and by electron microscopy that disorganization and loss of function takes place much earlier in the chloroplasts than in the mitochondria. The fact that the mitochondria can be self-sufficient for the energy requirements of protein synthesis in darkness may account for this difference in sequentially senescent leaves in which shading is a factor. There are also other major differences in the protein synthesizing machinery of the two organelles. For example both organelles contain specific and immunologically distinct elongation factors, EF - G and EF - Tu, as components of the protein synthesizing systems. Whereas EF - Gchl and EF - Tuchl are encoded and synthesized within the plastids, EF - Gmit, and EF - Tumit are encoded in the nucleus and synthesized on cytoplasmic ribosomes (112,113). Thus in circumstances where there is preferential shut-down of chloroplast metabolism at the chL-RNA polymerase-transcription level there will be attendant arrest of synthesis of the plastid-encoded elongation factors. If cytoplasmic protein synthesis is continuing at this time this could provide the necessary elongation factors and allied constituents of the mitochondria since these are encoded in the nucleus and synthesized in the cytoplasm.

(3) Control of Protein Degradation. If we turn now to the degradative side of the turnover process we find that the state of knowledge is slight. The most obvious thing to look at first is the activity of hydrolytic enzymes. I have culled from the literature a list of hydrolytic enzymes which have been reported to increase in activity in senescing leaves and for purposes of comparison a list of the enzymes which are active in the barley aleurone on stimulation with gibberellin, which may be viewed as a somewhat analogous senescing system (Table 2). The list of enzymes is notably similar in the two organs although in the case of leaves the list had to be compiled from widely scattered reports on a range of different species. It is a matter of regret that there does not appear to have been a systematic study of the spectrum of hydrolases in the leaves of a single species.

As a prelude to examination of the regulation of hydrolytic activity in senescing leaves it is well to ask whether the frequently-observed increases in activity are in fact necessary at all for senescence to occur. In many cases the hydrolase activity present before the onset of senescence is more than adequate for the observed rate of breakdown and the activity during senescence is increased only by a factor of two or three (114), which may be of much less significance than a change in compartmentation bringing enzymes and substrates into contact.

Table 2. Hydrolytic enzymes reported to show increased activity in senescing leaves. Included for purposes of comparison (see text) is a list of hydrolytic enzymes released from the aleurone layer of barley. For references see Woolhouse (131)

Leaves	
Protease	
Tobacco	Anderson & Rowan 1965
Mung bean (cotyledons)	Baumgartner & Chrispeels 1977
Pea	Storey & Beevers 1977
Wheat	Dalling et al. 1976
	Peoples & Dalling 1978
	Peoples, Frith & Dalling 1979
	Walters et al. 1980
	Feller 1978, 1979
	Nair et al. 1978
	Wittenbach 1978
Maize	Feller et al. 1977
Barley	Peterson & Huffaker 1975
	Miller & Huffaker 1979
Oats	Martin & Thimann 1972
	Drivdahl & Thimann 1977, 1978
Perilla	Kannangara & Woolhouse 1968
Lolium temulentum	Thomas 1978
Ribonuclease	
Rhoeo	De Leo & Sacher 1970
Rhoeo	Sacher & Davies 1974
Phaseolus vulgaris	Phillips & Fletcher 1969
Oat	Udvardy et al. 1967

Table 2. Continued

Acid Invertase		
Lolium temulentum	Pollock & Lloyd 1978	
Chlorophyllase		
Barley and oats	Sabater & Rodriguez 1978	
Esterase		
Festuca pratensis	Thomas & Bingham 1977	
Acid Phosphatase		
Perilla	Kannagara & Woolhouse 1968	
Aleurone layers of barley		
α-amylase	De novo synthesis	Varner 1975
protease	De novo synthesis	Varner 1975
ribonuclease	De novo synthesis (in part)	Varner 1975
β-glucosidase	De novo synthesis (in part)	Varner 1975
phosphatase	Release + de novo synthesis	Varner 1975
pentosanase	Release + de novo synthesis	Dashet & Chrispeels (1977)
peroxidase	Release	Varner 1975
esterase	Release	Varner 1975

Hydrolase regulation. The possible levels of control of the hydrolytic activity of senescing leaves, according to the present state of knowledge, are summarized in Figure 11. The diagram offers a vast array of possibilities but they may for convenience be broadly divided under three headings (i) *de novo* synthesis of the enzymes; reactions and factors a to k (ii) activation and *in situ* regulation; reactions m and n, and (iii) compartmentation; reactions o to p. It is important to consider briefly the evidence for these various levels of control.

De novo synthesis: Reaction (a) Several workers have treated excised leaves with actinomycin-D without affecting the rate of senescence (115,116,117) from which the conclusion has been drawn that a transcription control is not involved in senescence. It is not certain whether the inhibitor is able to gain access to the nuclei under the conditions of these experiments. Attempts to demonstrate qualitative differences in the RNA produced in young and old barley leaves by means of DNA:RNA hybridization have also proved negative (118) though here again uncertainties surround the efficiency of the technique employed.

Message processing: Reactions b and c. Virtually nothing is known concerning mRNAs for the hydrolytic enzymes in leaves, of whether they contain interrupted sequences and require processing or of whether they become polyadenylated. Maize seedlings have been shown to be an active source of RNA polydenylating enzyme and the polyadenylation inhibitor cordycepin has been shown to delay senescence in leaf discs of *Nicotiana* (119) providing an interesting lead worthy of further investigation.

Reactions and factors of the ribosome cycle: reactions d to k. The ribosome cycle in eukaryotes is still imperfectly understood but it seems reasonable to suppose that we are here concerned with a vast array of possible controls on protein synthesis which will include: availability of membrane surfaces upon which polysome assembly can occur, availability of intact ribosomal sub units, appropriate supplies of RNA and the corresponding aminoacyl tRNA synthetases and availability of active initiation, translocation and termination factors.

The cytoplasmic ribosome cycle in mature leaves has not been analyzed at this level of detail and so we have no direct evidence concerning rate-limiting reactions involving any of these factors. Callow et al (95) obtained active 80s-based polysome fractions from leaves of *Perilla* throughout maturation and senescence up to the time of abscission, long after the 70s-based plastid system had become inactive. Most of the evidence on the involvement of protein synthesis in senescence is however indirect and depends upon the

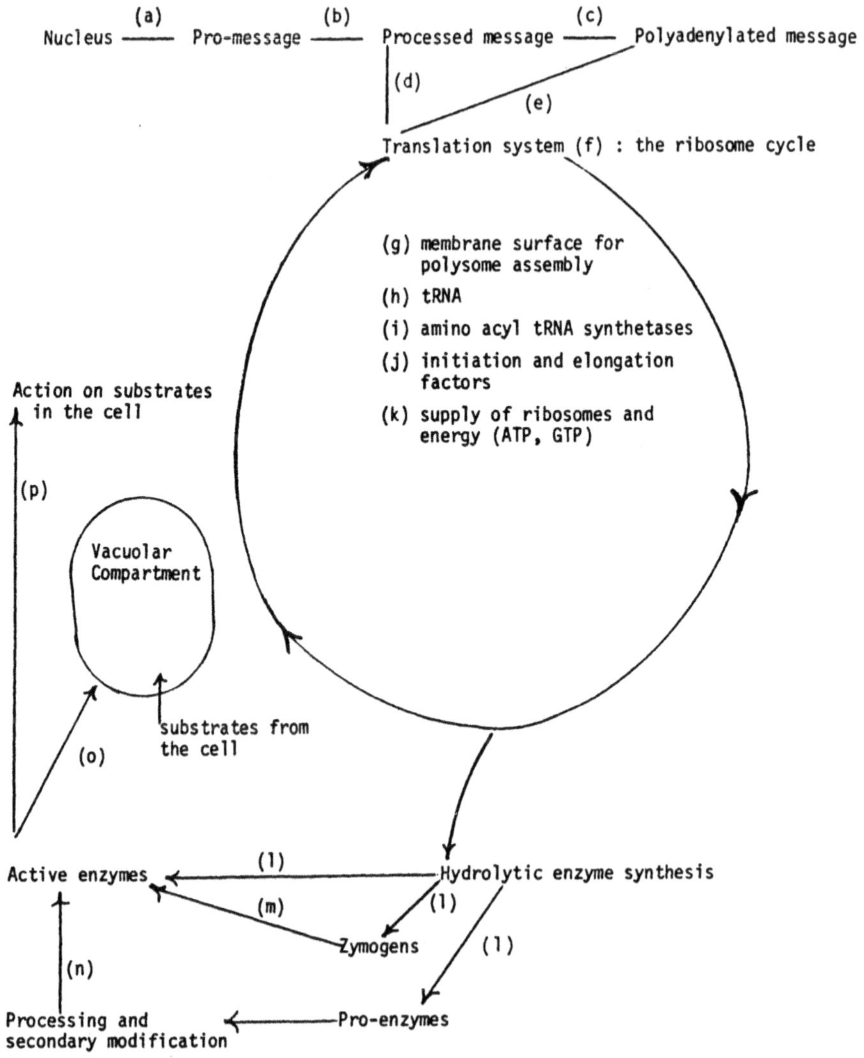

Fig. 11. Diagram indicating some of the multiple controls which may be involved in the regulation of the increased hydrolytic enzyme activity associated with the later stages of leaf senescence. Reactions and factors (a) to (l) are concerned with aspects of de novo synthesis of the enzymes; reactions m and n denote in situ regulation of the enzymes by processing, secondary modification or release from a zymogen; reaction (o) indicates enzyme regulation by compartmentation; reaction (p) denotes enzyme regulation at the substrate level.

credence which one is prepared to give to the results of applying inhibitors. It had been the general experience that inhibitors of protein synthesis such as MDMP (120) and cycloheximide, (121, 122, 123, 79) arrest the progress of senescence in excised leaf tissue. It is tempting to conclude that the key to this effect is the blocking of synthesis of degradative enzymes which mediate the autolysis of cell constituents; it may be born in mind however that these inhibitors will block the synthesis of _all_ enzymes being made on 80s ribosomes and this may be sufficient to arrest other aspects of metabolism whether involved in senescence or not, rather than just the synthesis of degradative enzymes.

The reader may cull from the literature numerous papers having a possible bearing upon the ribosome cycle in leaves; I refer to such matters as changing patterns of leucyl-tRNA and leucyl-tRNA synthetases, changes in lability of rRNA under electrophoresis and lowered protein: rRNA ratios in the ribosomes (124). In the opinion of this reviewer there is insufficient evidence on which to accord any of these observations a specific significance in the process of senescence.

The only conclusive evidence for regulation by _de novo_ synthesis, of a hydrolase which increases during leaf senescence concerns a ribonuclease. When leaf discs of _Rheo discolor_ are incubated in darkness there is a loss of RNA and protein accompanied by an increase in ribonuclease and acid phosphatase activities (125). The increase in activity of these enzymes is suppressed by IAA, kinetin and inhibitors of RNA and protein synthesis and is promoted by abscisic acid. Sacher and Davies (126) used a density labelling technique to show that the increased ribonuclease activity is accompanied by _de novo_ synthesis of the enzyme. The precise level at which this enzyme synthesis is being controlled is not known and there is clearly a great need to extend this type of work to the other enzymes which show increased activity in the course of leaf senescence.

Hydrolase control by compartmentation. As may be seen from Table 2, the reports of hydrolases associated with senescence in leaves are becoming legion; likewise we have seen that the substrates for these enzymes, the nucleic acids and proteins in particular, are being lost at different rates within a given organelle and the time course of losses is different for different organelles. Difficult questions therefore arise as to the localization these hydrolytic enzymes in the cell. Do the enzymes move to the components which are to be degraded or _vice-versa_? If the former then why is it that the hydrolase having arrived in a particular organelle degrades one protein relatively quickly but not its neighbour?

The currently popular view of leaf senescence urges that > 95% of the proteases present in leaves are contained in the cell vacuoles which function as lysosomes which degrade the proteins or indeed whole organelles which pass into them (Figure 12) (127). At the final dissolution of the cell the tonoplast membrane will be ruptured when these proteases pass from the vacuoles along with the rest of the lysosomal hydrolases bringing about the final dissolution of the cell contents. For a detailed discussion of the vacuolar hydrolase system see Woolhouse (73). It is my purpose here however to suggest that this represents far too crude a system to achieve the degree of control implied by the widely differing rates of turnover of the various proteins in the cell. I wish to suggest that the vacuolar system does not account for the observed facts of orderly events in leaf senescence and to propose in its stead an alternative hypothesis invoking a precisely regulated proteolytic system. Figure 13 shows quasi-independent protein cycles within the cytoplasm and organelles. In each of these cycles the numbers 1 and 2 refer to the generally accepted view of protein synthesis, being the activation of an amino acid (AA) to form aminoacyl tRNA and 2 the subsequent incorporation via the ribosomal cycle. Steps 3 and 4 indicate a regulated degradation of proteins in which a protein to be removed is first "recognized" and converted to an "activated" form which is susceptible to proteolytic attack (step 3) and the activated protein is then broken down (step 4).

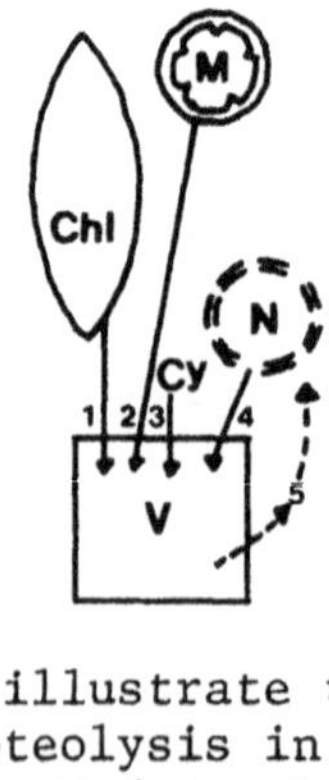

Fig. 12. A simple model to illustrate the autophagic vacuole concept in relation to proteolysis in plant cells. CY (cytoplasm): Chl (chloroplast) ; M (mitochondrion) ; N (nucleus) ; V (Vacuole). The numbers 1 - 4 denote transport pathways of proteins from the organelles to the vacuole. Number 5 denotes the pathway of release of amino acids formed by proteolysis in the vacuole. The difficulties in equating a model of this kind with the variety of half-lives of proteins in an organelle is evident (see text).

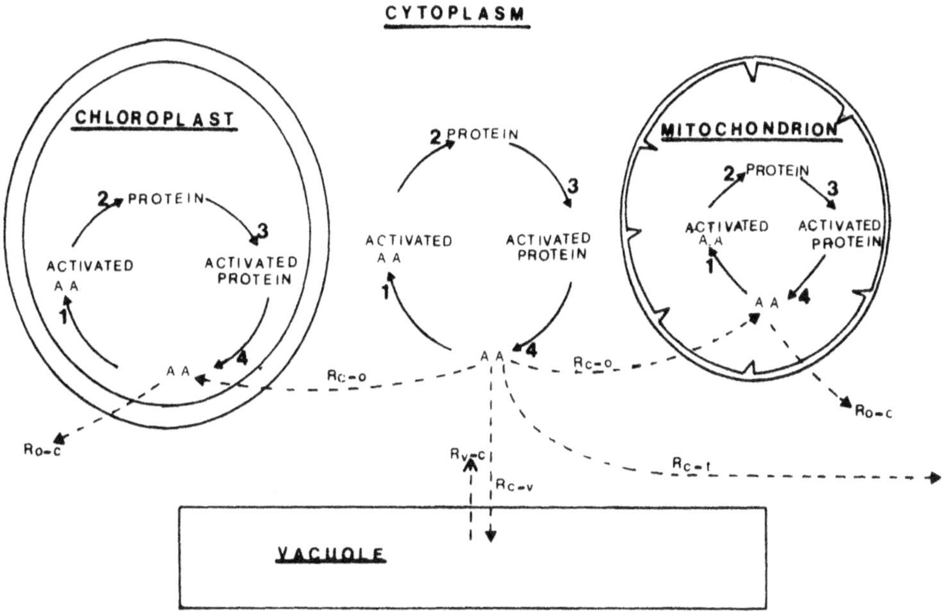

Fig. 13. Scheme representing the three protein cycles which co-exist in leaf-cells and aspects of their interrelationships. The numbered steps in each cycle indicate 1, amino acid incorporation into protein, 3, energy-dependent protein activation prior to degradation (see test and Fig. 14) and 4. hydrolytic breakdown of protein.

The dotted lines bearing the symbol R represent possible route along which amino acids or peptides may be transported, thereby linking the substrate pools of the 3 amino acid cycles as follows: Ro-c organelle to cytoplasm; Rc-o cytoplasm to organelle; Rc-v cytoplasm to vacuole, Rv-c vacuole to cytoplasm and Rc-t cytoplasm to the phloem transport system mediating export from the leaf.

Regulation of degradation is envisaged as occurring particularly at stage 3, which is an energy-dependent process. Step 3 is based on the findings in animal and microbial cells, that there is a low molecular weight protein APF-1 (128,129) which combines with the protein which is to be degraded, in an ATP-dependent reaction, to form a covalently bonded APF_1-protein ecomplex in which form the protein moiety is now in a conformational state in which it is susceptible to proteolytic degradation. A more detailed picture of the regulation of step 3 is presented in Figure 14.

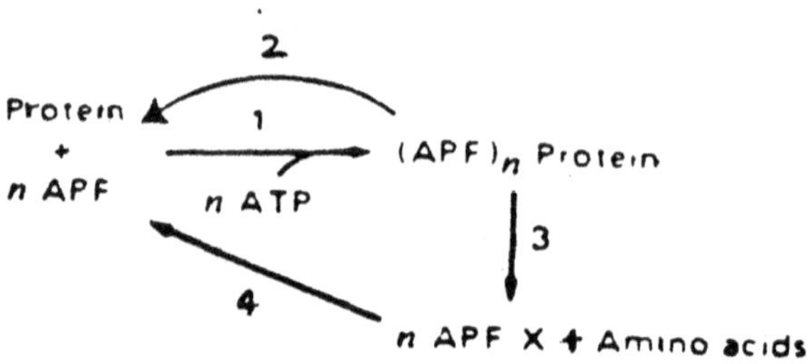

Fig. 14. A suggested model for energy-dependent breakdown of proteins. 1. APF-1-protein amide synthetase (acting on lysine ε - NH_2 groups); 2, Amidase that allows correction when n + 1 or 2; 3. Peptidases which act preferentially on (APF-1)n derivatives, when n > 1 or 2; 4, Amidase for APF-1-x. X is lysine or a small peptide. (From Hershko et al. (130). For explanation see text.

In advancing this hypothesis here I am seeking primarily to point to new areas in which the study of plant senescence appears to be leading. Considerations of space preclude the inclusion of full discussion of compartmentation and energy-dependent protein degradation which are required here, for details see references (131,132).

CONCLUSION

I have had in mind two main objectives in this introductory review of plant senescence. Firstly, in Part I, I have surveyed the various manifestations of senescence as they are to be found in all phases of the life cycles of plants and have attempted to show how the events of senescence are embodied in the genetic programme of development. Secondly in Part II, I have considered just one facet of the regulation of cell metabolism, nucleic acid and protein turnover, in order to exemplify the deeper questions which must be answered as we seek to explain how the senescence processes are regulated. It is important to emphasize that I have made a deliberate selection by way of an example.

An equally cogent case could have been made for considering the regulation of liquid turnover and its role in altering membrane functions in senescing tissues. Likewise I have said nothing of hormonal controls which impinge upon many of the processes we have mentioned (133). Thus we see that even in its basic aspects the study of plant senescence is a vast and complex study; on these grounds alone it is surely worthy of our serious attention but to this may be added its enormouse economic importance in agriculture and horticulture with which we shall be much concerned in the later stages of this symposium.

REFERENCES

1. P. B. Medawar, A unsolved problem of biology. Reprinted in "The Uniqueness of the Individual," Methuen, London (1957).
2. H. W. Woolhouse. Senescence processes in the life cycle of flowering plants. Bioscience, 28:25 (1978).
3. C. W. Minot, The Problem of Age, Growth and Death, G. P. Putnam's Sons, London. (1908).
4. W. Pfeffer, The Physiology of Plants, Volume 2 Transl. by A. J. Ewart, Clarenden Press, Oxford (1903).
5. D. W. Thompson, On Growth and Form. 2nd Edition Cambridge University Press, (1942).
6. P. F. Wareing and A. K. Seth, Aging and Senescence in the whole plant. Symposium of Society for Experimental Biology 21: 543 (1967).
7. D. H. Maggs, The distance from tree base to shoot origin as a factor in shoot and tree growth. J. Hort. Sci. 39:298 (1964).
8. M. Möbius. Beiträge zur lehre von der Fort - pflanzung der Gewäsche. Gustav Fischer, Jena, Germany (1897).
9. K. Sax, Aspects of Ageing in Plants. Ann. Rev. Plant Physiol. 13:489 (1962).
10. N. P. Krenke, Regeneratsiya rastenii. Izd. Akad. Nauk. USSR, Moscow-Leningrad (1950).
11. O.V.S. Heath, Ageing in higher plants. In: The Biology of Ageing. ed., W. B. Yapp and G. H. Bourne, Symposium of the Institute of Biology 6:9 (1957).
12. V. C. Wynn-Edwards, Animal dispersion in relation to social behaviour. Oliver and Boyd, Edinbergh and London (1962).
13. A. Comfort, The Biology of Senescence. Routledge and Kegan Paul, London (1956).
14. A. Comfort, The Biology of Senescence. Routledge and Kegan Paul, London (1956).
15. J. Maynard Smith, The causes of ageing, Proceedings of the Royal Society, London (B), 157:115 (1962).
16. A. Comfort, Ageing: The Biology of Senescence, Routledge and Kegan Paul, London (1965).
17. T. King. See Reference 13 (1933).
18. R. B. Setlow and W. L. Carrier, The disappearance of thymine dimers from DNA: an error-correcting mechanism. Proc. Natl. Acad. Sci. USA 51:226 (1964).
19. R. P. Boyce and P. Howard-Flanders, Release of Ultraviolet light-induced thymine dimers from DNA in E.coli K12, Proc. Natl. Acad. Sci. USA 51:293 (1964).
20. R. W. Hart and R. B. Setlow, Correlation between deoxyribonucleic acid excission-repair and life-span in a number of mammalian species, Proc. Natl. Acad. Sci. USA 71:2169 (1974),
21. R. E. Rasmussen and R. B. Painter, Radiation-stimulated DNA-synthesis in cultured mammalian cells. J. Cell Biol. 29:11 (1966).

22. R. B. Setlow, J. D. Regan and W. L. Carrier, Biophysical Society Abstracts 12:19a (1972).
23. J. E. Cleaver and J. E. Trosko, Absence of excision of ultraviolet-induced cyclobutane dimers in Veroderma pigmentosum, Photochem Photobiol 11:547 (1970).
24. J. E. Cleaver, G. H. Thomas, J. E. Trosko, and J. T. Lett Excision repair (dimer excision, strand breakage and repair replication) in primary cultures of eukaryotic (Bovine) cells, Exp. Cell Res. 74:67 (1972).
25. H. Sober (ed.) Handbook of Biochemistry, Chemical Rubber Company, Cleveland, Ohio (1970).
26. J. D. Regan and R. B. Setlow, In: Chemical Mutagens, ed. A. Hollaender, Plenum, New York 3:151 (1973).
27. C. W. Ferguson, Bristlecone Pine: Science and Esthetics, Science 159:839 (1968).
28. T. Tanada and S. B. Hendricks, Photoreversal of ultraviolet effects in soybean leaves. American J. Bot. 40:634 (1953).
29. T. A. Skokut, J. H. Wu and R. S. Daniel, Retardation of ultraviolet light accelerated chlorosis by visible light or by benzyladenine in Nicotiana glutinosa leaves: changes in amino acid content and chloroplast ultrastructure. Photochem. Photobiol. 25:183 (1977).
30. D. C. Swinton and P. C. Hanawalt, Absence of ultraviolet-stimulated repair replication in the nuclear and chloroplast genomes of Chlamydomonas rheinhardii, Biochem. Biophys. Acta 294:385 (1973).
31. S. Wolff and J. E. Cleaver, Absence of DNA replication after chemical mutagen damage in Vicia faba Mut. Res. 20:71 (1973).
32. G. D. Small and C. S. Griemann, Repair of pyrimidine dimers in ultraviolet irradiated Chlamydomonas, Photochem. Photobiol. 25:183 (1977).
33. G. P. Howland, Dark-repair of ultraviolet-induced pyrimidine dimers in the DNA of wild carrot protoplasts, Nature (London) 254:160 (1975).
34. T. M. Sonneborn, Genetic studies on Stenostomium incaudatium n. sp. 1. The nature and origin of differences in individuals formed during vegetative reproduction. J.Exp. Zool. 57:57 (1930).
35. E. Wangermann and E. Ashby, Studies in the morphogenesis of leaves VIII. Part I. Effects of light intensity and temperature on the cycle of ageing and rejuvenation in the vegetative life history of Lemna minor. New Phytol. 50:186 (1951).
36. E. Ashby and E. Wangermann, Studies in the morphogenesis of leaves VII. Part II. Correlative effects of fronds in Lemna minor, New Phytol. 30:200 (1951).
37. E. Wangermann, Studies in the morphogenesis of leaves VIII. A note on the effects of length of day and of removing daughter fronds on ageing of Lemna minor, New Phytol. 51:355 (1952).
38. E. Wangermann and H. J. Lacey, Studies in the morphogenesis of

leaves IX. Experiments on Lemna minor with adenine, tris-iodobenzoic acid and ultraviolet radiation. New Phytol. 52: 298 (1953).
39. E. Wangermann and H. J. Lacey. Studies in the morphogenesis of leaves X. Preliminary experiments on the relation between nitrogen nutrition, rate of respiration and rate of ageing of fronds of Lemna minor, New Phytol. 54:182 (1955).
40. F. C. Steward, M. O. Mapes, A. E. Kent and R. D. Holsten, Growth and development of cultured plant cells, Science 145:20 (1964).
41. F. C. Steward, Physiological aspects of organization In: Trends in Plant Morphogenesis, ed. E. G. Cutter, Longmans, Green & Co., London (1966).
42. F. A. L. Clowes and B. E. Juniper, Plant Cells, Blackwell Scientific Publication, Oxford (1968).
43. A. W. Robards, Plásmodesmata in Higher Plants. In: Intercellular communications in plants: Studies on plasmadesmata, ed. B. E. S. Gunning and A. W. Robards, Springer-Verlag, Berlin (1976).
44. R. F. M. Van Steveninck, Cytochemical evidence on ion transport through plasmodesmata, In: Intercellular communication in plants: Studies on plasmodesmata, ed. B.E.S. Gunning and A. W. Robards, Springer-Verlag, Berlin (1976).
45. A. Gibbs, Viruses and Plasmodesmata. In: Intercellular communications in plants: Studies on plasmodesmata, ed. G.E.S. Gunning and A.W. Robards, Springer-Verlag, Berlin (1976).
46. D. M. Prescott, The cell cycle and the control of cellular reproduction, Adv. Genetics 18:99 (1976).
47. R. Braun and K. Behrens, A ribonuclease from Physarum: biochemical properties and synthesis in the mitotic cycle, Biochem. Biophys. Acta 195:87 (1969).
48. M. D. Scharff and E. Robbins, Polyribosome disaggregation during metaphase, Science 151:992 (1966).
49. D. L. Steward, J. R. Shaeffer and R. H. Humphrey, Breakdown and assembly of polyribosomes in synchronized Chinese hamster cells, Science 161:791 (1968).
50. L. D. Hodge, E. Robbins and M. D. Scharff, Persistance of messenger RNA through mitosis in HeLa cells. J. Cell Biol. 40:497 (1969).
51. H. Fan and S. Penman, Regulation of synthesis and processing of nucleolar components in metaphase-arrested cells. J. Mol. Biol. 59:27 (1971).
52. R. M. Groyan and A. Kniazeff, Vaccinia virus infection of synchronized pig kidney cells, J. Virol. 1:1255 (1967).
53. G. S. Stein and D. E. Matthews, Non-histone chromosomal protein synthesis: utilization of pre-existing and newly transcribed messenger RNA's Science 181:71 (1973).
54. J. Heslop-Harrison, The cytoplasm and its organelles during meiosis. In: Pollen: Development of and Physiology. ed. J. Heslop-Harrison, Butterworths, London (1971).
55. J. H. Taylor, Autoradiographic studies of nucleic acids and

proteins during meiosis in Lilium longiflorum, Am. J. Bot. 46:477 (1959).
56. L. Albertini, Étude autoradiographique des synthesès d'acide ribonucléique (RNA) au cours de la microsporogenèse chez le Rhoes discolor Hanie. C.R.helsd. Séanc. Acad. Sci. Paris, 260:651 (1965).
57. N. K. Das, Inactivation of the nucleolar apparatus during meiotic prophase in corn anthers, Exp. Cell Res. 40:360 (1965).
58. J. J. Sauter, Istoautoradiographische Untersuchung der Protein-synthese während der Meiosis bei Paeonica tenuifolia L. Naturwiss. 55:187 (1968a).
59. J. J. Sauter, Istoaudioradiographische untersuchungen zur Ribonucleinsaure-synthese wahre d der Meiosis bei Paeonina tenuifolia, Naturwiss. 55:236 (1968b).
60. A. Mackenzie, J. Heslop-Harrison and H. G. Dickinson, Elimination of ribosomes during meiotic prophase. Nature, Lond. 215:997 (1967).
61. K. Maruyama, Electron microscopic observations of plastids and mitochondria during pollen development in Tradescantia paludosa, Cytologia 33:482 (1968).
62. H. G. Dickinson and J. Heslop-Harrison, The ribosome cycle, nucleoli, and cytoplasmic nucleoids in the meiocytes of Lilium, Protoplasma 69:187 (1970).
63. I. Borodin, Bot. Jahrb. 4:919 (1876).
64. E. Schulze, Über zersetzang und Neubildung von Eiweisstoffen in Lupinekeimlingen. Handw. Jb. 7:411 (1878).
65. K. Mothes, Die vakuuminfiltration in ernahrungsversuch (Dargestellt au untersuchungen über die assimilation des ammoniaks), Planta 19:117 (1933).
66. F. G. Gregory and G. K. Sen, Physiological studies in plant nutrition. VI. The relation of respiration rate to the carbohydrate and nitrogen metabolism of the barley leaf as determined by nitrogen and potassium deficiency. Ann. Bot. 1:521 (1937).
67. A. Meyer, Eiweisstoffwechsel und vergilben der laublatter von Tropaeolum majus, Flora (Jena) 111:85 (1918).
68. G. Michael, Über die Beziehungen zwischen Chlorophyll und Eiweissabbau vergelbenden Laublatt von Tropaeolum. Z. Bot. 29:385 (1936).
69. B. Parthier, Untersuchungen über den Aminosäureeinbau in die Blatteiweiss des Tabaks, Flora, Jena 151:368 (1961).
70. K. Hardwick and H. W. Woolhouse, Changes in the composition of leaves of Perilla frutescens during foliar senescence. New Phytol. 66:545 (1967).
71. H. W. Woolhouse, The nature of senescence in plants, Symp. Soc. Expt. Biol. 21:179 (1967).
72. H. W. Woolhouse, Longevity and senescence in plants. Sci. Prog. Oxford 61:223 (1974).
73. H. W. Woolhouse, Cellular and metabolic aspects of senescence in higher plants. In: Biology of Ageing, ed. J. Behnke, C.

Finch and J. Moment, Plenum Press (1978).
74. D. D. Davies, The measurement of protein turnover in plants. Adv. Bot. Res. 8:65 (1981).
75. H. W. Woolhouse and T. Batt, The nature and regulation of senescence in plastids. In: Perspectives in Experimental Biology, Vol. 2, ed. N. Sunderland, Pergamon Press, Oxford, (1976).
76. J. W. Friedrich and R. C. Huffaker, Photosynthesis and protein degradation in senescing barley leaves. I: Leaf resistances and ribulose-1, 5-bisphosphate carboxylase, Plant Physiology (1979).
77. K. Hardwick, M. E. Wood and H. W. Woolhouse, Photosynthesis and respiration in relation to leaf age in Perilla frutescens (L) Britt. New Phytol. 67:79 (1968).
78. L. W. Peterson, G. E. Kleinkopf and R. C. Huffaker, Evidence for lack of turnover of ribulose 1-5-diphosphate carboxylase in barley leaves, Plant Physiol. 51:1042 (1973).
79. L. W. Peterson and R. C. Huffaker, Loss of ribulose, 1,5-diphosphate carboxylase and increase in proteolytic activity during senescence of detached primary barley leaves. Plant Physiol. 55: 1009 (1975).
80. E. Simpson, R. J. Cooke and D. D. Davies, Measurement of protein in leaves of Zea mays using [^{3}H] acetic anhydride and tritiated water. Plant Physiol. 67:1214 (1981).
81. T. Batt and H. W. Woolhouse, Changing activities during senescence and sites of synthesis of photosynthetic enzymes in leaves of the labiate, Perilla frutescens (L) Britt. J. J. Exp. Bot. 26:569 (1975).
82. G. C. Kannangara and H. W. Woolhouse, The role of carbon dioxide, light and nitrate in the synthesis and degradation of nitrate reductase in Perilla frutescens, New Phytol. 66:553 (1967).
83. H. Brown, Personal communication (1981).
84. R. D. Butler and E. W. Simon, Ultrastructural aspects of senescence in plants, Adv. Gerontol. Res. 3:73 (1971).
85. B. D. McKersie, J. E. Thompson and J. K. Brandon, x-ray diffraction evidence for decreased lipid fluidity in senescent membranes from cotyledons. Can.J.Bot. 54:1074 (1976).
86. B. D. McKersie and J. E. Thompson, Phase behaviour of chloroplast and microsomal membranes during leaf senescence. Plant Physiol. 61:639 (1978).
87. J. A. Sacher, Senescence: effects of auxin and kinetin on RNA and protein synthesis in subcellular fractions of fruit and leaf tissue sections. In: Biochemistry and physiology of plant growth substances, ed. F. Wightman and G. Setterfield Runge Press, Ottawa (1967).
88. J. A. Sacher and S. O. Salminen, Comparative studies of effect of auxin and ethylene on permeability and synthesis of RNA and protein, Plant Physiol. 44:1371 (1969).
89. B. W. Pooviah and A. C. Leopold, Deferral of leaf senescence with calcium. Plant Physiol. 52:236 (1973).

90. A. R. J. Eaglesham and R. J. Ellis, Protein synthesis in chloroplasts. I: Light-driven synthesis of membrane proteins by isolated chloroplasts. Biochim. Biophys. Acta. 335-396 (1974).
91. M. Edelman and A. Reisfeld, Synthesis, processing and functional probing of P-32,000, the major membrane protein translated within the chloroplast. In: Geome organization and expression in plants, ed. C. J. Leaver, Plenum Press, London (1980).
92. S. Ochoa and C. de Hars, Regulation of protein synthesis in eukaryotes. Ann. Rev. Biochem. 48:549 (1979).
93. G. I. Jenkins and H. W. Woolhouse, Photosynthetic electron transport during senescence of the primary leaves of Phaseolus vulgaris L. I: Non-cyclic electron transport, J. Exp. Bot. 32:467 (1981).
94. N-H. Chua and N. W. Gilham, The sites of synthesis of the principal thylakoid membrane polypeptides in Chlamydemonas reinhardii J. Cell Biol. 74:441 (1977).
95. J. A. Callow, M. E. Callow and H. W. Woolhouse, In vitro protein synthesis, ribosomal RNA synthesis and polyribosomes in senescing leaves of Perilla, Cell Differentiation 1: 79 (1972).
96. W. Bünger and J. Feieraband, Capacity for RNA synthesis in 70S ribosome-deficient plastids of heat-bleached rye leaves, Planta 149:163 (1980).
97. R. J. Ness and H. W. Woolhouse, RNA synthesis in Phaseolus chloroplasts. 1: Ribonucleic acid synthesis in chloroplast preparations from Phaseolus vulgaris L. leaves and solubilization of the RNA polymerase, J. Exp. Bot. 31:223 (1980a).
98. P. J. Ness and H. W. Woolhouse, RNA synthesis in Phaseolus chloroplasts. II. Ribonucleic acid synthesis in chloroplasts from developing and senescing leaves, J. Exp. Bot. 31:235 (1980b).
99. C. J. Brady and N. S. Scott, The persistance of plastid polyribosomes and Fraction 1 protein synthesis in ageing wheat leaves. In: Colloques internationaux de centre nationale de la recherch scientifique No. 261. Acides nucleiques et synthese des proteines chez les végêtaux. ed. L. Bogorad and J. H. Weil, Paris, France (1976).
100. R. W. Brougham, N. Z. Jl. Agric. Res. 1:707 (1959).
101. K. V. Thimann, The senescence of leaves. In: Senescence in plants, ed. K. V. Thimann, C.R.C. Press Boca Raton (1980).
102. D. A. Teffer and D. E. Fosket, Hormone-mediated translational control of protein synthesis in cultured cells of Glycine max, Develop. Biol. 62:486 (1978).
103. M. E. Callow and H. W. Woolhouse, Changes in nucleic acid metabolism in regreening leaves of Perilla, J. Exp. Bot. 24: 285 (1973).
104. S. G. Siddell and R. J. Ellis, Protein synthesis in chloroplasts. VI: Characteristics and products of protein

synthesis _in vitro_ in etioplasts and developing chloroplasts from pea leaves, Biochem. J. 146:675 (1975).
105. J. Silverthorne and R. J. Ellis, Protein synthesis in chloroplasts. VIII: Differential synthesis of chloroplast protein during spinach leaf development. Biochem. Biophys. Acta 607: 319 (1980).
106. L. Bogorad, S. O. Jolly, G. Kidd, G. Link and L. McIntosh, Organization and transcription of maize chloroplast genes. In: Genome organization and expression in plants. ed. C. J. Leaver, Plenum Press (1980).
107. J. R. Bedbrook, R. Kolodner and L. Bogorad, _Zea mays_ chloroplast ribosomal RNA genes are part of a 22,000 base pair inverted repeat, Cell 11:739 (1977).
108. B. L. Jenni and E. Stutz, Physical mapping of the ribosomal DNA region in _Euglena gracilis_ Eur. J. Biochem. 88:127 (1978).
109. N. Chu and K. K. Tewari, Arrangement of the ribosomal RNA genes in the restriction endonuclease map of pea chloroplast DNA. Citation by T. A. Dyer, J. R. Bedbrook, Genes coding for chloroplast ribosomal RNA. In: Genome organization and expression in plants, ed. C. J. Leaver, Plenum Press (1979).
110. G. Burkard, J. Canady, E. Crouse, P. Guillemant P. Imbault, G. Keith, M. Keller, M. Mubumbila, L. Osorie, V. Sarantoglou, A. Steinmetz and J. H. Weil, Transfer RNAs and aminoacyl-tRNA synthesis in plant organelles. In: Genome organization and expression in plants. ed. C. J. Leaver, Plenum Press (1980).
111. K. Hardwick, M. E. Wood and H. W. Woolhouse, Photosynthesis and respiration in relation to leaf age in _Perilla frutescens_ (L) Britt. New Phytologist 67:79 (1968).
112. O. Tiboni, G. Di Pasquale and C. Cifferi, Purification of the elongation factors present in spinach chloroplasts, Eur. J. Biochem. 92:471 (1980).
113. O. Ciferri, G. Di Pasquale and O. Tiboni, Chloroplast elongation factors are synthesized in the chloroplast. Eur. J. Biochem. 102:331 (1979).
114. S. P. Waters, M. B. Peoples, R. J. Simpson and M. J. Dalling, Nitrogen redistribution during grain growth in wheat (_Triticum aestivum_ L.) 1: Peptide hyrolase activity and protein breakdown in the flag leaf, glumes and stem. Planta 148:422 (1980).
115. G. J. von Abrams, An effect of ornithine on degradation of chlorophyl and protein in excised leaf tissue. Z. Pflanzenphysiol. 72:410 (1974).
116. H. Thomas, Leaf growth and senescence in grasses, Annual Rep. Welsh Plant Breeding Station, Aberystwyth (1975).
117. H. Thomas and J. L. Stoddart, Biochemistry of leaf senescence in grasses. Ann. Appl. Biol. 89:461 (1977).
118. B. I. S. Srivastava, RNA-DNA hybridization competition studies

on senescing barley leaves, New Phytol. 71:93 (1972).
119. T. Jakezami and K. Yoshida, Remarkable retardation of the senescence of tobacco leaf discs by cordycepin, an inhibitor RNA polyadenylation Pl. Cell Physiol. 16:1163 (1975).
120. H. Thomas, Delayed senescence in leaves treated with the protein synthesis inhibitor MDMP Plant Sci. Lett. 6:369 (1976).
121. J. S. Knypl and W. Mazurczyk, Arrest of yellowing in senescing leaf discs of maize by growth retardents, coumarin and inhibitors of RNA and protein synthesis, Biol. Plant 12: 199 (1972).
122. S. N. Makovetzki and E. E. Goldschmidt, A requirement for cytoplasmic protein synthesis during chloroplast senescence in the aquatic plant Anacharis canadensis, Pl. Cell Physiol. 17:859 (1976).
123. C. Martin and K. V. Thimann, The role of protein synthesis in the senescence of leaves. 1: The formation of protease, Pl. Physiol. 49:64 (1972).
124. H. Thomas and J. L. Stoddart, Leaf senescence, Ann. Rev. Pl. Physiol. 31:83 (1980).
125. P. De Leo and J. A. Sacher, Control of ribonuclease and acid phosphatase by auxin and abscisic acid during senescence of Rhoes leaf sections, Pl. Physiol. 46:806 (1970).
126. J. A. Sacher and D. D. Davies, Demonstration of de novo synthesis of RNAase in Rheo leaf sections by deuterium oxide labelling. Pl. Cell Physiol. 15:157 (1974).
127. P. Matile, Biochemistry and function of vacuoles, Ann. Rev. Pl. Physiol. 29:193 (1978).
128. A. Ceichanover, H. Heller, S. Elias, A. L. Haas and A. Harsko, ATP-dependant conjugation of reticulocyte proteins with the polypeptide required for protein degradation, Proc. Natl. Acad. Sci. USA 77:1365 (1980a).
129. A. Ciechanover, S. Elias, H. Heller, S. Ferber and A. Hershko, Characterization of the heatstable polypeptide of the ATP-dependant proteolytic system from reticulocytes, J. Biol. Chem. 255:7525 (1980).
130. A. Hershko, A. Ciechanover, H. Heller, A. L. Haas and I. A. Rose, Proposed role of ATP in protein breakdown: Conjugation of proteins with multiple chains of the polypeptide of ATP-dependent proteolysis, Proc. Natl. Acad. Sci. USA 73:1783 (1980).
131. H. W. Woolhouse, Biochemical and Molecular Aspects of Plant Senescence. In: Molecular Biology of Plant Development, ed. H. Smith and D. Grierson Blackwells, Oxford,(1981).
132. H. W. Woolhouse and G. I. Jenkins, Physiological responses, metabolic changes and regulation during leaf senescence (1981).
133. H. W. Woolhouse, Hormonal control of senescence-related to reproduction in plants. In: Strategies of plant reproduction, Beltsville Symposia in Agricultural Research VI, ed. W. J. Meudt, US Dept. Agric. Beltsville (1981).

134. R. Wollgiehn, S. Lerbs and D. Munsche, Synthesis of ribosomal RNA in chloroplasts from tobacco leaves of different age. Biochem. Physiol. Pflanzen. 170:381 (1976).

CONTROL OF RIBONUCLEIC ACID AND ENZYME SYNTHESIS DURING FRUIT RIPENING

Don Grierson

Department of Physiology and Environmental Studies
University of Nottingham, School of Agriculture
Sutton Bonington
Loughborough, LE12 5RD, U.K.

INTRODUCTION

When fruits ripen they undergo major changes in composition which affect their attractiveness, storage life and nutritional value. This is a consequence of alterations in the physiological and biochemical processes occurring in the cells of the fruit. A list of some of the major changes that occur in tomatoes is given below. Similar changes occur in many other types of fruits.

1. Chlorophyll degradation
2. Synthesis of pigments such as lycopene and β-carotene.
3. Changes in organic acids and amino acids.
4. Starch degradation and formation of sugars.
5. Synthesis of compounds contributing to flavour and aroma.
6. Cell wall degradation leading to softening.

Ripening usually takes place after a certain stage of maturity has been reached. The timing is not directly related to chronological age, however, but is determined by a developmental sequence which is under genetic control. It is, therefore, misleading to consider ripening as a type of "ageing" and it is probably more correct to view it as a series of coordinated metabolic events switched on at a specific stage in development. In climacteric fruits, such as tomatoes, apples and bananas, the initiation of these changes is associated with increased synthesis of carbon dioxide and ethylene by the fruit tissue. Ethylene is regarded as a ripening "hormone" because unripe fruits can be stimulated to ripen by an exogenous supply of the gas and natural ripening can sometimes be retarded if ethylene produced by ripening fruits is removed from the surrounding atmosphere.

Early theories invoked alterations in membrane permeability and the release of degradative enzymes to explain the process of ripening. However, although such events may contribute to the later stages of ripening, there is now little doubt that the synthesis or activation of specific enzymes is required for normal ripening to occur. For example, the solubilization of cell-wall pectins that take place during ripening of tomato is at least partly due to the enzyme polygalacturonase. There is virtually no polygalacturonase activity in mature-green tomatoes but enzyme activity appears as fruits begin to ripen and continues to increase as ripening progresses (5,7,21). Invertase activity also increases dramatically during ripening (5,10) and other enzymes, such as pectinmethylesterase, may be present at the mature-green stage but subsequently increase substantially in activity (5).

Although many aspects of ripening appear to involve changes in enzyme synthesis or activity, including ethylene synthesis itself, which seems to depend on the appearance of enzyme ACC(1-aminocyclopropane-1-carboxylic acid) synthase, this does not mean that all enzyme changes occur by the same mechanism. Indeed this would be surprising given the range of synthetic and degradative processes which occur in such diverse cell compartments as the plastids, the cytoplasm and the cell walls. Treatment of tomatoes with gibberellins, for example, permits the respiratory climacteric to proceed while tomatoes remain green (3) and in the "greenflesh" mutant of tomato the plastids tend not to lose their chlorophyll although softening and lycopene synthesis appear to continue as normal (Grierson, unpublished observations). Furthermore, under controlled-atmosphere storage of green tomatoes, starch degradation occurs although there is little or no softening or colour change (Goodenough, Tucker, Grierson and Thomas, unpublished results). These observations suggest that some aspects of ripening can proceed independently of others and they do not all require exactly the same conditions to be manifested. It therefore follows that there may be more than one type of control mechanism operating during ripening although there may be one common initiating event.

For those interested in understanding the control of ripening the major challenge is, therefore, to learn how changes in enzyme synthesis and activity are brought about and to determine whether, and if so how, ethylene initiates these changes. In this article I shall discuss the role of RNA synthesis and protein synthesis in controlling enzyme changes during fruit ripening.

EVIDENCE THAT RIPENING INVOLVES CHANGES IN GENE EXPRESSION

A number of enzymes have been shown to change in activity during ripening (18) including malic enzyme (9), polygalacturonase (7) and invertase (10). Although such changes could be brought about by a variety of mechanisms, as discussed below, there is good evidence that specific stimulation of protein synthesis occurs during ripening. Hulme (8) showed that there was a net accumulation of protein during the ripening of apples and Hansen (6) showed ethylene caused an increase in protein nitrogen in pears. Furthermore, a stimulation of the incorporation of radioactive amino acids into protein has been demonstrated during the ripening of avocado (15), pears (4), apples (15), tomatoes (19) and bananas (17). In their experiments with pears Frenkel et al. (4) showed that there was enhanced incorporation of ^{14}C-phenylalanine into malic enzyme and other proteins during the early stages of ripening. These observations, and the finding that cycloheximide, which inhibits protein synthesis on 80S ribosomes, retards or prevents ripening in pears (4), citrus (16) and bananas (1), indicate that protein synthesis is an important event in ripening fruit.

RNA synthesis also seems to be associated with ripening. There is increased incorporation of radioactive uridine into RNA during the ripening of tomatoes (13,19) and ethylene has been shown to stimulate RNA synthesis in apples (9) and figs (11) and induce changes in the relative amounts of certain iso-accepting tRNAs in tomatoes (12).

The involvement of both nucleic acid and protein synthesis in the process of fruit ripening provides biochemical evidence that there is a requirement for gene expression in order for ripening to occur. There is also separate genetic evidence from the study of various tomato mutants implicating specific nuclear genes in the control of ripening (2,20). For example, in the honozygous condition the rin (ripening inhibitor) gene on chromosome 5 results in fruit which do not exhibit a respiratory climacteric or produce a burst of ethylene synthesis. Such fruit do not ripen normally but slowly lose their chlorophyll and turn yellow. They produce no polygalacturonase (21) and do not soften to any great extent. In contrast, fruit from plants homozygous for the Nr (Never ripe) gene on chromosome 9 show a reduced climacteric and ethylene evolution, turn orange but not red, and soften slowly, probably due to the production of low amounts of only one of two isoenzymes of polygalacturonase(21). Both these mutations produce multipe phenotypic effects whereas the gf (green flesh) mutation on chromosome 8 results in fruit which appear to ripen normally except for the fact that they do not lose their chlorophyll.

POSSIBLE CONTROL POINTS IN GENE EXPRESSION

Figure 1 gives a brief outline of the steps involved in the espression of genetic information in the form of proteins. The number of active molecules of a given enzyme could be governed by regulatory processes occurring at any one of the steps 1 - 7. Although RNA and protein synthesis (steps 1-3) do seem to be important for ripening, other regulatory mechanisms apart from transcription and translation may also play a role (see article by

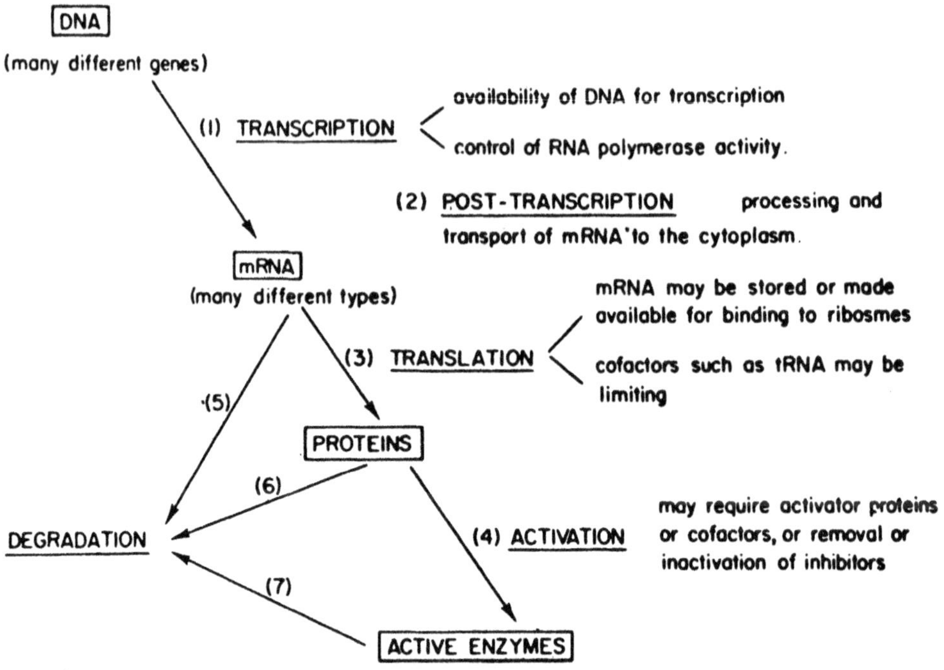

Fig. 1. Possible control points in gene expression. (From Grierson et al. (5)).

Rhodes in this volume). Figure 2 gives a general outline of the steps involved in the transcription, processing and translation of mRNA and the production of active enzymes. After transcription by RNA polymerase II many plant mRNAs are modified at the 5' and 3' ends by specific enzymes. Modification at the 5' end may involve the addition of a special "cap" sequence starting with a 7-methyl guanine linked via a 5'--5' triphosphate bridge to the second nucleotide in the chain. Modification at the 3' end of some, but not necessarily all, mRNAs involves the addition of a poly(A) sequence by poly(A) polymerase. In addition to being modified at each end the initial RNA transcript may be cut at specific points by nucleases. Following this type of processing reaction, only certain parts of the RNA sequence are retained and spliced together to produce functional mRNA molecules. Once formed, mRNAs may be made available for polyribosome formation and protein synthesis immediately or they may be stored until such time as their translation is allowed to proceed. It has been suggested that changes in isoaccepting tRNAs may be important for translational control (12).

The regulatory implications of the various RNA processing steps are enormous. Insufficient information is available at the present time, however, to allow a clear statement of the molecular mechanisms

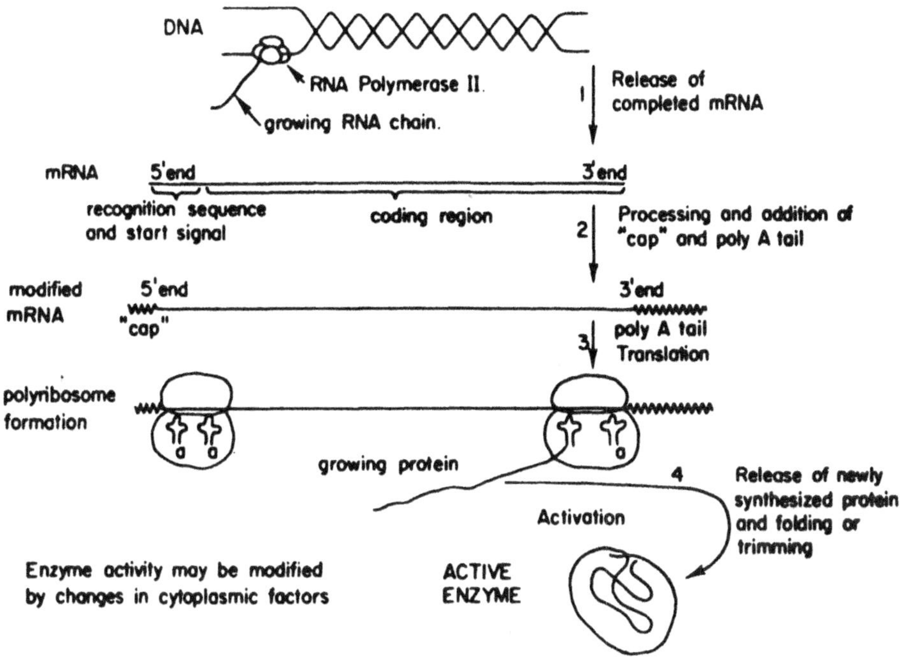

Fig. 2. The synthesis, processing and translation of messenger RNA. (From Grierson et a. (5)).

that determine whether or how a specific mRNA sequence is produced. Despite the complexity of the situation this nettle will have to be grasped if we are to understand how gene expression is controlled.

RNA SYNTHESIS DURING FRUIT GROWTH AND RIPENING

Figure 3 shows the change in nucleic acid content of two varieties of tomato fruits during growth and ripening. Plants were grown in a glasshouse and individual flowers tagged at anthesis. Fruits were sampled at various stages of growth and fresh weight, nucleic acid content and lycopene were measured. There is very little DNA synthesis or cell division during fruit growth and the increase in nucleic acid is due largely to RNA synthesis. The accumulations of RNA slows down or ceases at the mature-green stage (1-2 weeks before ripening) and RNA content may actually decline just before or during ripening (13) (measured by lycopene production).

Measurements of RNA synthesis, based on the incorporation of [5-^{3}H]-uridine into RNA by tomato segements, are shown in Figure 4. The results indicate that there is a decline in the rate of RNA synthesis as fruits approach maturity. Subsequently a burst of incorporation into RNA occurs which reaches a peak at or just before the initiation of ripening. This does not seem to be due to changes

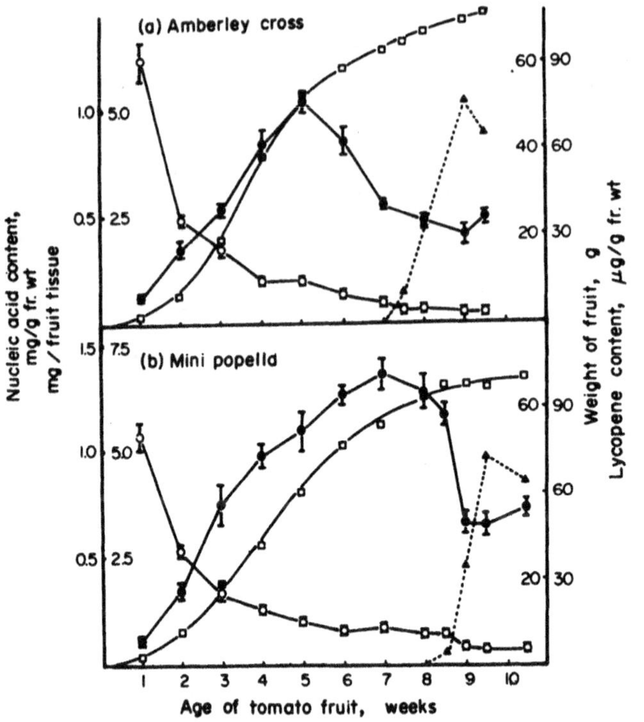

Fig. 3. Changes in nucleic acid content during the growth and ripening of tomatoes. Results are shown for a quick-ripening (Amberley Cross, 3a) and a slow-ripening (Minipopella, 3b) variety. Fresh weight of fruit (◘); nucleic acid in mg per fruit (●); nucleic acid in mg per g of fruit tissue (o); lycopene content (▲). There is virtually no cell division or DNA synthesis during fruit development and ripening and the changes in nucleic acid are due largely to RNA. Ripening is indicated by lycopene synthesis. (From Rattanapanone et al. (13)).

in tissue permeability and uptake of [5-^{3}H]-uridine (13) and is therefore interpreted as a genuine increase in RNA synthesis. Thus, at the time when total RNA content is decreasing there is a large stimulation of RNA turnover as fruits approach ripening. It is not clear whether this is an essential part of the ripening process but evidence discussed below does indicate that the production of new types of mRNA occurs as fruits ripen.

Polyacrylamide gel electrophoresis has shown that during growth of the fruit there is synthesis of 25S and 18S cytoplasmic rRNA and poly(A)-containing RNA (Figs. 5, 6 and reference 13). The poly(A)-

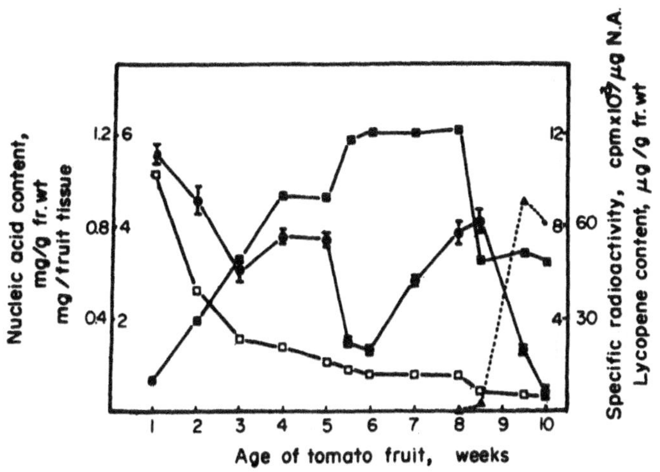

Fig. 4. Rate of labelling of RNA in segments of tomato fruit at different developmental stages. The Minipopella variety was used. Nucleic acid per fruit (■); nucleic acid per g (□); sp. radioactivity of RNA after 24 h incubation in [5-^{3}H]-uridine (●); lycopene content (▲). (From Rattanapanone et al.(13)).

containing RNA, purified by oligo(dT)-cellulose chromatography, stimulates protein synthesis in vitro. It can therefore properly be described as mRNA (Figs. 7 and references 13, 14). There is also synthesis of plastid 23S and 16S rRNA during tomato fruit growth but on a quantitative basis this is a relatively small part of the total RNA(13). As fruits ripen the cells retain the capacity for synthesis of high molecular weight 25S and 18S rRNA and soluble RNA (Fig. 5). They also synthesis poly(A)-containing RNA, similar to that shown in Figure 6. Thus the burst of RNA synthesis associated with ripening leads to the production of both stable rRNAs and mRNA. The ratio of total RNA to poly(A)-RNA does not change much during the ripening period, suggesting that the production of mRNA may actually be coupled to rRNA synthesis in some way (13).

Although in ripe fruit there is substantial breakdown of total cell RNA, giving rise to a decrease in high molecular weight molecules and an accumulation of small products, this only occurs during the later stages of ripening (13). This may be a consequence of the general release of hydrolytic enzymes associated with the advanced stages of senescence. In the early stages of ripening, although there may be a controlled turnover of RNA molecules, the cells have not lost their biosynthetic capacity and there is good reason to believe that many RNA molecules are functioning in protein synthesis.

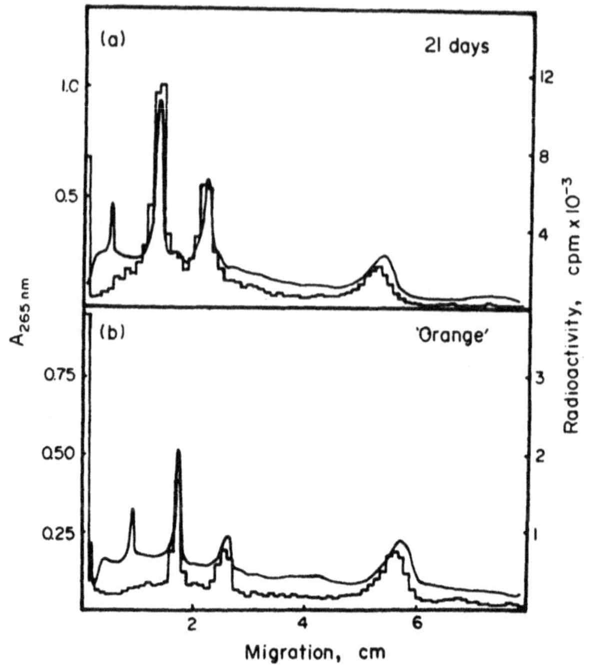

Fig. 5. Incorporation of [5-^{3}H]-urdine into different RNA fractions in unripe and ripening fruits. Pericarp segments from unripe (5a) or ripening (5b) fruit were labelled for 24 h with 1 µ Ci/µl [5-^{3}H]-urdine. RNA was extracted, purified and fractionated in polyacrylamide gels. Electrophoresis was from left to right. The optical density scan is shown by the smooth curve. The peaks are, from left to right; DNA, 25s and 18s ribosomal RNA and a mixture of 5s and transfer RNAs. The histogram shows the distribution of radioactivity in each 1 mm gel slice. (From Rattanapanone et al. (13)).

The most interesting question is whether new types of mRNA are transcribed or accumulated at the onset of ripening. The answer to this question seems to be yes. Purified poly(A)- containing RNA from green and ripening fruits directs the synthesis of a number of major polypeptides in vitro. A comparison by gel electrophoresis of the proteins synthesized in response to different tomato mRNA samples has shown that during ripening at least two extra translatable mRNAs appear (Figure 7 and reference 14). These experiments provided the first direct evidence that there are changes in fruit mRNA coding for specific proteins occurring during ripening. More recently, similar evidence has been obtained for avocado (23). Experiments of this type most readily detect changes in those mRNAs which accumulate in relatively large quantities. Furthermore, they only relate to the poly(A)-containing mRNA fraction. Quantitatively minor changes or those associated with the poly(A)-minus mRNA would have gone undetected. It therefore seems probable that further work

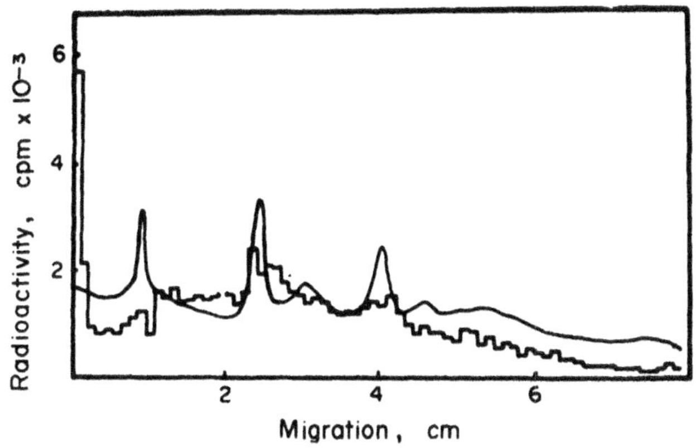

Fig. 6. Gel electrophoresis of poly(A)-containing RNA from tomatoes. Segments of pericarp from 1-week-old fruit were labelled with [5-^{3}H]-uridine for 24 h. Radioactive RNA was extracted from the tissue and poly(A)-containing RNA was purified by oligo(dT)-cellulose chromatography mixed with unlabelled carrier RNA (smooth curve) and fractionated by gel electrophoresis as in Fig. 5. Unlike the radioactive RNA in Fig. 5 the poly(A)-containing RNA is not present as distinct peaks but has a broad size distribution, shown by the histogram. Similar poly(A)-containing RNA fractions were obtained from fruits at all stages of development and ripening. (From Rattanapanone et al. (13)).

may reveal other mRNAs that alter in amount during ripening. Never-the less, the present indications are that those changes that do occur involve only a small number of new mRNAs. We are not yet certain which proteins they code for. Work which may help to solve this question is discussed below.

POLYGALACTURONASE SYNTHESIS DURING RIPENING

As stated earlier there is a lot of evidence for a general stimulation of protein synthesis during fruit ripening and there are quite a few examples where an increase in activity of a specific enzyme has been demonstrated. Generally speaking, however, there is much less information about the mechanism of such enzyme increases (whether by *de novo* synthesis or activation) and virtually no evidence about mRNA coding for specific enzymes. Without such information it is difficult to build up a proper picture of the complexity of ripening-related changes and impossible to determine where the regulatory factors, such as ethylene, operate. In order to achieve this latter objective it is necessary to study specific cases in much more detail than hitherto. We have chosen to investigate the problem of polygalacturonase production in tomato.

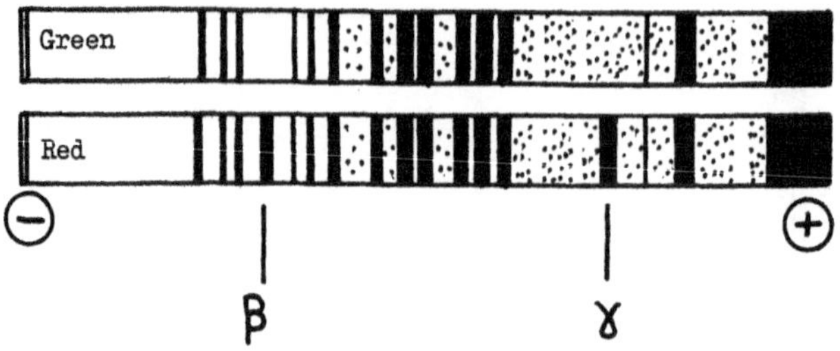

Fig. 7. Comparison of *in vitro* translation products of messenger RNAs from green and ripening fruit. The radioactive protein products produced by RNA translation in an *in vitro* protein synthesis system from wheat germ were fractionated in 15% polyacrylamide gels under denaturing conditions and then detected by fluorography. Electrophoresis was from left to right. This is a drawing of the original fluorograph. The approximate molecular weights of the proteins marked β and γ, estimated from the mobility of added marker proteins fractionated in the same gel, are 46,000 and 20,000. (From Rattanapanone et al. (14)).

Mature-green tomatoes contain extremely low amounts of extractable polygalacturonase (PG). As fruits begin to change colour enzyme activity appears and there is a massive increase during the later stages of ripening (21). There are at least two isoenzyme forms of tomato polygalacturonase which have been purified and studied. They have different physical properties (i.e. molecular weight, heat stability, density) but they both seem to be composed of the same basic polypeptide (21). Isoenzyme 2 is the major form of the enzyme which accumulated during ripening (Figure 8 and reference 21). Purifield PG2 can be converted into PG1 in vitro, confirming the belief that the enzymes are related, and it is probable that PG1 is a dimer of PG2 (22). The *rin* mutant produces no polygalacturonase (Figure 10 and reference 21) and the *Nr* mutant normally produces only PG1 (21).

There are several lines of evidence that suggest that polygalacturonase is synthesized *de novo* during ripening. Firstly, incubation of tomato fruit tissue with deuterium oxide leads to the production of enzyme with an increase in density compared to that of the water control (Figure 9). Secondly, radioimmunoassay (21) of tomato extracts, using an antibody raised against PG2, indicates that there is a proportional relationship between enzyme activity measured by conventional assay and the number of immunologically active protein molecules. Furthermore, no evidence

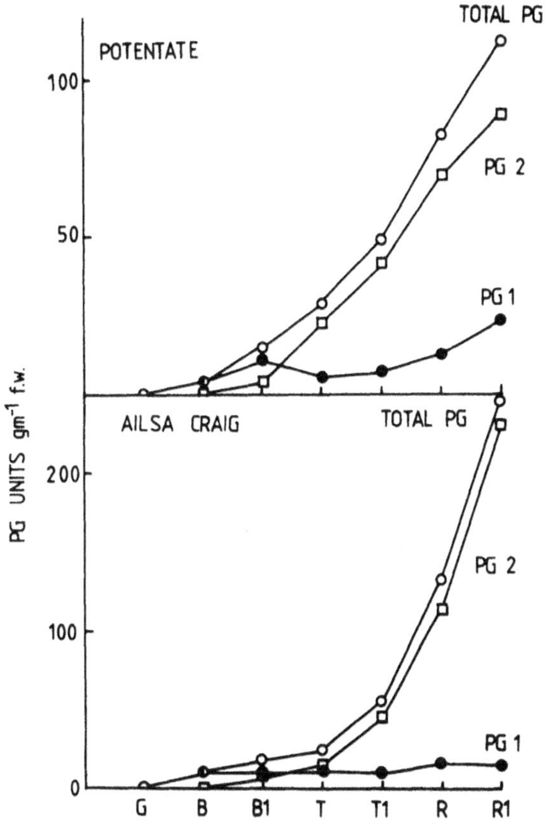

Fig. 8. Total polygalacturonase activity and isoenzyme levels extractable from individual fruit at various stages of ripeness. A slow ripening (Potentate) and fast-ripening (Ailsa Craig) variety were used. Isoenzyme activities were determined by the heat stability method. After heating for 5 min at 65°C the remaining activity can be attributed to 90% of the original polygalacturonase 1. Measurements were made on green fruit (G); breaker fruit showing the appearance of colouration (B); late breaker, with half the fruit coloured (B1); fruit turning orange (T): totally orange fruit (T1); firm red fruit (R) and slightly soft red fruit (R1). Total polygalacturonase activity (o); polygalacturonase 1 (●); polygalacturonase 2 (□). (From Tucker et al. (21)).

of an immunologically active protein was obtained by assaying extracts from green tomatoes (Tucker, Robertson and Grierson, unpublished). Thirdly, electrophoresis of protein extracts from green tomatoes in polyacrylamide-dodecyl sulphate gels shows no

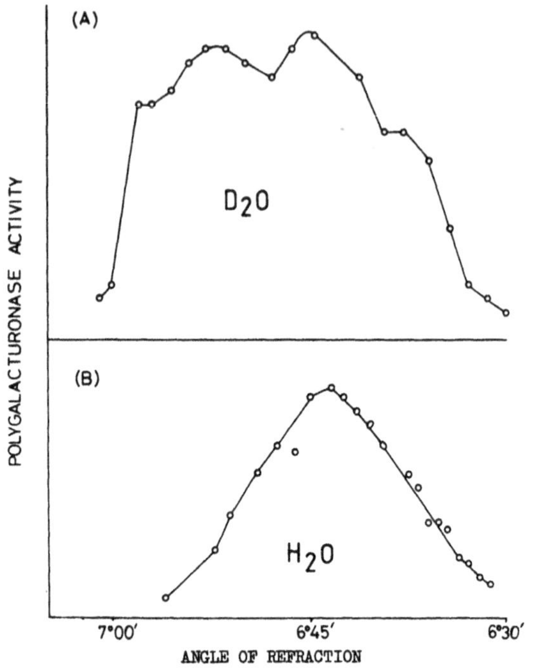

Fig. 9. Demonstration of de novo polygalacturonase synthesis by density labelling. Individual fruits were taken 35 h after the onset of ethylene production, the locule contents removed and the hollow pericarp filled with an approximately equal weight of either (A) 70% deuterium oxide or (B) water, for 18 h. Total cell-wall-bound proteins were then extracted and fractionated on a CsCl gradient. Fractions were assayed for polygalacturonase 1 activity after heating at 65°C for 5 min to remove any polygalacturonase 2 activity. (Unpublished result of N.G. Robertson and D. Grierson).

detectable protein running at the position of purified PG. In contrast, extracts from ripe tomatoes contain a very prominent protein band at the position corresponding to PG (Figure 10). On the basis of these three separate lines of evidence it seems reasonable to conclude that the enzyme is synthesized during ripening.

RELATIONSHIP BETWEEN THE SYNTHESIS OF ETHYLENE AND POLYGALACTURONASE

It has been suggested that in tomatoes the synthesis of polygalacturonase precedes ethylene production and that the action of PG

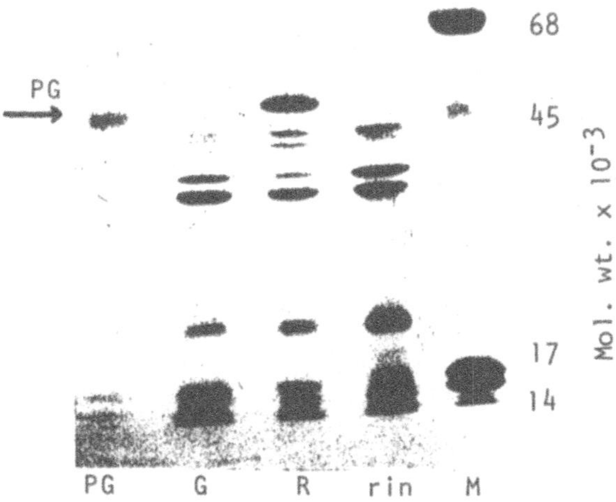

Fig. 10. Polyacrylamide gel profiles of cell-wall-bound protein. Protein samples were suspended in a buffer (40 mM Tris/HCl pH 9.18 containing 50% w/v sucrose, 10% w/v sodium dodecylsulphate and 5% v/v mercaptanoethanol) and boiled for 5 min. Samples containing 50 to 100 μg were then fractionated in a 10 to 15% gradient polyacrylamide slab gel under denaturing conditions in 0.1% sodium dodecylsulphate. The left hand track contains partially purified polygalacturonase 2 (PG). The remaining tracks contain total cell-wall-bound proteins from green normal (G), ripe normal (R) and yellow (rin)fruit respectively. The right-hand track contains marker proteins Mr = 68,000, 45,000, 17,900 and 14,300. Gels were stained in 0.1% Coomassie blue, 40% methanol, 7% acetic acid and destained in 30% methanol, 7% acetic acid. (Unpublished result of G.A. Tucker and D. Grierson).

on the cell walls of the fruit releases wall-bound enzymes important for ethylene synthesis and other aspects of ripening (20). This seems an improbable mechanism for the general control of ripening since some fruits do not produce PG. Furthermore, in tomato the kinetics of ethylene evolution and polygalacturonase synthesis do not support this view. We have made careful measurements of the timing of natural ethylene synthesis and polygalacturonase synthesis in individual tomatoes. The results clearly show that the enhanced ethylene production normally associated with ripening is initiated some 20-30 h before polygalacturonase activity can be detected

(Grierson, Robertson and Tucker, unpublished). We conclude that ethylene synthesis precedes polygalacturonase synthesis. A similar conclusion has recently been reached for cucumber (24). The mechanism for switching on ethylene production is unknown.

IS THERE CONTROL AT THE LEVEL OF mRNA?

As shown in Figure 7, tomato mRNA stimulates the synthesis of a number of polypeptides _in vitro_ and there are specific mRNA changes occurring during ripening. One of the mRNAs which increases greatly in quantity during ripening codes for a protein with a molecular weight of 46,000 (Figure 7). This is only slightly larger than the molecular weight of the polygalacturonase protein in polyacrylamidedodecyl sulphate gels (21). Furthermore, some of the radioactive protein molecules synthesized _in vitro_ in response to mRNA from ripening tomatoes reacts with antibody raised against polygalacturonase 2 (Figure 11). This provides an assay for measuring the amount of mRNA for polygalacturonase. Preliminary

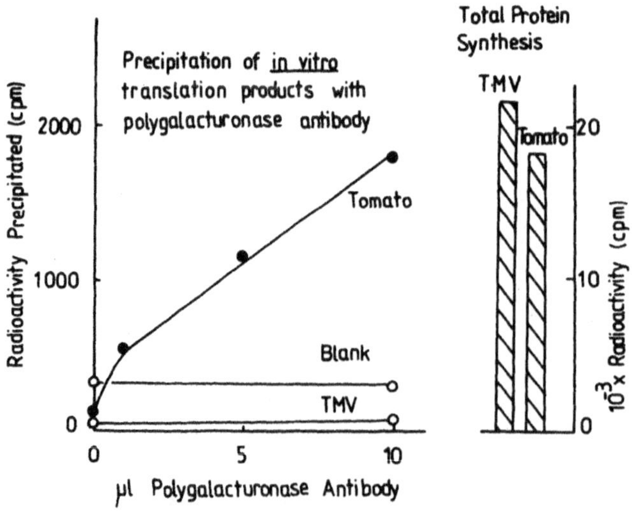

Fig. 11. Immunoprecipitation of protein synthesized in vitro in response to tomato RNA, using an antibody raised against polygalacturonase 2. The total incorporation into protein in response to RNA from orange tomatoes or tobacco mosaic virus is shown on the right. The results on the left show the titration of the _in vitro_ translation products with increasing concentration of PG 2 antibody. (From Grierson et al. in "Biochemistry of Fruits and Vegetables," J. Friend, ed., Academic Press, in press).

results suggest that there is an increase in this mRNA as tomatoes ripen. Further measurements will be necessary before it can be determined whether there is any mRNA for polygalacturonase before ethylene synthesis begins and whether ethylene induces the accumulation of polygalacturonase mRNA.

ACKNOWLEDGEMENT This work was supported by a grant from the Agricultural Research Council.

REFERENCES

1. C. J. Brady, J. K. Palmer, P.B.H. O'Connell and R. M. Smillie, An increase in protein synthesis during ripening of the banana fruit, Phytochemistry 9:1037 (1970).
2. L. A. Darby, Isogenic lines of tomato fruit colour mutants, Hort. Res. 18:73 (1978).
3. H. C. Dostal and A. C. Leopold, Gibberellin delays ripening of tomatoes, Science 158:1579 (1967).
4. C. Frenkel, I. Klein and D. R. Dilley, Protein synthesis in relation to ripening of pome fruits, Plant Physiol. 43:1146 (1968).
5. D. Grierson, G. Tucker and N. G. Robertson, The regulation of gene expression during the ripening of tomato fruits, in Long Ashton Symposium on 'Quality in Stored and Processed Vegetables and Fruit,' pp 179-191, P. W. Goodenough and R.K. Atkin, eds., Academic Press, London (1981).
6. E. Hansen, Ethylene stimulated metabolism of immature Bartlett pears, Proc. Am. Soc. Hort. Sci. 91:863 (1967).
7. G. E. Hobson, Polygalacturonase in normal and abnormal tomato fruit, Biochem J. 92:324 (1964).
8. A. C. Hulme, Studies in the nitrogen metabolism of apple fruits. The climacteric rise in respiration in relation to changes in equilibrium between protein synthesis and breakdown, J. Exptl. Bot. 5:159 (1954).
9. A. C. Hulme, M. J. C. Rhodes and L. S. C. Woodtorton, The relationship between ethylene and the synthesis of RNA and protein in ripening apples, Phytochemistry 10:749 (1971).
10. K. Iki, K. Sekiguchi, K. Kurata, T. Tada, H. Nakagawa, N. Ogura and H. Takehana, Immunological properties of β-fructofuranosidase from ripening tomato fruit, Phytochemistry 17:311 (1978).
11. N. Marei and R. Romani, Ethylene-stimulated synthesis of ribosomes, ribonucleic acid and protein in developing fig fruits, Plant Physiol. 48:806 (1971).
12. I. J. Mettler and R. J. Romani, Quantitative changes in tRNA during ethylene-induced ripening (ageing) of tomato fruits, Phytochemistry 15: 25 (1976).

13. N. Rattanapanone, D. Grierson and M. Stein, Ribonucleic acid metabolism during the development and ripening of tomato fruits, Phytochemistry 16:629 (1977).
14. N. Rattanapanone, J. Speirs and D. Grierson, Evidence for changes in messenger RNA content related to tomato fruit ripening, Phytochemistry 17:1485 (1978).
15. A. E. Richmond and J. Biale, Protein and nucleic acid metabolism in fruits. 1. Studies of amino acid incorporation during the climacteric rise in respiration of avocado, Plant Physiol 41: 1247 (1966).
16. J. Riov, S. P. Monselise and R. S. Kahan, Ethylene-controlled induction of phenylalanine ammonia lyase in citrus fruit peel, Plant Physiol. 44:631 (1969).
17. J. A. Sacher, Permeability characteristics and amino acid incorporation during senescence of banana tissue, Plant Physiol. 41:701 (1966).
18. J. A. Sacher, Senescence and postharvest physiology, Ann. Rev. Plant Physiol. 24:197 (1973).
19. G. H. De Swardt, J. H. Swanepoel and A. J. Duvenage, Relations between changes in ribosomal RNA and total protein synthesis and the respiration climacteric in pericarp tissues of tomato, Z. Pflanzenphysiol. 70:358 (1973).
20. E. G. Tigchelaar, W. B. McGlasson and R. W. Buescher, Genetic regulation of tomato fruit ripening, J. Hort. Sci. 13:508 (1978).
21. G. A. Tucker, N. G. Robertson and D. Grierson, Changes in polygalacturonase isoenzymes during the 'ripening' or normal and mutant tomato fruit, Europ. J. Biochem. 112:119 (1980).
22. G. A. Tucker, N. G. Robertson and D. Grierson, The conversion of tomato fruit polygalacturonase isoenzyme 2 into isoenzyme 1 _in vitro_, Europ. J. Biochem. 115:87 (1981).
23. R. E. Christoffersen, E. Warm and G. C. Laties, Gene expression during fruit ripening, Plant Physiol. Suppl. 47:73 (1981).
24. M. E. Saltveit, Jr. and R. F. McFeeters, Polygalacturonase activity and ethylene synthesis during cucumber fruit development and maturation, Plant Physiol. 1019:66 (1980).

RESPIRATION AND ENERGY METABOLISM IN SENESCING PLANT TISSUES

Theophanes Solomos

Department of Horticulture
University of Maryland
College Park, Md. 20742

INTRODUCTION

The senescence of detached plant organs is associated with cataclysmic physiological, histochemical and biochemical changes (9,16,17,64,90,100). Although the process is predominantly catabolic, anabolic reactions are nevertheless necessary for senescence to proceed (91). The prevention of senescence by inhibitors of anabolic processes bears witness to the above (43,71,74). Furthermore, overt catabolic reactions, such as the conversion of starch to sucrose, require at least one mole of ATP per mole of sucrose formed (118). The transport of inorganic and organic nutrients from senescing attached leaves to the parent plant must also be a sink of metabolic energy. In detached senescing organs respiration is the sole source of energy, since the photosynthetic capacity of certain detached organs is minimal and moreover senescence of detached leaves proceeds faster in the dark than in the light (69). In this presentation we shall examine the rate of production, regulation, and physiological significance of metabolic energy in senescing detached plant organs.

RESPIRATORY DRIFTS OF DETACHED PLANT ORGANS

Fruit

It was observed by Kidd and West (56) that the rate of respiration of detached apple fruits decreased initially and after reaching a minimum, suddenly increased rapidly, peaked and then declined. Kidd and West (56) named this sudden rise in respiration during fruit ripening the climacteric. It was subsequently shown that this phenomenon occurs during ripening of other fruits besides apples

(16,17,90). Fig. 1 shows the changes in the rates of O_2 uptake and CO_2 output of avocado fruits in the course of ripening. However this burst of respiration during ripening is not a sine qua non for all fruits. Thus the rate of respiration of detached citrus fruits decreases gradually during storage (16,17), whereas the rate of respiration of strawberries does not show pronounced changes during ripening (58). Biale (16) divided fruits into climacteric and non-climacteric groups, based on whether their respiration increased or gradually declined during ripening. Table 1 shows some fruits representative of both groups.

Leaves

The changes in respiration during senescence have been studied in both detached and attached leaves. Blackman was the first to carry out extensive studies with senescing detached leaves (see James, 53). The respiratory drift was divided into six phases as shown in Table 2. Thus the respiratory drift of senescing leaves shares many common features with that of climacteric fruits. The most important is the rise in respiration during yellowing (phase 4).

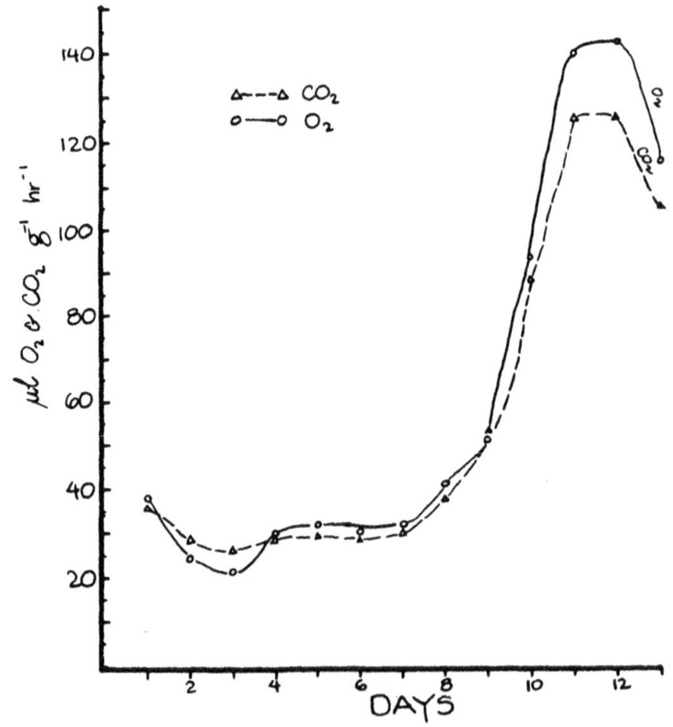

Fig. 1. Respiratory drift of detached avocados.

Table 1. Classification of Fruits

Climacteric	Non-climacteric
Apple	Cherry
Apricot	Citrus (grapefruit,lemon,orange)
Avocado	Melon
Banana	Pineapple
Peach	Strawberry
Pear	
Plum	
Tomato	

A rise in respiration during the later stages of leaf senescence has been observed with barley leaves (53), tobacco leaves (4,67,66,103), and oat leaves (69). On the other hand, work with pea leaves (104) shows that respiration declines steadily during senescence. Woolhouse (124) studied the photosynthetic efficiency and rate of respiration of attached leaves of *Perilla frutescens*. It can be seen in Fig. 2 that there is a climacteric-like increase in respiration prior to senescence.

Flowers

The respiratory behavior of detached flowers has not been studied as extensively as that of fruits and leaves. However, the available reports indicate that cut flowers show both climacteric and non-climacteric types of respiratory drifts (33,91,93). It has been shown that cut carnations exhibit a typical climacteric repiratory drift (39,80). Figure 3 shows that the rate of CO_2 output of cut carnations decreases sharply after cutting, to increase again during the wilting stage of the petals. The rate of respiration of cut roses, on the other hand, declines steadily during senescence (55).

CAUSES OF THE CLIMACTERIC RISE IN RESPIRATION

Since its discovery, the climacteric rise in respiration during fruit ripening has been attributed to a decrease in "organization" resistance of the cell (19), presence of natural uncouplers (77), enhancement of protein synthesis (50), and the synthesis of specific "ripening" enzymes (90). The climacteric occurs in fruits which differ widely with respect to growing conditions, composition and physiology. The only common metabolic feature is their ability to produce ethylene and to respond to exogenous ethylene with an increase in respiration (16,17,90). Moreover, ethylene enhances the rate of respiration of non-climacteric fruits such as citrus (17). The difference between citrus and climacteric fruits lies

Table 2. Phases of Leaf Senescence[a]

Phase	Duration in days	Slope	Color of Leaves
1	1	high level	green
2	6	rapid fall	green
3	6	low level	green-pale green
4	c.20	steady rise	pale green yellow
5	c.20	steady fall	yellow-brown
6		sharp rise	brown

[a]James (53)

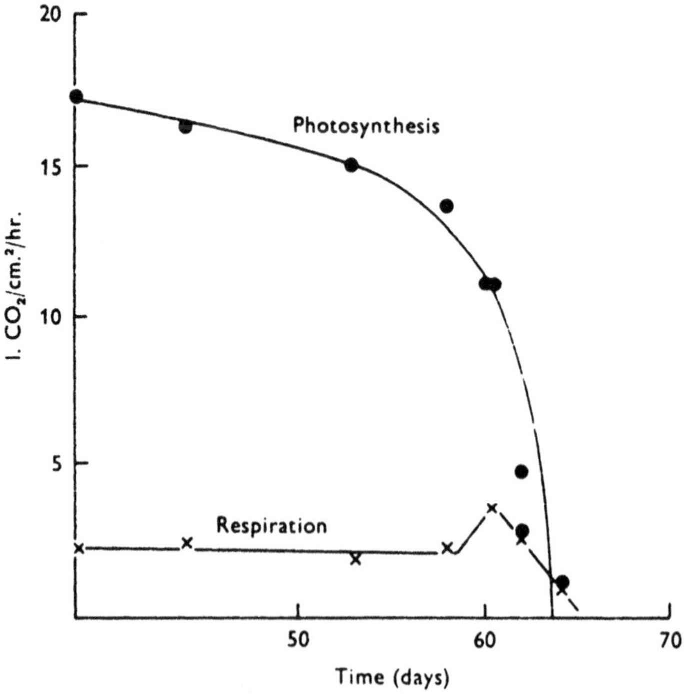

Fig. 2. Changes in the rates of respiration and photosynthesis of attached Perilla leaves during the later stages of development (Woolhouse, 124).

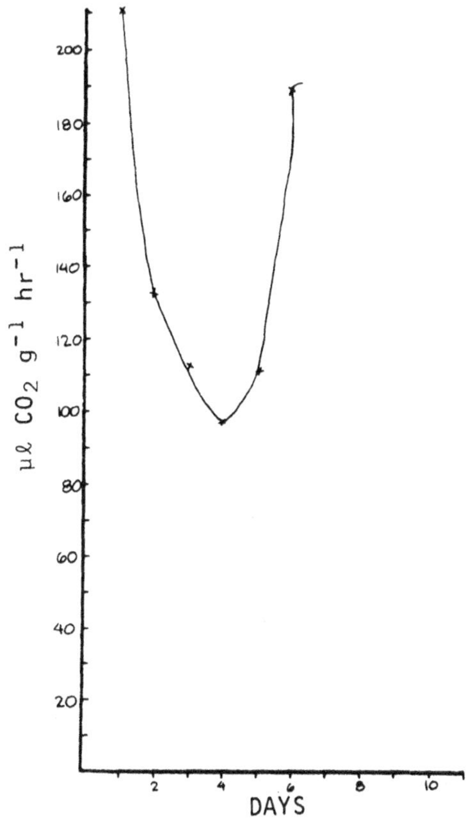

Fig. 3. Rate of CO_2 output of cut carnation flowers.

in the inability of the former to produce endogenous ethylene naturally or in response to the application of exogenous ethylene. Citrus fruits, however, can produce ethylene in response to stresses (32,72) or if they are harvested at an immature stage (3). On the other hand, ethylene is not produced by citrus fruits during normal ripening. McMurchie et al. (75) treated bananas, a climacteric fruit, and oranges, a non-climacteric fruit, with propylene, an ethylene analog, and found that the rate of respiration was enhanced in both fruits. Propylene failed to induce the production of ethylene in oranges. To use their own terminology, oranges lack system II, which the authors consider to be associated with the induction of the autocatalytic production of ethylene. The non-ripening tomato mutants further demonstrate the close relationship between the rise in respiration and ethylene. Thus, in the rin mutant, which shows only a limited degree of ripening, there is no increase in either the production of endogenous ethylene or respiration during storage (48). These authors have shown that application of exogenous ethylene enhances the rate of respiration. Moreover, interrupted applications of exogenous ethylene enhance respiration sequentially.

In addition, ethylene enhances respiration in tissues other than fruits where ripening is not at issue (89, 108-110). In Table 3 several tissues are shown whose respiration is enhanced by ethylene. In senescing leaves the association of the rise in respiration with ethylene production has not yet been established unequivocally. It has been shown that senescing tobacco leaves showed both an increase in respiration and ethylene production (4, 103). Tetley and Thimann (113) claimed that there is no correlation between senescence of oat leaves and ethylene. It has recently been observed that the addition of 1-aminocyclopropane-1-carboxylic acid (ACC), the immediate precursor of ethylene (2), to the bathing solution of senescing oat leaves enhances respiration and senescence (Thimann, personal communication). We have found recently that when ivy leaves (Hedera helix) senesce on the parent plant, they produce about 2-3 nl C_2H_4/g/h. The rate of respiration of green ivy leaves kept in the dark increases in response to ethylene treatment (Fig. 4). The data indicate that, as in the case of fruits, the respiratory upsurge during leaf senescence may be associated with an increase in the endogenous production of ethylene.

Certain cut flowers produce ethylene during senescence (26). Moreover the rise in respiration in carnations coincides with a sharp increase in ethylene production (39, 73). We have found

Table 3. The Effect of Ethylene on the Respiration Rate of a Variety of Plant Tissues

Tissue	Respiration Rate			
	Control		C_2H_4	
	O_2	CO_2	O_2	CO_2
	μl/g/h			
Apples	6	6	6	6
Avocados	35	36	150	140
Cherimoyas	35	35	160	150
Lemons	7	7	16	15
Grapefruit	11	11	30	29
Beetroot	11	11	22	22
Carrots	12	10	20	19
Potatoes	2.5	2.5	14	11
Sweet Potatoes	18	17	22	22
Rutabagas	9	9	18	15

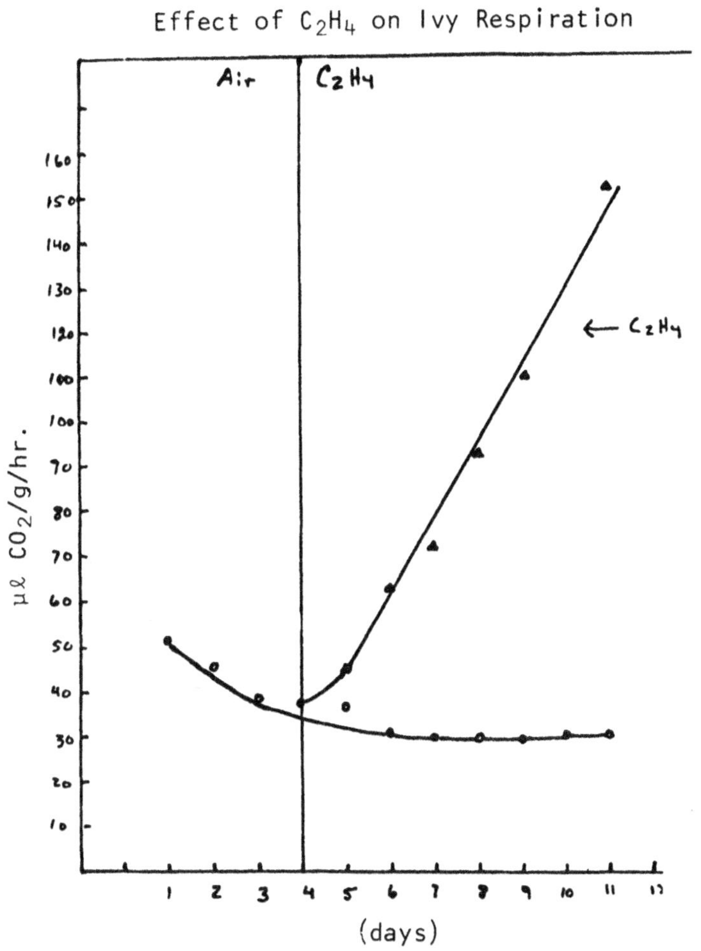

Fig. 4. Rate of respiration of detached ivy leaves kept in the dark in air and 10 p.p.m. ethylene

that, with cut carnations, the addition to the holding solution of aminoethyoxyvinyl glycine, an inhibitor of ethylene production (64), prevents both the rise in respiration and ethylene formation (Fig. 5). In addition, the application of exogenous ethylene enhances the rate of respiration of roses, which do not usually show a climacteric type of respiratory drift (55).

In summary the evidence indicates that the rise in respiration during senescence of fruits, cut flowers and possibly leaves is the result of ethylene action. In this respect the division of fruits

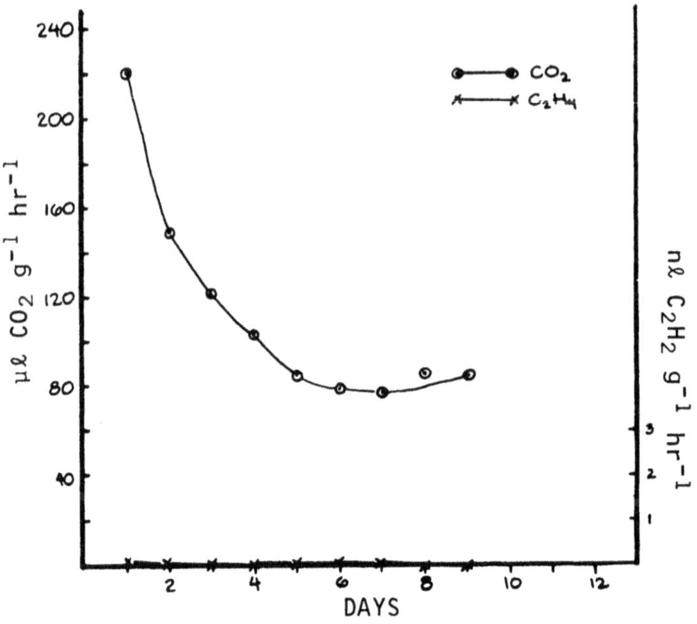

Fig. 5. Effect of AVG on the rates of respiration and ethylene production of cut carnation flowers.

into climacteric and nonclimacteric should be based on their ability to produce sufficient quantities of ethylene rather than on their respiratory behavior.

MODE OF ETHYLENE ACTION ON PLANT RESPIRATION

Ethylene induces a wide variety of physiological responses in higher plants (1,24,64,78,86). Theoretically the rise in respiration can be the result of de novo synthesis of new respiratory enzymes (coarse control), or the activation of pre-existing enzymes (fine control) or a combination of the two. Several lines of evidence indicate that with the possible exception of pome fruits (38,52), the respiratory potential of preclimacteric fruits is sufficient to sustain rates of respiration observed at the climacteric peak. Thus Millerd et al. (77) reported that the rate of respiration of preclimacteric avocado slices treated with uncouplers of oxidative phosphorylation was similar to that at the climacteric peak. In addition, the glycolytic potential of preclimacteric avocados is adequate to handle the carbon flux required during the climacteric (107). The initial rate of respiration of freshly cut carnations equals or even exceeds the respiratory peak during senescence (Fig. 3). Moreover the fact that inhibitors of protein synthesis prevent ripening of intact pears and banana slices even though they do not decrease the rise in respiration supports the above proposition (43,74).

REGULATION OF PLANT RESPIRATION

In senescing fruits, carbohydrates are the principal source of energy. Even in avocados, which may contain as much as 15% fat, carbohydrates are the main respiratory substrates (18). On the one hand, fat content does not change with ripening and avocados lack the enzymes involved in the first steps of the mobilization of fats (18). On the other hand, the stable carbon isotope composition of respiratory CO_2 resembles that of cellular carbohydrates at all stages of ripeness of avocado fruits (Table 4) (Solomos and Laties, unpublished). In the case of leaves, proteins may contribute to respiration at later stages of senescence (53).

The respiratory pathways of higher plants are the same as those of other higher organisms. They include glycolysis, the pentose pathway, tricarboxylic acid cycle and the electron transport chain. The existence of a net flux and the necessity for a rapid control in response to changes in the energy requirements necessitate the displacement from equilibrium of the overall reaction of the respiratory pathways (5a, 12, 79). It has been observed that certain enzymatic steps of a metabolic sequence are greatly displaced from equilibrium and that these steps are usually regulatory in nature. The attributes of regulatory enzymes are: (a) the reactions they catalyse are located at metabolic crossroads, (b) these reactions are displaced from equilibrium, (c) the enzymes are highly allosteric, and (d) there is an inverse relationship between the changes in the overall flux and the level of substrate. Thus, during a transition from a lower to a higher overall flux, the concentration of substrate decreases. This pattern of changes indicates that the activity of the enzyme under consideration has been increased to a greater extent than that of both the immediately preceding and the subsequent steps (5a,29,118). For a reaction to be considered non-equilibrium its observed mass action ratio must differ by 20-fold from the equilibrium constant (79). Although the kinetic control of enzymatic reactions involved in energy production is the most advantageous and prevalent (5a), in certain cases equilibrium reactions contribute to metabolic regulation(60).

Glycolysis

Of the glycolytic enzymes, phosphofructokinase (PFK) and pyruvate kinase (PK) are the key ones which fulfill the attributes of regulatory enzymes. They are located at strategic metabolic points (118), are highly allosteric, their activities being affected by several metabolites (Table 5), their observed mass action ratios are greatly displaced from equilibrium (Table 6), and crossover changes occur under conditions where the rate of glycolysis changes from one steady state to another (5,11,41,49,54,59,76,79).

The rate of glycolysis of pea seeds increases by two-fold with transfer to N_2 (11). This increase is associated with a character-

Table 4. Carbon Isotope Values of Carbohydrate and Protein Enriched Fractions, Total Fats and CO_2 in the Course of Ripening of Avocado Fruits.[a]

Stage of ripeness	$\delta^{13}C/^{12}C$ (0/00)			
	Carbohydrate	Proteins	Total Fatty Acids	CO_2
Preclimacteric	-23.31	-24.15	-31	-23
Climacteric	-23.40	-24.08	-30.7	-23.2
Postclimacteric	-23.54	-24.29	-32.1	-23.5

[a] Solomos and Laties, unpublished

Table 5. Modulators of PFK and PK[a]

Enzyme	Modulators	
	Positive	Negative
PFK	P_i, K^+, NH_4 Mg^{2+}	PEP, 2PGA, 3PGA, ATP, citrate gluconate-6-P
PK	K^+, Mg^{2+}	Ca^{2+}, ATP, citrate

[a] Turner and Turner (118)

istic pattern of changes in the glycolytic intermediates, which reveal the existence of crossover changes at the levels of PFK and PK (Table 7). It can be seen from the data of this table that the changes in the known modulators of PFK and PK are in the expected direction. Thus phosphoenolpyruvate (PEP), a strong inhibitor of pea PFK (118), decreases sharply. Similarly the inhibitor ATP, 2-phosphoglycerate (2-PGA) and 3-phosphoglycerate (3-PGA) also decrease. In the case of anaerobic castor bean endosperms there is in addition a decrease in the total adenylate pool, which would further contribute to the activation of PFK (59). It is known that citrate, an inhibitor of both PFK and PK, decreases in anaerobic pea

Table 6. Mass Action Ratios of Certain Glycolytic Enzymes in Air and N_2 of Pea Seeds[a]

Enzyme	Mass Action Ratio		
	Air	N_2	K_{eq}
Glucose phosphate isomerase	0.19	0.20	0.42
Phosphofructokinase	0.016	5.09	1.05×10^3
Phosphoglycerolmutase	0.841	0.816	0.15
Enolase	0.288	0.216	3.7
Pyruvate kinase	6.90	13.60	11×10^3

[a]Barker et al. (11)

Table 7. Changes in the sugar phosphates, adenosine nucleotides and phosphoenol pyruvate of green pea seeds in air and N_2.[a]

Compounds[b]	Treatment	
	Air	30 min in N_2
	μ μmoles/100 gr	
G6P	55.5	55.5
F6P	10.50	11
F,1,6diP	0.88	17.50
DHAP	2.10	10.57
3PGA	32.00	8.53
2PGA	26.90	6.96
PEP	7.75	1.50
AMP	2.30	13.70
ADP	18.67	64.33
ATP	96.30	20

[a]Barker et al. (11)

[b]G6P (glucose 6-phosphate); F6P(fructose 6-phosphate); F,1,6diP(Fructose 1,6di phosphate); DHAP (dihydroxy-acetone phosphate); 3PGA (3 phosphoglyceric acid); 2PGA(2 phosphoglyceric acid); PEP (phosphoenolpyruvate)

seeds (121). In addition, there is a 16-fold increase in the lactate content of anaerobic pea seeds (121), which may decrease the pH of the cytosol and hence enhance the activity of PFK (118). In short, the changes in the levels of the modulators of PFK and PK are those anticipated for the enhancement of glycolysis.

Crossover changes at the PFK and PK levels have been observed during the climacteric with bananas (10,101), avocados (Fig. 6) (107) and tomatoes (28). Because of the analytical difficulties there are no data concerning the simultaneous changes of all factors involved in the regulation of glycolysis either during the climacteric or during the rise in respiration of senescing detached leaves and flowers. However, from the fragmented data it is clear first that glycolysis is enhanced during the climacteric rise in bananas, avocados and tomatoes, and second that this increase in gycolytic activity is caused by an activation of PFK and probably PK. The activation of these enzymes, however, cannot be completely explained from the associated changes of the most important modulators. Thus, apart from PEP, which decreases sharply, ATP increases in avocado, cantaloupe, cherimoya and tomato fruits and in senescing leaves (Table 8). In bananas, citrate either does not change (105) or increases with the climacteric (82). It should be pointed out that the concentration of citrate in bananas is such (ca. 15 mM) that it is more than sufficient to completely inhibit PK in certain plant tissues (118). It is thus obvious that in vacuolated plant cells, cellular compartmentalization may alter the actual concentrations of the enzyme modulators, especially organic acids, at the site of the enzyme (13,81). Chalmers and Rowan (28) proposed that the activation of PFK during the climacteric of tomatoes is caused by an increase in the cytosolic concentration of inorganic phosphate, a positive modulator of PFK, in several plant tissues (117).

Avocado PFK is inhibited by ATP and stimulated by P_i (66). Salminen and Young (101) have shown that the kinetic characteristics of banana PFK change in the course of the climacteric. In particular the enzyme exhibited negative co-operativity with respect to the substrate, F6P. However, the negative cooperativity was much more stringent with the enzyme isolated from prelcimacteric than from climacteric bananas. Furthermore the $S_{0.5}$ changed from 5.8 mM at the preclimacteric stage to 1.72 mM at the climacteric peak. Also, the inhibition by the adenine nucleotides changed in the course of the climacteric. The most notable was that with AMP, which lost its inhibitory effect at the climacteric. P_i and NH_4 ions alone had no effect on the banana PFK. However, both ions reduce the inhibition caused by ADP. On the basis of their electrophoretic data, Salminen and Young (101) concluded that first there was no production of a new species of the PFK in the course of the climacteric, and second the enzyme remained oligomeric at all stages of ripeness.

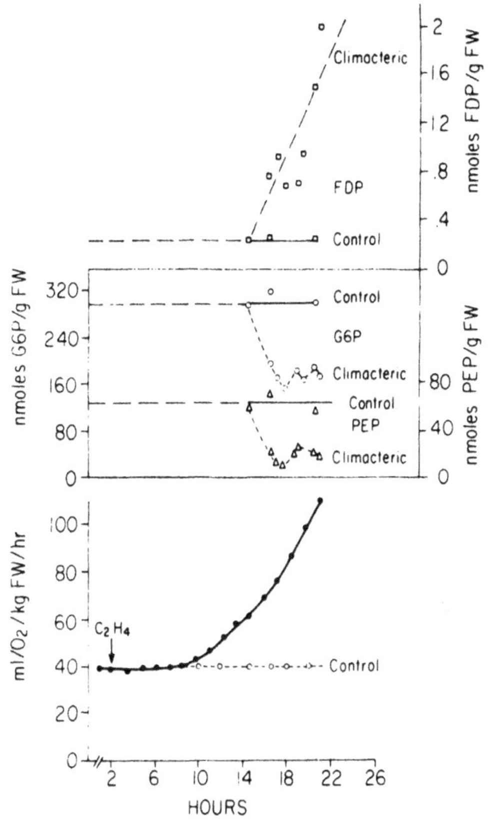

Fig. 6. Course of respiration and the levels of PEP, G6P, and FDP during the climacteric of intact avocado. Fruit treated with 30 µl/l ethylene where indicated

Isaac (1980) (reported by Rhodes (91)), on the other hand, found that PFK isolated from tomato fruits was oligomeric up to the climacteric peak, while the enzyme extracted at the post-climacteric stage was present in both oligomeric and low molecular-weight species. Further, under certain conditions the oligomeric form could be converted to smaller fractions. This form showed negative cooperativity with respect to substrate, the degree of which was determined by the Mg^{2+}/ATP ratio, whereas the small molecular weight species exhibited classical Michaelis-Menten kinetics. P_i converted the oligomeric species to smaller molecular weight species which in turn showed hyperbolic kinetics.

It has already been pointed out that PK shows all the characteristics of a regulatory enzyme (118). However, no information is yet available concerning the kinetic properties of fruit PK. As in the case of PFK the changes of the negative modulators, ATP

Table 8. Changes in the Levels of Adenine Nucleotides in Senescing Fruits and Leaves

Tissue	Stage	AMP	ADP	ATP
		nmoles/gr		
Avocado[1]	Preclimacteric	-	207	89
	Climacteric	-	124	140
Cherimoyas[2]	Preclimacteric	-	-	48
	Climacteric	-	-	115
Cantaloupes[3]	Preclimacteric			110
	Climacteric			350
	Controls			40
Tobacco leaves[4]	Senescing	-		90
	Control			111
Barley[5]	Senescing			218
	Control	18	40.8	45
Ivy leaves[6]	Senescing	23.9	39.7	74

[1]Young and Biale, (126)

[2]Solomos and Laties, (109, 110)

[3]Rowan et al., (95)

[4]Sisler and Pian, (103)

[5]Malik and Thimann,(69)

[6]Warman and Solomos, 1981, unpublished

and citrate, call for a decrease in PK activity in the course of the climacteric, since ATP increases in most climacteric fruits, and in the case of bananas, citrate may also increase with the climacteric rise. However the presence of K^+ ions is a requirement for PK and it also alleviates the inhibition of negative modulators (117). Thus, if the cytosolic concentration of K^+ increases during the climacteric, as has been reported for tomatoes (120), then PK would be expected to be activated. In short the present analytical data indicate that the enhancement of glycolysis during the climacteric is caused by an activation of PFK and PK and that in bananas and tomatoes a change in the kinetic properties of PFK in the course of the climacteric may be the most important factor in bringing about the enhancement of the glycolytic flux.

Pentose Pathway

The measurements of Ashihara and Komamine (6) show that of the enzymes of the pentose pathway, only the overall reaction of

G6P dehydrogenase is a non-equilibrium reaction. Further, this enzyme is shown to be under both coarse and fine controls in various plant tissues (118). The latter authors consider the NADPH/NADP ratio the most important factor in the regulation of G6P-dehydrogenase in vivo in higher plants. With respect to fruits there is no information available concerning the changes in reactants and modulators of this enzyme in the course of the climacteric.

Tricarboxylic acid cycle

Mitochondria capable of linking the oxidation of Krebs cycle intermediates to phosphorylation have been isolated from fruits at various stages of ripeness (37,51,61,94). Further, there seems to be no change in the mitochondrial activities with ripening. Lance et al. (61) stated: "In the presence of cofactors, taking state 3 as the most reliable criterion, no major changes in the oxidative abilities of mitochondria from avocados during the ripening process have been found." With isolated mitochondria the availability of ADP controls the rate of oxygen uptake. In vivo, although oxidative phosphorylation exerts a strong control on respiration, since the addition of uncouplers always elicits an increase in the rate of respiration (see Beevers, 12), this control cannot be thought of as a state-4-to-state-3 transition. On the one hand, ADP is always present in sufficient quantities in plant tissues. On the other, ADP decreases or shows marginal changes in the course of the climacteric of avocados (126), and during the rise in respiration of ivy leaves treated with ethylene (Table 8). Hence the control on plant respiration of oxidative phosphorylation is not exerted by the level of ADP per se, but rather by its feedback control on substrate mobilization, namely glycolysis and the pentose phosphate pathways and possibly on some of the enzymatic steps of the Krebs cycle (5a,117,123).

Several of the enzymes involved in the TCA cycle are located at metabolic crossroads, and are highly allosteric, hence potential control points. Wiskich (123) has summarized the most important regulatory metabolites of the enzymes of the Krebs cycle (Table 9). The studies of the changes in the levels of the intermediates of the Krebs cycle under different respiratory regimes and during the transition from one steady state to another are limited and inconclusive. In fact the pattern of changes in the TCA cycle intermediates in the air-to-N_2 and N_2-to-air transition of green peas led Wager (121) to propose that in pea seeds the TCA cycle is not a major pathway contributing to respiration. However, it should be kept in mind that measurements of overall concentrations may not reflect the concentration present inside the mitochondria and also that the availability of cofactors may distort the interpretation of the results of the changes in the levels of the TCA cycle intermediates. In addition, the metabolite transport systems, the presence of malic enzyme, and the action of transaminase reactions

Table 9. Metabolites Affecting the Activities of Enzymes of the Krebs cycle[a].

Enzyme	Modulators	
	Positive	Negative
Pyruvate dehydrogenase		NADH, CoA
Citrate synthetase		ATP
Isocitrate dehydrogenase		NADH
α-oxoglutarate dehydrogenase	AMP	
Succinate dehydrogenase		oxaloacetate
Malic enzyme	CoA	NADH
Malic acid dehydrogenase		NADH, ATP

[a]Wiskich (123)

would be expected to drastically alter the assumed cyclic utilization of the TCA cycle intermediates. In banana fruits most of the TCA cycle acids increase with ripening (Fig. 7) (10,82), and this increase results in the decrease of the pH of the pulp (44). In the case of apples, however, malic acid decreases during ripening due to the synthesis of malic enzyme (38, 52). It is obvious that on the basis of analytical data neither the activity of the TCA cycle nor the possible regulatory steps can be ascertained in senescing plant tissues.

Metabolite Transport

The following metabolite transport systems have been identified in plant mitochondria: pyruvate/OH^-, P_i/OH^-, dicarboxylate, tricarboxylate, glutamate, α-oxoglutarate, adenine nucleotides and adenosine diphosphate (122). Several lines of evidence indicate that the metabolite transport system functions *in vivo*. For instance in fresh potato slices, malonate inhibits the stimulation of respiration by uncouplers, notwithstanding the absence of the functioning of the TCA cycle (62). Further, the α-oxidation of fatty acids, which are the respirable substrates of fresh potato slices, is inhibited by rotenone (125), indicating that the electron transport chain is the terminal electron carrier. Wu and Laties (125) attributed the inhibition of the respiration of uncoupled fresh potato slices by malonate to the impairment by the acid of the malate-oxaloacetate shuttle, which is considered to be the vehicle for the transport of reducing equivalents from the NADH generated in the microsomes during the α-oxidation of fatty acids to mitochondria. Although the operation *in vivo* of the metabolite transporters is not in doubt, their regulatory role has not

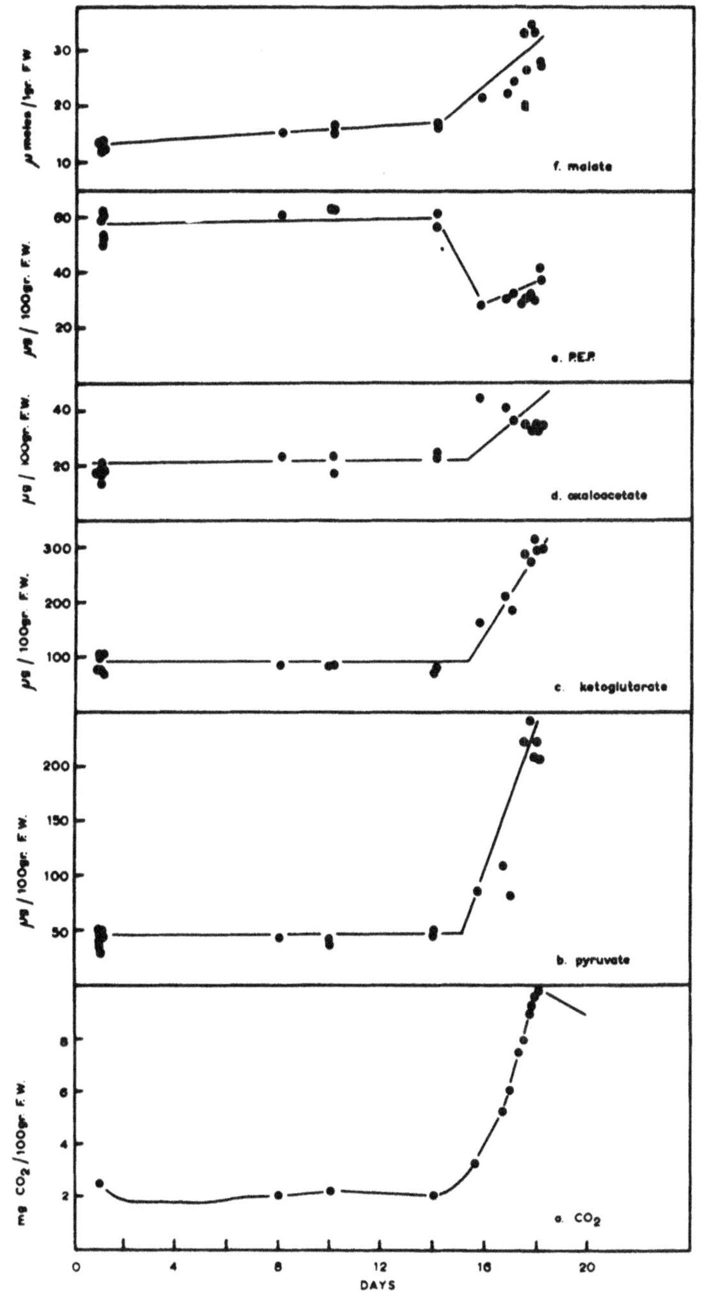

Fig. 7.a. The continuous line curve shows the changes in rate of CO_2 output of a single banana finger ripened at 18°C. The circles show the rates of respiration of other single fingers taken for analysis at different stages of ripeness.

7.b,c,d,e,f. Changes in content of pyruvate, α-ketoglutarate, oxalacetate, PEP, and malate, respectively, in banana in which the rates of CO_2 output are shown in a.

been established. Wiskich (123) considers it unlikely to be of major importance in the regulation of the TCA cycle.

Electron Transport

Although the composition of the electron transport chain in senescing fruits has not been determined explicitly, there is nevertheless no reason to believe either that it differs from other plant tissues or that it changes in the course of ripening (61,94). Intact fruits and mitochondria isolated therefrom are resistant to cyanide (94,109). It is well documented that cyanide resistant oxidase branches from the main electron transport chain at the level of ubiquinone (14,36,47,106,111).

It was observed that (a) HCN induces the climacteric rise in respiration and eventual ripening of avocado fruits, (b) the changes in the glycolytic intermediates of avocado and potato tubers treated with HCN are similar to those induced by ethylene,(c) ethylene and HCN induce in potato tubers identical changes in respiration and sugar content, (d) ethylene enhances respiration in tissues where cyanide acts similarly, while having no effect on tissues whose respiration is inhibited by cyanide (45,107,108,109,110). On the basis of the above observations it was proposed that for ethylene to enhance plant respiration the presence of cyanide-resistant respiration is necessary. Two crucial questions have to be examined experimentally: first, is ethylene-enhanced respiration prevented by inhibitors of the alternate oxidase? and second, is alternate oxidase functioning in the absence of inhibitors of the cytochrome path?

To answer the first question, we treated with ethylene thick banana slices which had first been infiltrated and were constantly supplied with a 3 mM solution of SHAM, an inhibitor of the alternate oxidase (102). It can be seen in Figure 8 that the inhibition of the alternate oxidase did not prevent the rise in respiration in response to ethylene (Tucker and Solomos, unpublished). Similar data were obtained with ivy leaves (Solomos, unpublished observations). Thus ethylene enhances plant respiration independent of the nature of the terminal oxidase.

To answer the second question, the mechanism which regulates the apportioning of electrons between the cytochrome and alternate paths will first be examined. Bahr and Bonner (7,8) have shown that the diversion of electrons to the alternate path is independent of the redox state of the cytochromes, since antimycin-A, which inhibits the electron transport at the level of the b cytochromes, is as effective in diverting electrons to the alternate path as is cyanide. Also, the energy state of the mitochondria is not a controlling factor, since electrons are diverted to the alternate oxidase in both coupled and uncoupled mitochondria. Bahr and Bonner

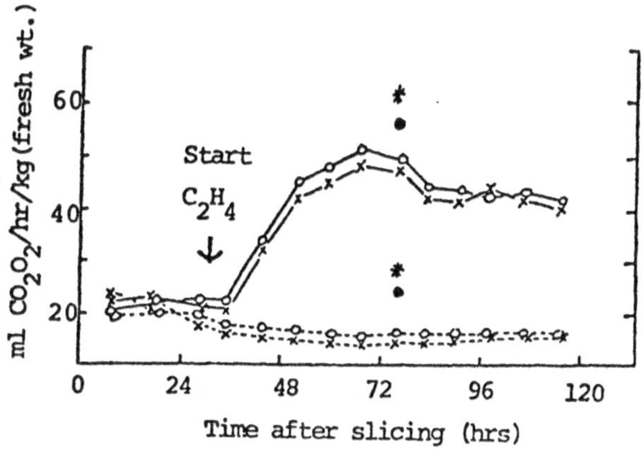

Fig. 8. Effects of SHAM on respiration of thick banana slices not treated with C_2H_4 and treated continuously with 5 ppm C_2H_4. Mannitol 0.3M was used as an osmotic stabilizer in all samples. Each point is the mean of two composite samples each of four 1.5 cm thick slices. The symbols indicate the following: 5 ppm C_2H_4(-); no C_2H_4(---); CO_2 output (x,o); O_2 uptake (*, o); control (x, *); and 3mM SHAM (o, •).

(8) derived an equation by which the fraction of the alternate oxidase which is engaged in the absence of inhibitors of the cytochrome path can be calculated. For this, mitochondria were titrated with inhibitors of the alternate oxidase in the presence of cyanide. These values of oxygen uptake were called g(i). The authors carried out similar titrations in the absence of cyanide. Assuming that the degree of inhibition by the hydroxamic acid derivatives, inhibitors of the alternate oxidase (102), is the same irrespective of the presence of cyanide, then the total oxygen uptake $V_T = \rho g(i) + V_{cyt}$, where V_{cyt} is the flux through the cytochrome path and ρ is a number between 0 and 1. If $\rho = 1$ then the full capacity of the alternate path is realized, whereas if $\rho = 0$ all electron transport is carried out via the cytochrome oxidase. Bahr and Bonner (7) concluded that the apportioning of electrons between the two paths is controlled by an equilibrium mechanism. They proposed that both paths share a common component B and the first component of the alternate path is A which is in ready equilibrium with B. The standard redox potentials of these two electron carriers differ, with B being more positive, so that it allows A to be fully oxidized while B is partially reduced.

From the Nernst equation, $E' = \frac{RT}{nF} \ln \frac{A^{ox}}{A^{red}} \cdot \frac{B^{red}}{B^{ox}}$.

Bahr and Bonner calculated that the difference in the redox potentials of the above carriers would be 35 mV. Storey (111) showed that ubiquinone is the carrier share by the two paths (component B) and FPma (Flavo-protein, mediumpotential, absorbing light) is the first carrier of the alternate path (component A). The E values of ubiquinone and FPma are +60 and +20 respectively (112) which meets the requirement of a difference of 35 mV proposed by Bahr and Bonner (8). The corollary of these observations is that the flux through the cytochrome path alone determines the degree of the diversion of electrons to the alternate path. In other words as long as the capacity of the cytochrome path is sufficient to handle the electron flux the alternate path will not be engaged.

Theologis and Laties (114-116) tested the above hypothesis with a variety of tissue slices, including bananas and avocados at various stages of ripeness. With the possible exception of avocado slices, in all other tissues ρ was zero; i.e. the alternate path is not engaged unless the cytochrome path is either inhibited or saturated by the electron flux, as in coupled slices.

We have attempted to ascertain the contribution of the alternate oxidase to the respiration of banana and sweet potato slices by relying on the fact that the apparent K_m for O_2 of the alternate oxidase is 5-8 fold higher than that of the cytochrome oxidase (36,106,112). If the alternate oxidase is engaged and the mechanism of controlling the apportionment of electrons is operative, the concentration of oxygen below which the rate of oxygen uptake ceases to be of zero order will differ in control slices and SHAM-treated slices, being larger in the former case. It can be seen from Figure 9 that the time versus rate of oxygen uptake of the oxygen electron trace departs from linearity at the same point for both control and SHAM-treated slices (13-19 μM), while in slices treated with KCN the rate of oxygen uptake becomes dependent on the oxygen concentration at about 78-120 μM of external oxygen concentration.

The data therefore indicate that, first, in the absence of cyanide the alternate oxidase is not engaged. Secondly, the ratio of the apparent K_m for O_2 of the alternate oxidase to that of the cytochrome oxidase is about 6-8. It can be seen from Figure 9 that when the external oxygen concentration decreases to 10.5 μM the rate of O_2 uptake is reduced by 30%. If the slices are kept at this concentration for 90 min, lactate accumulates (Table 10). Hence when the concentration of oxygen becomes limited to the cytochrome oxidase a Pasteur effect is induced.

It is a well established fact that reduction of external oxygen tensions causes a decrease in the rate of respiration in several plant tissues in general and in fruits in particular (42,53,57).

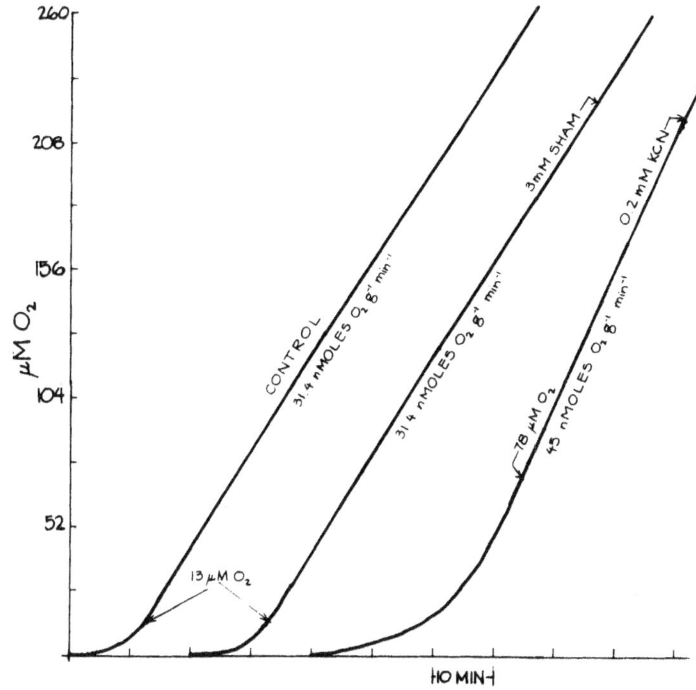

Fig. 9. Effects of 3 mM SHAM and .2 mM KCN compared to control, on the rate of O_2 uptake of 1 mm thick sweet potato slices

Table 10. Effect of Oxygen Concentration on the Levels of Pyruvate and Lactate in 1 mm Thick Sweet Potato Slices

Treatment	nmoles/g FW	
	Pyruvate	Lactate
Air	350	47
60 min in 10.5 μM O_2	522	356
90 min in 10.5 μM O_2	578	642.4

Mapson and Burton (70) proposed that in intact potato tubers two terminal oxidases with a different affinity for oxygen are functioning. Chevillote (30) explained the decrease in respiration caused by lowering the oxygen tension in terms of constraints of the oxygen diffusion into the cells.

The beneficial effects of the low O_2 concentrations on the longevity of fruits and flowers preclude the possibility that the decrease in the rate of respiration is the result of a diminished activity of the cytochrome oxidase, since, as has been pointed out earlier, this would be expected to lead to the accumulation of glycolytic end products, such as lactate, ethanol and acetaldehyde, which are known to be highly detrimental to stored fruits.

In order to avoid the complications of the interactions between O_2 and C_2H_4 action (27) we have studied the effect of low oxygen concentrations on the rate of CO_2 output in sweet potato roots. Figure 10 shows that the gradual decrease in external O_2 concentration causes a drop in the rate of CO_2 output until the concentration of the O_2 drops to about 1.5 - 2%, where CO_2 output begins to increase. The rate of CO_2 output continues to increase as the O_2 concentration decreases to zero.

If the roots are kept for 12 h. at 1.4% O_2, a concentration at which the rate of CO_2 output stops falling and instead begins to rise, lactate accumulated (Table 11). Hence the data indicate that not until the O_2 concentration is decreased below 2.2% does the supply of O_2 begin to limit cytochrome oxidase. In order to further eliminate the possibility that the rate of O_2 diffusion was not the factor limiting respiration during the gradual decrease of O_2 concentration, we treated sweet potato roots with C_2H_4 at 10% oxygen. Figure 11 shows that ethylene raises the rate of CO_2 output by about 100% above that of the air level. Collectively the data indicate that the decrease in the rate of respiration with decreasing O_2 concentrations is caused by the curtailment of an "oxidase" with a lower affinity for O_2 than cytochrome oxidase. Furthermore, the ratio of the apparent K_m for O_2 of the low affinity oxidase, ca. 8.5%, to that of cytochrome, ca. 0.90%, is about 9, which is close to that, 6-8, found with the slices where the respective apparent Km could be ascertained more accurately. Thus, in sweet potatoes about 30-40% of the terminal electron transport is probably mediated via the alternate oxidase.

It has been pointed out earlier than the diversion of electrons to the alternate oxidase is regulated by the flux of electrons through the cytochrome path. Unless the latter is inhibited or saturated, the alternate oxidase is not expected to function. In intact fruits and storage roots only a small fraction of the cytochrome capacity is engaged. For instance, the rate of oxygen uptake of intact sweet potato roots and uncoupled fresh slices at 25°C is 18-20 and 100-120 μl/g/h respectively. Thus the contribution of the alternate oxidase to tissue respiration should be non-existent. However, the experimental evidence cited earlier indicates that an "oxidase" with an affinity for oxygen lower than that of the cytochrome oxidase, presumably alternate oxidase, contributes substantially to the respiration of intact fruits and stor-

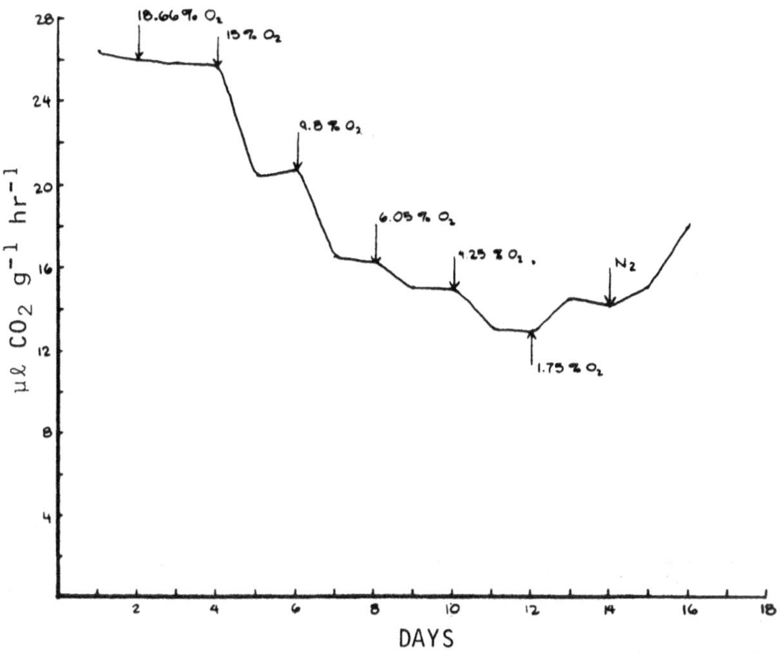

Fig. 10. Effects of O_2 concentration lower than air on the rate of CO_2 output of intact sweet potato roots.

Table 11. Effect of O_2 Concentration on the Level of Lactate in Intact Sweet Potato Roots

	Lactate nmoles/g FW
Air	< 10
10% O_2	< 10
1.4% O_2 12 h	260

age organs. It should be mentioned that (a) the low rates of oxygen uptake of certain preclimacteric fruits and storage roots and stems, (b) the high affinity for O_2 of the cytochrome oxidase ($K_m \sim 0.5$ μM), and (c) the absence of the accumulation of glycolytic end products at O_2 tensions lower than air all preclude the possibility that cytochrome oxidase is limited by the rate of oxygen diffusion. These observations raise two crucial questions. If

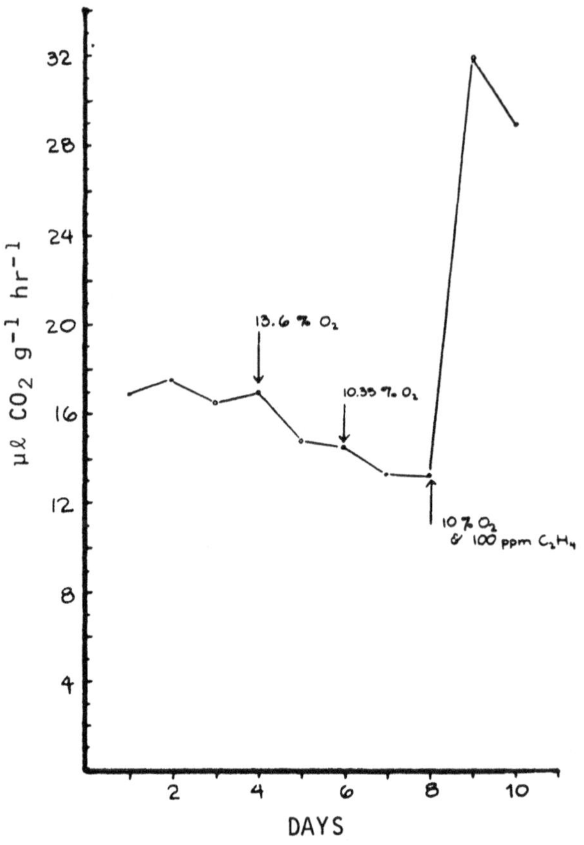

Fig. 11. Effect of 100 p.p.m. ethylene on the rate of CO_2 output of intact sweet potato roots kept at 10% oxygen.

the availability of ADP is the main factor which restricts cytochrome oxidase in vivo, how then does site I of oxidative phosphorylation function, in preference to the other two sites, allowing the operation of the alternate oxidase? Secondly, since there is no accumulation of respiratory intermediates at oxygen tensions lower than those in air, what is the feedback mechanism by which the diminution of the activity of the alternate oxidase restricts the initial steps of glucose oxidation?

Since it is very unlikely that preferential disengagement of oxidative phosphorylation at site I is involved, then the electrons from the oxidation of the NAD-linked substrates must bypass this step. Experimental evidence for this argument comes from avocado slices, where amytal stimulated slice respiration by about

50% and this stimulation was sensitive to inhibitors of the cytochrome path (see 106). Further work with isolated plant mitochondria suggests that site I can be bypassed either by the action of malic dehydrogenase (83,84) or through the transmembrane hydrogen transport (34,123). Thus it appears that in plant mitochondria site I of oxidative phosphorylation can be bypassed, thereby allowing the operation of the alternate oxidase.

The mechanism by which the curtailment by low O_2 of the low affinity oxidases leads to the decrease in the mobilization of food reserves is not clear.

Residual Oxidase

The combined application of cyanide and hydroxamic acid derivatives to slices of storage organs and fruits does not completely inhibit the rate of oxygen uptake (36,114). The nature of these residual oxidases is completely unknown. We have recently observed that the inhibition by the simultaneous application of inhibitors of both the cytochrome and alternate paths decreases with increasing oxygen tension (Figure 12). Thus while 63% O_2 has no effect on the rate of résspiration of either control or cyanide-treated slices, it enchances by more than three-fold the rate of O_2 uptake of slices treated both with cyanide and SHAM together (Table 12). In the case of avocado slices, the increase in oxygen concentration to 100% completely eliminates the inhibition caused by the simultaneous application of cyanide and SHAM (Tucker and Laties, personal communication). Since the rate of oxygen uptake of isolated plant and fruit mitochondria is negligible when KCN and SHAM are added simultaneously, the residual "oxidase(s)" must be cytosolic in origin. The above observations raise the following crucial questions: What is the pathway of glucose oxidation in the presence of both KCN and SHAM? Is the Krebs cycle partially functioning and are the reducing equivalents then transferred to cytosolic enzymes by the mechanism proposed by Day and Wiskich (34) for isolated mitochondria? Or is glucose being oxidized exclusively by the pentose phosphate shunt? The answers to these questions must remain unanswered until detailed analytical data become available concerning the changes of respiratory intermediates in tissues inhibited by both KCN and SHAM. It should be noted that in short-term experiments with tissue slices, like the present one, respiratory electrons are first channelled to the cytochrome path and only the inhibition of the latter engages the alternate path. Similarly since the increase in oxygen tension has no effect on oxygen uptake when cyanide is present, and since the rate of oxygen uptake in the presence of KCN is zero order up to concentrations of O_2 equal to twice the apparent K_m for oxygen of the alternate oxidase, the alternate oxidase appears to be the main terminal electron acceptor. Only when both paths are inhibited, are respiratory-reducing equivalents transferred to O_2 by cytosolic

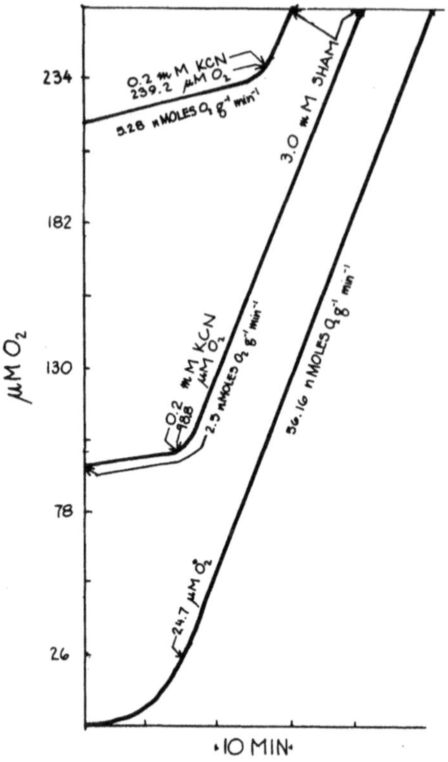

Fig. 12. Effect of O_2 concentrations on the respiration of 1 mm thick sweet potato slices treated with 3 mM SHAM and 0.2 mM KCN simultaneously

Table 12. Rate of Oxygen and Uptake of Sweet Potato Slices Treated with KCN, KCN and SHAM, and of the Control at Different O_2 Concentrations

μMO_2	$\mu l\ O_2/g/min$			
	Control	3 mM SHAM	0.2 mM KCN	3mM SHAM + .2 mM KCN
260	0.92	0.92	1.13	0.20
780	0.91	0.92	1.15	0.69

oxidases whose affinity for O_2 is much smaller than that of the alternate path.

Cellular Organization

Blackman and Parija (19) speculated that the increase in the rate of respiration during the climacteric was the result of the "decrease in cellular organization resistance." It was visualized that with the onset of senescence, the permeability of cellular membranes increases, thus enhancing the availability of substrates to the enzymes. There is little doubt that membrane permeability increases in senescing tissues. Sacher (99) found that the free space of banana slices increased prior to the climacteric rise in respiration. Brady et al. (21) found an increase in membrane permeability of ripening bananas before the increase in respiration. However, when aseptically prepared slices were treated with ethylene, a respiratory rise preceded the increase in amino acid leakage. Hanson and Kende (46) also observed enhanced ion efflux in the rib tissue of morning glory flower (*Ipomoea tricolor*). The authors concluded that ethylene increases membrane permeability in selective cells of the rib tissue. However this increase is probably the result rather than the cause of senescence. In the first place it is difficult to measure changes in the membrane permeability of bulky tissues like fruits. Secondly, it has been pointed out by Burg (25) and Vickery and Bruinsma (120) that ion leakage, the method used to determine changes in membrane permeability, may reflect changes in the ions available for leakage. An increase in the cytosolic ion concentration will be expected to show enhanced efflux. The increase in the apparent free space observed during ripening in bananas and avocados (15,99) may overestimate the actual changes which occur *in vivo* (25).

It must be stressed that changes in the cellular distribution of regulatory metabolites and ions need not require gross changes in membrane permeability. It is to be expected that the compartmentalization of regulatory metabolites, particularly ions, are under metabolic control (88). A case in point is the increase in the cytosolic concentration of K^+ in the course of the ripening of tomato fruits (120). Since K^+ is a requirement for PK and a positive modulator of PFK, this increase in the cytosolic K^+ will be expected to contribute to the enhancement of respiration.

Hence the available experimental evidence indicates that the rise in respiration during senescence of certain fruits, cut flowers and detached leaves is caused by ethylene. Although ethylene induces cyanide-resistant respiration in potato tubers (35,98), it enhances respiration independent of the nature of the terminal oxidase. The respiratory rise is associated with an increase in glycolysis which in turn is caused by an activation of PFK and PK. The activation of banana PFK appears to be due in part to changes in its kinetic properties in the course of the climacteric and to the decrease in the concentration of PEP. The probable increase in the cytosolic concentration of ions, which activate both PFK and PK, could also contribute to the enhancement of respiration. Further, the results discussed above indicate that in tissue slices, and probably intact tissues, the bulk of the respiratory electrons is channelled via the cytochrome oxidase. However, in intact sweet potato roots and probably other bulky organs, as well as preclimacteric fruits, a substantial portion of the terminal electron acceptors (ca. 30-40% in sweet potatoes) is handled by an oxidase other than cytochrome oxidase, presumably alternate oxidase.

A note of caution is in order. It is obvious that the curtailment by relatively low O_2 concentrations of the non-phosphorylating oxidation of glucose will be beneficial to the extension of the storage life and quality of commodities where both of these properties are critically dependent on sugars or other food reserves. However the action of low O_2 in delaying senescence cannot be thought of as being exerted through the diminution of respiration. In the first place, in fruits with large food reserves, such as bananas, cherimoyas, avocados, etc, the shortage of respiratory substrates is not at issue. The delaying effect of relatively low O_2 concentrations on senescence must be exerted through the diminution of the activities of oxygen-utilizing enzyme(s) with relatively low affinity for oxygen and which are involved in the induction, synthesis (2) and action of ethylene (25) and/or other unknown processes (64). Alternatively, it has been proposed that C_2H_4 and high O_2 concentrations generate active forms of oxygen, namely peroxides, which in turn enhance senescence (23,31,97).

PHYSIOLOGICAL SIGNIFICANCE OF THE RISE IN RESPIRATION

It has already been pointed out that for senescence to proceed, metabolic energy is required. Teleologically an increase in energy demand will be expected to decrease ATP concentration and increase that of ADP. However, enhancement of anabolic activities would require an increase in the energy charge (5a). The available analytical data concerning the changes in the energy charge in the course of senescence of fruits and detached leaves are scanty. In avocados, energy charge increases (126), while in senescing ivy leaves, the increase in the energy charge is not very pronounced. It can be calculated from the data of Table 8 that the energy charge in ivy leaves during senescence increases from 0.51 to 0.6. The available analytical results show that the rise in respiration in response to ethylene treatment is associated with a large increase in ATP levels (Table 8). Further, in cases where energy charge remains relatively constant, this increase in ATP is associated with a net increase in the total pool of the adenine nucleotides. In view of the inhibitory effects of these metabolites on PFK, one would anticipate a decrease in the rate of respiration. Therefore, ethylene must decrease the regulatory restraints of the adenine nucleotides on respiration, allowing a simultaneous enhancement in respiration, elevation of ATP and/or of total adenine nucleotide concentration.

On the basis of the present state of knowledge regarding the nature and magnitude of the sinks of metabolic energy in senescing tissues, it is impossible to calculate the energy demand with any degree of precision. At the climacteric peak of bananas at 20°C the rate of respiration is 60-70 µl O_2/g/h and the level of ATP is 90-100 nmoles/g/h (Solomos, unpublished observation). If we assume that glucose is oxidized via glycolysis and the Krebs cycle, and assuming tight coupling of glucose oxidation to phosphorylation, about 18,750 nmoles of ATP are synthesized per gram of tissue in one hour (63). Hence the turnover of the ATP pool per hour is approximately 200. Therefore, for respiration to proceed for any length of time, ATP must be utilized very rapidly.

In ripening bananas the conversion of starch to sucrose can be a sizable sink for ATP. It can be calculated from the data presented by Gane (44) that at the climacteric peak the rate of sucrose formation is about 1,900 nmoles/g/h (Table 13). If starch is broken down by starch phosphorylase, then one ATP is required for each molecule of sucrose synthesized (118). Hence, about 2,000 nmoles ATP/g/h are required for sucrose synthesis, a rather small fraction of the total ATP which could be synthesized.

Brady and O'Connell (20) calculated that in ripening bananas the rate of protein turnover represents the replacement of 25-50% per day of the total proteins. If it is assumed that (a) the protein content of bananas is 2%, (b) the average molecular weight of the proteins is 60,000, (c) the average weight of the individual amino-acids is 135, (d) four ATP are required for the synthesis of a peptide bond (63), and (e) 50% of the total protein is replaced in 24 h, then 12,350 nmoles ATP/g/h are required to sustain the above rate of protein synthesis. Even this rather large overestimation

Table 13. Calculated rates of ATP Production and Utilization at the Climacteric Peak of Ripening Bananas

µl CO_2/g /h	nmoles ATP /g/h	nmoles sucrose synthesized /g/h	nmoles ATP/g/h required for protein synthesis
70	18,750	1,900	12,350

of protein synthesis together with the formation of sucrose could be satisfied from the point of view of energy with a smaller increment of respiration than that extent at the climacteric peak (Table 13). Moreover, ripening of bananas can proceed in the absence of any perceptible rise in respiration (87). Further, in most fruits the overt changes associated with ripening occur after the climacteric peak.

It has been proposed that the upsurge of respiration in climacteric fruits reflects an increase in the energy requirements for the ripening process, and further that the slow ripening process of nonclimacteric fruits does not require as much energy as do climacteric ones, hence the lack of a respiratory rise (86,90,92).

However, nonclimacteric fruits such as strawberries ripen rapidly with no concomitant increase in respiration (58). Vendrell (119) has shown that the infiltration of banana slices with auxin solutions delays senescence caused by exogenous ethylene but does not prevent the initial rise in respiration. Dostal and Leopold (40) observed that application of GA to green tomatoes delays the breakdown of chlorophyl but has no effect on either respiration or ethylene production. Further, the application of exogenous ethylene does not alter the effect of GA. Pratt and Goeschl (85) have found that in ripening cantaloupes, on the one hand 0.1 p.p.m. ethylene induced softening and carotenoid synthesis and on the other hand the enhancement of respiration required 3 p.p.m C_2H_4. Moreover, the rate of protein and nucleic acid synthesis in several fruits diminishes at the later stages of the climacteric rise (22,91,92).

In summary, the rise in respiration during senescence of fruits, leaves and cut flowers may not reflect the energy requirements of the senescence process but it is a facet of ethylene action and not of senescence *per se*. To paraphrase Churchill, senescence is a riddle wrapped in a mystery inside an enigma, but the parcel is mainly tied up with ethylene.

ACKNOWLEDGEMENTS

This is scientific article no. A-3020, contribution no. 6083 of the Maryland Agricultural Experimental Station and the Department of Horticulture, University of Maryland, College Park. The research was supported in part by a grant from the National Science Foundation No. PCM 76-21560.

REFERENCES

1. Abeles, F. B., 1973, "Ethylene in Plant Biology," Academic Press: New York.
2. Adams, D. O., and Yang, S. F., 1979, Ethylene biosynthesis. Identification of 1-aminocyclopropane-1-carboxylic acid as an intermediate in the conversion of methionine to ethylene, Proc. Nat. Acad. Sci. U.S.A., 76:170.
3. Aharoni, Y., 1968, Respiration of oranges and grapefruits at different stages of development, Plant Physiol., 43:92.
4. Aharoni, N., and Lieberman, M., 1979, Ethylene as a regulator of senescence in tobacco leaf discs, Plant Physiol., 64:801
5. apRees, T., Wright, B. W., and Fuller, W. A., 1977, Measurements of starch breakdown as estimates of glycolysis during thermogenesis of spadix of Arum maculatum, Planta, 134:53.
5a. Atkinson, D., 1977, "Cellular Energy metabolism and its regulation," Academic Press: New York.
6. Ashihara, H., and Komamine, A., 1974, Enzyme and metabolite profiles of the pentose phosphate pathway of Phaseolus mung seedlings, Plant Sci. Lett. 2:331.
7. Bahr, J. T., and Bonner, W. D., Jr., 1973, Cyanide-insensitive respiration. I. The steady state of skunk cabbage spadix and bean hypocotyl mitochondria, J. Biol. Chem., 248:3441.
8. Bahr, J. T., and Bonner, W. D., Jr., 1973, Cyanide-insensitive respiration. II. Control of the alternate pathway, J. Biol Chem., 248:3446
9. Bain, J. M., and Mercer, F. V., 1964, Organization resistance and the respiration climacteric. Aust. J. Biol. Sci., 17:78.
10. Barker, J., and Solomos, T., 1962, The mechanism of the climacteric rise in respiration in banana fruits, Nature, 69: 180.
11. Barker, J., Khan, A. A., and Solomos, T., 1967, The mechanism of the Pasteur effect in peas, New Phytol., 66:577.
12. Beevers, H., 1974, Conceptual developments in metabolic control 1924-1974, Plant Physiol., 54:437.
13. Beevers, H., Stiler, M. L., and Butt, V. S., 1966, Metabolism of organic acids, in: "Plant Physiology," F. C. Stewards, ed., Vol. 4A, Academic Press: New York.
14. Bendall, D. S., and Bonner, D. W., Jr., 1971, Cyanide-insensitive respiration in plant mitochondria. Plant Physiol. 47: 236.
15. Ben Yehoshua, S., 1964, Respiration and ripening of discs of the avocado fruit, Physiol. Plant., 17:71
16. Biale, J. B., 1960, Respiration of fruits, in: "Encyclopedia of plant physiology," vol. XII/2, Springer Verlag: Berlin.
17. Biale, J. B., 1960, The postharvest biochemistry of tropical and subtropical fruits, Adv. Food Res., 10:293.
18. Biale, J. B., and Young, R. E., 1970, The avocado pear, in: "The Biochemistry of fruits and their products," vol. 2, A. C. Hulme, ed., Academic Press: London-New York.

19. Blackman, F. F., and Parija, P., 1928, Analytical studies in plant respiration. I. The respiration of a population of senescent ripening apples, Proc. Roy. Soc. B. Biol. Sci., 103:412.
20. Brady, C. J., and O'Connell, P. B. H., 1976, On the significance of increased protein synthesis in ripening banana fruits, Aust. J. Plant Physiol., 3:301
21. Brady, C. J., O'Connell, P. B. H., Smydzuk, J., and Wade, N.L., 1970, Permeability, sugar accumulation and respiration rate in ripening banana fruits, Aust. J. Biol. Sci., 23:1143.
22. Brady, C. J., Palmer, J. K., O'Connell, P. B. H., and Smillie, R. M., An increase in protein synthesis during ripening of the banana fruit, Phytochemistry, 9:1037.
23. Brennan, T., and Frenkel, C., 1977, Involvement of hydrogen peroxide in the regulation of senescence in pears, Plant Physiol., 59:411.
24. Burg, S.P., 1962, The physiology of ethylene formation, Ann Rev. Plant Physiol., 13:265.
25. Burg, S.P., 1968, Ethylene, plant senescence and abscision, Plant Physiol., 43:1503.
26. Burg, S. P., 1973, Hypobaric storage of cut flowers, Hort.-Sci., 8:202.
27. Burg, S. P., and Burg, E. A., 1967, Molecular requirements of the biological activity of ethylene, Plant Physiol., 42:144.
28. Chalmers, D. J., and Rowan, K. S., 1971, The climacteric in ripening tomato fruit, Plant Physiol., 48:235.
29. Chance, R., Holmes, W., Higgins, J., and Connelly, C. M., 1958, Localization of interaction sites in multicomponent transfer systems, Nature 182:1190.
30. Chevillote, P., 1973, Relation between the reaction of cytochrome oxidase, oxygen and oxygen uptake in cells in vivo. The role of diffusion, J. Theor. Biol., 39:277.
31. Chin, C. K., and Frenkel, C., 1977, Upsurge in respiration and peroxide formation in potato tubers as induced by ethylene, propylene and cyanide, Plant Physiol., 54:515.
32. Cooper, W. C., Rasmussen, G. K. Roger, B. J., Reece, P. C., and Henry, W. H., 1968, Control of abscission in agricultural crops and its physiological basis, Plant Physiol., 43:1560
33. Coorts, G.P., 1973, Internal metabolic changes in cut flowers, HortSci., 8:195.
34. Day, D.A., and Wiskich, J.T., 1978, Pyridine nucleotide interactions with isolated plant mitochondria, Biochem. Biophys. Acta, 501:396.
35. Day, D.A., Arron, G. P., Chritoffersen, A.E., and Laties, G. G., 1978, Effect of ethylene and carbon dioxide on potato metabolism. Stimulation of tuber and mitochondrial respiration, and inducement of the alternate path, Plant Physiol., 62:820
36. Day, D. A., Arron, J. P., and Laties, G. G., 1980, Nature and control of respiratory pathways in plants. The inter-

action of cyanide-resistant respiration with cyanide-insensitive pathways, in: "The Biochemistry of Plants. A Comprehensive Treatise," Vol. 2, D. D. Davies, ed., Academic Press: New York.

37. Dickinson, D. B., and Hanson, J. B., 1965, Comparison of mitochondria from tomato fruits at various stages of ripeness, Plant Physiol., 40:161
38. Dilley, D. R., 1962, Malic enzyme activity in apple fruit, Nature, 196:337.
39. Dilley, D. R., and Carpenter, W. J., 1978, The role of chemical adjuvants and ethylene synthesis on cut flower longevity, Acta Horticulturae. Post Harvest Physiology of Cut Flowers, 41:117.
40. Dostal, H. C., and Leopold, C. A., 1967, Gibberellin delays ripening of tomatoes, Science, 158:1579.
41. Faiz-ur-Rahman, Trewavas, A. J., and Davies, D. D. 1974, The Pasteur effect in carrot root tissue, Planta, 118:195.
42. Fidler, J. C., Wilkinson, B. G., Edney, K. L., and Sharpless, R. O., 1973, The biology of apple and pear storage, Research Bulletin No. 3, Commonwealth Bureau of Horticulture and Planation Crops, East Malling, Maidstone, Kent.
43. Frenkel, C., Klein, I., and Dilley, D. R., 1968, Protein synthesis in relation to ripening of pome fruits, Plant Phy siol., 43:1146.
44. Gane, R., 1936, A study of respiration of bananas, New Phytol. 35:383.
45. Hanes, C. S., and Barker, J., 1931, The physiological action of cyanide. I. The effects of cyanide on the respiration and sugar content in potato tubers at 15°C., Proc. Roy. Soc. B, 108:95.
46. Hanson, A.D., and Kende, H., 1975, Ethylene-enhanced ion and sucrose efflux in morning-glory flower tissue, Plant Physiol., 55:663.
47. Henry, M. F., and Nyns, E. J., 1975, Cyanide-insensitive respiration. An alternate mitochondrial pathway, Sub-Cell Biochem., 1.
48. Herner, R. C., and Sink, K. C., Jr., 1973, Ethylene production and respiratory behavior, Plant Physiol., 52:38.
49. Hess, G. M., and Meeuse, B. J. D., 1968, Factors contributing to the respiratory flare up in the appendix of sauromatum (Araceae), Kon. Neder. Acad. Wet. Amsterdam Ser. C. 71:456.
50. Hulme, A. C., 1948, Studies in the nitrogen metabolism of the apple fruit. Changes in the nitrogen of the apple during the normal and ethylene induced rise in rate of respiration, Biochem. J., 43:343.
51. Hulme, A. C., Jones, J. D., and Wooltorton, L. S. C., 1963, The respiration climacteric in apple fruits. Proc. Roy. Soc. Ser. B., 158:514.
52. Hulme, A. C., Rhodes, M. J. C., Galliard, T., and Wooltorton, L. S. C., 1968, Metabolic changes in excised fruit tissue,

IV Changes occurring in discs of apple peel during the development of the respiration climacteric, Plant Physiol., 43:1154.
53. James, W. O., 1953, "Plant respiration," Oxford U.P.
54. Johnson, T. F., and Meeuse, B. J. D., 1972, The phosphofructokinase of the sauromatum appendix (Araceae). Purification, and activity-regulation in vitro, Kon. Neder. Akad. Weten, series C., 75:1.
55. Kaltaler, R. E. L., and Steponkis, P. L., 1976, Factors affecting respiration in cut roses, J. Am. Soc. Hort. Sci., 101: 352.
56. Kidd, F., and West, C., 1924, The course of respiratory activity throughout the life of an apple, Gt. Brit. Dept. Sci. Ind. Food Research Invest. Board Rept., 1924:27.
57. Kidd, F., and West, C., 1927, Gas storage of fruits, Spec. Report, Fd. Invest., DSIRO.
58. Knee, M., Sargent, J. A., and Osborne, D. J., 1977, Cell wall metabolism in developing strawberry fruits, J. Exp. Bot., 28:377.
59. Kobr, M., and Beevers, H., 1971, Gluconeogenesis in castor bean endosperms. Changes in the glycolytic intermediates, Plant Physiol., 47:48.
60. Krebs, H. A., 1969, The role of equilibria in regulation of metabolism, Curr. Top. Cell. Regul., 1:45.
61. Lance, C., Hobson, G. E., Young, R. E., and Biale, J. B., 1965, Metabolic processes in cytoplasmic particles of the avocado fruit. VII Oxidative and phosphorylative activity throughout the climacteric cycle. Plant Physiol., 40:1116.
62. Laties, G. G., 1978, Development and control of respiratory pathways in slices of plant storage organs, in: "Biochemistry of wounded plant storage tissues," G. Kahl, ed., de Gruyter: Berlin.
63. Lehninger, A. L., 1976, "Biochemistry: The molecular basis of cell structure and function," Worth Publishers, Inc: New York.
64. Lieberman, M., 1979, Biosynthesis and action of ethylene, Ann Rev. Plant Phys. 30:533.
65. Lieberman, M., and Baker, J. E., 1965, Respiratory electron transport, Ann. Rev. Plant Physiol., 16:343
66. Lowry, D. H., and Passoneau, J. W., 1964, A comparison of the kinetic properties of phosphofructokinase from bacterial plant and animal sources, Naunyn-Schmiedeborgs, Arch. Exp. Path. Pharmakol., 248:185.
67. Macnicol, P. K., 1973, Metabolic regulation in the senescing tobacco leaf. II. Changes in the glycolytic metabolic levels in detached tobacco leaf, Plant Physiol., 51:798.
68. Macnicol, P. K., Young, R. E., and Biale, J. B., 1973, Metabolic regulation in the senescing tobacco leaf. I. Changes in the pattern of ^{32}P incorporation into leaf disc metabolites, Plant Physiol., 65:855.

69. Malik, N. S. A., and Thimann, K. V., 1980, Metabolism of oat leaves during senescence, VI. Changes in ATP levels, Plant Physiol., 65:855
70. Mapson, L. W., and Burton, W. G., 1962, The terminal oxidases of potato tuber, Biochem. J., 82:19
71. Martin, C., and Thimann, K. V., 1972, The role of protein synthesis in the senescence of leaves. I. Formation of protease. Plant Physiol., 49:64.
72. Maxie, E. C., Eaks, I. L., Sommer, W. F., Rae, H. L., and El-Batal, S., 1965, Effect of gamma radiation on rate of ethylene and carbon dioxide evolution by lemon fruit, Plant Physiol., 40: 407.
73. Mayak, S., Vaadia, Y., and Dilley, D. R., 1977, Regulation of senescence in carnations (*Dianthus caryophyllus*) by ethylene: Mode of action, Plant Physiol., 59:591.
74. McGlasson, W. B., Palmer, J. K., Vendrell, M., and Brady, C. J., 1971, Metabolic studies with banana fruits. II. Effect of inhibitors on respiration, ethylene production and ripening, Aust. J. Biol. Sci., 24:1103.
75. McMurchie, E. J., McGlasson, W. B., and Eaks, J. L., 1972, Treatment of fruit with propylene gives information about the biogenesis of ethylene, Nature, 237:235.
76. Meeuse, B. J. D., 1975, Thermogenic respiration in aroids, Plant Physiol., 26:117.
77. Millerd, A., Bonner, J., Biale, J. B., 1953, The climacteric rise in fruit respiration as controlled by phosphorylative coupling, Plant Physiol., 28:521
78. Morgan, P. W., 1976, Effects on ethylene physiology, in: "Herbicides: Physiology, Biochemistry, Ecology," L. J. Audies, ed., Academic Press: New York.
79. Newsholme, E. A., and Start, C., 1979, "Regulation in metabolism," John Wiley & Sons: Chichester, New York, Toronto.
80. Nichols, R., 1973, Senescence of cut carnation flowers: Respiration and sugar status, J. Hort. Sci., 48:111
81. Oaks, A., and Bidwell, R. G. S., 1970, Compartmentation of intermediary metabolites, Ann. Rev. Plant Physiol., 21:43.
82. Palmer, J. K., 1970, The banana, in:"Biochemistry of fruits and their products," Vol. 2, A.C. Hulme, ed., Academic Press: New York.
83. Palmer, J. M., 1973, The organization and regulation of electron transport in plant mitochondria, Ann. Rev. Plant Physiol., 27:133.
84. Palmer, J. M., and Arron, G. P., 1976, The influence of exogenous nicotinamide adenine dinucleotide on the oxidation of malate by Jerusalem artichoke, J. Exp. Bot., 27:418.
85. Pratt, H. K., and Goeschl., J. D., 1968, The role of ethylene in fruit ripening, in: "The Biochemistry and physiology of plant growth substances," Wightman and Setterfield, ed., The Runge Press Ltd.: Ottawa.

86. Pratt, H. K., and Goeschl, J. D., 1969, Physiological roles of ethylene in plants, Ann. Rev. Plant Physiol., 20:541.
87. Quazi, M. H., and Freebairn, H. T., 1970, The influence of ethylene, oxygen and carbon dioxide on ripening of bananas, Bot Gaz., 131:5
88. Raven, J., Smith, A., 1979, The chemiosmotic view point, in: "Plant Membrane Transport: Current Conceptual Issues," R.M. Spanswick, W. J. Lucas, and J. Dainty, eds., Elsevier-North Holland: Amsterdam.
89. Reid, M. S., and Pratt, H. K., 1972, Effects of ethylene on potato tuber respiration, Plant Physiol., 49:252
90. Rhodes, M.J.C., 1970, The climacteric and ripening of fruits, in: "The biochemistry of fruits and their products," Vol. 1, A.C. Hulme, ed., Academic Press: New York.
91. Rhodes, M.J.C., 1980, Respiration and senescence of plant organs, in: "The biochemistry of plants: A comprehensive review," vol. 2, D. D. Davies, ed., Academic Press: New York.
92. Richmond, A., and Biale, J. B., 1966, Protein and nucleic acid metabolism in fruits. I. Studies of aminoacid incorporation during the climacteric rise in respiration of avocado, Plant Physiol., 41:1247.
93. Rogers, M., 1973, An historical and critical review of post harvest physiology research on cut flowers, HortSci., 8:189.
94. Romani, R., 1975, Long term maintenance of mitochondria function in vitro and the course of cyanide-insensitive respiration, in: "Plant Mitochondria," G. Ducet, and C. Lance, eds., Elsevier-North-Holland: Amsterdam.
95. Rowan, K. S., McGlasson, W. B., and Pratt, H. K., 1969, Changes in adenosine phosphates in cantaloupe fruit ripening normally and after treatment with ethylene, J. Exp. Bot., 20:145.
97. Rychter, A., Janes, H. W., and Frenkel, C., 1978, Cyanide-resistant respiration in freshly cut potato slices, Plant Physiol., 610:667
98. Rychter, A., Janes, H. W., and Frenkel, C., 1979, Effect of ethylene and oxygen on the development of cyanide-resistant respiration in whole potato mitochondria, Plant Physiol., 63:149.
99. Sacher, J. A., 1967, Studies of permeability RNA and protein turnover during ageing of fruit and leaf tissues, Symp. Exp. Biol. XXI:269.
100. Sacher, J. A., 1973, Senescence and post-harvest physiology, Ann. Rev. Plant Physiol., 24:197.
101. Salminen, S. O., and Young, R. E., 1975, The control properties of phosphofructokinase in relation to the respiratory climacteric in banana fruit, Plant Physiol., 55:45.
102. Schonbaum, G. R., Bonner, W. D. G., and Storey, B. T., 1971, Specific inhibition of the cyanide-insensitive respiratory pathway in plant mitochondria by hydroxamic acids, Plant Physiol., 47:124

103. Sisler, E., and Pian, A., 1973, Effect of ethylene and cyclic olefins on tobacco leaves, Tob. Sci., XVII:68.
104. Smillie, R. M., 1962, Photosynthetic and respiratory activities of growing pea leaves, Plant Physiol., 37:716
105. Solomos, T., 1963, Studies of the respiratory metabolism of bananas, Ph.D. thesis, University of Cambridge, U.K.
106. Solomos, T., 1977, Cyanide-resistant respiration in higher plants, Ann. Rev. Plant Physiol., 28:279.
107. Solomos, T., and Laties, G. G., 1974, Similarities between the actions of ethylene and cyanide in initiating the climacteric and ripening of avocados, Plant Physiol., 54:506.
108. Solomos, T., and Laties, G. G., 1975, The mechanism of ethylene and cyanide action in triggering the rise in respiration in potato tubers, Plant Physiol., 55:73
109. Solomos, T., and Laties, G. G., 1976, Induction by ethylene of cyanide-resistant respiration, Biochem. Biophys. Res. Comm., 70:663.
110. Solomos, T., and Laties, G. C., 1976, Effects of cyanide and ethylene on the respiration of cyanide-sensitive and cyanide-resistant plant tissues, Plant Physiol., 55:47.
111. Storey, B. T., 1976, Respiratory chain of plant mitochondria, XVIII, Point of interaction of the alternate oxidase with the respiratory chain, Plant Physiol., 55:73.
112. Storey, B. T., 1980, Electron transport and energy coupling in plant mitochondria, in: "The Biochemistry of Plants: A Comprehensive treatise," Vol. 2, D. D. Davies ed., Academic Press: New York.
113. Tetley, R. M., and Thimann, K. V., 1974, The metabolism of oat leaves during senescence. I. Respiration, carbohydrate metabolism, and action of cytokinins, Plant Physiol., 54:294
114. Theologis, A., and Laties, G. G., 1978, Relative contribution of cytochrome-mediated and cyanide-resistant electron transport in fresh and aged potato slices, Plant Physiol., 62: 232.
115. Theologis, A., and Laties, G. G., 1978, Cyanide-resistant respiration in fresh and aged sweet potato slices, Plant Physiol. 62:243.
116. Theologis, A., and Laties, G. G., 1978, Respiratory contribution of the alternate path during various stages of ripening of avocado and banana fruits, Plant Physiol., 62: 255,
117. Turner, J. F., and Turner, D. H., 1975, The regulation of carbohydrate metabolism, Ann. Rev. Plant Physiol., 26:159.
118. Turner, J. F., and Turner, D. H., 1980, The regulation of glycolysis and the pentose pathway, in "The Biochemistry of Plants: A comprehensive treatise," Vol. 2, D. D. Davies, ed., Academic Press: New York.
119. Vendrell, M., 1969, Reversion of senescence: Effects of 2.4 dichlorophenoxyacetic acid and indolacetic acid on respiration, ethylene production, and ripening of banana fruit

slices, Aust. J. Biol. Sci., 22:601.

120. Vickery, R. S., and Bruinsma, J., 1973, Compartments and permeability of potassium in developing fruits of tomato (Lycopersicon esculentum Mill), J. Exp. Bot., 24:1261.

121. Wager, H. G., 1961, The effect of anaerobiosis on acids of the tricarboxylic cycle in peas, J. Exp. Bot., 12:34.

122. Wiskich, J. T., 1977, Mitochondrial metabolite transport, Ann. Rev. Plant Physiol., 28:45.

123. Wiskich, J. T., 1980, Control of the Krebs cycle, in: "The Biochemistry of Plants. A Comprehensive Treatise," Vol. 2, p. 243, D. D. Davies, ed., Academic Press: New York.

124. Woolhouse, H. W., 1967, Nature of senescence in plants, in: Symp. Soc. Expt. Biol., Vol. XXI, Cambridge U.P., Cambridge.

125. Wu, S.-Y., and Laties, G. G., 1978, A malate OAA shuttle as a link between α-oxidation and electron transport in uncoupled potato slices, Plant Physiol., 61:S70:386 Abst.

126. Young, R. E., and Biale, J. B., 1967, Phosphorylation in avocado fruit slices in relation to the respiratory climacteric, Plant Physiol., 42:1359.

ENZYME ACTIVITIES AND POST-HARVEST CHANGE

M.J.C. Rhodes

Agricultural Research Council Food Research Institute
Colney Lane
Norwich, England

INTRODUCTION

A complex series of metabolic adjustments occur in horticultural crops after harvest which are influenced by the dislocation of supply of nutrients, minerals and hormones from the parent plant to the harvested plant organ. Basic respiratory processes are necessary to supply the energy required to maintain the essential repair reactions upon which the integrity and independent existence of the organ depend. Over and above this maintenance metabolism, the harvested fruit or vegetable is able to continue its normal development which may lead to ripening and senescence in the case of fruits and leaves or to dormancy and regrowth in tubers and seeds. In addition the harvested organ has the ability to respond metabolically to the environment in which it is stored, to wounding and to microbial attack. Such basic biochemical and physiological changes lead to major changes in the chemical composition and physical structure of the tissue and have an important effect on the quality of the horticultural crop as food whether it is eaten directly or following some food processing operation.

Underlying these biochemical changes are major changes in the relative activities of metabolic pathways in the tissue. Figure 1 illustrates some of the visible changes that occur in ripening fruits and the metabolic pathways which are likely to be responsible for these changes. Changes in the flow of metabolites through the pathways involved must be reflected in changes in the *in vivo* activity of their constituent enzymes. Modern ideas on the regulation of metabolic pathways stress the concept of control being a function of the system as a whole with each of the constituent enzymes of the pathway making some contribution to overall control

COLOUR	
1. Loss of green colour	Lipolysis Proteolysis
2. Formation of red/blue colours	Carotenoid and phenylpropanoid synthesis
FLAVOUR	
1. Loss of acidity	TCAC and decarboxylation
2. Increase in sweetness	Starch hydrolysis
3. Flavour volatiles	Biosynthesis of alcohol, esters etc.
TEXTURE	
1. Softening	Pectolytic activity Cellulose breakdown

Fig. 1. Metabolic Pathways Associated with Fruit Ripening.

(13). However, the contributions of the individual enzymes are not equal and it is frequently possible to identify steps in a pathway catalysed by particular enzymes which make major contributions to control. We need to concentrate on such enzymes in order to understand how the metabolic adjustments which occur during the post harvest period are achieved. The present paper will discuss the mechanisms available for regulating enzyme activities in cells and will then discuss the present state of our knowledge of the regulation of two diverse areas of metabolism, both of which are of importance in harvested organs.

Factors Affecting the Activity of Enzymes in vivo

The *in vivo* activity of an enzyme is determined by factors which affect either the number of active enzyme molecules available or the degree to which the potential catalytic activity of these available enzyme molecules is expressed. The first group of factors lead to a relatively slow acting gross control while the second group involve fast acting, fine tuning responses of an existing enzyme system. The actual number of active molecules of a particular enzyme in the cell is a function of the relative rates of its synthesis and degradation and may also be affected by mechanisms which either activate or inactivate the enzyme reversibly. Increased rate of synthesis or decreased rate of degradation provide a powerful means of controlling the number of available enzyme molecules. There is evidence in both plants and animals that different enzymes turnover at vastly different rates. For instance in rat liver the range of half lives of individual proteins in the cell can vary from a few minutes up to 15-25 days (20). Similar variations occur among higher plant proteins (Woolhouse, Chapter 1).

In plants the nature of the proteases responsible for the proteolytic breakdown of enzymes is unknown. In animals, degradation is under the control of a range of proteases of varying specificity. There are group-specific proteases which recognize the co-enzyme binding sites of certain enzymes and split the apoprotein at this site. Group specific proteases for pyridoxal dependant, NAD dependant and FAD containing proteins have been described (31,34). These enzymes appear to attack only a few peptide bonds near the cofactor binding site and subsequent degradation is by other less specific proteolytic enzymes. In higher plants the presence of a protease specific for nitrate reductase has been proposed (60) but wider aspects of the specificity of this protease were not studied in detail. However, enzymes such as the insulin specific protease of rat skeletal muscle, (15) specific for a particular protein do exist in animal cells but these appear to be the exception rather than the rule. Kahl (30) has discussed the factors which tend to promote proteolysis and concluded that factors tending to promote denaturation promote proteolysis while those tending to preserve the structural integrity of proteins protected against it. The binding of ligands or attachment to cellular structures such as ribosomes or membranes protect enzyme proteins from degradation. It seems very likely that conformational change is the first step in the breakdown of an enzyme molecule and it is suggested by Goldberg and St John (20) that these initial conformational changes in protein structure are the rate limiting step in enzyme degradation rather than the presence of specific proteases. However, the factors responsible for these structural changes are unknown. In enzymes attacked by the group-specific proteases the loss of the co-enzyme must be an early event in these changes since only the apo-enzyme is degraded and the presence of co-factor protects against proteolysis. A possibly important but as yet unexplained feature of intracellular protein degradation is its sensitivity to inhibitors of energy metabolism such as cyanide or dinitrophenol. This is an intriguing finding since it is unexpected on thermodynamic grounds, the hydrolysis of the peptide bond being exergonic and none of the proteases described require high energy intermediates (20). It could be that the requirement for ATP is an indirect one but it perhaps indicates that proteolysis is under strict control.

In the synthesis of enzymes major structural modifications occur following the release of nascent polypeptide chains from polyribosomes leading to the formation of the complete enzymically active enzyme structurally located and operating in a pathway. The rate of synthesis of polypeptide chains is sensitive to a number of factors including the availability of the relevant mRNA species and these factors will be discussed elsewhere in this symposium (Grierson, Chapter 2). The development of secondary and tertiary structure in the newly formed polypeptide chains is a function of the amino acid sequence determined during translation. The folding of these chains, the formation of disulphide bridges and other interactions between chains determines the overall shape of subunits of the enzyme molecule (11). The formation of quaternary structure in more complex proteins involves interactions between subunits and their assembly can be complex. The small and large subunits of ribulose phosphate carboxylase:oxygenase are formed respectively on cytoplasmic and chloroplast ribosomes. The small subunit is transported across the chloroplast limiting membrane before the complete molecule of the enzyme can be assembled within the chloroplast (6). The small subunit bears a polypeptide tail which is removed by proteolysis during the process. Similarly, the mitochondrial genome can code for only 8-10 polypeptides. These polypeptides are components of proteins which contain polypeptide constituents made in the cytoplasm and transported to the mitochondrion (38). Thus assembly of complex proteins may involve diverse sites of synthesis of subunits and their transport to the site of assembly. Transport and assembly must be important with most enzymes since their sites of action, which may be within organelles or on cellular membranes, are distant from their sites of synthesis. Possible mechanisms for incorporation of proteins into a membrane system have been reviewed by Wickner (61). Many enzymes require the covalent attachment of a metal or prosthetic group to achieve catalytic activity and enzyme synthesis may be coupled with the synthesis of the prosthetic group and its incorporation into the protein molecule may well itself be catalyzed by specific enzymes. Many plant enzymes are glycoproteins and their synthesis involves post translational transfer of sugar units to the protein. Plant glycoproteins are generally glycosylated either by N-glycosidic linkages to the amide N of asparagine frequently involving the C-1 of N acetyl-D-glucosamine to which D mannose residues are attached, or by O-glycosidic linkages to hydroxyproline, serine or threonine involving L-arabinose or D-galactose (51). In animals and plants (27) groups to be linked N-glycosidically are first assembled, from nucleotide sugar precursors, on a C_{80-100} saturated isoprenyl phosphate (dolichol phosphate) carrier and then transferred en bloc to the protein acceptor. Oligosaccharides, to be linked by O-glycosidic bonds, are assembled by direct transfer of sugars from the nucleotide sugar precursor to the protein acceptor. Firestone and Heath (19) studied the de novo synthesis in culture mouse cells of a glycoprotein having alkaline phosphatase activity. They showed that if protein glycosylation was inhibited using tunicamycin, the formation of enzyme protein was also inhibited. It appeared that glycosylation protected the nascent protein from proteolytic attack. These findings indicate that glycosylation is an integrated step in the synthesis of the enzyme. The function of the carbohydrate in such enzymes is not fully understood but may play a role either protecting against proteolysis or in the structural localization of the enzyme in the cell (51). It can be seen from the above discussion that there are many steps in the synthesis of enzymes which have to be integrated with other pathways such as those leading to the formation of prosthetic groups or carbohydrate. These provide possibilities for control but their possible significance *in vivo* is not known.

Another factor in addition to synthesis or degradation which may affect the concentration of active enzyme protein in the cells is the activation or inactivation of pre-formed enzyme protein. A number of animal digestive enzymes are synthesized and stored as inactive precursors (pro-enzymes) and these may be converted into the active form by specific irreversible limited proteolysis. For instance, the conversion of trypsinogen into the active trypsin involves the proteolytic cleavage of a hexa peptide at the N-terminal end and a change in conformation of the protein molecule. In other cases such as the conversions of procarboxypeptidase A to the active enzyme in which nearly two-thirds of the molecule is lost, the proteolysis is less 'limited' (63). In plants, there are a number of cases reported of the presence of specific protein inhibitors for particular enzymes. These include invertase (40) and phenylalanine ammonia lyase (8,12). Inactivation is thought to occur by formation of stable protein:protein complexes between the inhibitor and enzymes. Such a mechanism provides for a pool of inactive enzymes which can rapidly be activated in response to metabolic demands. Such protein:protein interactions will be discussed further in a later section.

For a given amount of enzyme protein there is a maximum potential catalytic activity. There is scope for regulation of the *in vivo* activity of enzymes by the extent to which this potential activity is actually expressed. Many enzymes in the cell operate probably under conditions of limiting substrate and cofactors and in the presence of potentially inhibitory products. Under these conditions they are susceptible to regulation by factors which either increase or decrease the supply of substrates or cofactors, increase or decrease the rate of removal of products or which change the effectiveness with which they are able to utilize a given level of substrate or cofactor. The pH at which the enzyme operates *in vivo* is an important factor controlling activity. The supply of substrate for a given enzyme may depend on previous steps in the pathway or may in addition involve changes in cell membrane permeability and the release of stored substrate from the vacuole or from organelles. Of special significance in the overall regulation of metabolic sequences is the action of small molecular weight compounds (allosteric regulators) which, while not structurally related to either the substrates or products of the enzyme, may effect the activity of regulatory enzymes (35). These regulatory enzymes are characteristically complex both in quaternary structure and in their kinetic behaviour towards their substrates. Allosteric regulators bind to regulatory enzymes and change both the quaternary structure of the protein and its catalytic activity. The effector may either promote or inhibit the activity of the enzyme at a given concentration of substrate by affecting the efficiency of binding between substrate and enzyme. Such regulation may allow very rapid reversible adjustments of the rate of flow of intermediates through a regulatory step and appear to be of wide importance in cellular regulation.

In addition to allosteric regulation, some enzymes typically regulatory ones, are subject to reversible covalent modifications which can have important effects on their catalytic and regulatory properties. Figure 2 shows examples of the type of covalent modifications occurring with animal enzymes which include carbamylation, acylation, oxidation and reduction of thiol groups, adenylation and phosphorylation. The most studied example of covalent modification of proteins is the role of phosphorylation/dephosphorylation in the control of glycogen phosphorylase which catalyses the breakdown of glycogen in muscle. This reaction is controlled by the combined action of non-specific protein kinases, a highly specific phosphorylase kinase and phosphatase (36). The protein kinases, which are activated by cAMP, phosphorylate proteins using ATP as phosphate donor. These activate an inactive form of phosphorylase kinase by phosphorylating specific serine residues in the molecule and causing a subsequent conformational change in the enzyme protein. The activated kinase in turn phosphorylates and activates an inactive dimeric form of the glycogen phosphorylase, phosphorylase-b to form the fully functional phosphorylase-a which is tetrameric. The dephosphorylation of phosphorylase a and of the activated form of phosphorylase kinase are carried out by specific protein phosphatases and the system of phosphorylation/dephosphorylation allows for very rapid activation or deactivation of the glycogen degrading enzyme (44).

The importance of such reactions in higher plants is far from clear and covalent modification of a regulatory plant enzyme has yet to be conclusively demonstrated. The plant enzyme equivalent to glycogen phosphorylase i.e. starch phosphorylase does not appear to have regulatory properties and there is no evidence that it can undergo phosphorylation (5). This lack of regulatory mechanism may be related to the relatively slow changes in the rate of utilization of starch in plants compared with the rapid responses in glycogen breakdown in animals during strenuous activity. Protein kinases exist in plants (1,57) but these are not cAMP dependant and in fact the presence of cAMP in plants is still open to doubt (4). The phosphorylation of proteins occurs in chloroplasts and the phosphory-

MODIFICATION	DONOR	ACCEPTOR ENZYME	EFFECT ON ENZYME
Phosphorylation	ATP	Glycogen Phosphorylase (Serine)	Activation
Carbamylation	Carbamylphosphate	Glutamic Dehydrogenase	Inactivation
Adenylation	ATP	Glutamic Synthetase (Tyrosine)	Inactivation
ADP-Ribosylation	NAD^+	Translocase	Inactivation
S-H/S-S	-	Pyruvate Lyase	Activation

Fig. 2. Covalent Modification of some Animal and Bacterial Enzymes.

lated products include the light harvesting chlorophyll:protein complex. Phosphorylation of this complex has been proposed as a regulatory mechanism by which photosynthetic systems adapt to changing wavelengths of light (2). Covalent modifications allow for relatively rapid responses in enzyme function to external factors and can be seen as conserving metabolic energy as compared with enzyme synthesis.

The foregoing discussion of the potential mechanisms for the control of both enzyme formation and enzyme activity illustrate the limited information available for plant as opposed to animal systems. Yet even in animals, evidence for the functional significance of such mechanisms in vivo is often incomplete. The second part of this paper will discuss the state of our knowledge of the regulation of two enzymes operating in fruit tissues. One is the role of phosphofructokinase (PFK) in the control of glycolysis, a pathway of primary metabolism and the other is the role of phenylalanine ammonia-lyase (P.A.L.) in the control of the synthesis of a group of secondary products, the phenylpropanoids.

The Role of Phosphofructokinase in the regulation of Glycolysis in Plants

Glycolysis is a constituent metabolic pathway in all living plant cells which can respond rapidly to respiratory and synthetic demands made upon it. Davies (13) pointed out that, although the control of glycolysis is a property of the whole integrated system, the interaction between two of its constituents enzymes PFK and pyruvate kinase was likely to be the major feature of this control in higher plants. PFK, which catalyses the essentially irreversible phosphorylation of fructose 6-phosphate (F6P) in the presence of ATP to form fructose 1:6 diphosphate, is the first enzyme of the sequence entirely committed to glycolysis. PFK is a high molecular weight protein which exists as an oligomer. It shows non-linear kinetics towards one of its substrates, F6P, and its activity is affected by a wide range of compounds structurally unrelated to either its substrates or products. It thus has the characteristics expected of a regulatory enzyme (58). PFK is sensitive to allosteric effectors which are intermediates in glycolysis (2 phosphoglyceric acid, 3-phosphoglyceric acid and phosphoenol pyruvate (PEP)), the pentose phosphate pathway (6-phosphogluconate), and the tricarboxylic acid cycle (malate and citrate), as well as to intermediates in energy exchange reactions (adenine nucleotides and P_i). This coupled with its strategic position as the first step in the pathway suggests an important role for PFK in the metabolic control of glycolysis.

There is evidence from a number of different fruits including the banana (64) and the tomato (10) that there is an increased flux of carbon through the glycolytic pathway during the period of the climacteric. A detailed discussion of the regulation of glycolysis in plants has recently been published (13, 59) and it would not be appropriate to discuss the overall regulation of the pathway here. However, I would like to discuss some of our work on the properties of PFK, in relation to its role in relation to the control of glycolysis. In work in this laboratory PFK was isolated under carefully controlled conditions as we have shown that both the catalytic and regulatory properties of PFK are sensitive to the molecular state of the extracted enzyme (28). Table 1 shows estimates of the maximum catalytic activity of PFK isolated from tomatoes harvested at four stages of ripening and this shows that there are no significant changes in the extractable activity of the enzyme. This confirms other workers' findings that even though there is a major increase in flux through glycolysis during ripening the maximum catalytic activity of PFK remains constant (10,47). In studies of avocado slices, Solomos and Biale (54) concluded that the glycolytic potential of pre-climacteric fruit was sufficient to sustain the rates of respiration found in fruit at the climacteric peak. These findings suggest that there is potential of the maintenance of widely different glycolytic fluxes by regulation of the activity of pre-existing glycolytic enzymes without changes in the level of available enzyme protein.

PFK was purified from the tomato (29) using affinity chromatography and it was shown that the pH of isolation was critical in retaining the structural and regulatory features of the enzyme. Raising the pH during isolation promoted the breakdown of the oligomeric form of the enzyme (M.Wt. 180,000 daltons) giving progressively more of a lower molecular weight form (35,000 daltons) which may be a monomer. This loss of subunit structure was associated with a loss of regulatory properties. The oligomer showed negative cooperativity towards its substrate, F-6-P in the presence of MgATP (28) while the dissociated form showed simple Michaelis-Menten kinetics.

PEP was the most potent allosteric inhibitor of PFK found; with the tomato enzyme 50% inhibition was caused by 10 μM PEP (see Fig. 3). The inhibition was sigmoidal in nature and if the data was analyzed as a Hill Plot, the Hill coefficient was 1.5 showing marked deviation from hyperbolic kinetics (28). Such inhibition by PEP could explain interactions between PFK and pyruvate kinase since PEP is the substrate of the latter enzyme. The presence of PEP reduced the binding of enzyme of both its substrates, F6P and MgATP. However, the inhibiting effect of PEP could be reversed by addition of P_i to the assay; 7-8 mM P_i being required to give 50% relief of inhibition in the presence of 20 μM PEP (see Fig. 4).

Chalmers and Rowan (10) proposed that an enhanced rate of leakage of P_i from the vacuole to the cytoplasm was important in

Table 1. Activity of PFK in Tomatoes at Various Stages of Ripening

Stage	PFK Activity nKat/g fresh wt	PFK Activity nKat/ g protein	Vmax. nKat/ g protein
Mature-Green	191.8	0.085	0.2
Breaker	145.6	0.076	0.19
Orange	122.7	0.073	0.17
Red	118.2	0.073	0.17

(after 28)

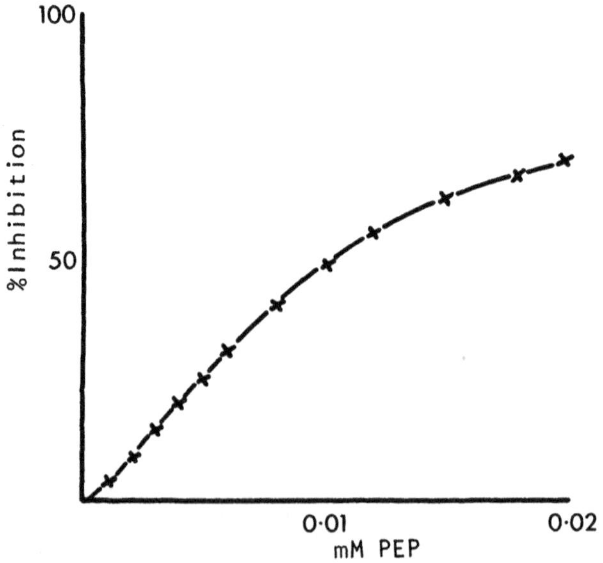

Fig. 3. Effect of PEP on the activity of purified tomato PFK at pH 7.5 (after 28).

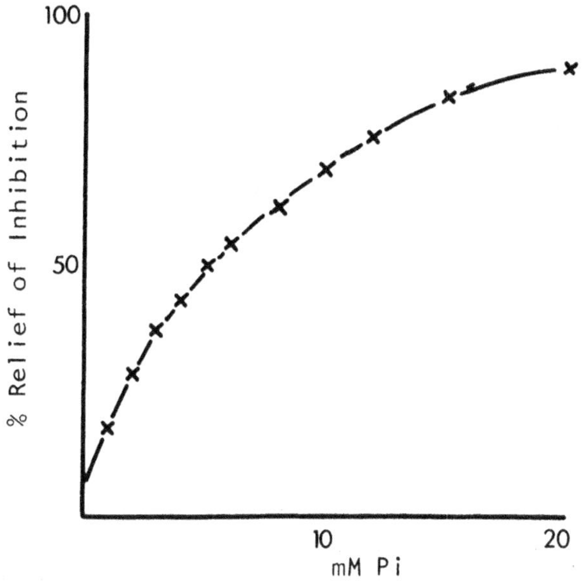

Fig. 4 Effect of P_i in relieving the inhibition of activity of purified tomato PFK at pH 7.5 by 20 μM PEP (after 28).

controlling the rate of glycolysis in tomatoes. They showed a positive correlation between the rate of respiration of tomato fruits and their P_i contents. More recently, Woodrow and Rowan (62) have studied the rise in respiration in slices cut from the pericarp of mature-green tomatoes and incubated under conditions in which colour changes occurred similar to those in the whole ripening fruit. They measured changes in the fluxes of P_i and found that the flux from vacuole to cytoplasm increased as the rate of respiration rose. The effect of P_i could be due to its effect on PFK activity.

A number of workers (10,28,58) have shown that P_i stimulates the activity of isolated PFK. Isaac (28) showed that P_i stimulated PFK activity by up to 2 fold at concentrations up to 20 mM (see Fig. 5). P_i caused a shift in the pH optimum of the enzyme from 8.0-8.25 to 7.5-7.8 and tended to promote degradation of the oligomeric structure with loss of the non linear kinetics. The oligomeric form of PFK shows negative co-operativity towards its substrate, F6P and P_i has the effect of both increasing the activity of the enzyme and changing its kinetics towards its substrate to simple Michaelis-Menten hyperbolic relationship. Such negative co-operative kinetics towards substrate in the absence of P_i have been found in bananas by other workers (46). Those authors considered that the effect of

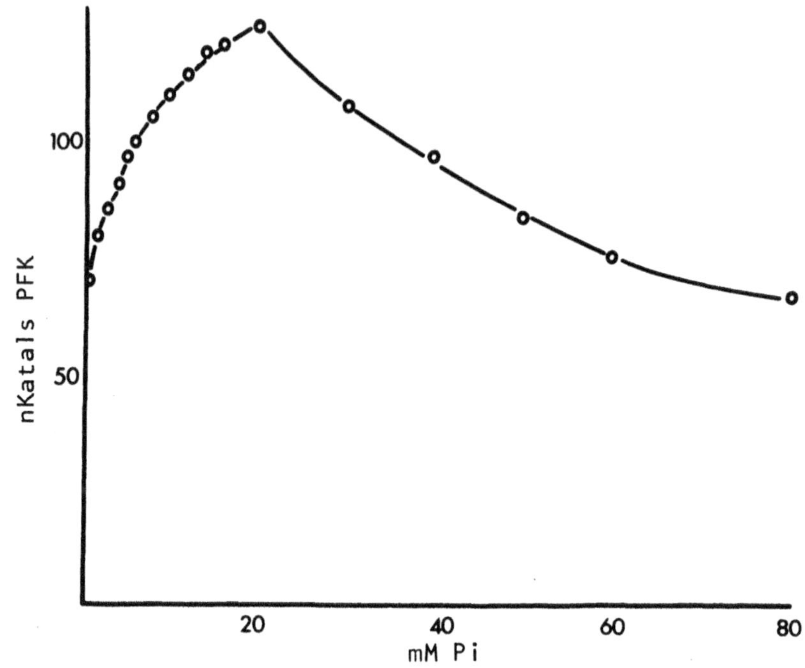

Fig. 5. Effect on P_i on the activity of purified tomato PFK at pH 7.5 (after 28).

the negative co-operativity was for the binding of one substrate molecule to make the binding of a second molecule more difficult so as to maintain the metabolism of carbohydrates at low levels in spite of large changes in substrate level. The effect of P_i by promoting deaggregation of the enzyme complex with resulting non-interactive kinetics, could be to relieve the constraint imposed by the negative co-operativity and thus stimulate the enzyme activity. In the controlling interaction between PFK and pyruvate kinase an additional effect of P_i would be to relieve the constraint on PFK activity due to inhibition of the enzyme by PEP. Malate and citrate, the major storage acids of the tomato can inhibit PFK allosterically but less effectively than PEP and here again P_i can relieve the inhibition. It is thus possible to propose a model that increased permeability of vacuolar membranes leading to an increased cytoplasmic level of P_i could stimulate glycolysis and the climacteric by relieving the constraints on PFK imposed by substrates and negative effectors.

Such a model is based on in vitro enzymological experiments. It assumes that the isolated enzyme is both structurally and kinetically identical to the enzyme in vivo. We know very little of the substrate and product concentrations, and of structural relations between enzymes which actually occur in the cell. We know average concentrations of metabolites but neither their actual concentrations in the organelles in which they act nor their rates of change in concentration. In the case of the possible interactions between PEP and P_i in controlling PFK, the concentrations required to give 50% responses (10 μM and 7-8 mM respectively) are at least of the same order as the average concentrations of these compounds in tomato tissues (10-20 μM and 2-3 mM respectively (10). However, it may be too simplistic to think of effectors acting singly and, in all probability, complex interactions between groups of positive and negative effectors may be important in the regulation of such a complex enzyme as PFK. It has been calculated that for a single enzyme with 8 reactants and effectors over 1.6×10^6 assays would be required to establish the complete rate law (13) and many more assays would be required for the complete analysis of a complex enzyme such as PFK. There is clearly a gap between our relatively detailed knowledge of the properties of high purified isolated enzymes and our ignorance of the behaviour of enzymes *in vivo*.

Some aspects of the Regulation of Phenylpropanoid Metabolism in Plants

The metabolism of phenylpropanoids is important in a number of aspects of the post-harvest physiology of horticultural crops. These compounds contribute to the colour, texture and flavour of fruits and vegetables and are important in some responses to damage and stress. Anthocyanins are responsible for the red and blue colours of many fruits and vegetables while leucoanthocyanins are important in the astringent taste of some fruits. Lignin formation occurs in the endocarp of stone fruits and contributes to the texture such as pears by its presence in stone cells. In addition, o-diphenols such as caffeic and chlorogenic acids and flavonoids such as catechins are endogenous substrates for the browning reaction of fruit and vegetables following damage which depend on the enzyme, o-diphenol oxidase. Frequently the synthesis of phenylpropanoids is stimulated in response to stress in plants and some fruit and vegetables respond to low temperature stress by increased metabolism of chlorogenic acid (42).

All of these phenylpropanoid compounds are synthesized from the aromatic amino acid, phenylalanine which in turn arises from erythrose 4-phosphate and PEP by the shikimate pathway. Figure 6 shows that the route from phenylalanine to the various products is highly branched and its regulation must involve control of the total flow of carbon through the pathway and the distribution of the flow to the particular branches. It is probably generally true that not all branches operate in a given tissue and in some cases the flow may be almost exclusively to a single product, as in lignifying tissues. The pathway as outlined in Fig. 6 has a common core consisting of the formation cinnamic acid and its ring substituted

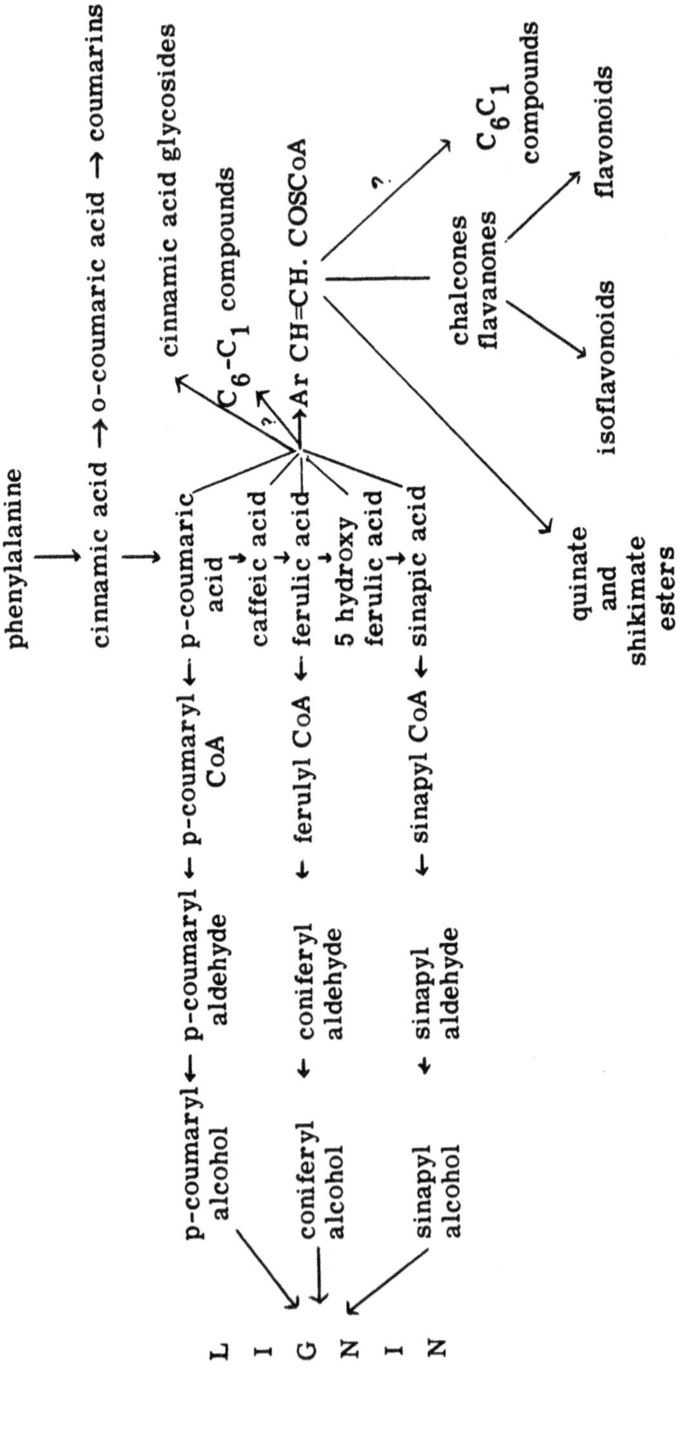

Fig. 6. The Pathway of Phenylpropanoid Biosynthesis.

derivatives from phenylalanine (p-coumaric, caffeic, ferulic and sinapic acids), and the activation of the cinnamic acid derivatives to form their CoA thioester derivatives which are precursors for a wide range of phenylpropanoid products (24). In the regulation of this pathway it appears that the enzymes catalyzing the steps from phenylalanine to cinnamic acid, phenylalanine ammonia-lyase (P.A.L.), from t-cinnamic acid to p-coumarate, cinnamic acid 4-hydroxylase (C.A.4H.) and from cinnamic acid derivatives to their CoA thioesters, hydroxycinnamate CoA ligase (CL) are most important.

P.A.L., which catalyses the stereospecific deamination of phenylalanine to form t-cinnamic acid, is a protein of molecular weight 300-320,000 composed of two subunits each consisting of two separate polypeptide chains (9,25). It displays non-linear kinetics with apparent negative co-operativity towards its substrate (43) and thus in its structural and kinetic complexity has the character of a regulatory enzyme. The reaction it catalyses is reversible. Havir and Hanson (26) demonstrated the formation of phenylalanine from cinnamate but this required the presence of a very high concentration of ammonium ions (0.374M) and even this was insufficient to saturate the reverse reaction. Such ammonium ion concentrations are not likely to occur under physiological conditions and since neither free cinnamate nor its derivatives normally accumulate in plant tissues it is likely that in vivo the P.A.L. reaction is effectively irreversible.

P.A.L. is mainly found in the 'soluble' fraction of tissue homogenates yet a small but significant fraction is bound to a membrane fraction, consisting of fragments of the endoplasmic reticulum and containing the second enzyme of the pathway, C.A.4H. C.A.4H is a mixed function oxidase which is linked to a membrane bound electron transport system in which the flow of electrons from $NADPH_2$ to cytochrome P450 is coupled to the hydroxylation of t-cinnamic acid in the position 4 of the aromatic ring to form p-coumaric acid (45). Kindl (32) has proposed that P.A.L. in vivo exists as a membrane bound complex associated with C.A.4H. and suggested that the complex between P.A.L. and the membrane bearing C.A.4H involves a weak interaction which is largely destroyed during homogenation under aqueous conditions. Kindl (32) showed that in the complex phenylalanine was a better precursor for p-coumarate formation than was t-cinnamic. The presence of such a complex in allowing restricted exchange of intermediates, vectoral transport of intermediates and to high local concentrations of substrates would offer major advantages. Since the hydroxylation reaction is irreversible, the channelling of cinnamate to the membrane bound system ensures that the conversion of phenylalanine to p-coumaric acid by the complex is irreversible. It is likely that the complex plays a major role in controlling the pathway of phenylpropanoid biosynthesis. However, much of the work in the past has concentrated on P.A.L. since C.A.4H. as a complex multienzyme system is much more difficult to handle experimentally. Another possible control point is the ligase step, CL in which substituted cinnamic acids are activated in the presence of ATP, CoA and Mg^{2+} to form thioester derivatives. These reactions are reversible and since they are inhibited by the ADP and AMP (33,41) may be subject to control depending on the availability of high energy carriers.

In many tissues and in response to different environmental and hormonal stimuli there is a rise in the production of phenylpropanoid products (9) and this is associated with co-ordinated increases in the extractable activity of P.A.L., C.A.4H and CL. An example of such a response is shown in Fig. 7 in which increases in the extractable activities of P.A.L., C.A.4H and CL in soya bean suspension cultures following exposure to light (24a).

Measurement of levels of extractable enzyme activity involve a number of difficulties which include the quantitative extraction of enzyme proteins which are structurally and catalytically intact from tissues often containing high levels of acidity and interfering compounds, and the assay of the extracted enzyme. With suitable precautions, the first of these two problems may frequently be overcome but measurements made of enzyme activities _in vitro_ may not be relevant to the conditions pertaining _in vivo_. To take P.A.L. as an example it is assayed _in vitro_ at the pH optimum (8.9) in the presence of excess substrate (13-14 mM, i.e. about x20 the Km). It is unlikely that the physiological pH is as high as 8.9 and it is interesting that all the other enzymes in this region of the pathway have optima between 7.0-7.5 and this is probably nearer the physiological range found in plants (52). In this context it is interesting that P.A.L. is inhibited by both its product, cinnamic acid and by p-coumaric acid and this inhibition is more potent at pH 7.0 than at the pH optimum (25). Margna (39) pointed out that for a range of plant tissues, the average level of phenylalanine was about 0.1-0.2 mM while the Km value for P.A.L. was in the range 0.2-0.7 mM. Thus it is unlikely that the maximum catalytic potential of the enzyme would be fully expressed _in vivo_. Methods for _in vivo_ assay of enzymes including one for P.A.L. (3) have been developed but these have not been widely used and have their own inherent problems. Margna (39) pointed out that since the maximum catalytic potential of P.A.L. is not always realized _in vivo_ there was considerable scope, at least in some cases, for the control of the pathway by increasing the availability of the substrate, phenylalanine.

Although it is not possible to generalize, there are now some well documented cases in which it is clear that the observed rise in P.A.L. activity involves an increase in the level of available P.A.L. protein rather than increased availability of substrate to an existing enzyme system. The best example does not relate to post-harvest behaviour but to the effect of light on the formation

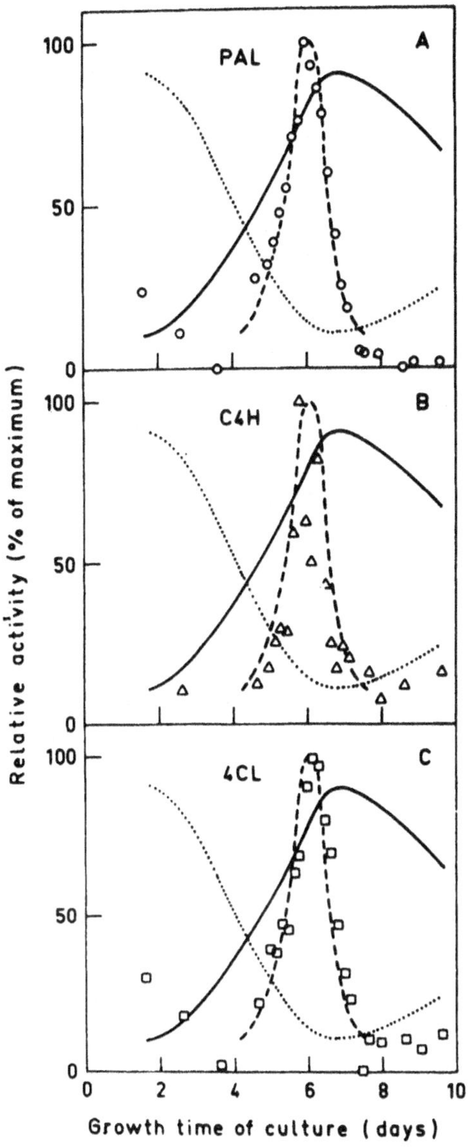

Fig. 7. Changes in P.A.L., C.A.4H and CL activities during the growth of soybean cell suspension cultures (after 24a - with permission).

of two flavonoids, apiin and its methoxy derivative in cell suspension cultures of parsley. Hahlbrock and his colleagues (22) showed that the formation of apiin and methoxyapiin was preceded by a co-ordinated increase in a number of enzymes including P.A.L., C.A.4H. and CL. The increases in these enzymes showed a similar time course to a peak 15 hours after the start of illumination (21). Hahlbrock (23) studied the rates of synthesis and degradation of P.A.L. during the period of rising activity and proposed a model to explain the observed pattern of changes in P.A.L. activity in terms of large changes in the rate of synthesis coupled with a more or less constant rate of degradation of the enzyme. The *de novo* synthesis of the enzyme was indicated by its sensitivity to inhibitors of both transcription and translation (49) but more definitive evidence has come from experiments with antisera raised against purified P.A.L. Schroder et al. (49) measured rates of formation of P.A.L. *in vitro* using polysomes prepared from cultured cells after varying periods of illumination and measuring the incorporation of ^{35}S-methionine into a protein product which was precipitable by the antiserum to purified P.A.L. The immunoprecipitate which was enzymically inactive (7) was analyzed by SDS-PAGE and among the products were recognizable labelled fragments of P.A.L. In *in vivo* experiments in which ^{35}S-methionine was fed to cells kept in the dark or after various periods following illumination, only in light grown tissues was there significant incorporation into the immuno precipitate. Subsequent analysis by SDS-PAGE showed that the radioactivity was associated with the 83,000 dalton subunit of P.A.L. The time course experiments indicated that the maximum rate of P.A.L. synthesis was reached 5-6 hours after the start of illumination and corresponded with the time of maximum rate of increase of P.A.L. activity. These findings confirmed the model proposed by Hahlbrock (23) that the rise in P.A.L. activity observed was due to the effect of light in promoting its synthesis. Subsequent work (50) showed that the effect of light was mediated through an increase in mRNA activity specific for P.A.L. It is, however, not clear whether this involves *de novo* transcription of mRNA or the activation of mRNA molecules or a combination of these two effects.

There are a number of other cases in which it is likely that the observed rise in P.A.L. activity is due to an increased rate of synthesis of the enzyme but the evidence is less complete than that for parsely (14,37,48). The evidence is based on the use of inhibitors of transcription and translation, density labelling with heavy isotopes or on *in vitro* or *in vivo* labelling of the enzyme.

In other cases, however, mechanisms other than *de novo* synthesis of the enzyme appear to operate. In a number of responses of different plants to light, it has been shown that there is an initial rise in P.A.L. activity to a peak value, which frequently appears to be due to enhanced enzyme synthesis, followed by a period of decline in activity to a lower steady level. It has been shown that application of inhibitors of protein synthesis to either potato disks (65) or gherkin hypocotyls (16) at the point of maximum P.A.L. activity inhibited subsequent decline in P.A.L. activity. Similarly, if light stimulated gherkin hypocotyls, which had been left at 4° for 24 hours during which time the P.A.L. activity fell back to control levels, were subsequently transferred back to 25° in darkness, there was a secondary rise in P.A.L. activity which was insensitive even to a very high concentrations of inhibitors of protein synthesis (17). Smith (53) showed a stimulation of P.A.L. activity in cold treated gherkin hypocotyls which was insensitive to cycloheximide. Subsequent density labelling experiments strongly suggested that de novo synthesis was not responsible for this rise in P.A.L. activity (53). Such experiments are taken as evidence for the involvement of proteineous inhibitors playing a role in the activation/inactivation of P.A.L. in vivo. In subsequent work an inhibitor was isolated from gherkin (8) which was present in both the soluble and membrane fractions of aqueous extracts. It could be 'solubilized' from the membrane fraction by treatment with chaotropes. The inhibitor is apparently proteineous being both thermolabile and sensitive to proteolytic attack. It has an apparent molecular weight of 19,000 and inhibits P.A.L. in a reversible manner. It is interesting that it also inhibits C.A.4H. but not a range of other enzymes tested. Evidence for the presence of other inhibitors of P.A.L. have been given by other workers (12,55,56) but very little work has been done to purify these inhibitors and to study their kinetic and mechanistic properties in any detail. There is a limited amount of evidence at a physiological level attempting to correlate changes in inhibitor level with observed changes in P.A.L. activity.

There has been some work on the enzymic factors responsible for the control of the accumulation of the red anthocyanin pigments of the skin of apples which is known to be stimulated by exposure to low temperatures and to light. Faragher and Chalmers (18) found that in whole Johanthan apples, light stimulated anthocyanin formation and a rise in P.A.L. to a peak followed by a secondary decline. When cycloheximide was applied as a single drop to a whole fruit, it stimulated P.A.L. activity and this may be interpreted as an inhibition of the inactivation of P.A.L. Tan (5,56) proposed that a P.A.L. inactivating system (P.A.L.-IS) plays a role in the regulation of P.A.L. activity in vivo. Apples stored in constant light accumulate higher levels of P.A.L. and more anthocyanin when given alternating periods at 6° and 25° compared with fruit stored continuously at 25°. The alternating temperature regime also led to lower levels of P.A.L.-IS compared to 25° throughout. In general, P.A.L. activity was found to be negatively correlated with the level of P.A.L.-IS. The nature of the P.A.L.-IS was studied by Creasy (12) who showed it to be associated with a heat labile high molecular weight compound and that D-phenylalanine protected P.A.L.

against P.A.L.-IS. P.A.L.-IS differs from the inhibitor of Billet et al. (8) in being irreversible in its action. This prompted Smith (53) to suggest it might be acting proteolytically on P.A.L. but there is no clear evidence on how it functions.

In general, it appears that the responses of the pathway of phenylpropanoid metabolism to external or internal stimuli are to a large extent mediated by changes in the available level of P.A.L. protein whether this be by de novo synthesis or by activation/inactivation. This is the possibility of fine control of the activity of the individual enzyme since P.A.L. and C.A.4H. are sensitive to both their immediate products (25,45) or to end products of the pathway (53) and CL is susceptible to control by energy charge (33,41). However, although it is likely that such control may be important under some circumstances, the major control is on the level of available enzyme protein.

It is important to return to the point made by Margna (39) who questioned why it was necessary to invoke the production of new enzyme proteins when the existing system does not appear to be fully saturated. It is possible to criticize this approach in relating measured P.A.L. activities to production of particular phenylpropanoid products. Since the pathway is highly branched it is important to compare the catalytic capacity of P.A.L. with the total throughput of the pathway by measuring all products (including lignin which is frequently overlooked because of the difficulty of its analysis but which is often quantitatively the most important product of the pathway). While the estimate of the carbon flow via P.A.L. may be too low it is almost certain that the in vitro estimates of its activity are too high. The pH optimum is probably well above the physiological pH of its cell compartment and its kinetic behaviour may well be very different at pH's close to 7.0 when it forms part of a membrane associated complex than in free solution at pH 9.0. This points to a general problem that confronts enzymologists and physiologists. The enzymologist carries out his in vitro experiments measuring initial rates of reactions in the presence of excess substrate under conditions in which the concentration of products is essentially zero. The activity of an enzyme may be large when the concentration of products is low but may be much lower when the substrate and products are present at close to their equilibrium values. Thus the information the enzymologist provides may not always be very useful to the physiologist who wants information on the operation of enzymes in situ. It seems likely that the reaction catalyzed by P.A.L. may under physiological conditions be limited by its catalytic potential in vivo rather than by the availability of substrate or cofactors.

In conclusion, this paper has set out some of the potential mechanisms available for regulation of enzyme activity in vivo. Our knowledge of such regulation is at a very elementary level;

many of the potential mechanisms have not yet been demonstrated with plant enzymes and even with those that have been shwon to occur it is very difficult to obtain evidence that they actually operate in vivo. The two pathways we have considered represent reactions or primary and secondary metabolism which play an important role in the post-harvest physiology of fruits and vegetables. The present state of knowledge suggests that very different mechanisms operate in those two pathways. It is tempting to suggest the first type of mechanism to be expected in obligatory systems such as glycolysis while induction and inactivation may be appropriate to a pathway which may only operate in certain cells for limited periods of their existence. At present, one can only speculate on the relative importance of these two types of mechanism of control.

REFERENCES

1. Alfonso, R., Nelson, N., and Racker, E., 1980, Plant Physiol., 65:730.
2. Allen, J. F., Bennett, J., Steinback, K. E., and Arntzen, C. J., 1981, Nature, Lond. 291:25.
3. Amrhein, N., Gödeke, K-H., and Gerhaudt, J., 1976, Planta, Berl., 131:33
4. Amrhein, N., 1977, Ann. Rev. Plant Physiol., 28:123.
5. Ap. Rees, T., 1974, in: 'Plant Biochemistry', D. H. Northcote, ed., Butterworths University Park Press, London, p. 89.
6. Barraclough, R., and Ellis, R. J., 1979, Europ. J. Biochem., 94:165.
7. Betz, B., and Hahlbrock, K., 1979, FEBS Letters, 107:233.
8. Billett, E. E., Wallace, W., and Smith, H., 1978, Biochim. Biophys. Acta, 524:219.
9. Camm, E. L., and Towers, G. H. N., 1973, Phytochemistry, 12: 961.
10. Chalmers, D. J., and Rowan, K. S., 1971, Plant Physiol., 48:235.
11. Chou, P. Y., and Fasman, G. D., 1978, Ann. Rev. Biochem., 47:251.
12. Creasy, L. L., 1976, Phytochemistry, 15:673
13. Davies, D. D., 1978, in: 'Biochemistry of Wounded Plant Tissues', G. Kahl, ed., W. de Gruyter and Co., Berlin, p. 309.
14. Duchesne, M., Fritig, B., and Hirth, L., 1977, Biochim. Biophys. Acta, 485:465.
15. Duckworth, W. C., Heinemann, M. A., Kitabchi, A. E., 1972, Proc. Nat. Acad. Sci. USA, 69:3698.
16. Englsma, G., 1967, Natwwissenschaften, 54:319.
17. Englsma, G., 1970, Planta, Berl., 91:246.
18. Faragher, J. D., and Chalmers, D. I., 1977, Aust. J. Plant Physiol., 4:133

19. Firestone, G. L., and Heath, E. C., J. Biol. Chem., 256:1404.
20. Goldberg, A. L., and St John, A. C., 1976, Ann. Rev. Biochem., 45:747.
21. Hahlbrock, K., Ebel, J. Ortmann, R., Sutter, A., Wellman, E., and Grisebach, H., 1971, Biochim. Biophys. Acta, 244:745.
22. Hahlbrock, K.,and Wellman, E., 1973, Biochim. Biophys. Acta, 304:702.
23. Hahlbrock, K., 1976, Europ. J. Biochem., 63:137.
24. Hahlbrock, K., and Grisebach, H., 1979, Ann. Rev. Plant Physiol., 30:105.
24a Hahlbrock, K., Schroder, J., and Vieregge, J., 1980, Advances Biochem. Engin., 18:40.
25. Hanson, K.R., and Havir, E. A., 1972, Recent Adv. Phytochem., 4:45.
26. Havir, E. A., and Hanson, K. R., 1968, Biochemistry, 7:1904
27. Hemmings, F. W., 1978, Phil. Trans. Royal Soc. Lond., B284:559.
28. Isaac, J., 1978, PhD Thesis, University of East Anglia
29. Isaac, J., and Rhodes, M.J.C., 1981, Phytochemistry, (in press).
30. Kahl, G., 1978, in: 'Biochemistry of Wounded Plant Tissues,' G. Kahl, ed., W. de Gruyter and Co., Berlin, p. 347.
31. Katsunuma, T., Schott, E., Elsasser, S., and Holzer, H., 1972, Europ. J. Biochem., 27:520.
32. Kindl, H., 1978, in: 'Regulation of Secondary Product and Plant Hormone Metabolism,' M. Luckner, and K. Schneber, ed., FEBS Symp. 58, Vol. 55, p. 49.
33. Knobloch, K., and Hahlbrock, K., 1977, Arch. Bioch. Biophys., 184:237.
34. Kominami, E., Banno, Y., Chichibee, K., Shiotani, T., Hamaguchi, Y., and Katsunuma, N., 1975, Europ. J. Biochem., 52:51.
35. Koshland, D. E., and Neet, K. E., 1968, Ann. Rev. Biochem., 37:359.
36. Krebs, E. G., and Beavo, J. A., 1979, Ann. Rev. Biochem., 48:923.
37. Lamb, C. J., and Dixon, R. A., 1978, FEBS Letters, 94:277.
38. Leaver, C. J., 1975, Proc. Ann. Phytochem. Soc. Symp., 11:137.
39. Margna, U., 1977, Phytochemistry, 16:419.
40. Pressey, R., 1967, Plant Physiol., 42:1780.
41. Rhodes, M.J.C., and Wooltorton, L.S.C., 1973, Phytochemistry, 12:2381.
42. Rhodes, M.J.C., and Wooltorton, L.S.C., 1977, Phytochemistry, 16:655.
43. Ricard, J., Nari, J., Buc, J., and Meunier, J-C., 1977, Proc. Ann. Phytochem. Soc. Symp., 14:156.
44. Rubin, C. S., and Rosen, O. H., 1973, Ann. Rev. Biochem., 44:831.
45. Russell, D. W., 1971, J. Biol. Chem., 246:3870
46. Salminen, S. O., and Young, R. E., 1974, Nature, Lond., 247: 389.

47. Salminen, S. O., and Young, R. E., 1975, Plant Physiol., 55: 45.
48. Schopfer, P., 1977, Ann. Rev. Plant Physiol., 28:223.
49. Schroder, J., Betz, B., and Hahlbrock, K., 1976, Europ. J. Biochem., 67:527
50. Schroder, J., Betz, B., and Hahlbrock, K., 1977, Plant Physiol., 60:440.
51. Selvendran, R. R., and O'Neill, M. A., 1981, Encyclopaedia of Plant Physiol., Vol. II, (in press).
52. Smith, F. A., and Raven, J. A., 1979, Ann. Rev. Plant, Physiol., 30:289.
53. Smith, H., 1973, Proc. Ann. Phytochem. Soc. Symp., 9:303.
54. Solomos, T., and Biale, J. B., 1974, in 'Facteurs et Regulation de la Maturation des Fruits,' R. Ulrich, ed., Colloq. Int. C.N.R.S., No. 238, p. 221.
55. Tan, S. C., 1979, J. Amer. Soc. Hort. Sci., 104:581.
56. Tan, S. C., 1980, Aust. J. Plant Physiol. 7:159.
57. Trewavas, A., 1976, Ann. Rev. Plant Physiol., 27:349.
58. Turner, J. F., and Turner, D. H., 1975, Ann. Rev. Plant Physiol., 26:159.
59. Turner, J. F., and Turner, D. H., 1980, in: 'The Biochemistry of Plants,' Vol. 2, D.D. Davies, ed., Academic Press, New York, p. 279.
60. Wallace, W., 1977, Proc. Ann. Phytochem. Soc. Symp., 14:177.
61. Wickner, W., 1980, Science, 210:861.
62. Woodrow, I., and Rowan, K. S., 1979, Aust. J. Plant Physiol., 6:39.
63. Yamasaki, M., Brown, J. R., Cox, D. J., Greenshield, R. N. Wade, R. N., and Neurath, H., 1963, Biochemistry, 2:859.
64. Young, R. E., Salminen, S., and Sornsrivichai, P., 1974, in: 'Facteurs et Regulation de la Maturation des Fruits,' R. Ulrich, ed., Colloq. Int. C.N.R.S., No. 238, p. 271.
65. Zucker, M., 1968, Plant Physiol., 43:365.

PLANT MEMBRANE LIPIDS : CHANGES AND ALTERATIONS DURING AGING AND SENESCENCE

Paul Mazliak

University of Paris
4 place Jussieu, 75005 Paris

INTRODUCTION

Some types of lipid molecules (phospholipids or galactolipids), because of their amphiphilic properties, are essential constituents of all biomembranes. Changes occurring in the lipid composition of these membranes will certainly modify their permeability, their energy transduction capacity, the activities of their membrane-bound enzymes, before and after harvest, when plants of economical value are concerned. Furthermore, after harvest, many alterations are likely to affect membrane lipids, thus playing a major role in the senescence or the post-harvest physiology of plant crops. It seems therefore reasonable to pay some attention to plant membrane lipids when discussing the factors allowing good crop preservation.

LIPID COMPOSITION OF PLANT CELL MEMBRANES

Many articles have dealt recently with the lipid composition of plant cell membranes (see for instance 1,2) and it will be sufficient, for this short review, to present rapidly the lipid composition of some well-defined plant cell membranes (Tables 1 and 2).

Some general considerations can be drawn from the data accumulated these last years. To summarize, one can say that:

(1) Polar lipids (glyco- and phospholipids) are the main lipid components of all plant cell membranes. The amphipathic properties of these molecules confer to membrane lipids a self-organization (or auto-aggregation) tendency in aqueous mediums, due to the thermodynamically favourable hydrophobic interactions occurring between the non-polar parts of membrane lipids when they are immersed in water.

Table 1. Lipid composition of some well defined plant cell membranes

Membranes (origin)	% Total lipids											
	N.L*	PC	PE	PGP	PA	PI	PS	PG	DGDG	MGDG	SL	References
Potato tuber plasmalemma	32	12.6	17.5	1.1	-	7.2	-	-	12.5	17.1	-	(3)
Potato tuber mitochondria :												
- outer membrane	-	52.6	25	12.1	-		10.3		-	-	-	(4)
- inner membrane	-	27	29	19.5	-		24.5		-	-	-	
Spinach chloroplast												
- envelope	-	27	-	-	-	1.3	-	8.4	32	22	5.0	(5)
- thylakoids	-	3.2	-	-	-	1.4	-	9.1	26	51	7.1	
Potato tuber microsomes	36.5	39	14.4	0.6	1.2	6.2	0.6	-				(6)
- Sunflower hypocotyl nuclear membranes	35.8	13.1	13.0	-	-	3.8	(6.1)**	15.3	6.1	6.8	-	(7)
- Castor bean endosperm glyoxysome	-	49.0	31.4	2.4	11.4	6.1	-	-		not measured		(8)
- Potato tuber peroxisomes	22	16.3	41	-	-	10.9	-	-	10	traces	-	(9)

*Abbreviations : N.L = neutral lipids, PC = phosphatidylcholine, PE = phosphatidylethanolamine, PGP = diphosphatidylglycerol, PA = phosphatidic acid, PI = phosphatidylinositol, PS = phosphatidylserine, PG = phosphatidylglycerol, DGDG = digalactosyldiacylglycerol, MGDG = monogalactosyldiacylglycerol, SL = sulfolipid.

** = lysophospholipids.

Table 2. Main fatty acid composition of some plant cell membrane lipids

Membranes (origin)	% Total fatty acids								
	16:0	16:1	16:1 Δ^3*trans*	18:0	18:1	18:2	18:3	Double bond index*	References
Potato tuber plasmalemma	28.5	-	-	7.5	4.5	41	17.5	3.7	(3)
Potato tuber mitochondria									
- outer membrane	24.0	-	-	3.5	1.8	58.0	12.7	5.7	(4)
- inner membrane	12.5	-	-	2.3	0.8	67.4	16.8	12.6	(4)
Vicia faba chloroplasts									
- envelope	13.3	1.8	0.0	4.3	5.8	11.5	63.3	12.5	(10)
- thylakoids	5.9	0.0	1.2	1.6	3.2	5.2	82.9	33.8	(10)
Potato tuber microsomes	31	-	-	7.0	0.5	49.5	12.0	3.5	(11)
Sunflower hypocotyl nuclear membranes	28.4	4.9	-	12.3	12.1	33.8	5.8**	2.3	(12)
Potato tuber peroxisome	22.0	-	-	2.1	4.3	56	13.3	5.9	(9)

*Double bond index = % Monounsaturated + 2 x % Diunsaturated + 3 x % Triunsaturated fatty acids/ % saturated fatty acids.

**Eicosanoic acid (2.7 %) has also been found.

(2) Chloroplast thylakoid lipid composition is unique in several aspects: i) galactolipids (MGDG, DGDG and SL) form the main part of thylakoid lipids whereas phospholipids are predominating in other cell endomembranes (the chloroplast envelope is somehow hybrid between these two classes of biomembranes, although rich in galactolipids, one can see that phospholipids (PC and PG) do represent important percentages (more than 35%) of total envelope lipids, ii) a second characteristic of chloroplast thylakoids, always observed in all types of green tissues, is the high fluidity of these membranes due to very high percentages of linolenic acid ($C_{18:3}$) among total fatty acids, which gives to thylakoid lipids a double-bond index far greater than any other cell membrane lipid (Table 2).

(3) In cell membranes other than thylakoids, the major phospholipids are always phosphatidylcholine and phosphatidylethanolamine, phosphatidylinositol being often a third important lipid component. Once again, the chloroplast envelope is in a curious position, phosphatidylcholine is predominating but phosphatidylglycerol (which is important in thylakoids) is also important in the envelopes. All other known phospholipids (PS, PA, PGP) are minor components, envisioned most frequently as intermediary metabolites.

CURRENT SCHEMES FOR THE BIOGENESIS OF PLANT CELL MEMBRANE LIPIDS

Thanks to the joined efforts of many laboratories throughout the world, the biosynthetic pathways of cell membrane lipids is beginning to be known (see 13, 14). The striking point in this research area is that the biosynthesis of all cell membrane lipids necessitates an active cooperation between different cell compartments. This necessary cooperation implies some transfer of acyl residues or entire phospholipid molecules within and between cell organelles. Those transfers of mainly hydrophobic molecules through the aqueous cytosol cannot be realized without the participation of some kinds of acyl carriers, for instance acyl carrier proteins within the plastids (15), phospholipid transfer proteins within the cytosol (16); alternatively, membrane vesicles budding from the endoplasmic reticulum and membrane flow towards cell organelles can also be considered following the conceptions of Morré et al (17). There is no doubt that CO_2 photosynthetically fixed from the atmosphere, is the ultimate precursor of all acyl residues of plant cell membrane lipids. However, the biosynthetical process in chloroplasts stops with the production of palmitic and oleic acids which are then exported towards the cytosol in an unknown manner. The oleoyl residues are then desaturated in the membranes of endoplasmic reticulum. Linoleoyl-phosphatidylcholine is the product of the desaturation. As linolenoyl-galactolipids are the main lipid components of the photosynthetic apparatus some transfer back to the chloroplast of linoleoyl-phosphatidylcholine must be envisioned to allow the last desaturation step to occur in the plastidial compartment.

CHANGES INVOLVING CELL-MEMBRANE LIPIDS DURING AGING AND SENESCENCE

We have described some natural turn-over of membrane lipid during the "aging" of adult plant cells of two quiescent organs; abortive cauliflower inflorescence and dormant potato tubers (waiting for some months the break of dormancy, in natural conservation at 10-12°C). I must point out that this kind of research upon the natural turn-over of plant membrane constituents is very rare, as far as I know, although similar researches are rather numerous in animal tissues (18).

Natural Turn-over of Membrane Lipid in Plant Organs

Half-lives of membrane lipids from potato or cauliflower cells have been calculated, in our laboratory (19), by measuring, over several days, the decrease in specific activities of phospholipids, following a prior incubation of tissue slices in ^{14}C-acetate solution. After a 3 hrs incubation period, the slices were transferred to a medium without any radioactive precursor. In both tissues, the phospholipid showed different turn-over rates in the different cellular fractions. In cauliflower floret slices, where the interpretation of the results was not complicated by a superimposed lipid synthesis due to artificial aging, the measured half-lives for phospholipids in the various subcellular fractions varied from 3 to 22 days which is very similar to the corresponding data obtained with animal tissues (18). The greatest turn-over rates were found in microsomal membranes and the lowest in mitochondria (although nuclear membranes had not been examined). These results agree well with the rates of direct labelling of the different PC or PE pools present in the various subcellular fractions of cauliflower floret tissues. It has been always found that 2 hrs pulses of ^{14}C-acetate labelled the phospholipids of the microsomal fractions more rapidly and more intensively than those of mitochondria or any other fractions (11).

Similar results have been obtained with onion stem slices (20) labelled with ^{14}C-choline and the following order of decreasing labelling has been found: microsomes > dictyosomes > smooth E.R. > nuclei > mitochondria-proplastids.

Pulse labelling and dual substrate (^{14}C-acetate and ^{3}H-glycerol) labelling were used by Wilson and Rinne (1976) (21) to evaluate lipid turn-over in soybean cotyledons. These studies indicated first-order decay kinetics for phosphatidic acid, phosphatidylinositol, phosphatidylcholine, phosphatidylethanolamine, N-acyl phosphatidylethanolamine and diacylglycerols. On the other hand, triacylglycerols accumulated continuously in cotyledons, thirty days after flowering. Phosphatidic acid and diacylglycerol turned over rapidly (half-times of 6 min), which simply resulted from their role of intermediary metabolites in the biosynthetic pathway of triacylglycerols. Half-times of 30-60 min were calculated for PC or PE and were found similar for the acyl groups and the glycerol parts of these molecules. On the contrary, in N-acyl-phosphatidylethanolamine, acyl groups were replaced independently of the glycerol moiety (half-life: 3 hrs).

Moore (1977) (22) examined phospholipid turn-over in soybean tissue cultures and found much longer half-times than Wilson and Rinne. Phosphatidylcholine, labelled from ^{14}C-choline, had a half-life of 36 hrs and phosphatidylethanolamine, labelled from ^{14}C-ethanolamine, decayed in three phases with half-lives of 12, 34 and 136 hrs. Those data, obtained without distinguishing the different pools of a peculiar phospholipid, engaged for instance in several organelle membranes, point out the difficulty of measuring turn-over half-lives without separating first, the different organelles or membranes of the plant-cells under examination. Cumming and Osborne (23) recently demonstrated a continuous turn-over of membrane phospholipids in imbibed dormant embryos of the wild oat by carrying pulse-chase experiments with ^{3}H-glycerol. The suggestion was made by the authors that membrane turn-over was necessary to maintain the viability of dormant imbibed seeds and that the capacity for continuous membrane replacement may determine the longevity of the wild oat-seed under field conditions.

Changes During Aging

It must be admitted, from the rare data collected in the literature, that, if the storage conditions chosen for various crops (either seeds or fruits, or tubers) are adequate (relatively low temperatures, low relative humidity, darkness, slight anoxia), those crops can survive several months after harvest, in apparently good physiological conditions, without displaying important changes in the composition of their membrane lipids (24,25). That does not mean, in any way, that metabolic activity in stored plant tissues is nil: turn-over of membrane lipids can be measured with radioactive precursors in those tissues, but this process does not lead to any noticeable change in lipid composition of stored organs. It must be pointed out here however, that a recent paper by Pearce and Samad (26) showed a decrease in the total lipid content of peanuts stored for 38 months at 5°C, without any change in the fatty acid composition of the various neutral or polar lipids. If this aging of the seeds was realized in an environment of 38°C and 90% relative humidity, the content of polar lipid fractions fell to almost 10% of the original, in only 28 days.

It has been generally observed that a long storage period at low temperature leads to an increase in the degree of unsaturation of membrane fatty acids and this fact is usually interpreted as the result of an adaptation of cells from chilling resistant plants to low temperatures. It is assumed that such increase in lipid unsaturation would increase membrane fluidity and allow physiological processes dependent on fluid lipid structure to be maintained at lower temperatures.

Jerusalem artichoke tuber is a good example of such a chilling tolerant organ (27). Artichoke tubers develop and mature during spring and autumn but have to survive subzero soil temperatures during winter before sprouting next spring. Succinate-oxidase activity of the mitochondria extracted from these tubers, at various times of the year, has been studied, as well as the changes in the energy activation of succinate-oxidation reaction at various temperatures. The Arrhenius plots of succinate-oxidase activity thus obtained, in various seasons, all displayed two breaks (T_f and T_s) corresponding to two major changes in the molecular ordering of mitochondrial membrane lipids; this was shown with coupled spin-label motion studies. For instance at the end of autumn, the T_f and T_s for mitochondrial membrane lipids of Jerusalem artichoke are 22 and 3°C, respectively. Both T_f and T_s of tuber membranes decrease during winter and during this time there is a general increase in membrane fluidity. This phenomenon is probably part of the mechanism adopted by tubers to withstand freezing but cannot be correlated, in that peculiar case, with any significant change in the unsaturation of the fatty acids of membrane lipids. Thus, many plants which survive freezing temperatures, show negligible increases in fatty acid unsaturation when exposed to low temperature (Siminovitch et al. 28). Other mechanisms of the adaptation of plant cells to a low temperature stress can be envisioned: changes in the position of esterification of unsaturated fatty acids on the glycerol moiety of complex lipids, changes in the relative proportion of membrane phospholipids, etc...

Lipid Breakdown During Senescence

Few studies are concerned with the changes occurring in membrane lipids during senescence. However, all the data which can be collected in the literature point out the same tendency: in every senescent plant organ, whatever it is, there is a general lipid breakdown. This phenomenon is generally accompanied by some disorganization of membranes in senescent cells or organelles. Sometimes the appearance of plastid globuli, probably built with membrane lipid materials, is noticed, in chloroplasts for instance (29).

Newmann has thoroughly studied the evolution of chloroplast lipids in the leaves of Cucurbita maxima, during development and senescence: a noticeable reduction in percentage of linolenic acid was observed in senescent leaves(30) (Table 3). The same author has compared the lipid composition of the chloroplasts from leaves of various trees, either green or yellow(senescent). He has found again a noticeable reduction in the percentages of polyunsaturated fatty acids (Table 4). In the same time, short chain fatty acids tended to accumulate in senescent leaves.

Table 3. Evolution of the fatty acid composition of chloroplast lipids during development and senescence of Cucurbita maxima leaves (Newmann 30).

Leaf age	Percent of total fatty acid weight		
	Palmitic acid	Oleic acid	Linolenic acid
Cotyledons	10.3	2.0	66.9
First internode leaves	4.5	0.8	79.0
Second internode leaves	10.1	1.3	75.5
Sixth internode leaves	7.9	3.3	73.8
Tenth internode leaves	12.7	5.0	70.9

Table 4. Comparison between the fatty acid composition of the chloroplasts of either green or yellow, senescent leaves from two trees (Newmann 30).

Species	Percent of total fatty acid weight							
	$C_{12:0}$	$C_{14:0}$	$C_{16:0}$	$C_{16:1}$	$C_{18:0}$	$C_{18:1}$	$C_{18:2}$	$C_{18:3}$
Cladastris lutea								
Green leaves	2.1	5.9	13.5	-	2.0	-	7.7	64.1
Yellow leaves	9.9	4.3	10.2	1.8	7.8	14.5	4.9	28.0
Ginkgo biloba								
Green leaves	10.6	9.1	23.9	4.6	4.3	8.9	4.0	34.7
Yellow leaves	12.8	19.2	17.2	1.3	6.4	9.5	4.6	7.0

These pioneering observatons were largely confirmed by subsequent researches. In senescent leaves, rapid losses of those lipids which are components of chloroplast thylakoids are seen (31): the glycolipids are quickly deacylated (32) and the liberated linolenic acid is particularly susceptible to lipoxygenase-catalyzed attack. Accumulating unesterified fatty acids have also been noted to react with xanthophylls and various alcohols in senescing leaves of Acer platanoides (33).

The degradation of linolenic acid has been observed in senescent apple fruits (34), with the noticeable development of osmiophilic granules in the yellowing peels of senescent fruits.

An unbalance between biosynthetic and degradative pathways of lipid metabolism is probably the simplest explanation which can be offered to this lipid breakdown during senescence. As it will be indicated below, peroxidation of polyunsaturated fatty acids is probably overtaking the biosynthesis of the same compound at the onset of senescence, and this precise reversing of metabolic equilibrium could mark the involvement of plant tissues towards an irreversible disorganization.

Confirming this idea, changes in fatty acid metabolism were studied in soybean cotyledons during senescence as well as in cotyledons which had been caused to regreen by removal of the epicotyl from the seedling (35). The activities of the enzymes acetyl-CoA synthetase and fatty-acid synthetase in plastids isolated from the cotyledons decreased during senescence but increased in response to regreening. These changes in enzyme activities followed the same pattern as changes in the quantities of chlorophyll and polyunsaturated fatty acids in these tissues. Similarly, the galactosyl transferase activity of chloroplast envelopes decreases in senescent soybean cotyledons (36).

The loss of polyunsaturated fatty acids, during senescence has obviously deep consequences on the fluidity of biomembranes in senescing organs. This question has been studied by McKersie and Thompson (37) in senescent cotyledons from bean. Lipid transition temperatures for rough and smooth microsomal membranes isolated from bean cotyledons at various stages of germination were determined by wide angle X-ray diffraction. For rough and smooth microsomes from 2-day-old tissues, the transition occurred at 0 and 3°C, respectively. By the 4th day of germination, the transition temperatures had increased to 32°C for smooth microsomes and 35°C for rough microsomes. During the later stages of germination, the transition temperatures of smooth microsome lipids continued to rise to 56°C, by day 9, at which time the cotyledons were extensively senescent and beginning to abscise. The data indicate that substantial amounts of membrane lipids are in a crystalline or pseudo-crystalline state in senescing membranes. The onset of senescence in the cotyledons of germinating beans is marked by a general metabolic decline (decreased ATPase activity, ATP-dependent cation transport, etc...) correlating well with a structural disassembly of all membranes (38).

Finally, some relationship between phospholipid metabolism, senescence and ethylene production has been described in senescing tissues of the morning glory flowers (Ipomoea tricolor Cav.) (39). Rib segments excised from flower buds passed through the same phases of senescence as the respective tissues on the intact plant. It was found that the phospholipid contents of rib segments had already started to decline before rolling up of the ribs, a visible sign of senescence. As the segments began to roll up and to produce ethylene, the rate of phospholipid loss accelerated sharply.

During the same period, the level of fatty acids esterified to phospholipids also fell by 40 per cent. No qualitative changes in any lipid component could be detected during senescence. Exogenously applied ethylene accelerated the loss of phospholipid and the senescence of rib segments while benzyladenine retarded both of these processes. Two metallic ions had also contrary effects: Ag^+ inhibited the rolling up of flower segments but did not affect ethylene generation or phospholipid loss; in contrast, Co^{2+} reduced both processes.

Those observations are interpreted in the following way: the onset of senescence is triggered by some degradation of membrane lipids resulting in increased membrane permeability. This loss in cellular compartmentation is thought to cause mixing of previously sequestered metabolites of the ethylene-generating system, thereby stimulating ethylene evolution. Once formed, ethylene enhanced its own synthesis and accelerated aging by further increasing membrane permeability and perhaps, by further reducing phospholipid biosynthesis (40). As one can see from the preceding results and interpretations, membrane lipid breakdown emerges as a crucial step in the onset of senescence.

MEMBRANE LIPID PEROXIDATION: A STEP TOWARDS SENESCENCE

One of the good ideas which has been advanced concerning the senescence of living tissues is that lipid peroxidation is greatly involved in the phenomenon. The obvious advantage of this idea is that experimental research can be done and has been done, indeed, on the matter : the data collected so far strengthen the hypothesis. On another hand, it cannot be masked that lipid peroxidation is also of immense practical importance, for it leads to the deterioration of tissues *in vivo* and so to the spoilage of foods.

The Mechanism of Lipid Peroxidation

Lipid peroxidation can result both from purely chemical or photochemical processes as well as from biological reactions. In the former case catalysts can be either organic or metallic; in the latter case, the enzymes specialized in lipid peroxidation are named lipoxidases or lipoxygenases (41).

Table 5 shows the detailed reaction mechanisms which have been worked out for chemical or enzymatic lipid peroxidation. The first products of such oxidations, involving the capture of atmospheric oxygen, are always hydroperoxides which are very unstable products. The best substrates of lipid peroxidation are common polyunsaturated

Table 5. Lipid peroxidation

AUTOXIDATION

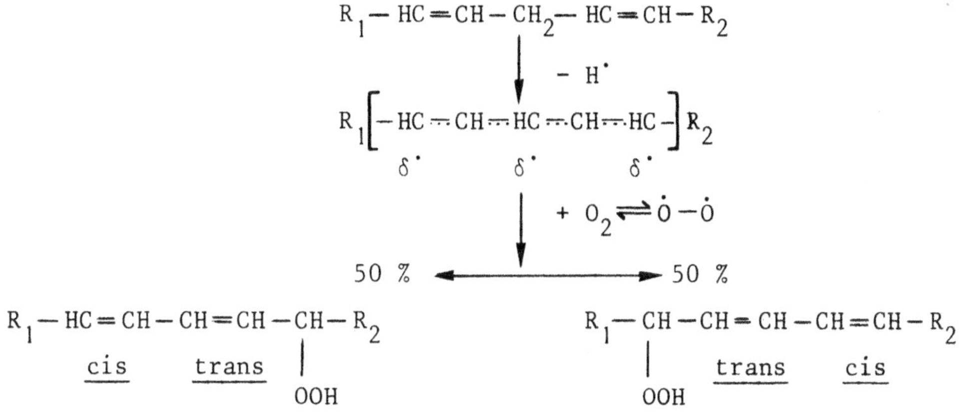

CHEMICAL CATALYSIS

- hematin compounds
- hemoglobin
- myoglobin
- cytochromes
- trace metal catalysts (Cu, Fe, etc...)

ENZYMIC CATALYSIS

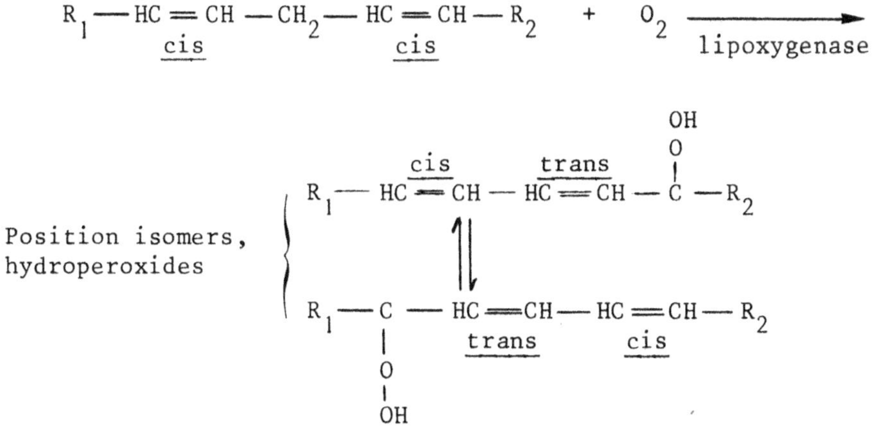

fatty acids (linoleic and linolenic), containing the reactive center *cis*, *cis*, 1,4, pentadiene ($-CH=HC-CH_2-CH=CH-$). The hydroperoxides formed constitute a great number of possible isomers and a major difference between chemical lipid peroxidation (also called autoxidation) and biological peroxidation is that, generally, the chemical process leads to different isomers appearing roughly in equal amounts while the biological process gives rise preferentially to one peculiar isomer. Free radicals and chain reactions are characteristic of lipid peroxidation.

Lipid peroxidation is indeed accelerated or inhibited by a series of physical or chemical factors (Table 6); the lipid peroxides, as indicated previously, are very unstable compounds and their decomposition leads to many volatile end products (Table 7) (many of which are responsible for the spoiling of foods). Lipid peroxidation is often coupled to other compound oxidation, which, again contributes to food deterioration.

Volatile products : the scavenging of lipid peroxides

Many of the flavours associated with vegetables (such as beans and cucumbers (42)) are formed by oxidative enzymic breakdown of unsaturated tissue lipids. The sequence is begun by lipoxygenase. The formed hydroperoxide cleaves to liberate volatile aldehydes that are responsible for the aroma characteristics of many vegetables (Fig.1).

After cleavage of the hydroxyperoxides and departure of the short volatile products, there remains always a portion of the acyl chain which could well be at the origin of traumatic acid, the so called "wound hormone," the structure of which is *trans*-10-$C_{12:1}$ dibasic acid.

A very recent finding deserves some attention : among the cleavage products of hydroperoxides many aldehydes are found. Some of them could be reduced *in vivo* to short chain alcohols (butanol, hexanol) : it appears that these short chain alcohols have senescence retardant effects (43). This would allow considering the relatively rare reductions of short chain aldehydes as feed-back reactions controlling negatively the onset of senescence (see below)

Table 6. Factors affecting lipid oxidation (41)

ACCELERATION BY	INHIBITION BY
High temperature	Refrigeration
Light (U.V. and blue)	Opaque or colored
Ionizing radiation (α,β,γ)	Containers or wrappers
Peroxides	Exclusion of oxygen
Lipoxygenases	Blanching
Organic iron catalysts	Antioxidants
Trace metal catalysts	Metal deactivators

Table 7. Hydroperoxides and volatile aldehydes formed in lipid autoxidation(41)

Fatty acids	Isomeric hydroperoxides	Aldehydes formed by decomposition
Oleic	11-OOH-9-ene	octanal
	9-OOH-10-ene	2-decenal
	8-OOH-9-ene	2-undecenal
	10-OOH-8-ene	nonanal
Linoleic	13-OOH-9,11-diene	hexanal
	11-OOH-9,12-diene	2-octenal
	9-OOH-10,12,diene	2,4-decadienal
Linolenic	16-OOH-9,12,14-triene	propanal
	14-OOH-9,12,15-triene	2-pentenal
	12-OOH-9,13-15-triene	2,4-heptadienal
	13-OOH-9,11,15-triene	3-hexenal
	11-OOH-9,12,15-triene	2,5-octadienal
	9-OOH-10,12,15-triene	2,4,7-decatrienal

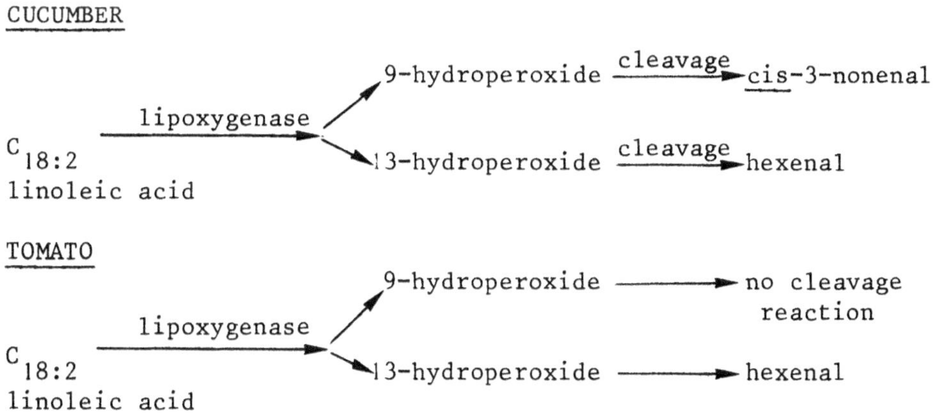

Fig. 1. Comparison of the enzymic formation of volatile fragments by hydroperoxide cleavage in cucumber and tomato fruits (42)

A Step Towards Senescence

Lipid peroxidation products on one side, hydroperoxides (lipoxygenase products) on the other side are susceptible to accumulate *in vivo* under certain circumstances. It must be stressed first that lipoxygenase activity, for instance, is very different from one species to another, as shown in Table 8. Secondly, the physiological state of development of tissues is mostly important when considering *in vivo* lipoxygenase activities. Accumulation of lipid peroxidation products during senescence of fruit has been demonstrated (45). It must be admitted that it is not easy, presently, to differentiate *in vivo* enzymic and non-enzymic lipid peroxidation systems. Gardner (46) has discussed important aspects of both enzymic and non enzymic formation and breakdown of lipid hydroperoxides. But whatever the formation pathway of these compounds, both their biosynthesis and their ultimate cleavage may produce free radicals which undoubtedly are susceptible to greatly damage plant cell organelles and membranes. These free radicals would react at random with other compounds through hydrogen removal and a variety of addition reactions. Through these reactions a wide variety of membrane lipid components would be chemically damaged. The loss of some fatty acids, carotene, pigments, etc...would appear of great significance for the physiology of plant cell organelles and membrane bound-enzymes.

Table 8. Oxidation of linoleic acid by aqueous extracts of edible plants (44)

Tissue	Lipoxygenase activity* (µl O_2 consumed x min^{-1} x g fresh matter)
Lettuce leave	120
Spinach leaves	80
Strawberry	40
Pear	0
Pumpkin	120
Avocado	720
Egg plant fruit	4320
Date	0
Carrot root	0
Beet root	0
Turnip root	0
Potato tuber	4560
Celery stalk	120
Artichoke heart	3360
Cauliflower florets	1440

*linoleic acid was used as a substrate.

For instance, damage to membranes may simply result in breaking the diffusion barrier, which would allow cell contents to leak out. But damage to membranes may be more complex as in the case of energy transducing membranes where electron-transport systems or coupling factors are greatly inhibited by lipid peroxidation, either in chloroplast or mitochondria. Damages to lysosome membranes or tonoplasts would be particularly devastating.

CONCLUSIONS

Let us propose in conclusion the following hypothesis which is not entirely original (47). Senescence could be the result of biomembrane damages within plant cells. At a certain period in the life cycle of mature plant cells, there would cease to be the efficient repression of the lipoxygenase peroxidation system by natural antioxidants or superoxide dismutase. Free radicals would appear more and more numerous formed from all membrane lipids, particularly those containing polyunsaturated fatty acids which are so numerous. As a consequence, tonoplast or lysosome membrane integrities, for instance, could not be maintained and many transducing energy functions in membranes would be hampered. Death, then, seems unavoidable.

In other words, one could say that lipid, in living cells, has been an obligatory component since the origin of life, because of the necessity to isolate the protoplasma from the external aqueous medium or to divide the aqueous protoplasma into several distinct functional compartments. The unfortunate and accompanying consequence of this necessity is that membrane lipids, susceptible as they are to oxygen attack, represent truely a potential danger for life itself. Membrane lipid alteration could well be the first sign of senescence and the leading pathway to death.

REFERENCES

1. P. Mazliak, Glyco- and phospholipids of biomembranes in higher plants, pp. 48-74 in "Lipids and lipid polymers in higher plants," M. Teveni and H.K. Lichtenthaler, ed., Springer-Verlag, Heidelberg (1977).
2. J.K. Raison, Membrane lipids : structure and function, pp 57-83 in "The Biochemistry of Plants, vol 4 : Lipids : Structure and Functions," P.K. Stumpf, ed., Academic Press, New York, (1980).
3. C. Demandre, "Les membranes cytoplasmiques du tubercule de Pomme de terre (*Solanum tuberosum*, L.)," Ph. D. thesis, University of Paris (1976).
4. D. Meunier, P. Mazliak, Differences de composition lipidique entre les deux membranes des mitochondries de Pomme de terre, C.R. Acad. Sc., 275:213 (1972).
5. R. Douce, R.B. Holz, A.A. Benson, Isolation and properties of the envelope of spinach chloroplasts. *J.Biol.Chem.*, 248:7215 (1973).

6. A. Ben Abdelkader, "Biogenese des lipides membranaires pendant la "survie" de tranches de tubercules de Pomme de terre." Doctorate thesis, Paris, (1972).
7. A.M. Justin, D. Robert, P. Mazliak, Préparation de membranes nucléaires d'hypocotyles de Tournesol (Helianthus annuus, L.). Analyse des lipides, C.R. Acad. Sc., 288:81 (1979).
8. R.P. Donaldson, N.E. Tolbert, C. Schnarrenberger, A comparison of microbody membranes with microsomes and mitochondria from plant and animal tissue, Arch. Biochem. Biophys., 152:199 (1972).
9. F. Tchang, "Biosynthèse des lipides contenus dans les peroxysomes du tubercule de Pomme de terre (Solanum tuberosum, L.)." Ph. D. Thesis, University of Paris, (1977).
10. L.R.O. Mackender, R.M. Leech, The galactolipid, phospholipid and fatty acid composition of the chloroplast envelope membranes of Vicia faba. Plant Physiol., 39:262 (1974).
11. A. Ben Abdelkader, P. Mazliak, Echanges de lipides entre mitochondries, microsomes et surnageant cytoplasmique de cellules de Pomme de terre ou de Chou-fleur, Eur. J. Biochem., 15:250, (1970).
12. A.M. Justin, "Les lipides des noyaux et membranes nucléaires d'hypocotyles de Tournesol (Helianthus annuus, L.)." Doctorate Thesis, University of Paris (1979).
13. P.K. Stumpf, ed., "The Biochemistry of plants, vol 4 : Lipids, Structure and Function Academic Press, New-York (1980).
14. P. Mazliak, P. Benveniste, C. Costes, R. Douce, "Biogenesis and Function of Plant Lipids." Elsevier-North-Holland, Amsterdam (1980).
15. P.K. Stumpf, Biosynthesis of saturated and unsaturated fatty acids, p. 177-204 in "The Biochemistry of Plants, vol 4." P.K. Stumpf, ed., Academic Press, New-York (1980).
16. P. Mazliak, J.C. Kader, Phospholipid exchange systems, p. 283-300, in "The Biochemistry of Plants, vol 4." P.K. Stumpf, ed., Academic Press, New-York (1980).
17. D.J. Morré, J. Kartenberg, W.W. Franke, Membrane flow and interconversions among endomembranes, Biochim. Biophys. Acta, 559: 71 (1979).
18. P. Mazliak, A. Ben Abdelkader, A. Trémolières, Le renouvellement des lipides dans les membranes animales et végétales, Cah. Nut. Diet., 7:51 (1972).
19. A. Ben Abdelkader, P. Mazliak, Renouvellement des lipides dans diverses fractions cellulaires de parenchyme de Pomme de terre ou d'inflorescence de Chou-fleur, Physiol. Vég., 9:227 (1971).
20. M.J. Montague, P.M. Ray, Phospholipid-synthesizing enzymes associated with golgi dictyosomes from pea tissue. Plant Physiol., 59:225 (1977).
21. R.F. Wilson, R.W. Rinne, Studies on lipid synthesis and degradation in developing soybean cotyledons. Plant Physiol., 57: 375 (1976).

22. T.S. Moore, Phospholipid turnover in soybean tissue cultures, Plant Physiol., 60:754 (1977).
23. A.C. Cumming, D.J. Osborne, Membrane turn-over in imbibed dormant embryos of the wild oat (Avena fatua, L.) II Phospholipid turn-over and membrane replacement, Planta, 139:219 (1978).
24. P. Mazliak, Le métabolisme des lipides au cours de la maturation des Pommes, Qualitas Plant., 19:19 (1969).
25. A. Cherif, A. Ben Abdelkader, Analyse quantitative des acides gras présents dans différentes régions du tubercule de Pomme de terre ; variation au cours de la conservation à 10°C, Potato Res., 13:284 (1970).
26. R.S. Pearce, I.M.A. Samad, Changes in fatty acid content of polar lipids during ageing of seeds of peanut (Arachis hypogea, L.), J. Exp. Bot., 31:1283 (1980).
27. E. Chapman, L.C. Wright, J.K. Raison, Seasonal changes in the structure and frunction of mitochondrial membranes of artichoke tubers, Plant Physiol., 63:363 (1979).
28. D. Siminovitch, J. Singh, I.A. de la Roche, Studies on membranes in plant cells resistant to extreme freezing. I. Augmentation of phospholipids and membrane substance without changes in unsaturation of fatty acids during hardening of black locust bark, Cryobiology, 12:144 (1975).
29. J.L. Bailey, A.G. Whyborn, The osmiophilic globules of chloroplasts II. Globules of the spinach-beet chloroplast, Biochim. Biophys. Acta, 78:163 (1963).
30. D.W. Newman, Chloroplast fatty acid transformations in nitrogen deficient and senescent tissues, Plant Physiol., 41:328 (1966).
31. F. Fong, R.L. Heath, Age dependent changes in phospholipids and galactolipids in primary bean leaves (Phaseolus vulgaris), Phytochem., 16:215 (1977).
32. S.R. Draper, Lipid changes in senescing cucumber cotyledons, Phytochem., 8:1641 (1969).
33. E.C. Grob, L. Csupor, Zur Kenntnis der Blattlipide von Acer platanoides L. während der herbstlichen Vergilbung, Experientia, 23:1004 (1967).
34. P. Mazliak, A.M. Catesson, Métabolisme des lipides dans le parenchyme externe de l'épiderme des Pommes. Fruits, 23:247 (1968).
35. M. Place, M.S. Morgan, A. Rutkoski, D.W. Newman, J.G. Jaworski, Fatty-acid metabolism in senescing and regreening soybean cotyledons, Planta, 147:247 (1979)
36. D. Dalgarn, P. Miller, T. Bricker, N. Speer, J.G. Jaworski, D.W. Newmann, Galactosyl transferase activity of chloroplast envelopes from senescent soybean cotyledons, Plant Sc. Lett., 14:1 (1979).
37. B.D. McKersie, J.E. Thompson, Lipid crystallization in senescent membranes from cotyledons, Plant Physiol., 59:803 (1977).
38. B.D. McKersie, J.E. Thompson, Patterns of cytoplasmic membrane senescence in cotyledon tissue, Phytochem., 14:1485 (1975).

39. P. Beutelmann, H. Kende, Membrane lipids in senescing flower tissue of Ipomoea tricolor, Plant Physiol., 59:888 (1977).
40. R.F. Irvine, D.J. Osborne, The effect of ethylene on (1-^{14}C) glycerol incorporation into phospholipids of etiolated pea-stems, Biochem. J., 136:1133 (1973).
41. M. Keeney, Secondary degradation products, p. 79-89 in "Lipids and their oxidation." H.W. Schultz, E.A. Day, R.O. Sinnhuber ed., Avi publishing Cy, Westport, Connecticut (USA) (1962).
42. T. Galliard, Lipolytic and lipoxygenase enzymes in plants and their action in wounded tissues, p. 155-201 in "Biochemistry of wounded plant tissues", G. Kahl ed., W. de Gruyter, Berlin (1978).
43. S.O. Satler, K.V. Thimann, The influence of aliphatic acohols on leaf senescence, Plant Physiol., 66:395 (1980).
44. A. Pinsky, S. Grossmann, M. Trop, Lipoxygenase content and antioxidant activity of some fruits and vegetables, J. Food Sci., 36:571 (1971).
45. Y.P. Maguire, N.F. Haard, Fluorescent product accumulation in ripening fruit, Nature, 528:599 (1975).
46. H.W. Gardner, Decomposition of linoleic acid hydroperoxides. Enzymic reactions compared with non-enzymic, J. Agric. Food Chem., 23:129 (1975).
47. J.L. Marx, Aging research : cellular theories of senescence, Science, 186:1105 (1974).

HORMONAL REGULATION OF SENESCENCE, AGEING, FADING, AND RIPENING

Johan Bruinsma

Agricultural University
Wageningen
The Netherlands

INTRODUCTION

Senescence is generally looked upon as an irreversible process of gradual degeneration and desintegration, eventually leading to the death of the organism. Such a form of senescence, which may proceed uncontrolled, occurs in most perennial plants, e.g. trees, that after many years of flowering and fruiting (polycarpy) gradually die off.

However, senescence need not occur as a more or less pathological event, but may proceed as a process under strict control as is the case with any other stage of development. This can be seen in monocarpic plants, mostly annuals or biennuals, in which senescence starts in the course of the single reproductive state of the plant. The lowermost plant parts, that were formed first, may commence to senesce and the process moves acropetally.

Next to the senescence of organisms, such organs as leaves, flowers and fruits also senesce. With polycarpic plants this may often be to the benefit of the plant individual and species. For instance, the dying off of leaves of woody and bulbous plants at the approach of the unfavourable season is a means in the strategy for survival of the perennial plant during the adverse period.

Detachment of plant organs usually advances the onset of senescence. The artificially induced senescence of isolated plant parts is called ageing. Without special precautions, cut leaves will yellow sooner than when they remain attached. Similarly, cut flowers will wilt, and picked fruit will ripen earlier than when left on the plant. Insofar as the senescence of these organs proceeds controlled, manipulation of the regulating mechanism(s) may affect the longevity and shelf life of postharvest commodities.

It will be demonstrated that the senescence of monocarpic plants is under the control of plant hormones. Subsequently, we will consider how far the ageing of isolated plant parts, too, is regulated by these hormones. The present state of our knowledge of the course of hormone levels during yellowing, fading, and ripening, and of the effects of treatments with exogenous plant growth regulators during these processes, will also be surveyed. First of all, however, the plant hormones themselves will be briefly introduced.

PHYTOHORMONES

Plant hormones are endogenously produced molecules that can act at the site of their synthesis and also after translocation to other plant parts. They exert their effects at low concentrations, in the part per million (ppm) and part per billion (ppb) ranges, mainly by interacting with the genetically controlled protein synthesis, with the enzymatic activities of these proteins, and with membrane-bound processes. In these respects, phytohormones are similar to the hormones known from vertebrate physiology. They differ from the animal ones in their chemical structures and in that their actions are not confined to specific target organs. On the contrary, phytohormones can be produced in different organs and also exert their manifold effects in various plant parts. In this way, locally synthesized hormones can interact with each other and with those derived from elsewhere. This leads to shifting patterns of agonistic and antagonistic components, which patterns specifically regulate the enzymatic composition and metabolic activity of each organ at every state of development. With such a pattern of promoting and inhibiting factors a much more precise regulation can be obtained than with either one alone.

In order to be effective, the hormone molecules need to be recognized by receptor molecules in the cell. Insensitivity to a hormone may mean a reduced amount or availability of its receptor molecules. That the amount of receptor molecules may strongly vary during development was shown for the cytoplasmic auxin receptor in cultured tobacco-pith callus (Fig. 1). Hormones that antagonize other hormones, might also do so by keeping receptor molecules at a low level or unavailable for these other hormones.

At present, five groups of phytohormones have been established with certainty. Moreover, other naturally occurring plant substances may participate in the hormonal pattern and some of them may possibly be recognized as hormones before long. Among these are phenolic, particularly steroid compounds (Grove et al., 1979, Hewitt et al., 1980). Polyamines are known to retard leaf senescence (Kaur-Swahney and Galston, 1979), whereas a specifically senescence-promoting substance, (-)-methyl jasmonate, has recently been isolated from wormwood, Artemisia absinthum L. (Ueda and Kato, 1980).

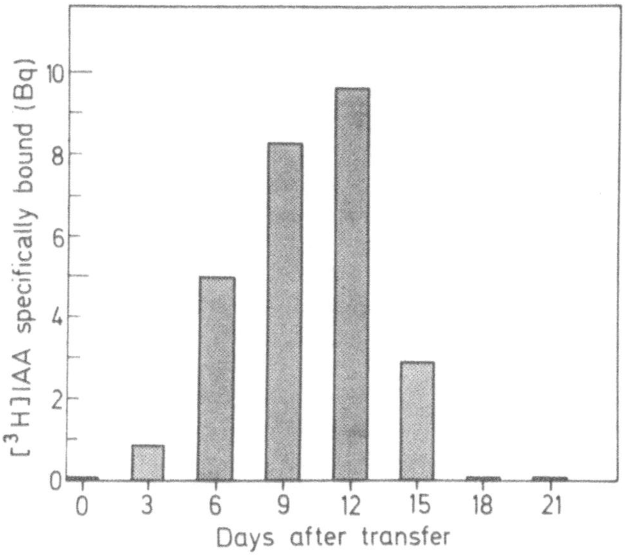

Fig. 1. Course of amount of soluble auxin receptor during culture in vitro of tobacco-pith callus (after Oostrom et al., 1980).

The plant hormones are derived either from isoprene units or from amino acids. The natural cytokinins are adenine compounds containing an isoprene side chain (Fig. 2). Cytokinins are highly effective in delaying the onset of ageing in isolated plant parts used as a bioassay to test senescence-promoting and -inhibiting substances. The use of organ sections, e.g. leaf segments incubated on test solutions in darkness, prevents interference by hormonal substances from other organs of the plant. It is particularly the deprivation of the roots that renders isolated plant parts liable to advanced senescence because the roots are the main site of cytokinin biosynthesis. On the contrary, enabling an overground plant organ, e.g. a leaf, to form adventitious roots prevents the leaf to senesce because of re-establishment of its source of cytokinin. This is the 'special precautions' referred to in the previous section when discussing the advanced senescence of detached organs.

The two other groups of hormones, derived from three and four isoprene units, respectively, are the abscisins and the gibberellins (Fig. 2). The gibberellins are not very active in senescence bioassays, although they may retard ageing of leaf discs in specific *Rumex* and *Taraxacum* species, and also counteract some ripening phenomena. The abscisins, on the contrary, strongly promote the onset and rate of senescence processes. The main representative, abscisic acid (ABA), is known to enhance the synthesis and release of hydrolytic enzymes that play a role in senescence, and to decrease membrane integrity. It specifically interferes with the biosynthesis of cytokinin nucleotides (Miernyk, 1979). Gibberellins and abscisins are formed both in roots and in leaves, especially in the chloroplasts of the leaf parenchyma (Hartung and Abou-Mandour, 1980, Heilmann et al., 1980, Milborrow, 1979). They can be released from the chloroplasts into the cytoplasm to exert their hormonal functions. In the cytoplasm, ABA is gradually inactivated, partly into conjugate(s), partly into (dihydro)phaseic acid (King, 1979).

The natural auxins are synthesized from amino acids (Fig. 3), mainly in the young, overground plant parts. Also synthetic chemicals with auxin activity have been developed that can replace the natural auxins. When these growth regulators are applied at moderate concentrations, they usually retard senescence. High auxin dosages, however, have the opposite effect, they promote senescence by inducing the biosynthesis of the other amino acid-derived hormone, ethylene (Fig. 3) (Aharoni et al., 1979). Ethylene is the most potent accelerator of senescence phenomena and, according to its ability to induce *de novo* enzyme synthesis, a true hormone (Lieberman, 1979). It is the rate-limiting step in the production of ethylene from methionine, *viz.* the formation of 1-aminocyclopropane-1-carboxylic acid (ACC), which is strongly promoted by auxins (Yu et al., 1979). The same step can also be inhibited by aminoethoxyvinylglycine (AVG) and by α-aminooxyacetic acid (AOA). These artificial regulators are thus potent inhibitors of senescence phenomena (Boller et al., 1979; Amrhein and Schneebeck, 1980). The conversion of ACC into ethylene requires oxygen and can be inhibited by ethanol (Heins and Blakely, 1980) and by cobaltous ions (Yu and Yang, 1979), whereas silver ions prevent the action of ethylene, probably by interfering with its receptor (Aharoni and Lieberman, 1979; Veen et al., 1980). During its action, ethylene may be oxidized to carbon dioxide or converted into ethylene glycol and its glucoside (Beyer, 1979, Blomstrom and Beyer, 1980). Ethylene can be replaced by the, less active, propylene that is converted in a similar way.

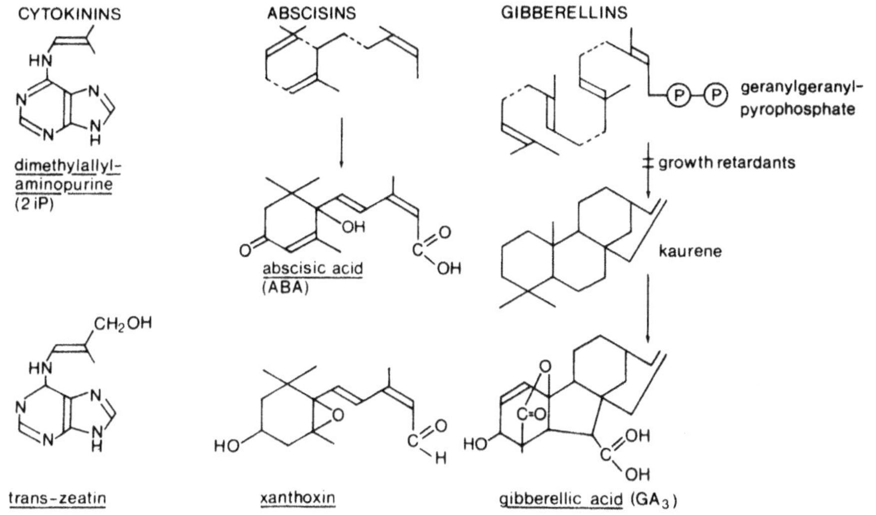

Fig. 2. Isoprenoid phytohormones.

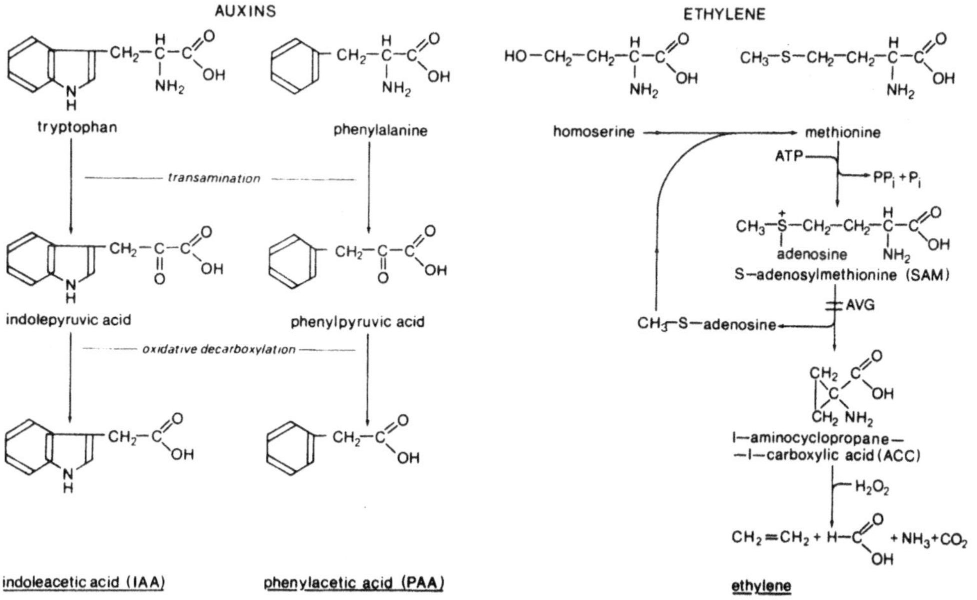

Fig. 3. Amino acid-derived phytohormones.

The physiological effects of plant growth regulators, both artificial ones and the naturally occurring hormones, can be studied by their application and following the course of the developmental process under study, e.g., senescence. Also the level of hormones during development can be followed by repeated sampling and determination of hormone contents by bioassay or, preferably, by physical methods (Hillman, 1978). It is by such methods that phytohormones have been found to be involved in the regulation of, e.g., monocarpic senescence.

MONOCARPIC SENESCENCE

The overall senescence of monocarpic plants, which occurs during or after their single flowering phase, often coincides with the approach of the adverse season. However, monocarpic plants do not die because of unfavourable environmental conditions but rather in response to internal changes. This is shown by individuals that fail to flower and survive the adverse season (Leopold, 1975). The coincidence illustrates the ecological adaption of the species.

Every gardener knows that removal of withered flowers prolongs the longevity of his plants. This prolonged viability is not so much due to the removal of potential sinks for nutrients but rather of producers of senescence-inducing substance(s) (Noodén and Leopold, 1978). For example, removal of developing soybean pods reduces the abortion of younger fruitlets and allows vegetative development to continue; in fruitlets and shoot buds, where the ABA content is lower than with intact plants, ABA application inhibits growth of fruits and shoots (Tames et al., 1979 a, b). Therefore, the developing pods probably produce this inhibitor. However, because ABA application fails to induce senescence in depodded plants, it is probably not the 'death hormone' itself but an accompanying factor. Very little research has been done yet to identify other hormonal factors released from developing fruit.

On the other hand, the application of cytokinin and auxin to intact soybean plants largely prevents the yellowing and nutrient export of the leaves without reducing the growth of the pods (Noodén et al., 1979). Apparently these regulators are able to counteract the senescing factor(s) from the fruit. In cereal plants, the ABA content of the ears increases during grain filling (King, 1979). Here, too, leaf senescence during this period is retarded by cytokinins from the roots (Biswas and Choudhuri, 1980).

When the basal, mature leaves of a tobacco plant start yellowing, removal of the top of the plant will not only halt this but the leaves will turn green again. This regreening demonstrates that senescence, at least in its initial stage, is a reversible process. That the reversion is brought about by root-produced cytokinins that had otherwise moved to the top, is indicated by the regreening of basal leaves at intact plants when sprayed with a cytokinin solution (Dyer and Osborne, 1971). A similar reversibility of senescence is shown by the recovery of wheat seedlings after two days of darkness. After four days in the dark the plants are unable to recover completely, unless they had been treated with cytokinin (Wittenbach, 1977).

Generally, the response of plants to environmental conditions is mediated through hormonal messengers. As long as the soil conditions are favourable, the roots produce cytokinins and gibberellins that promote the growth and development of the overground parts. However, when the roots perceive drought or lack of oxygen or of mineral nutrients, they decrease the production of these promoting hormones and increase that of abscisins, so that the life of the plant tends to be shortened by advancing senescence (Atkin et al., 1973). For example, the leaf yellowing under nitrogen deficiency may not directly be caused by shortage of this nutrient but rather by lack of cytokinin from the root system (Wagner and Michael, 1969).

It can be concluded that plants, that cannot run away from adverse conditions, adapt their life cycle to their environment. They regulate their physiological functions, including senescence and death with monocarpic plants, by way of their hormonal patterns. Developing fruits alter these patterns, either directly or by influencing the hormonal production of the roots which is an important factor in monocarpic senescence. In this form of senescence, a role of ethylene has not yet been established.

LEAF AGEING

We now turn to the question of how plant organs react when they are removed from the plant. To what extent is the inevitable senescence of these isolated plant parts still under hormonal control?

Although the ageing of leaf sections is by far the most used bioassay to test for senescence activity, yet the role of hormones in the yellowing of leaf discs is still incompletely understood, largely because the course of endogenous hormones during ageing has hardly been studied. It is no doubt the removal from the source of cytokinin supply which predisposes the leaf section to accelerated ageing, but the mechanism by which cytokinins are able to prevent the leaf tissue from senescing is not clear. Cytokinins are known to activate anabolic pathways and perhaps they sustain the metabolic state of the leaf cells to such an extent that it remains above some threshold below which catabolic reactions become predominant (Thomas and Stoddart, 1980). Ethylene and ABA, on the other hand, which advance and accelerate leaf ageing (Hall, 1977; Noodén and Leopold, 1978), might do so by stimulation of catabolic pathways and reduction of membrane stability (e.g. Collins and Morgan, 1980). Thimann and Satler (1979) recently observed that treatments that cause stomatal opening of oat leaf sections in darkness also retard ageing and, conversely, treatments that induce stomata to close in the light promote ageing. These treatments in the light can be considered to induce ABA as a response to stress, particularly in the absence of root-supplied cytokinin. Exogenous cytokinin can, indeed, prevent this ABA synthesis (Gepstein and Thimann, 1980). In prolonged darkness, these leaf sections also accumulate ABA that may play a role in accelerated ageing under these circumstances (Gepstein and Thimann, 1980). This need not necessarily involve a direct causal relationship with the stomatal closure in darkness. The ABA accumulation may well result from release of the hormone from the inactive, cytokinin- deprived chloroplasts followed by further ABA synthesis in these organelles. Radin (1981) showed ageing and stomatal closure to be not directly related in cotton leaves, both processes responding apparently independently to endogenous ABA levels.

Whatever the mode of action of hormones in leaf ageing may be, the shelf life of leafy commodities can be enhanced by keeping the levels of ABA and ethylene as low as possible; cytokinins are not permitted to be applied to foodstuffs for toxicological reasons.

FADING OF CUT FLOWERS

The senescence phenomena of cut flowers are studied more thoroughly than those of leaf sections, on the one hand because of their practical importance, on the other hand because of focusing upon the changes in endogenous hormone levels in flowers. As early as 1970, Mayak and Halevy reported that rose varieties have a longer vase life when their petals have a higher cytokinin content. Immersion of rose buds in cytokinin solution prolonged their longevity. Later these authors reported that the endogenous cytokinin level in the petals rapidly drops after flower opening and that subsequently the ABA level rises. This occurs more rapidly in roses with a shorter vaselife than in those with a better keepability. They also state that the rise in ABA is provoked by ethylene. ABA should accelerate senescence but suppress further ethylene biosynthesis (Mayak et al., 1972; Mayak and Halevy, 1972). Although inhibition of ethylene evolution by ABA has been occasionally reported (Wright, 1980), it is more frequently observed that ABA promotes ethylene that thereupon triggers the senescence process (e.g. Sagee et al., 1980).

The crucial phenomenon in flower ageing, the fading process, results from leakiness of petal-cell membranes, particularly of the tonoplast surrounding the vacuoles. Such osmotically active substances as sugars may preserve membrane stability and in the spike of Gladiolus exogenously applied gibberellin (GA_3) may stimulate starch breakdown and thus give a more full and prolonged flowering by increased supply of soluble sugars (Ramanuja Rao, 1979). Also the protecting effect of cytokinin seems to be exerted through prolonged membrane integrity, and this effect can be mimicked by synthetic benzthiadiazoles (Apelbaum and Katchansky, 1977). Cut flowers lack a cytokinin supply from the roots.

Ethylene, on the contrary, strongly promotes membrane permeability (Asen and Lieberman, 1963; Kende and Baumgartner, 1974; Hanson and Kende, 1975). This is the cause of the wilting of fertilized flowers: after pollination of, e.g., orchid blossoms, ethylene is evolved from the pistil and reaches the petals that subsequently start producing the hormone (Burg and Dijkman, 1967, Nichols, 1977, Suttle and Kende, 1978). Consequently, inhibition of either the biosynthesis of ethylene with AVG or ethanol (Suttle and Kende, 1978, Heins and Blakely, 1980), or its action with silver ions (Beyer, 1976), effectively extends the vase life of cut flowers. If, in particular, the silver is complexed in the thiosulphate ion (STS complex) it effectively enhances the flower longevity (Veen and Van de Geijn, 1978). When added to the vase water of cut carnations, the complex was found to be deposited on the plasmalemma of the receptacle cells (Veen et al., 1980), which membrane may be involved in ethylene biosynthesis (Lieberman, 1979). Isolated protoplasts from morning glory flowers (Ipomoea tricolor Cav.) are still able to convert ACC into ethylene but the AVG-sensitive synthesis of ACC from methionine is no longer possible, one of the essential enzymes probably being located at the plasmalemma (Konze et al., 1980).

It must be questioned, however, as to how far ethylene can be considered as the essential trigger for flower fading in general. Although AVG completely suppresses the ethylene evolution from Tradescantia hybrid petals, it only partially reduces the loss of membrane semipermeability, indicating that other factors may be involved in the initiation of membrane instability (Fig. 4). Moreover, in the petal cells of morning glory, the amounts of membrane phospholipids decline, and fading starts, well before the increase in ethylene production (Fig. 5). On the contrary, in carnation flowers AOA suppresses not only ethylene synthesis but also completely removes the normal senescence of the flowers, characterized by the rolling of the petals. Instead the flowers ultimately senesce by desiccation and necrosis of the petals (Bufler et al.,1980).

Therefore, ethylene seems to exert two different functions. On the one hand, in relatively long-living flowers it may well be the real trigger for senescence, particularly upon pollination. On the other hand, in such ephemeral flowers as Tradescantia and morning glory, which start fading a definite number of hours after opening already as determined by some endogenous clock, ethylene may act as a regulator that accelerates the loss of compartmentation induced

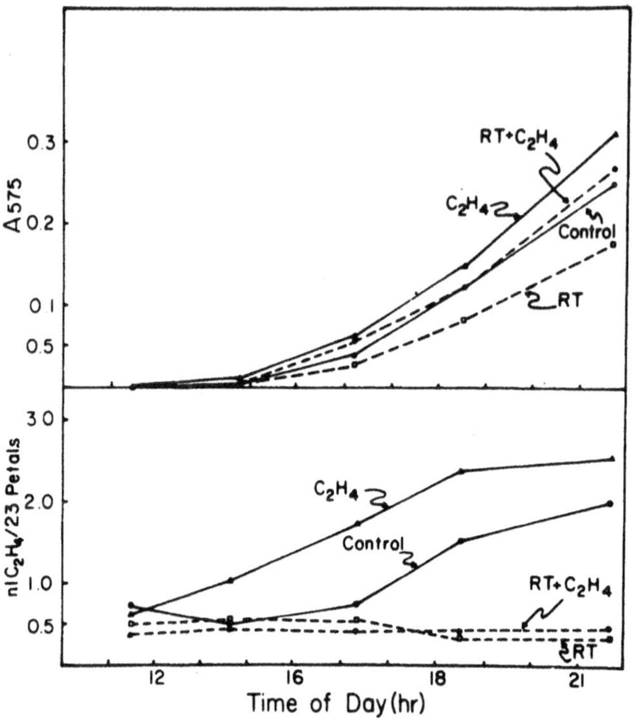

Fig. 4. Differential effect of AVG (RT) on ethylene evolution and pigment efflux, measured at 575 nm, from <u>Tradescantia</u> petals (after Suttle and Kende, 1978).

by some other, clock-determined factor(s), thus speeding up rather than triggering the terminal deterioration of the flower.

FRUIT RIPENING

The senescence of fruits starts with the ripening of the mature fruit. Because fruit ripening involves the biosynthesis and activity of several enzymes and requires large amounts of energy and the prolonged integrity of membranes, some prefer to consider only the deterioration of the fruit in its over-ripe stage as senescence. However, since some of these enzymes are similar to those occurring in other senescence phenomena such as abscission, and since the hormonal regulation of ripening is also very similar, it is justifiable to consider the ripening process itself as a senescence phenomenon.

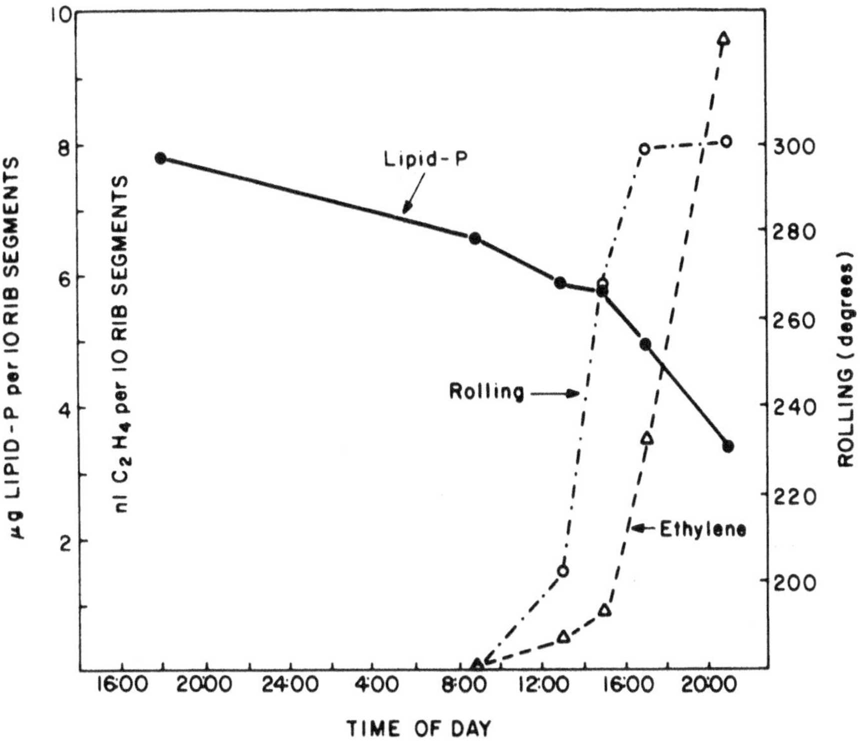

Fig. 5. Time course of change in phospholipids, wilting (rolling), and ethylene evolution of petal sections of Ipomoea <u>tricolor</u> Cav. (after Beutelmann and Kende, 1977).

The onset of ripening is often advanced by the picking of the fruit. For apple, this has been known since 1925 (Hulme et al., 1969). It also occurs in tomato (Sawamura et al., 1978), while such tropical fruits as mango and avocado hardly ripen pre-harvest probably because of the presence of some hormonal, ripening-inhibiting substance derived from the vegetative parts (Burg and Burg, 1964). On the one hand, roots may be involved in retardation of fruit ripening since cytokinins have been reported to delay the onset or decrease the rate of ripening (Bruinsma et al., 1974, Varga and Bruinsma, 1974, Lieberman et al., 1977). On the other hand, a mango fruit will ripen if a steam ring prevents supply to it from the leaves (Burg and Burg, 1964). The nature of this foliar, ripening-inhibiting hormone is unknown, but hormonal levels of auxin are able to keep fruits from ripening (Tingwa and Young, 1975, Inaba et al., 1976). In pear, endogenous auxin has to be removed by oxidation before ripening can occur (Frenkel and Dyck, 1973).

Gibberellins can exert a pronounced, though limited, inhibition of ripening by retardation, or even reversion, of the degreening process. They promote the re-assembly of functional chloroplasts from plastids in various degrees of degradation into carotenoid-filled chromoplasts, e.g. in citrus and banana peel (Bruinsma et al., 1974, Goldschmidt et al., 1977). Other aspects of the ripening process, however, are little affected by gibberellins.

Abscisic acid accumulates in maturing fruit and its exogenous application may advance the onset of ripening (Bruinsma et al., 1974; Inaba et al., 1976; Hale and Coombe, 1974). However, it may well act by inducing ethylene biosynthesis. Because ripening can be completely inhibited by preventing the biosynthesis or action of ethylene, a direct role for ABA is doubtful (Bangerth, 1980). Apparently, ethylene is the dominant hormone in fruit ripening and the pre- and postharvest regulation of ripening largely concerns the control of ethylene synthesis and action (e.g. Bangerth, 1978, Williams, 1980; Bramlage et al., 1980). The effect of the other hormones in fruit ripening may well be exerted by affecting either the biosynthesis of ethylene or the sensitivity of the tissue to this hormone, for instance, by keeping the ethylene receptor sites at a low level or inaccessible.

The sensitivity to ethylene gradually develops during maturation and particularly after picking (Burg and Burg, 1962b; Sawamura et al., 1978). Once the hormone becomes effective it also greatly stimulates its own biosynthesis so that a burst of ethylene evolution, similar to the respiratory climacteric peak, is a definite indication of the onset of ripening. This autocatalytic ethylene synthesis (Burg and Burg, 1965) can also be evoked by propylene, although such an artificial induction may lead to a changed sequence of the ripening events, an artificial lack of synchronization (Eaks, 1980). The increased ethylene production results from enhanced activities of what are probably newly synthesized, ACC-forming and -oxidizing enzymes. In over-ripe fruit ACC can accumulate considerably, possibly because of inactivation of the membrane-bound ACC peroxidase during the gradual loss of membrane integrity. This may contribute to the termination of ethylene evolution (Hoffman and Yang, 1980).

Non-climacteric fruits show no peak, either in respiratory activity or in ethylene evolution. Monogenic mutants of tomato that fail to ripen behave as non-climacteric fruits: these rin- and nor-mutants form stress ethylene but have no autocatalytic ethylene release. Because these mutants also lack the cell wall-softening enzyme, polygalacturonase, and exogenous ethylene alone is unable to induce its activity, Tigchelaar and McGlasson (1977) hypothesized that the biosynthesis or activation of this enzyme might be the primary, ripening-inducing process. By its action, alterations in compartmentation and/or release of cell wall-bound enzymes might occur, amongst other phenomena leading to ethylene production. However, there is no indication that these mutants would form ethylene upon supply of polygalacturonase. In addition, these mutants do ripen, and cell walls do undergo solubilization, in the presence of exogenous ethylene, provided the fruits either remain attached to the plants (Mizrahi et al., 1975) or are kept at a high partial oxygen tension (Frenkel and Garrison, 1976). This indicates a genetic block at the site of ethylene expression, including the stimulation of its own biosynthesis, rather than at the biosynthetic locus of one of the pectinolytic enzymes. Moreover, the appearance of polygalacturonase activity is not the first event in normal tomato-fruit ripening.

The difficulty in determining time sequences of events in fruit ripening is that because of lack of uniformity within fruit samples the distinctness of the phenomena decreases. Fig. 6 shows this for cucumber fruit; if the four individual fruits were measured together, no clear peak of ethylene evolution would be detected. Since such a peak is correlated neither with fruit weight nor with the number of days since anthesis, the only solution to determine the sequence of two events is to measure them in single fruits and to plot them, not against time, but against each other. Because the dose-response curve of ethylene in fruit ripening is logarithmic (Burg and Burg, 1962a), ethylene evolution should be plotted logarithmically. Such graphs show that in tomato, although ethylene and carbon dioxide evolutions may peak together, the endogenous ethylene level already increases when the carbon dioxide production is still low and, thus, the rise in ethylene precedes the climacteric rise in respiration (Fig. 7). Also the observation in honeydew melon that the climacteric peak only occurs if the endogenous level of ethylene surpasses a threshold value (Pratt and Goeschl, 1968) indicates that ethylene acts as the inducer of the increase in respiration. In tomato this threshold value is high in attached fruits, about 5 ppm on an average, but after picking it drops to about 1 ppm, probably because of decay of a ripening inhibitor (Sawamura et al., 1978).

Polygalacturonase activity rises in tomato at high ethylene levels only, without any interference from the ripening inhibitor (Fig. 8A). This renders it highly improbable that this enzyme plays a role in the triggering of the ripening process. Water-soluble pectin, the presumed product of cell wall-solubilization, increases even before polygalacturonase activity enhances appreciably (Fig. 8B), so that the _in vitro_ determination of enzyme activity may be deceptive. What should be determined is the activity of the polygalacturonase isoenzyme located within the cell walls and under the conditions prevailing there which may be less than optimal. However, the water-soluble pectin also increases only after an ethylene threshold has been surpassed (Fig. 9).

In some fruits, the development of ethylene sensitivity and accumulation does not seem to be the primary event in ripening. With avocado, for instance, the in vitro cellulase activity was found to increase 1.5 days prior to ethylene evolution (Awad and Young, 1979). However, particularly in avocado, both the sensitivity to ethylene and its production dramatically increase upon picking (Burg and Burg, 1962 b), so that a role of ethylene as a trigger of ripening is quite plausible in this fruit, too.

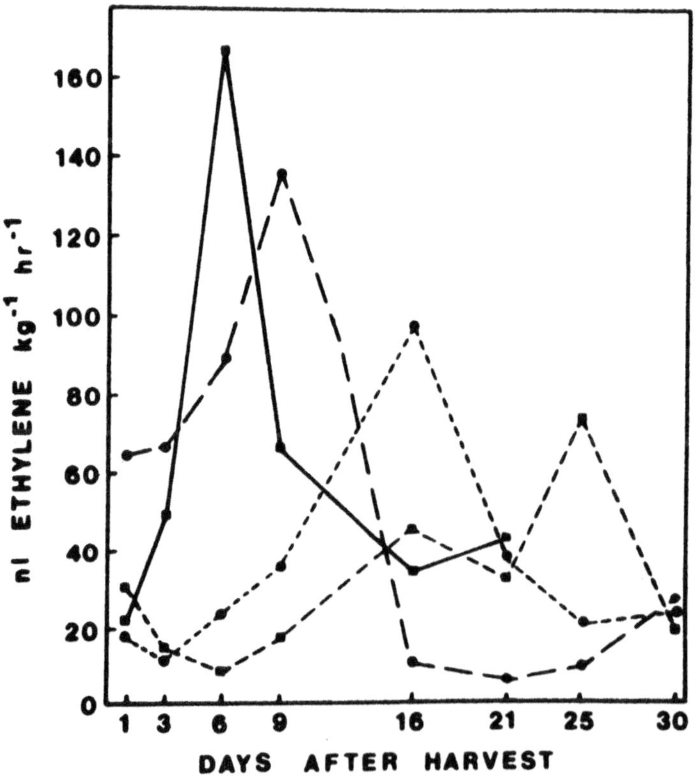

Fig. 6. Non-synchronous ethylene evolution from four harvested cucumber fruit (after Saltveit and McFeeters, 1980).

It can be concluded that ethylene is the primary hormone involved in fruit ripening but whether it acts as the trigger, as it does in the abscission process (Sagee et al., 1980), or only as one of the key factors accelerating a multicausal process (Hobson, 1979) remains to be established once the many methodological pitfalls have been overcome. Whatever the outcome may be, the control of ethylene sensitivity and biosynthesis is the main means of regulating the shelf life of harvested fruits.

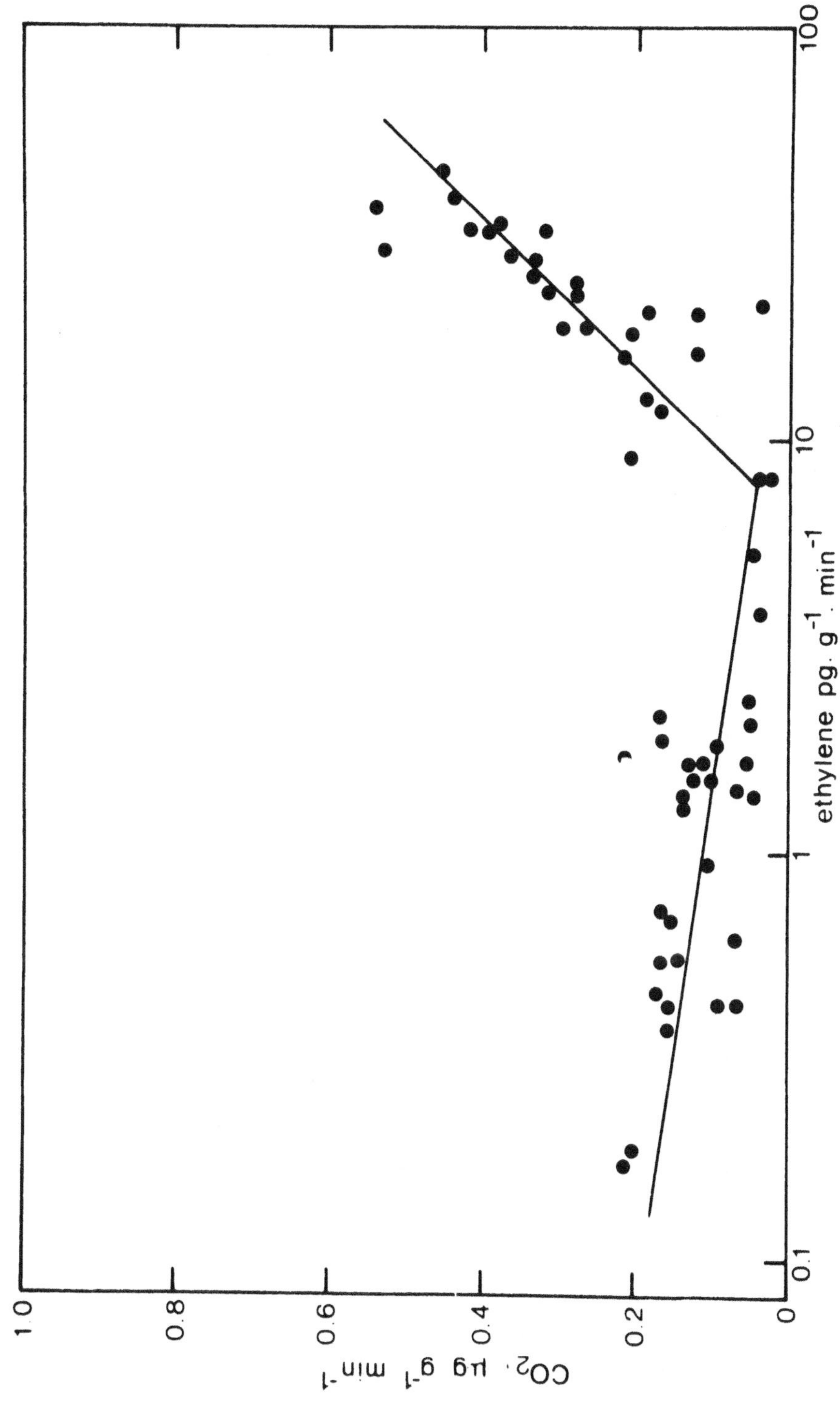

Fig. 7. Evolution of ethylene and carbon dioxide, measured together in the gas from single 'Sonato' tomato fruits at different stages of ripening.

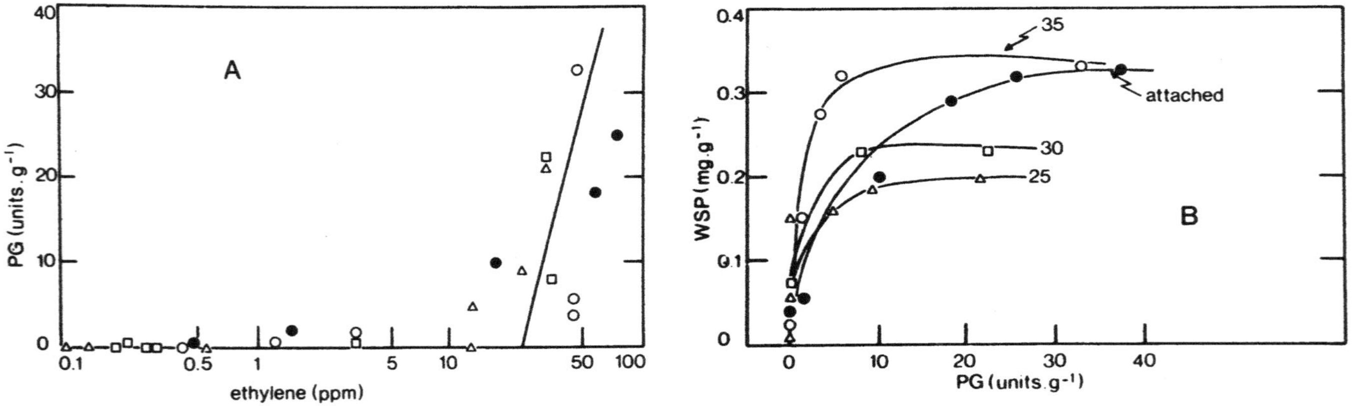

Fig. 8 (A), (B) Occurrence of polygalacturonase activity (PG), water-soluble pectin (WSP), and ethylene in 'Moneymaker' tomato fruits, detached 25, 30 or 35 days after fertilization, or left on the vine until analysis. (A) Relation between PG and internal ethylene. (B) Relation between PG and WSP (after Sawamura et al., 1978).

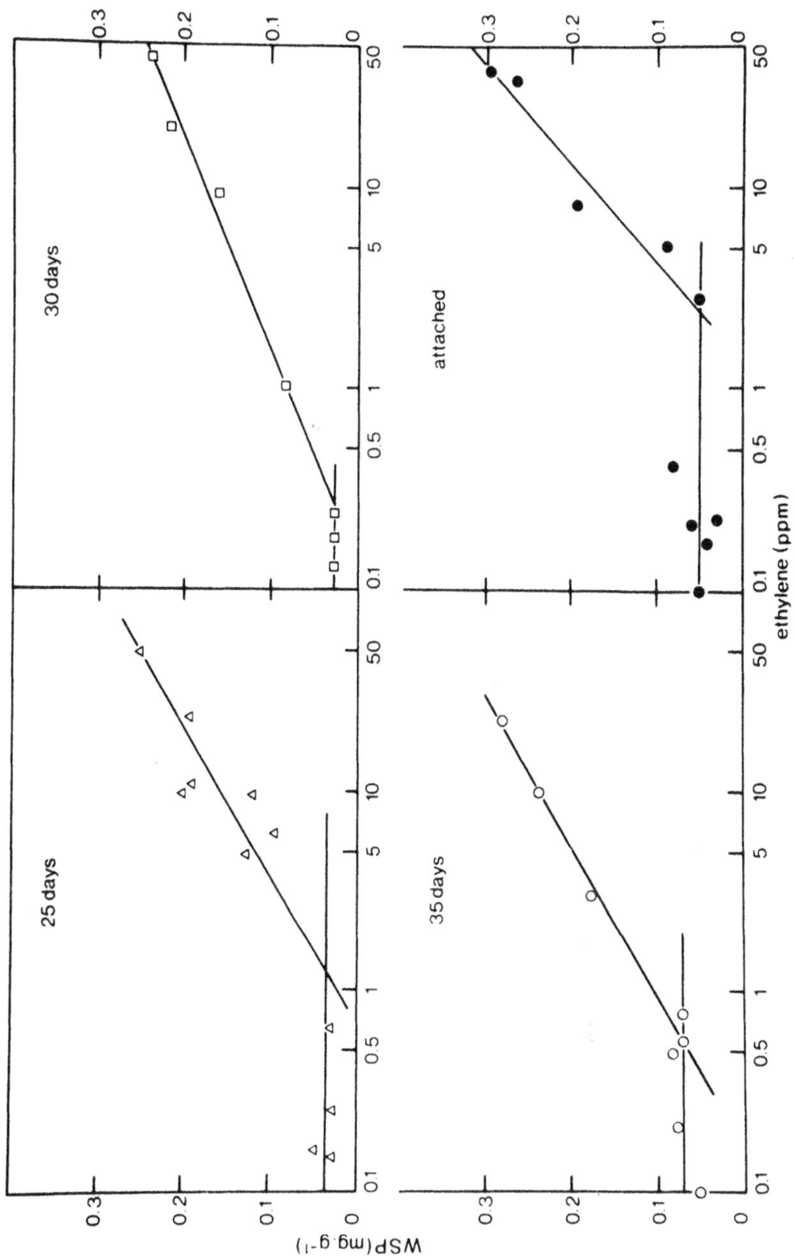

Fig. 9. Level of internal ethylene concentration and occurrence of water-soluble pectin (WSP) in individual 'Sonato' tomato fruits, detached 25, 30 or 35 days after fertilization or left on the vine until analysis (after Sawamura et al., 1978).

CONCLUSION

Senescence, also that of harvested plant parts, is regulated by a pattern of phytohormones. In this pattern cytokinins, also auxins, and to some extend gibberellins, postpone the occurrence of senescence and retard the process, whereas abscisins and ethylene advance and accelerate it. Pre- and postharvest treatments with plant growth regulators, as well as physical storage conditions, should aim at the promotion and maintenance of a hormonal pattern that benefits the shelf life of the harvested commodities. This concerns mainly the prevention of accumulation of abscisins and, particularly, the inhibition of the biosynthesis and action of ethylene.

ACKNOWLEDGEMENT

The critical reading of the manuscript by Drs. E. Knegt, L. C. van Loon, D. A. Priestley, A Varga, and H. Veen is highly appreciated.

REFERENCES

Aharoni, N., Anderson, J. D., and Lieberman, M., 1979, Production and action of ethylene in senescing leaf discs, Plant Physiol., 64:805.
Aharoni, N., and Lieberman, M., 1979, Ethylene as a regulator of senescence in tobacco leaf discs, Plant Physiol., 64:801.
Amrhein, N., and Schneebeck, D., 1980, Prevention of auxin-induced epinasty by α-aminooxyacetic acid, Physiol. Plant., 49:62
Apelbaum, A., and Katchansky, M., 1977, Improving quality and prolonging vase life of bud cut flowers by pretreatment with thiabendazole, J. Amer. Soc. Hort. Sci., 102:623.
Asen, S., and Lieberman, M., 1963, Ethylene oxide found to combat deterioration of carnations, Florists Rev., 132: 3398.
Atkin, R. K., Barton, G. E., and Robinson, D. K., 1973, Effect of rootgrowing temperature on growth substances in xylem exudate of Zea mays, J. Exp. Bot., 24:475.
Awad, M., and Young, R. E., 1979, Postharvest variation in cellulase, polygalacturonase, and pectinmethylesterase in avocado fruits in relation to respiration and ethylene production, Plant Physiol., 64:306.
Bangerth, F., 1978, The effect of a substituted aminoacid on ethylene biosynthesis, respiration, ripening and preharvest drop of apple fruits, J. Amer. Soc. Hort. Sci., 103:401.

Bangerth, F., 1980, Funktion der Abscisinsäure bei der Reife von Apfelfrüchten, Gartenbauwiss., 45: 224.
Beutelmann, P., and Kende, H., 1977, Membrane lipids in senescing flower tissue of Ipomoea tricolor, Plant Physiol., 59:888.
Beyer, E. M., 1976, A potent inhibitor of ethylene action in plants, Plant Physiol., 58:268.
Beyer, E. M., 1979, ^{14}C-ethylene metabolism during leaf abscission in cotton, Plant Physiol., 64:971.
Biswas, A. K., and Choudhuri, M. A., 1980, Mechanism of monocarpic senescence in rice, Plant Physiol., 65:340.
Blomstrom, D. C., and Beyer, E. M., 1980, Plant metabolize ethylene to ethylene glycol, Nature, 283: 66.
Boller, T., Herner, R. C., and Kende, H., 1979, Assay for and enzymatic formation of an ethylene precursor, 1-aminocyclopropane-1-carboxylic acid, Planta, 145:293.
Bramlage, W. J., Greene, D. W., Autio, W. R., and McLaughlin, J. M., 1980, Effects of aminoethoxyvinylglycine on internal ethylene concentrations and storage of apples, J. Amer. Soc. Hort. Sci., 105:847.
Bruinsma, J., Knegt, E., and Varga, A., 1974, The role of growth-regulating substances in fruit ripening, Colloques Internat. C.N.R.S., 238: 193.
Buescher, R. W., and Tichelaar, E. C., 1975, Pectinesterase, polygalacturonase, Cx-cellulase activities and softening of the rin tomato mutant, Hortscience, 10:624.
Bufler, G., Mor, Y., Reed, M. S., and Yang, S. F., 1980, Changes in 1-aminocyclopropane-1-carboxylic-acid content of cut carnation flowers in relation to their senescence, Planta, 150:439.
Burg, S. P., and Burg, E. A., 1962 a, Role of ethylene in fruit ripening, Plant Physiol., 37:179.
Burg, S. P., and Burg, E. A., 1962 b, Relationship between post-harvest changes in ethylene content and the respiration and ripening of fruits, Plant Physiol., 37:xviii.
Burg, S. P., and Burg, E. A., 1964, Evidence for a natural occurring inhibitor of fruit ripening, Plant Physiol., 39:x.
Burg, S. P., and Burg, E. A., 1965, Ethylene action and the ripening of fruits, Science, 148: 1190.
Burg, S. P., and Dijkman, M. J., 1967, Ethylene and auxin participation in pollen induced fading of Vanda orchid blossoms, Plant Physiol., 42:1648.
Collins, J. G., and Morgan, M., 1980, The influence of temperature on the abscisic acid stimulated water flow from excised maize roots, New Phytol., 84:19.
Dyer, T. A., and Osborne, D. J., 1971, Leaf nucleic acids. II. Metabolism during senescence and effect kinetin, J. Exp. Bot., 22:552.

Eaks, I. L., 1980, Respiratory rate, ethylene production, and ripening response of avocado fruit to ethylene or propylene following harvest at different maturities, J. Amer. Soc. Hort. Sci., 105:744.

Frenkel, C., and Dyck, R., 1973, Auxin inhibition of ripening in Bartlett pears, Plant Physiol., 51:6.

Frenkel, C., and Garrison, S. A., 1976, Initiation of ripening process in the rin tomato mutant and in chilling stressed tomatoes as influenced by oxygen and ethylene interactions, Plant Physiol. 57:S-26.

Gepstein, S., and Thimann, K. V., 1980, Changes in the abscisic acid content of oat leaves during senescence, Proc. Natl. Acad. Sci. USA 77:2050.

Goldschmidt, E. E., Aharoni, Y., Eilati, S. K., Riov, J. W., and Monselise, S. P., 1977, Differential counteraction of ethylene effects by gibberellic acid and N_6-benzyladenine in senescing citrus peel, Plant Physiol., 59:193.

Grove, M. D., Spencer, G. F., Rohwedder, W. K. Mandava, N., Worley, J. F., Warthen, J. D., Steffens, G. L., Flippen-Anderson, J. L., and Cook, J. C., 1979, Brassinolide, a plant growth-promoting steroid isolated from Brassica napus pollen, Nature, 281: 216.

Hale, C. R., and Coombe, B. G., 1974, Abscisic acid - an effect on the ripening of grapes, in: "Mechanisms of regulation of plant growth," ed. Bieleski, Ferguson and Crepwell, Roy. Soc. New Zealand, Wellington, 831.

Hall, M. A., 1977, Ethylene involvement in senescence processes, Ann. appl. Biol., 85:424.

Hanson, A. D., and Kende, H., 1975, Ethylene-enhanced ion and sucrose efflux in morning glory flower tissue, Plant Physiol., 55:663.

Hartung, W., and Abou-Mandour, A. A., 1980, Abscisic acid in root cultures of Phaseolus coccineus L., Z. Pflanzenphysiol., 97 265.

Heilmann, B., Hartung, W., and Gimmler, H., 1980, The distribution of abscisic acid between chloroplasts and cytoplasm of leaf cells and the permeability of the chloroplast envelope for abscisic acid, Z. Pflanzenphysiol., 97:67.

Heins, R. D., and Blakely, N., 1980, Influence of ethanol on ethylene biosynthesis and flower senescence of cut carnation, Scientia Hortic., 13:361.

Hewitt, S., Hillman, J. R., and Knights, B. A., 1980, Steroidal oestrogens and plant growth and development, New Phytol., 85: 329.

Hillman, J. R., 1978, "Isolation of plant growth substances," Cambridge University Press, Cambridge, U.K.

Hobson, G. E., 1979, What factors are involved in the onset of ripening in climacteric fruit?, Current Adv. Plant Sci., 37: 1.

Hoffman, N. E., and Yang, S. F., 1980, Changes of 1-aminocyclopropane-1-carboxylic acid content in ripening fruits in relation to their ethylene production rates, J. Amer. Soc. Hort. Sci. 105:492.

Hulme, A. C., Rhodes, M. J. C., Wolltorton, L. S. C., and Galliard, T., 1969, Biochemical changes associated with ripening in apples, Qual. Plant. Mat. Veget., 19:1.

Inaba, A., Ishida, M., and Sobajima, Y., 1976, Changes in endogenous hormone concentrations during berry development in relation to the ripening of Delaware grapes, J. Japan, Soc. Hort. Sci., 45:245.

Kaur-Swahney, R., and Galston, A. W., 1979, Interaction of polyamines and light on biochemical processes involved in leaf senescence, Plant, Cell & Environment, 2:189.

Kende, H., and Baumgartner, B., 1974, Regulation of aging in flowers of Ipomoea tricolor by ethylene, Planta, 116:279.

King, R. W., 1979, Abscisic acid synthesis and metabolism in wheat ears, Aust. J. Plant Physiol., 6:99.

Konze, J. R., Jones, J. F., Boller, T., and Kende, H., 1980, Effect of 1-aminocyclopropane-1-carboxylic acid on the production of ethylene in senescing flowers of Ipomoea tricolor Cav., Plant Physiol., 66:566.

Leopold, A. C., 1975, Aging, senescence, and turnover in plants, BioScience, 25:659.

Lieberman, M., Baker, J. E., and Sloger, M., 1977, Influence of plant hormones on ethylene production in apple, tomato, and avocado slices during maturation and senescence, Plant Physiol., 60:214.

Lieberman, M., 1979, Biosynthesis and action of ethylene, Ann. Rev. Plant Physiol., 30:533.

Mayak, S., and Halevy, A. H., 1970, Cytokinin activity in rose petals and its relation to senescence, Plant Physiol., 46: 497.

Mayak, S., and Halevy, A. H., 1972, Interrelationship of ethylene and abscisic acid in the control of rose petal senescence, Plant Physiol., 50:341.

Mayak, S., Halevy, A. H., and Katz, M., 1972, Correlative changes in phytohormones in relation to senescence processes in rose petals, Physiol. Plant., 27:1.

Miernyk, J. A., 1979, Abscisic acid inhibition of kinetin nucleotide formation in germinating lettuce seeds, Physiol. Plant., 45:63.

Milborrow, B. V., 1979, Antitranspirants and the regulation of abscisic acid content, Aust. J. Plant Physiol., 6:249-254

Mizrahi, Y., Dostal, H. C., and Cherry, J. H., 1975, Ethylene-induced ripening in attached rin fruits, a non-ripening mutant of tomato, Hortscience 10:414.

Nichols, R., 1977, Sites of ethylene production in the pollinated and unpollinated senescing carnation (Dianthus caryophyllus) inflorescence, Planta, 135:155.

Noodén, L. D., and Leopold, A. C., 1978, Phytohormones and the regulation of senescence and abscission, in: "Phytohormones and regulated compounds." II, ed. Letham, Goodwin and Higgins, Elservier, Amsterdam.

Noodén, L. D., Kahanak, G. M., and Okatan, Y., 1979, Prevention of monocarpic senescence in soybeans with auxin and cytokinin: and antidote for self-destruction, Science, 206:841.

Oostrom, H., Kulescha, Z., Vliet, Th. B. van, and Libbenga, K. R., 1980, Characterization of a cytoplasmic auxin receptor from tobacco-pith callus, Planta, 149:44.

Pratt, H. K., and Goeschl, J. D., 1968, The role of ethylene in fruit ripening, in: "Biochemistry and physiology of plant growth substances," ed. Wightman and Letterfield, Runge Press, Ottawa, 1295.

Radin, J. W., 1981, Water relations of cotton plants under nitrogen deficiency. IV. Leaf senescence during drought and its relation to stomotal closure, Physiol. Plant., 51:145.

Ramanuja Rao, I. V., 1979, Postharvest physiology of the spike and regulation of flower development in Gladeolus, Thesis, Delhi.

Sagee, O., Goren, R., and Riov, J., 1980, Abscission of Citrus leaf explants. Interrelationships of abscisic acid, ethylene, and hydrolytic enzymes, Plant Physiol., 66:750.

Saltveit, M. E., and McFeeters, R. F., 1980, Polygalacturonase activity and ethylene synthesis during cucumber fruit development and maturation, Plant Physiol., 66: 1019.

Sawamura, M., Knegt, E., and Bruinsma, J., 1978, Levels of endogenous ethylene, carbon dioxide, and soluble pectin, and activities of pectin methylesterase and polygalacturonase in ripening tomato fruits, Plant Cell Physiol., 19:1061.

Suttle, J. C., and Kende, H., 1978, Ethylene and senescence in petals of Tradescantia, Plant Physiol., 62:267.

Tamas, I. A., Ozbun, J. L., Wallace, D. H., Powell, L. E., and Engels, C. J., 1979 a, Effects of fruits on dormancy and abscisic acid concentration in the axillary buds of Phaseolus vulgaris L., Plant Physiol., 64:615.

Tamas, I. A., Wallace, D. H., Sudford, P. M. and Ozbun, J. L., 1979 b, Effect of older fruits on abortion and abscisic acid concentration of younger fruits in Phaseolus vulgaris L., Plant Physiol., 64:620.

Thimann, K. V., and Satler, S., 1979, Relation between senescence and stomatal opening: senescence in darkness, Proc. Nat. Acad. Sci. USA, 76:2770.

Thomas, H., and Stoddardt, J. L., 1980, Leaf senescence, Ann. Rev. Plant Physiol., 31:83.

Tigchelaar, E. C., and McGlasson, W. B., 1977, Tomato ripening mutants: a key role for polygalacturonase in fruit ripening?, Plant Physiol., 59:S-121.

Tingwa, P. O., and Young, R. E., 1975, The effect of indole-3-acetic acid and other growth regulators on the ripening of avocado fruits, Plant Physiol., 55:937.

Ueda, J., and Kato, J., 1980, Isolation and identification of a senescence-promoting substance from wormwood (Artemisia absinthum L.), Plant Physiol., 66:246.

Varga, A., and Bruinsma, J., 1974, The growth and ripening of tomato fruits at different levels of endogenous cytokinins, J. Hort. Sci., 49:135.

Veen, H., and Geijn, S. C. van de, 1978, Mobility and ionic form of silver as related to longevity of cut carnations, Planta 140:93.

Veen, H., Henstra, S., and Bruyn, W. C. de, 1980, Ultrastructural localization of silver deposits in the receptacle cells of carnation flowers, Planta, 148:245.

Wagner, H., and Michael, G., 1969, Cytokininbildung in Wurzeln von Sonnenblumen bei unterschiedlicher Stickstoffernährung und Chloramphenicol-Zusatz, Naturwiss., 56:379.

Williams, M., 1980, AVG cuts apple production time, boots harvest, Hortscience, 15:76.

Wittenbach, V. A., 1977, Induced senescence of intact wheat seedlings and its reversibility, Plant Physiol., 59:1039.

Wright, S. T. C., 1980, The effect of plant growth regulator treatments on the levels of ethylene emanating from excised turgid and wilted wheat leaves, Planta, 148:381.

Yu, Y. -B., Adams, D. O., and Yang, S. F., 1979, Regulation of auxin-induced ethylene production in mung bean hypocotyls, Plant Physiol., 63:589.

Yu, Y. -B., and Yang, S. F., 1979, Auxin-induced ethylene production and its inhibition by aminoethoxyvinylglycine and cobalt ion, Plant Physiol., 64:1074.

POST HARVEST PHYSIOLOGY OF SEEDS AS RELATED TO QUALITY AND GERMINABILITY

Daniel Côme

Laboratoire de Physiologie des Organes Végétaux
après Récolte, C.N.R.S.,
and Université Pierre et Marie Curie (Paris 6)
4ter, route des Gardes, 92190 Meudon, France

INTRODUCTION

Seeds are remarkable organs in many ways and are of considerable economic interest. Their role is essential in the everyday life of men and animals, since they ensure plant reproduction. In this repect they are used to produce nutritive and floral plants, as well as fruit and forest trees. They are also used as direct sources of food or in various industries such as oil manufacturing and malting.

Seeds are mostly highly dehydrated organs, naturally adapted to long survival after their release by the mother plant until, in an almost inert state, they find the necessary conditions for germination. The problems involved in their storage are therefore very different from those which usually apply to other plant organs.

Some seeds germinate very well at harvesting but may lose this property more or less quickly during storage. Others are totally unable to germinate or do so with great difficulty if they have not been stored for sufficiently long or have not been given special treatment.

BIOLOGICAL PECULIARITIES OF SEEDS AFTER HARVESTING

Seeds result from the development of the ovules after fertilization. Despite their morphological diversity, their constitution is fundamentally the same (1). The embryo is the essential seed structure; it germinates and gives rise to a new plant. The reserves are localized in the embryo cotyledons or in a particular tissue, the endosperm. The seed-coats protect the embryo and control its germination. In some cases, like dry indehiscent fruit, the pericarp and sometimes other parts of the flower adhere to the seed. In certain Graminaceae, the caryopsis is covered by bracts called glumellae.

Seeds are mostly organs of resistance living at a slowed rate since their maturation is a very special phenomenon ending in extreme dehydration. At maturity, they generally contain no more than 10-15% water. Some seeds, however, contain more.

Consequences of Dehydration

The very great loss of water by seeds at the end of maturation is accompanied by profound changes in their cell ultrastructure, condensation of reserves and a significant decrease in metabolic activity (2). During this dehydration period, the coats of certain seeds become waterproof. This is known to be the case for Leguminous plants (3), whose seed-coats have a very characteristic structure (4, 5,6). In addition, the hilum of these seeds plays an important part in their dehydration, since it enables them to lose water without taking it in again. In this connection, seeds have been shown to lose water, first through all their surface and then, when the seed-coat becomes waterproof towards the end of maturation, water escapes through the hilum which remains open and acts as a valve (7). In a dry atmosphere, the hilum opens in less than a minute, so the seed can lose water, but in a humid atmosphere, it closes just as quickly and prevents rehydration.

The very low metabolic activity of mature seeds due to dehydration slows down their life processes. However, except when seed-coats are waterproof, this low activity ceases as soon as the seeds are placed in the presence of water, since the metabolism of imbibing seeds is quickly reactivated.

The ability of seeds to survive in a dehydrated state enables them to withstand conditions which in active life would lead to their death and makes them highly resistant organs. As early as 1907, Becquerel,(8) showed that naturally dehydrated seeds could survive exposure to the temperature of liquid air. However, such resistance is only possible if seeds are sufficiently dehydrated. When imbibed, they resist no better than any other organ.

Seed Sensitive to Desiccation

Some seeds cannot withstand desiccation. Unlike most seed types, they contain a good deal of water, and their embryo remains active and dies of dehydration. This applies to various trees with large seeds and to numerous tropical and subtropical species (9,10, 11,12,13,14,15). Under natural conditions, such seeds germinate as soon as they fall to the ground. These, however, constitute special cases worth closer investigation. Although seed viability greatly varies from one species to another, most seeds display an ability to survive in the dehydrated state, which enables them to remain alive for very long periods.

ORIGIN OF SEED DEATH AND ITS RELEVANCE TO SEED PRESERVATION

Seed Longevity

The extraordinary longevity of some seeds has fascinated scientists for a long time. The earliest work on this subject dates back to the middle of the last century (16).

Ewart, (17) in a treatise entitled "On the Longevity of Seeds," divided seeds into three classes, according to their longevity under ambient conditions:

a) Macrobiotic seeds, capable of surviving for more than 15 years. Examples of seeds with a longevity exceeding 50 to 100 years are far from rare (18). Their seed-coats are often waterproof. In 1907 and 1934, Becquerel, (8,19) succeeded in making very old seeds from the collections of the Natural History Museum in Paris germinate (Table 1). Lotus seeds (*Nelumbo nucifera*) are also well-known for their longevity of several hundred years (20,21). Ødum, (22) obtained the germination of *Chenopodium album* and *Spergula arvensis* seeds about 1,700 years old. The record is held by *Lupinus arcticus* seeds which germinated perfectly even though they were more than 10,000 years old (23). However, it is quite untrue that wheat found in Egyptian tombs was made to germinate (13).

b) Mesobiotic seeds with a life-span of 3 to 15 years. The great majority of species are in this category, which includes wheat.

c) Microbiotic seeds which survive at most for 3 years. These often die quickly, since they cannot withstand desiccation. Harrington, (18) quotes many such cases. Under natural conditions, some seeds only remain viable for very short periods. Those of *Oxalis*, for instance, die within a few days.

Viability Tests

Numerous tests have been proposed for evaluating seed quality and vigour (13,24).

Loss of viability is chiefly shown by a decrease in germinability (25), i.e. lower rates and percentages of germination, slow growth of the seedlings obtained and production of abnormal plants. The degree of germinability is the most commonly applied criterion of quality (26, 27).

Seed deterioration is also shown by diminished resistance to

Table 1. Longevity of some seeds (from Becquerel,(19))

Macrobiotic species	Date of collection	Seeds growing in 1906(x)	Seeds growing in 1934(x)	Determined longevity (years)	Probable longevity (years)
Mimosa glomerata Forsk.	1853	5	5	81	221
Melilotus lutea Gueld.	1851	3	0	55	-
Astragalus Massiliensis Lam.	1848	0	1	86	100
Cytisus austriacus Linn.	1843	1	0	63	-
Lavatera Pseudo-olbia Desf.	1842	2	0	64	-
Dioclea Pauciflora Rusby	1841	1	2	93	121
Ervum Lens Linn.	1841	1	0	65	-
Trifolium arvense Linn.	1838	2	0	68	-
Leucaena leucocephala Linn.	1835	2	3	99	155
Stachys nepetifolia Desf.	1829	1	0	77	-
Cytisus biflorus L'Herit.	1822	2	0	84	-
Cassia bicapsularis Linn.	1819	3	4	115	199
Cassia Multijuga Rich.	1776	-	2	158	-

(x) 10 seeds of each species were sown, except for Cassia Multijuga, of which only 2 seeds were sown.

different types of stress. To determine the vigour of some seeds, a Cold Test is sometimes used, especially for maize, in which the idea is to simulate bad natural germination conditions. The vigorous seeds are those which resist this treatment best (28,29).

The tetrazolium test, very widely used to determine seed viability (30,31,32,33), is based on the reduction of colourless tetrazolium salts and on the formation of red formazan by the dehydrogenases in imbibed embryos. Only living zones colour intensely.

The measurement of respiratory activity during imbibition has been suggested as a viability test, but its validity is often limited (34).

Another way of judging seed viability and vigour is by the amount of inorganic salts or organic compounds which seeds release into a waterbath, since damaged seeds release more of these substances than vigorous ones (35). The electrolytes thus freed are usually evaluated by measuring the electrical conductivity of the waterbath (36,37).

X-ray analysis is used to determine the viability of seeds of conifers and other trees (38,39), and can also be applied to other species. Without any particular treatment, radiography permits detection of seed malformations, but does not enable viable seeds to be distinguished from non-viable ones. The method, based on the impregnation of seed-coats and dead tissues with a barium chloride solution followed by radiography, permits the dead parts, which are more highly contrasted, to be distinguished from the nonimpregnated living parts. Since neither the X-ray doses used nor the barium chloride damage the seeds, they can germinate after this treatment and their germinability can be correlated with the damage observed by X-ray.

Why do Seeds Die?

The numerous biochemical, cytological or genetic changes observed during seed ageing have given rise to various theories concerning their death (13,24,25,40,41,42,43,44). These are summarized in Figure 1, reproduced from Roberts, (41).

Contrarily to what various authors have assumed, it does not seem possible to impute the loss of viability to the degradation or exhaustion of seed reserves, to the destruction or denaturation of molecules essential to seed life such as vitamins, hormones and enzymes, or to the production of toxic substances. Although fungi accelerate the degradation of excessively humid seeds, they cannot be involved when seeds are stored under favourable conditions (41). The theories prevailing at present concern the accumulation of damage in the nucleus and cytoplasmic organelles.

During dry storage of pea, broadbean and barley seeds, loss of viability was found to be closely related to chromosome damage (45,46). This chromosomal abberration was clearly visible during the first divisions of the meristem cells at the end of the root of the surviving seeds. Such damage seems very general (41) and apparently leads to the production of abnormal plants by the old seeds. It sometimes persists for a very long time and is probably

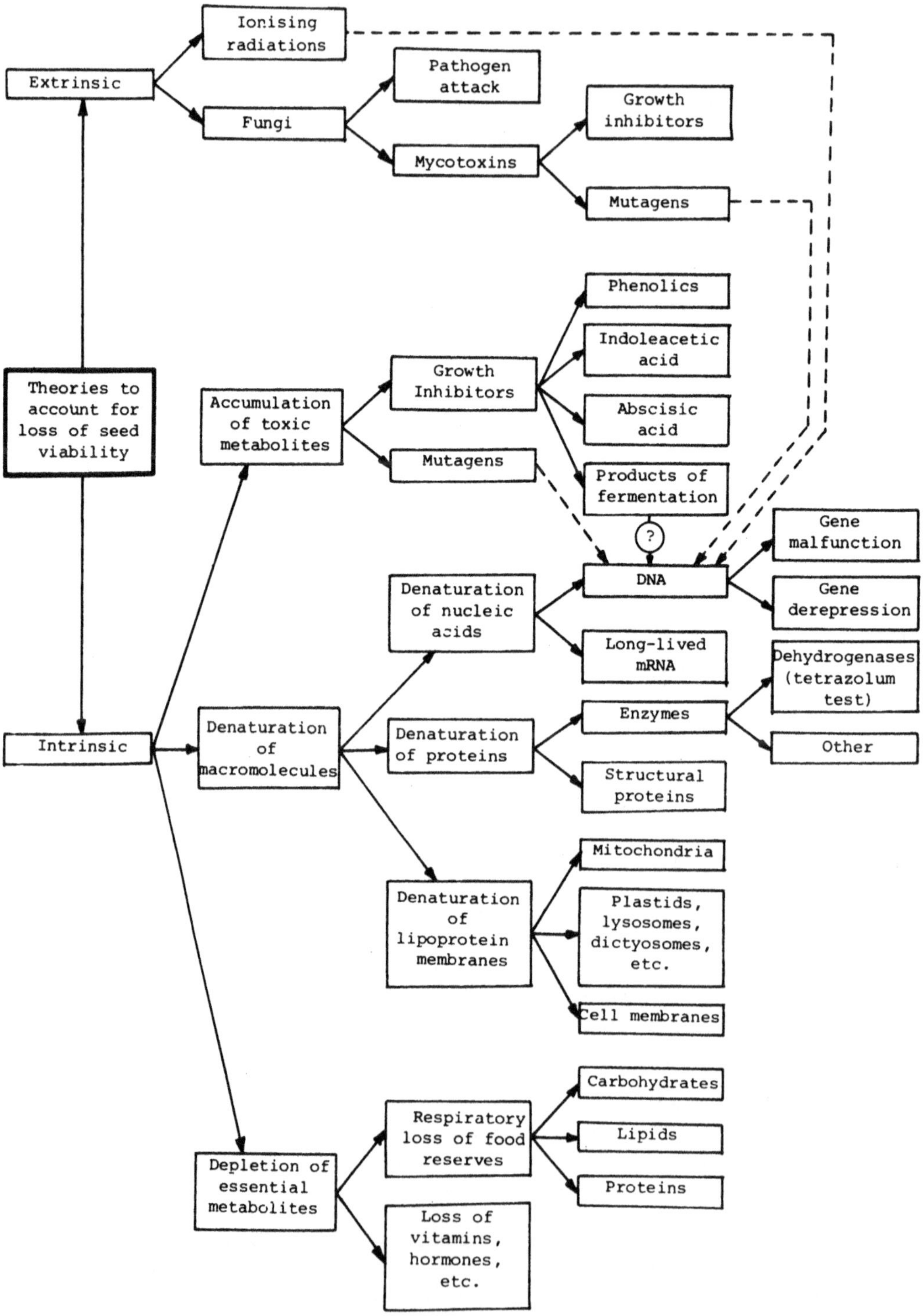

Fig. 1. Classification of viability theories. (from Roberts,(41) with permission).

responsible for the abortive pollen formation in plants grown from old seeds (47). Detailed studies of rye embryos have shown that DNA undergoes fragmentation during dry storage (48,49).

Severe disorders in the cytoplasmic organelles have been observed by electron microscopy in connection with loss of seed viability. Mitochondrial degeneration was shown by Abu-Shakra and Ching,(50) in Glycine max seedlings grown from 3-year-old seeds. The work of Berjak and Villiers, (51,52) on maize has shown that disorders occur in all cell organelles during ageing. Damage is reversible if ageing is not too advanced, but increases with the loss of viability and gradually becomes irreversible. The same types of degeneration were observed by Hallam, (53) in non-viable embryos. Such cell disorders are probably due to lipoprotein membrane deterioration and might explain why the metabolite leak into the germination medium is greater in old than fresh seeds (25,54,55).

Storage Problems

The maintainance of the nutritive value, viability and soundness of seeds is obviously an economic problem and considerable loss may result from germination, parasitic destruction or seed death during storage. Storage problems therefore depend on the end in view and the type of seed involved.

Numerous factors are likely to affect seed quality either before or during the harvest (18,56,57,58,59). We shall only consider here those which are active during storage; they are first of all temperature and moisture, and to a lesser degree, oxygen (60).

Prevention of germination - The problem of prevention germination only arises when seeds are stored in bulk in a very humid atmosphere or when they can only stand mild desiccation.

The nutritive seeds like cereals or leguminous seeds, germination and fungi growth may start if there is too much moisture in the atmosphere or if seeds have been harvested too wet. Such conditions cause appreciable heating of the seeds (61), thus favouring these developments.

Seed water content increases with the relative humidity of the atmosphere (15). For instance, wheat water content reaches 25% of its fresh weight at a relative humidity of 95%. Temperatures within the usual range affect the water vapour absorption rate of seeds but do not greatly change their water content at equilibrium. Refrigeration of seeds does not reduce their water content significantly but slows down germination appreciably. Below a given temperature, which varies depending on the species, germination is impossible. To prevent germination of seeds stored in silos or in large quantities, it is therefore necessary to keep them in a dry cold atmosphere.

Seeds which cannot stand much desiccation must be stored wet. Table 2 gives examples of such seeds. For some species like certain Citrus (11), essentially of tropical or subtropical origin, lowering of the temperature does not allow the seeds to stand more intensive dehydration. Such seeds can only be stored wet, at cold temperatures (about +5°C), but in that case the risks of their being degraded by fungi or bacteria are considerable. However, these difficulties can be overcome by various chemical treatments (63). Other seeds which cannot stand desiccation under natural conditions may, when desiccated and stored at cold temperatures, remain viable for long periods. This is true of Ulmus americana seeds, which die of dehydration within a year under usual conditions but can be stored for 15 years at -4°C if their water content is only 3% (12). The viability of other seeds known to be very sensitive to desiccation can also be prolonged by partial dehydration and storage at cold temperatures. This applies to the seeds of oak (Quercus), beech (Fagus), horse-chestnut (Aesculus), walnut (Juglans), chestnut (Castanea), willow (Salix) and poplar (Populus) (10,64,65,66,67).

Anti-parasitic action - Microflora activity depends much more on atmospheric humidity than on seed water content (60), but there is of course a direct relationship between the relative humidity of the atmosphere and the degree of seed hydration. The growth of fungi is totally inhibited when relative humidity is less than about 70% (68), and bacterial growth requires a relative humidity of at least 90% (69).

Table 2. Minimum water content necessary for survival, in several species of seed.

Species	Minimal water content (%)	Reference
Acer saccharinum	30 - 34	JONES (62)
Quercus alba	25,6	KORSTIAN (9)
Quercus rubra	8,7 - 13,6	KORSTIAN (9)
Citrus sinensis	25	BARTON (11)
Citrus grandis	52	BARTON (11)

A low temperature is just as important as a dry atmosphere in combatting the growth of microflora (68). Seeds which can stand desiccation should therefore be stored in a dry atmosphere at cold temperatures.

It is much more difficult to prevent the often considerable damage caused by insects, otherwise than by appropriate chemical treatment. Seed storage in a dry cold medium can, however, kill certain larvae such as those of weevils, which at 0°C die within two months (70).

In general, cold temperatures prevent germination and limit seed deterioration, but they must be combined whenever possible with dry storage conditions. It is also advisable to heat the seeds with dry air immediately they are removed from cold storage, so as to prevent water condensation on their surface.

Côme and Tissaoui, (71) succeeded in destroying the fungi growing on Robinia pseudacacia seeds thanks to an unusual kind of treatment. These seeds germinate with difficulty, since they are unable to imbibe and are rapidly overgrown with fungi. Both satisfactory germination and destruction of fungi were obtained as follows: seeds were plunged into water at room temperature for a day, allowed to dry in the open air for another day and plunged into liquid nitrogen for 15 minutes. As will be seen later, sudden freezing in liquid nitrogen permits good germination of hard seeds, i.e. seeds whose coats are waterproof. Although the immersion in water no doubt permits germination of the fungus spores retained by the seed-coats, these are subsequently killed by freezing.

Prolongation of viability - As the particular case of seeds which cannot withstand desiccation has already been dealt with, we shall consider here the great majority of seeds, which undergo normal dehydration.

The main factors affecting seed longevity during storage are temperature and seed water content (13,18,60,72,73). In most cases, the lower the seed water content and storage temperature, the longer the viability. Results concerning the role of oxygen are sometimes conflicting (60,74,74). However, an increase in partial oxygen pressure is generally admitted to have an unfavourable effect. In addition, Barton, (13,40) emphasized that the hermetical sealing of containers improved the preservation of germinability.

The very close relationship between seed longevity, water content and temperature has enabled quite satisfactory viability equations to be established (60). Harrington, (73) reported that seed longevity doubled when seed water content was reduced by 1% or when temperature was lowered by 5°C. However, according to Roberts and Ellis, (76) it would be more exact to say that longevity doubles when seed water content is reduced by 2.5% and temperature by 6°C.

At a given relative humidity, the water content of seeds in equilibrium varies with the species considered. It also depends on how seeds are brought to equilibrium. For the same relative humidity, they contain more water when equilibrated by dehydration than by rehydration (73). This phenomenon is known as hysteresis. In general, small seeds can stand very fast artificial dehydration. However, in large seeds, too sudden a dehydration may fracture their cotyledons or endosperm at germination (77).

Provided, of course, that seeds are not too greatly hydrated, very low temperatures (-20°C or less) are often very helpful in maintaining the viability of most species (13,60,78,79).

During the past 20 years, several countries have shown interest in prolonging seed storage as long as possible, since this is the best way of preserving plant genetic resources. The International Board for Plant Genetic Resources recommends storage at -18°C or less in hermetically sealed containers of seeds whose water content has been reduced to 5 ± 1% (76). Recent results suggest that some seeds could be stored for a very long time in liquid nitrogen or by lyophilization. These methods might be very useful for particularly fragile seeds. Table 3 gives the results obtained by Côme (unpublished data) for certain floral and vegetable seeds treated for 4 years by lyophilization. They seem excellent compared to results for storage under ambient conditions. However, it is still too early for a definite assessment of the lyophilization method.

Finally, it appears that naturally dehydrated seeds are not totally inert organs. Although much reduced, their metabolism is not nil, and in fact this residual metabolism seems to be at the origin of seed death. Any treatment resulting in prolonged seed survival reduces seed metabolism, as happens when the temperature is lowered or seeds are artificially dehydrated. In the most extreme case, total dehydration of the seeds by suitable methods should completely suppress their metabolism and enable them to remain viable almost indefinitely. Lyophilization might be one way of achieving this.

ACQUISITION OF GERMINABILITY AFTER HARVESTING

Germination requires certain external conditions to be met. The seed must first be able to become imbibed; next, the metabolic activity started by imbibition requires oxygen and lastly, as in all physiological processes, temperature is very important. Water, oxygen and temperature are therefore the three decisive factors in germination. Moreover, they are virtually indissociable (80, 81,82). Light may also play an important part in germination (80,83,84).

Table 3. Germinability of seeds stored by lyophilization (from Côme, unpublished data)

Species	% germination: at harvesting, non lyophilized	at harvesting, lyophilized	after 2 years, open air	after 2 years, lyophilized	after 4 years, open air	after 4 years, lyophilized
Lobelia sp.	89	79	50	75	30	80
Petunia hybrida	84	90	37	64	14	70
Mimulus sp.	92	90	100	100	0	100
Helichrysum bracteatum	83	80	0	64	0	70
Cineraria sp.	80	95	58	80	10	84
Ageratum mexicanum	92	95	80	90	40	80
Alyssum sp.	96	100	100	100	80	100
Clarkia sp.	96	93	88	90	40	100
Reseda sp.	96	95	52	72	40	73
Papaver sp.	92	90	87	100	80	100
Linum sp.	80	89	74	80	8	66
Bellis perennis (Daisy)	60	66	40	54	0	84
Rudbeckia sp.	84	66	7	65	0	59
Agrostis sp.	52	91	67	78	54	84
Polygonum fagopyrum (Buckwheat)	96	96	100	100	32	100
Phalaris canariensis	96	100	92	92	84	100
Festuca pratensis	92	92	78	100	11	100
Dactylis sp.	85	90	92	100	40	100
Nasturtium officinale (Water-cress)	92	86	20	92	8	84
Milium sp. (Millet)	92	95	80	100	72	100
Coriandrum sativum (Coriander)	95	70	47	47	29	47
Ocimum basilicum (Basilica)	52	90	80	70	48	72
Cichorium endivia	96	90	84	100	4	100
Asparagus officinale (Asparagus)	93	96	12	52	0	62
Foeniculum officinale (Fennel)	100	83	80	100	4	72
Allium cepa (Onion)	92	92	72	84	8	72
Valerianella olitoria (Corn salad)	88	96	68	97	4	100

However, many seeds are unable to germinate or germinate with great difficulty even under favourable conditions. This inability to germinate is commonly called "dormancy." Some dormant seeds germinate easily after storage, others need special treatment.

Inability to Germinate

When the term "dormancy" is used to designate the inability of a seed to germinate properly, it does not characterize the physiological mechanism involved. It may therefore be ambiguous, in which case it is preferable to speak of "inability to germinate."

Inability to germinate can be of two types (1,80): in some cases it is due to the embryo which cannot germinate properly even when stripped of its various covering structures (seed-coats and sometimes endosperm). This phenomenon is called embryo dormancy. In other cases, the naked embryo germinates perfectly, but the structures covering it inhibit its germination; this is known as inhibition of germination. Some seeds may display both types of inability.

Embryo dormancy is a very complex phenomenon and little is yet known about its mechanisms (80,85,86,87,88). Inhibition of germination by seed-coats is characterized by the fact that seed-coat removal or scarification leads to germination. Seed-coats often have an inhibitory effect by depriving the embryo of oxygen (80,81). They may also be waterproof, as in the well-known case of Leguminous plants (3,80). In some species, such as Chenopodium amaranticolor (89) and Fraxinus excelsior (90), they are permeable to water, but their mechanical resistance is too great for the embryo to break them. All these types of inability to germinate are very relative phenomena whose manifestations depend on many factors. Thus in some cases, such inability can be overcome by dry storage.

Effect of Dry Storage

Just after harvesting, the seeds of certain species germinate with great difficulty at the usual moderate temperatures, but germinate perfectly after various periods of dry storage under ambient conditions. This phenomenon is called after-ripening in dry storage (80, 91). Crocker and Barton, (92) mention 41 species with seeds capable of such after-ripening. Many photosensitive seeds gradually lose their photosensitivity during dry storage (93), but the most typical case is that of the Graminaceae (91,94), which we shall consider here.

As an example, Figure 2 shows that wheat caryopses germinate with difficulty at 20°C at harvest time, but germinate well after storage at room temperature in the open air for 2 months or a year. This particular dormancy of freshly harvested Graminaceae caryopses is in fact only an inability to germinate at relatively high temperatures, since germination occurs very easily at 10°C or less (95,96). During dry storage, the caryopses become capable of germinating at increasingly high temperatures (Fig. 3). The rate at which they acquire this ability to germinate within a wide temperature range depends on the storage temperature. Storage at 35 or 40°C for a few days may have the same effect as prolonged storage at 20°C (96,97). On the other hand, storage at very low temperature (-18°C) maintains caryopsis dormancy (95).

This phenomenon depends essentially on the external structures of the caryopsis. In barley, for instance, the glumellae adhere to the grain and are largely responsible for its bad germination at harvesting. During dry storage, the inhibiting effect of the glumellae gradually disappears (95).

Several explanations have been suggested to account for afterripening in dry storage. Some authors believe that volatile inhibitors are eliminated during storage (97), but it is more likely that the external structures of the caryopsis intervene by depriving the embryo of oxygen and that their effectiveness in acting as barriers to the passage of oxygen decreases during storage. This

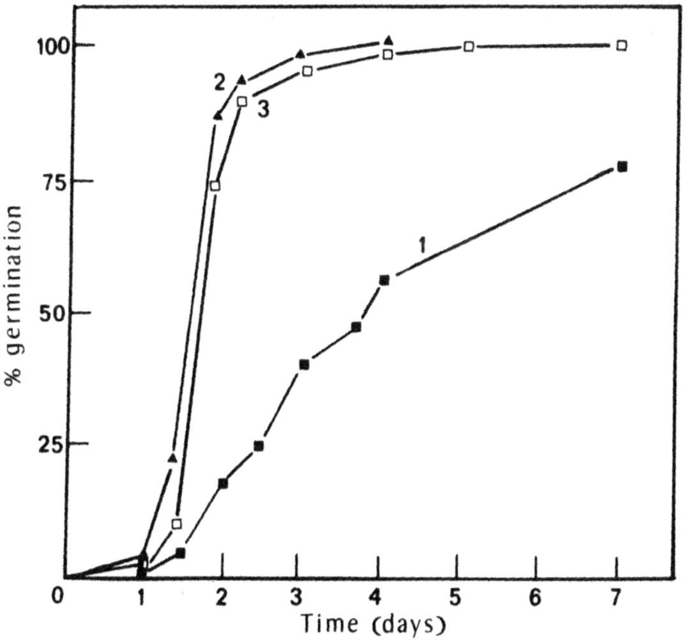

Fig. 2. Germination of wheat caryopses at 20°C at harvesting (1), or after dry storage at ambient conditions for 2 months (2) or one year (3). (from Corbineau et al., (95)).

is proved by the fact that barley and wheat caryopses clearly require less oxygen to germinate after dry storage than at harvesting (95,96) and that, in the case of barley, the presence of the glumellae enveloping the caryopsis increases its need for oxygen (Fig. 4). Phenolic compounds present in the seed-coats and glumellae might be responsible for oxygen absorption, as in many other seeds (81). However, it is also possible that, at harvesting, caryopsis envelopes are made up of living cells whose respiration might deprive the embryo of oxygen, and the death of these cells during dry storage would render them incapable of further oxygen absorption (98).

This inability of seeds to germinate easily as soon as they ripen is of considerable economic importance for cereals, since it prevents them from sprouting. However, it may be very inconvenient when seed germination is desired just after harvesting. This is especially the case of barley grown for malting.

Need for Special Treatment

The germination of seeds exhibiting embryo dormancy or whose seed-coats are waterproof is only possible at harvest time or after

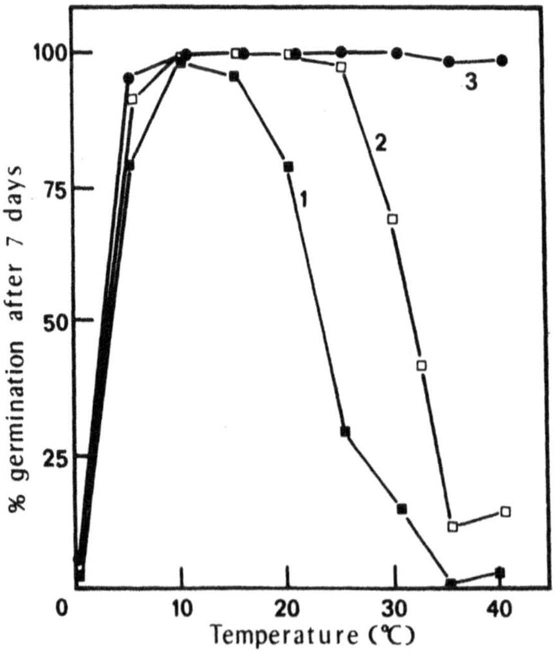

Fig. 3. Effect of temperature on the germination of wheat caryopses (Champlein variety). (1) in freshly harvested caryopses; (2) after 2 months of dry storage; (3) after one year of dry storage. (from Corbineau et al., (95)).

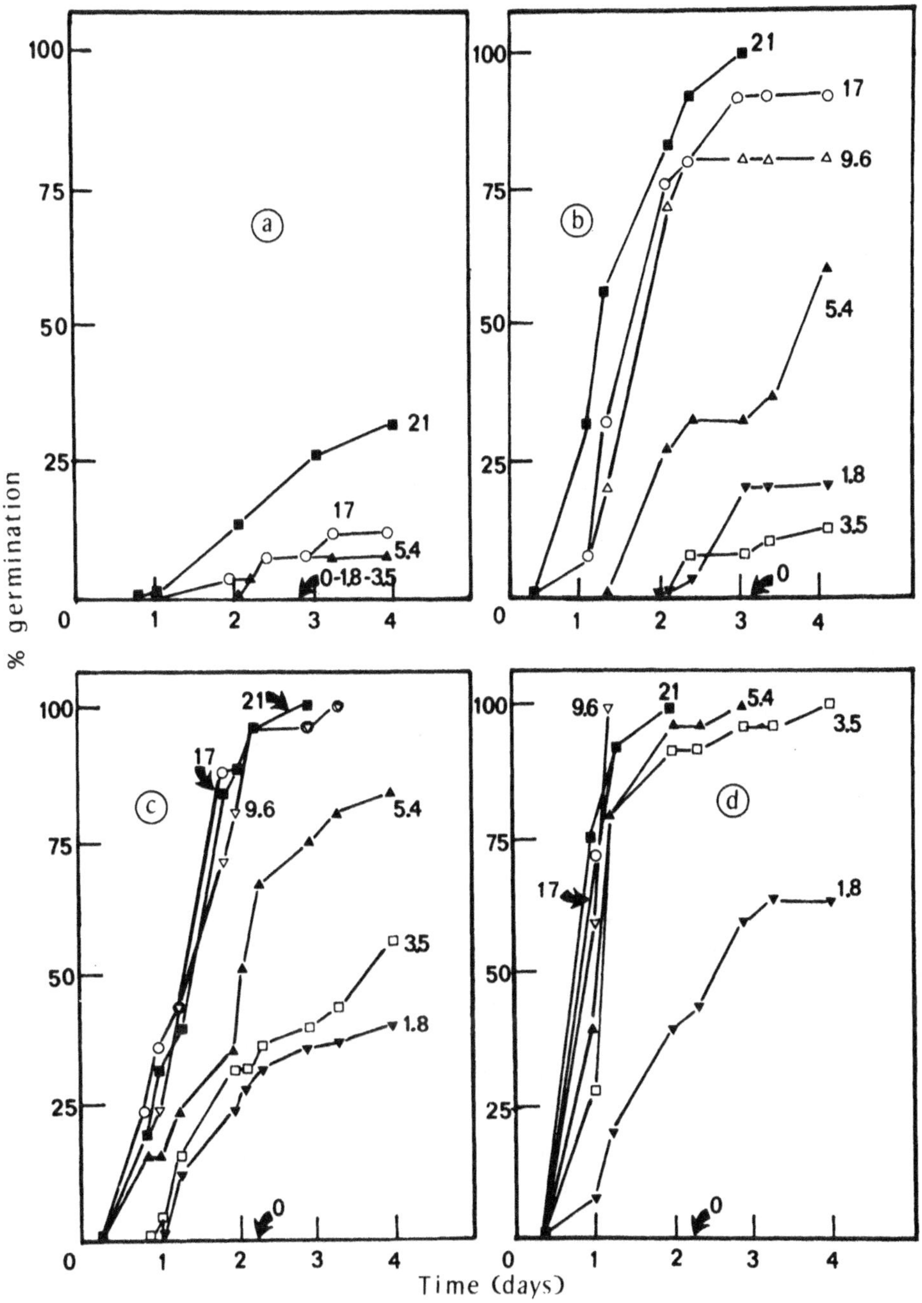

Fig. 4. Germination of barley caryopses (Sonja variety) at 20°C, in atmospheres containing 0, 1.8, 3.5, 5.4, 9.6, 17 or 21% of oxygen. (a) dormant caryopses with their glumellae; (b) non-dormant caryopses with their glumellae; (c) dormant caryopses stripped of their glumellae; (d) non-dormant caryopses stripped of their glumellae. (from Corbineau and Côme,(96)).

dry storage if they have undergone special treatment.

Embryo dormancy. Embryo dormancy is characteristic of Rosaceae seeds, but is also found in many other species. The subject has been treated in a large number of investigations, of which certain authors have made a synthesis (80,85,86,87,91,94,99).

Non-dormant embryos germinate easily at temperatures of the order of 20°C. Dormant embryos, on the contrary, do not germinate at these temperatures or do so very slowly. When they do, the seedlings they give - sometimes called dwarf seedlings - are abnormal, and are generally not viable. Their growth anomalies are more or less pronounced, depending on the state of dormancy of the embryos(80, 100).

Embryo dormancy is broken by placing the seeds at cold but positive temperatures in a wet medium. In temperate climates, this breaking process occurs naturally in winter. Cold temperatures only break embryo dormancy if the embryo is imbibed. If not they maintain its dormancy.

This property of cold temperatures has long been exploited by nurserymen who put the seeds into wet sand or soil in cold rooms, or simply plant them out in the open in winter. This method is known as stratification. Other cold treatments are also possible (86, 88).

The breaking of embryo dormancy always requires prolonged cold treatment. The evolution of this process is shown by the increasingly fast germination of the embryos when sown at moderate temperatures (Fig. 5) and by the regression of seedling growth anomalies. It is also characterized by a widening of the range of temperatures permitting germination (101) since, although dormant embryos can only germinate at low temperatures, they acquire the ability to germinate at higher temperatures during cold treatment (Fig. 6). Dormancy is regarded as completely broken when the embryos germinate rapidly within a wide temperature range, without producing abnormal seedlings.

There is a close relationship between temperatures and duration of treatment. The higher the temperature, the longer the treatment must last. For instance, the thermal optimum for breaking of apple embryo dormancy is between 0 and 5°C; temperatures above 12-15°C have no effect (Fig. 7). The need for cold temperatures varies with the species and variety considered, but in most cases a treatment of one or several months at temperatures around 5°C enables embryo dormancy to be broken.

Cold temperatures are not, however, absolutely necessary for breaking embryo dormancy, since other treatments have the same effect (86). For instance, apple embryo dormancy can be broken

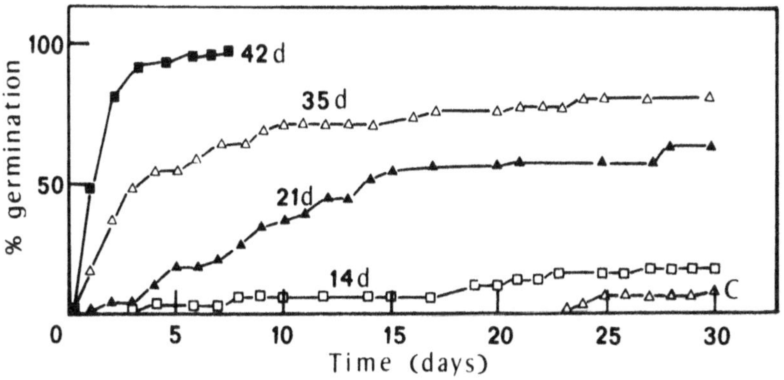

Fig. 5. Germination at 20°C of Golden delicious apple embryos from seeds treated at 5°C on wet cotton wool for 14, 21, 35 or 42 days. Curve C corresponds to control embryos which were not given cold treatment. (from Thevenot, (88)).

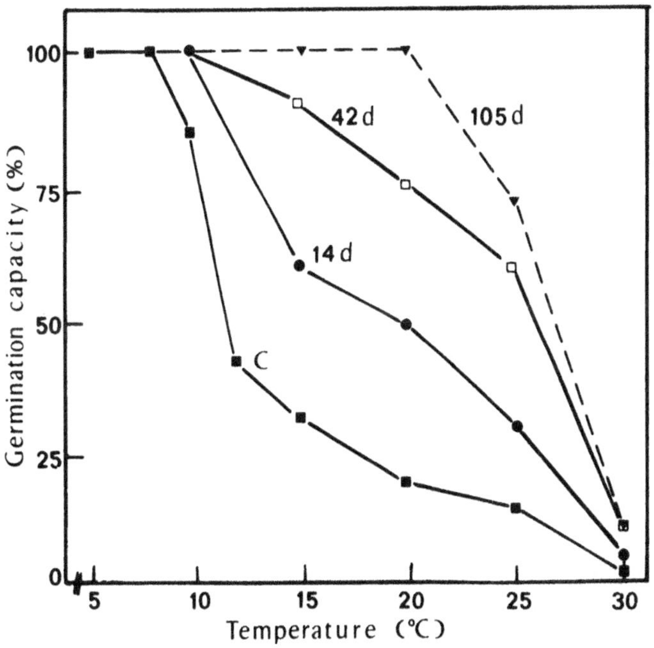

Fig. 6. Effect of temperature on the gerimination capacity of Golden Delicious apple embryos kept inside the fruit at 0°C for 14, 42 or 105 days. Curve C corresponds to control embryos which were not given cold treatment. (from Thevenot, (88) modified).

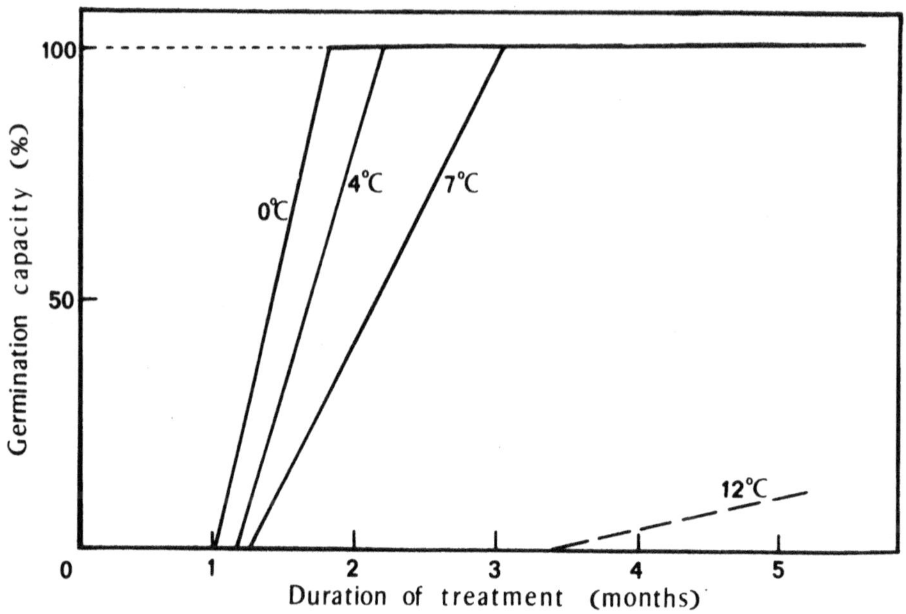

Fig. 7. Variation in the germination capacity, at 18°C, of Reinette du Mans apple embryos with duration of treatment in a wet medium, at 0, 4, 7 or 12°C (from Côme, (100).)

within 2 or 3 weeks when embryos are treated at 15 or 20°C in a wet medium without any oxygen (102). When apple seeds with dormant embryos are placed in the presence of water at 30 or 35°C, they do not germinate, but the dormancy of their embryos is gradually eliminated (103), since at these temperatures no oxygen reaches them under the seed-coats (100). Addition of gibberellins to the culture medium can also more or less replace cold treatment (104,105,106, 107,108).

Hard seeds. Some seeds are unable to germinate because their seed-coats are totally impermeable to water and their embryo cannot imbibe. Seeds which imbibe normally swell, soften and can easily be crushed. Impermeable seeds remain dry and resist crushing, and are therefore called hard seeds. This applies to leguminous seeds and to those of many other species (3). The percentage of hard seeds produced by a given species varies from one year to another. It depends on the place of culture and climatic conditions. There may also be a connection between the proportion of hard seeds and the colour of seed-coats (109).

Various treatments have been proposed to render hard seeds permeable to water (3). All of them partly destroy or crack the seed-coats. Hardness in cultivated species is often eliminated by violently projecting seeds against a metallic wall in order to break their seed-coats. However, the effectiveness of such treatments is limited as the seeds too may be broken. Immersion of seeds in concentrated sulfuric acid and their subsequent washing with water can give good results. However, this treatment is rather difficult to apply, since immersion must last long enough to destroy the seed-coats, but not too long, to avoid killing the embryos. In some cases brief immersion in boiling water is also possible. An alternative method, which is very effective, easily applied and does not damage the seeds consists of plunging them into liquid nitrogen (Fig. 8). The process can be repeated until satisfactory germination is obtained (71, 110). The enormous temperature difference between liquid nitrogen and the ambient air produces fine cracks in the seed-coats, which then become permeable to water.

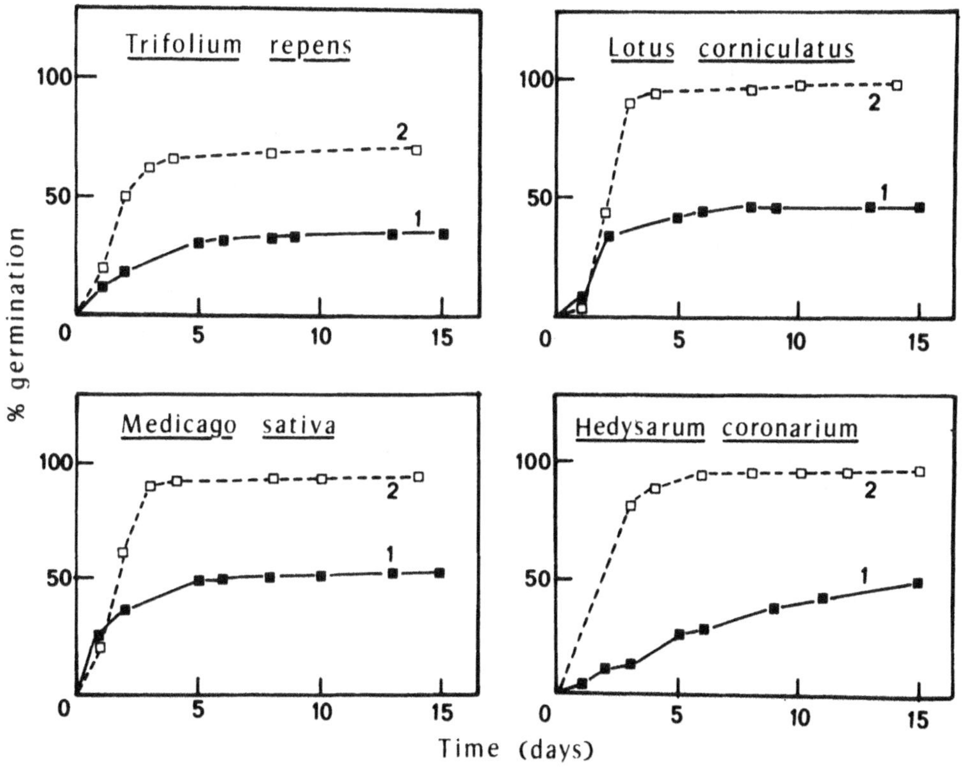

Fig. 8. Germination of Leguminous seeds at 20°C. (1) untreated seeds; (2) seeds frozen in liquid nitrogen for 15 minutes. (from Côme and Tissaoui, (71))

CONCLUSION

Seeds are very special organs whose postharvest physiology is quite different from that of other plant organs. Their extraordinary ability to survive in a dehydrated state enables them to subsist for a very long time until the conditions necessary to their germination are present, and facilitates their storage. Their germination is often delayed by numerous regulating systems. In wild species, inability to germinate is very beneficial since it helps to preserve these species when climatic or ecological conditions are difficult. For cultivated species, however, dormancy often causes serious problems which have to be solved by suitable treatments.

REFERENCES

1. D. Côme, Quelques problèmes de terminologie concernant les semences et leur germination, in: "La germination des semences," R. Chaussat and Y. Le Deunff, ed., Gauthier-Villars, Paris (1975).
2. H. Thomas, Control mechanisms in the resting seed, in: "Viability of Seeds," E.H. Roberts, ed., Chapman and Hall, London (1972).
3. L. V. Barton, Dormancy in seeds imposed by the seed coat, in: "Encyclopedia of Plant Physiology," W. Ruhland, ed., Springer-Verlag, Berlin, Heidelberg, New York, 15/2 (1965).
4. D. H. Hamly, Softening of the seeds of Melilotus alba, Bot. Gaz. 93:345 (1932).
5. A. Pitot, Le tégument des graines de Légumineuses, Bull. Soc. Bot. France 82:307 (1935).
6. L. Cavazza, Recherches sur l'impermeabilité des graines dures chez les Légumineuses, Ber. Schweiz. bot. Ges. 60:596 (1950).
7. E.O.C. Hyde. The function of the hilum in some Papilionaceae in relation to the ripening of the seed and the permeability of the testa, Ann. Bot. 18:241 (1954).
8. P. Becquerel, Recherches sur la vie latente des graines, Ann. Sci. Nat., Bot. 9:193 (1907).
9. C.F. Korstian, Factors controlling germination and early survival in oaks, Yale Univ. School of Forestry Bull. n° 19 (1927).
10. W. Crocker, Life span in seeds, Bot Rev. 4:235 (1938).
11. L.V. Barton, The storage of citrus seeds, Contr. Boyce Thompson Inst. 13:47 (1943).
12. L.V. Barton, Seed storage and viability, Contr. Boyce Thompson Inst. 17:87 (1953).
13. L.V. Barton, "Seed Preservation and Longevity," Leonard Hill, New York (1961).
14. G.D. Holmes, and G. Buszewicz, The storage of temperate forest tree species, Forestry Abstr. 19:nos 3 and 4 (1958).
15. A. Lang, Effects of some internal and external conditions on seed germination, in: "Encyclopedia of Plant Physiology," W. Ruhland, ed., Springer-Verlag, Berlin, Heidelberg,

New York, 15/2 (1965).
16. A. de Candolle, Sur la durée relative de la faculté de germer dans des graines appartenant à diverses familles, Ann. Sci. Nat., ser. 3, 6:373 (1846).
17. A. J. Ewart, On the longevity of seeds, Proc. Roy. Soc. Victoria 21:1 (1908).
18. J. F. Harrington, Seed storage and longevity, in: "Seed Biology," T.T. Kozlowski, ed., Academic Press, New York and London, III (1972).
19. P. Becquerel, La longévité des graines macrobiotiques, C.R. Acad. Sci. Paris 199:1662 (1934).
20. I. Ohga, On the longevity of the fruit of Nelumbo nucifera, Bot. Mag., Tokyo 37:87 (1923)
21. I. Ohga, On the structure of some ancient, but still viable fruits of Indian Lotus, with special reference to their prolonged dormancy, Jap. J. Bot. 3:1 (1926).
22. S. Ødum, Germination of ancient seeds. Floristical observations and experiments with archaeologically dated soil samples, Dansk. bot. Ark. 24:1 (1965).
23. A.E. Posild, C.R. Harrington, and G.A. Mulligan, Lupinus arcticus Wats. grown from seeds of the Pleistocene age, Science 158:113 (1967).
24. J. D. Maguire, Seed quality and germination, in: "The Physiology and Biochemistry of Seed Dormancy and Germination," A.A. Khan, ed., Elsevier/North-Holland Biomedical Press, Amsterdam (1977).
25. A. A. Abdul-Baki, and J.D. Anderson, Physiological and Biochemical deterioration of seeds, in: "Seed Biology," T.T.Kozlowski, ed., Academic Press, New York and London, II (1972).
26. ISTA (International Seed Testing Association), International rules of seed testing, Proc. Int. Seed Test. Ass. 31:1 (1966).
27. P. S. Wellington, Handbook for Seedling Evaluation, Proc. Int. Seed Test. Ass. 35:449 (1970).
28. D. Isely, The cold test for corn, Proc. Int. Seed Test. Ass. 16:299 (1950).
29. D. Isely, Vigor tests, Proc. Ass. Off. Seed Anal. 47:176 (1957).
30. G. Lakon, The topographical tetrazolium method for determining the germination capacity of seeds, Plant Physiol. 24:389 (1945).
31. R. P. Moore, Tetrazolium as a universally acceptable quality test of viable seed, Proc. Int. Seed Test. Ass. 27:795 (1962).
32. R. P. Moore, Tetrazolium staining for assessing seed quality in: "Seed Ecology," W. Heydecker, ed., Butterworths, London (1973).
33. H. Bulat, Das topographische Tetrazoliumverfahren in der Saatgutprüfung, in: "Hundert Jahre Saatgutprüfung," F. Ader, ed., Sauerländer, Frankfurt am Main (1970).
34. L. W. Woodstock, Biochemical tests for seed vigor, Proc. Int. Seed Test. Ass. 34:253 (1969).

35. K. Takayanagi, and K. Murakami, Rapid germinability test with exudate from seed, Nature 218:493 (1968).
36. S. Matthews, and W.T. Bradnock, Relationships between seed exudation and field emergence in peas and French beans, Hort. Res. 8:89 (1968).
37. W. T. Bradnock, and S. Matthews, Assessing field emergence potential of wrinkled-seeded peas, Hort. Res. 10:50 (1970).
38. M. Simak, Å. Gustafsson, and G. Granström, Die Röntgendiagnose in der Samenkontrolle, Proc. Int. Seed Test. Ass. 22:330 (1956).
39. R. Evrard, L'analyse de la qualité des semences par la méthode aux rayons, Ann. Gembl. 63:81 (1957).
40. L. V. Barton, Longevity in seeds and in the propagules of fungi, in: "Encyclopedia of Plant Physiology," W. Ruhland, ed., Springer-Verlag, Berlin, Heidelberg, New York, 15/2 (1965).
41. E. H. Roberts, Cytological, Genetical, and Metabolic Changes Associated with Loss of Viability, in: "Viability of Seeds," E. H. Roberts, ed., Chapman and Hall, London (1972).
42. T. A. Villiers, Ageing and the longevity of seeds in field conditions, in: "Seed Ecology," W. Hedydecker, ed., Butterworths, London (1973).
43. D. J. Osborne, Senescence in Seeds, in: "Senescence in Plants," K.V. Thimann, ed., CRC Press, Boca Raton, Florida (1980).
44. T.A. Villiers, Ultrastructural Changes in Seed Dormancy and Senescence, in: "Senescence in Plants," K.V. Thimann, ed., CRC Press, Boca Raton, Florida (1980).
45. E.H. Roberts, F.H. Abdalla, and R.J. Owen, Nuclear damage and the ageing of seeds with a model for seed survival curves, Symp. Soc. exp. Biol. 21:65 (1967).
46. F. H. Abdalla, and E.H. Roberts, Effects of temperature, moisture, and oxygen on the induction of chromosome damage in seeds of barley, broad beans, and peas during storage, Ann. Bot. 32:119 (1968).
47. F. H. Abdalla, and E.H. Roberts, The effect of temperature and moisture on the induction of genetic changes in seeds of barley, broad beans, and peas during storage, Ann. Bot. 33:153 (1969).
48. K. S. E. Cheah, and D.J. Osborne, DNA lesions occur with loss of viability in embryos of ageing rye seed, Nature 272:593 (1978).
49. D. J. Osborne, R. Sharon, and R. Ben-Ishai, Studies on DNA integrity and DNA repair in germinating embryos of rye (Secale cereale), Israel J. Bot. 29:259 (1980/81).
50. S. S. Abu-Shakra, and T.M. Ching, Mitochondrial activity in germinating new and old soybean seeds, Crop Sci 7:115 (1967).
51. P. Berjak, and T.A. Villiers, Ageing in plant embryo. I. The establishment of the sequence of development and senescence in the root cap during germination, New Phytol. 69:929 (1970).

52. P. Berjak, and T.A. Villiers, Ageing in plant embryos. II. Age-induced damage and its repair during early germination, New Phytol. 71:135 (1972).
53. N. D. Hallam, Fine structure or viable and non-viable rye and other embryos, in: "Seed Ecology," W. Heydecder, ed., Butterworths, London (1973).
54. T. M. Ching, and I. Schoolcraft, Physiological and chemical differences in aged seeds, Crop Sci. 8:407 (1968).
55. A. A. Abdul-Baki, and J. D. Anderson, Viability and leaching of sugars from germinating barley, Crop Sci. 10:31 (1970)
56. W. Heydecker, The "vigour" of seeds - a review, Proc. Int. Seed Test. Ass. 34:201 (1969).
57. W. Heydecker, Vigour, in: "Viability of Seeds," E.H. Roberts, ed., Chapman and Hall, London (1972).
58. R. B. Austin, Effects of Environment Before Harvesting on Viability, in: "Viability of Seeds," E.H. Roberts, ed., Chapman and Hall, London (1972).
59. R. P. Moore, Effects of Mechanical injuries on viability, in: "Viability of Seeds," E.H. Roberts, ed., Chapman and Hall, London (1972).
60. E.H. Roberts, Storage Environment and the Control of Viability, in: "Viability of Seeds," E.H. Roberts, ed., Chapman and Hall, London (1972).
61. A. Lynen, P. Schribaux, and J.M. Gatheron, Influence des caractéristiques d'enceintes differentes propres au stockage d'une masse de grains sur les équilibres entre calories émises et calories dispersées, C.R. Acad. Sci. Paris 237: 1028 (1953).
62. H. A. Jones, A physiological study of maple seeds, Bot. Gaz. 69:127 (1920).
63. J.F.L. Childs, and G. Henciar, A method of maintaining viability of citrus seed in storage, Proc. Florida State Hort. Soc. 61:64 (1949).
64. G. M. Buszewicz, The longevity of beechnuts in relation to storage conditions, Rep. For. Res., For. Comm., London 1961:117 (1962).
65. B. Suszka, Storage of beech (Fagus silvatica L.)seed for up to 5 winters, Arboretum Kórnickie 19:105 (1974).
66. B. Suszka, Cold storage of already after-ripened beech (Fagus silvatica L.) seeds, Arboretum Kórnickie 20:299 (1975).
67. M. Bonnet-Masimbert, and C. Muller, La conservation des faines est possible, Rev. Forest. Francaise 27:129 (1975).
68. C. M. Christensen, Microflora and Seed Deterioration, in: "Viability of Seeds," E.H. Roberts, ed., Chapman and Hall, London (1972).
69. G. Semeniuk, Microflora, in: "Storage of Cereal Grains and their Products," J.A. Anderson and A.W. Alcock, ed., Amer. Soc. of Cereal Chemists, St Paul, 152 (1954).
70. R. Ulrich, "La conservation par le froid des denrées d'origine végétale," Baillère, Paris (1954).

71. D. Côme, and T. Tissaoui, Nouveautés dans l'amélioration de la germination des graines par le froid, Rev. Gén. du Froid 3:287 (1973)
72. E. James, Preservation of seed stocks, Adv. Agron. 19:87 (1967).
73. J. F. Harrington, Problems of seed storage, in: "Seed Ecology," W. Heydecker, ed., Butterworths, London (1973).
74. E. H. Roberts, The viability of rice seed in relation to temperature, moisture content, and gaseous environment, Ann. Bot. 25:381 (1961).
75. J. Touzard, Influences de diverses conditions constantes de température et d'humidité sur la longévité des graines de quelques especes cultivees, Proc. 15th Intern. Hort. Cong. (Nice) I:339 (1961).
76. E. H. Roberts, and Ellis R.H., Prediction of seed longevity at sub-zero temperatures and genetic resources conservation, Nature 268:431 (1977).
77. M. Yamamoto, The decreased germination of dry seeds soaked in water, Jap. J. Ecol. 5:74 (1955).
78. G. Weibull, The cold storage of vegetable seeds, further studies, Rep. 14th Intern. Hort. Cong.: 647 (1955).
79. L. V. Barton, Effects of temperature and moisture on viability of stored lettuce, onion, and tomato seeds, Contr. Boyce Thompson Inst. 23:285 (1966).
80. D. Côme, "Les obstacles à la germination," Masson, Paris (1970).
81. D. Côme, Rôle de l'eau, de l'oxygène et de la température dans la germination, in: "La germination des semences." R.Chaussat and Y. Le Deunff, ed., Gauthier-Villars, Paris (1975).
82. D. Côme, and T. Tissaoui, Interrelated effects of imbibition, temperature and oxygen on seed germination, in: "Seed Ecology," W. Heydecker, ed., Butterworths, London (1973).
83. M. Evenari, Light and seed dormancy, in: "Encyclopedia of Plant Physiology," W. Ruhland, ed., Springer-Verlag, Berlin, Heidelberg, New York, 15/2 (1965).
84. W. Vidaver, Light and seed germination, in: "The Physiology and Biochemistry of Seed Dormancy and Germination," A. A. Khan, ed., Elsevier/North-Holland Biochemical Press, Amsterdam (1977).
85. D. Côme, Acquisition de l'aptitude à germer, in: "La germination des semences," R. Chaussat and Y. Le Deunff, ed., Gauthier-Villars, Paris (1975).
86. D. Côme, Problems of embryonal dormancy as exemplified by apple embryo, Israel J. Bot. 29:145 (1980/81).
87. S. Lewak, and R.M. Rudnicki, After-ripening in cold-requiring seeds, in: "The Physiology and Biochemistry of Seed Dormancy and Germination," A.A. Kahn, ed., Elsevier/North-Holland Biomedical Press, Amsterdam, 1977.
88. C. Thévenot, Corrélations entre les cotylédons et l'axe de l'embryon de Pommier. Interprétation de la dormance, These Doct. Sci. Nat. Paris (1980).

89. F. Lona, L'influenza delle condizioni ambientali, durante l'embriogenesi, sulle caratteristiche del sem e della pianta che ne deriva, in: "Lavori di botanica, volume pubbl. in occasione del 70e genetliaco del Prof. G. Gola (Padova 1947)," Rosenberg and Sellier, Torino (1947).
90. L. V. Barton, and W. Crocker, "Twenty years of Seed Research at Boyce Thompson Institute for Plant Research," Faber and Faber, London (1948).
91. L. V. Barton, Seed dormancy: General survey of dormancy types in seeds, and dormancy imposed by external agents, in: "Encyclopedia of Plant Physiology," W. Ruhland, ed., Springer-Verlag, Berlin, Heidelberg, New York, 15/2 (1965).
92. W. Crocker, and L.V. Barton, "Physiology of seeds. An introduction to the experimental study of seed and germination problems," Chronica Botanica, Waltham, Mass., U.S.A.(1953).
93. M. Evenari, Seed germination, in: "Radiation biology," A. Hollaender, ed., McGraw Hill, New York, vol. 3 (1956).
94. P. Stokes, Temperature and seed dormancy, in: "Encyclopedia of Plant Physiology," W. Ruhland, ed., Springer-Verlag, Berlin, Heidelberg, New York, 15/2 (1965).
95. F. Corbineau, A. Sanchez, D. Côme, and R. Chaussat, La dormance du caryopse de Blé (Triticum aestivum L., var. Champlein) en relation avec la température et l'oxygène, C.R. Acad. Agric. France, in the press (1981).
96. F. Corbineau, and D. Côme, Quelques caractéristiques de la dormance du caryopse d'Orge (Hordeum vulgare L., variété Sonja), C.R. Acad. Sci. Paris D-290:547 (1980).
97. G. T. Harrington, Forcing the germination of freshly harvested wheat and other cereals, J. Agric. Res. 23:79 (1923).
98. I. Gadd, On methods for the elimination of seed dormancy in seed control work, Proc. Int. Seed. Test. Ass. 11:108 (1939).
99. T. A. Villiers, Seed dormancy, in: "Seed Biology," T.T. Kozlowski, ed., Academic Press, New York and London, II (1972).
100. D. Côme, L'inhibition de germination des graines de Pommier (Pirus malus L.) non dormantes, Rôle possible des phénols tégumentaires, Ann. Sci. Nat., Bot. VIII:371 (1967).
101. C. Thévenot, and D. Côme, Manifestations de la dormance embryonnaire du Pommier (Pirus malus L.) en fonction des conditions thermiques de germination, Physiol. Vég. 11:151 (1973).
102. T. Tissaoui, and D. Côme, Levée de dormance de l'embryon de Pommier (Pirus malus L.) en l'absence d'oxygène et de froid, Planta 111:315 (1973).
103. C. Thévenot, and D. Côme, Levée de dormance des embryons de Pommier (Pirus malus L.) par traitement des graines à des temperatures elevees, C.R. Acad. Sci. Paris D-287:1127 (1978).
104. M. Mes, Gibberellic acid and chilling requirements of peach seeds, Nature 184:2034 (1959).

105. C. Bulard, and J. Monin, Action de l'acide gibbérellique sur des embryons dormants d'Evonymus europaeus cultivés in vitro, C.R. Acad. Sci. Paris D-250:2922 (1960).
106. B. Frankland, and P.F. Wareing, Hormonal regulation of seed dormancy in Hazel (Corylus avellana L.) and Beech (Fagus sylvatica L.), J. Exp. Bot. 17:596 (1966).
107. J. W. Bradbeer, and N.J. Pinfield, Studies in seed dormancy. III. The effects of gibberellin on dormant seeds of Corylus avellana L., New-Phytol. 66:515 (1967).
108. M. Durand, Influence de quelques régulateurs de croissance sur la germination et la dormance de l'embryon de Pommier (Pirus malus L.), Thèse 3ème cycle, Paris (1974).
109. D. Côme, and A. Semadeni, Dégazage des enveloppes séminales lors de leur imbibition. III. Application á l'étude da la dureté des graines d'Hedysarum coronarium L., Physiol. Vég. 11:171 (1973).
110. L. V. Barton, Special studies on seed coat impermeability, Contr. Boyce Thompson Inst. 14:355 (1947).

PHYSIOLOGY AND STORAGE OF BULBS: CONCEPTS AND NATURE OF DORMANCY IN BULBS

M. Le Nard

Institut National de la Recherche Agronomique
Station d'Amélioration de la Pomme de terre et
des Plantes à bulbes, 29207 Landerneau, France

INTRODUCTION

Bulbous plants constitute a group of a very great diversity, as much in their geographical origin as in their biological characteristics. Moreover, the term "bulbous plants" is generally used in a very broad sense and includes plants whose vegetative multiplication organs are true bulbs, corms, rhizomes and even tubers and tuberous roots.

Nevertheless, these "bulbs" have a point in common : after being lifted they can withstand being stored for certain periods, the duration depending on the species. Storage presents at least two advantages. First of all, it allows the bulbs to be withdrawn from adverse climatic conditions and thus to survive. Next it enables them to be placed in such conditions that their behaviour can be controlled and so to be prepared for a much more definite use : bulb production, early or late flowering, etc...

Contrary to seeds, bulbs are organs generally rich in water and show an active life after harvest, when they are sometimes apparently at rest. Their physiological evolution continues therefore during storage and it will be easily understood that the environmental factors will affect this evolution.

Another characteristic of bulbs is that it is generally possible to lift them at different degrees of maturity, then to store them. On the experimental level this is very interesting, for it is thus possible to have at one's disposal organs which are physiologically different and then to study the effects of environmental factors on their evolution. It must however be pointed out that it is often difficult to characterize adequately or define the degree of maturity of a bulb. From a physiological point of view, the date of harvest of a bulb is not therefore always a good reference.

There are at least three environmental factors which can influence the evolution of bulbs during storage : humidity, composition of the atmosphere in the store, and temperature.

Control of humidity during storage can be a very important factor in the case of bulbs, like lily bulbs, which are subject to dessication. In other cases a too high humidity can be harmful for it allows the growth of roots during storage. Similarly, by encouraging the development of certain parasites, chiefly fungi, humidity can have an indirect influence on the physiology of bulbs.

The composition of the atmosphere in the store can also strongly influence the physiological evolution of bulbs. The role of a gas like ethylene has especially been studied. Its effects can be varied and sometimes very great on some species (1,2,3,4,5).

Temperature is certainly the factor whose effects have been studied the most. This probably results from the fact that it very soon appeared that its influence on the behaviour of bulbs could be very great and that it was possible to benefit from this.

Extensive research concerning the influence of temperature on the behaviour of a great number of bulbous species has been carried out in the Netherlands under the direction of Blaauw. The results of this work, which has largely contributed to the improvement of our knowledge of the biology of bulbous plants, has already been summed up by numerous authors (6,7,8,9,10,11,12,13,14,15).

Blaauw and his co-workers have been mainly interested in the influence of temperature on the differentiation of the flower bud and on its elongation. The object was to determine the bases necessary for the control of flowering and thus to produce flowers of a good commercial quality out of season and, if possible, all the year round.

The results of Blaauw and his collaborators being henceforth classic, the purpose of this review is not to make an additional resume but, by providing complementary information, to try and propose for a more global approach to the role of storage temperature in the physiology of bulbs. Indeed, it must be kept in mind that bulbs are complex organs which, in addition to one or several flower buds, possess vegetative buds, roots and storage tissues, and that storage conditions affect the behaviour of this whole. To illustrate this, the effects of storage temperature will be studied in the case of two species possessing a true bulb with annual replacement, the tulip and the bulbous iris, and a species possessing an annual corm, the gladiolus (Gladiolus grandiflorus hort.)

The complexity of bulbs must also be taken into account when tackling a subject like dormancy. This term, sometimes controverted, is also sometimes used without precise definition and applied indifferently to the bulbs or to the buds of bulbs. A brief discussion of the problem of dormancy in bulbous plants will be made, referring more particularly to the behaviour of the tulip, bulbous iris and gladiolus.

TULIP

The study of the development of the tulip bulb under temperate climates, has shown that the initiation of the buds begins early, at the end of winter beginning of spring, as the bulb enlarges,(16, 17). The most outer buds are initiated first, but their development is then inhibited by the dominance exerted by the apical bud. At harvest, when the aerial organs of the mother-plant are mature, end of spring beginning of summer, a bulb of sufficient size possesses an apical bud which has already differentiated two or three leaf primordia. This differentiation will continue during storage and will end in the differentiation of a flower. The innermost vegetative bud, situated at the base of the flower bud, is the last to be initiated, during the differentiation of the flower bud.

Few observations have concerned the roots, but they seem to be similarly initiated at the beginning of storage (18).

In the weeks that follow the end of its formation, a tulip bulb thus possesses all the organs which will give rise to the daughter-plant : rooting in autumn-winter, growth of the aerial parts, flowering and enlargement of the daughter-bulbs in spring.

The rate of bud and root differentiation and their subsequent evolution are greatly influenced by the storage temperature of bulbs.

Effects of Storage Temperature on Bud and Root Differentiation

By exposing bulbs, immediately after lifting, to different temperatures, from 1,5°C to 35°C, kept constant, Blaauw et al.,(11) showed that flower bud differentiation occurred most rapidly at 17°-20°C. The further the temperature moves away from this, above or below, the more the differentiation is slowed down.

However, if instead of being stored at a constant temperature, the bulbs are subjected, immediately after lifting, to changing temperatures, it is possible to show that a treatment of short duration at high temperature leads simultaneously to a subsequent more rapid flower bud differentiation and a more rapid root differentiation which allows earlier rooting (19). The influence of a high temperature treatment is quantitative, in the sense that extending it, within certain limits, allows a more and more rapid subsequent flower bud differentiation and earlier and earlier rooting (Table 1). It has been noted elsewhere that extending the high-temperature treatment led to an increase in the flower bud height at the end of its differentiation (20).

As a direct effect of the high temperature is to delay considerably the flower bud differentiation, a subsequent more rapid differentiation can only be explained as a residual effect of the high-temperature treatment. This induces the possibility of a more rapid flower bud and root differentiation, but the induction can only be expressed when the bulbs are then placed at mean temperatures which allow the actual differentiation.

Besides, when the duration of a high-temperature treatment is extended beyond certain limits, the completion of the flower bud can be delayed because the acceleration caused by extending this treatment no longer allows compensation for the delay caused by moving the bulbs somewhat later into a temperature favorable to the manifestation of the process induced (Table 1 and Table 2). Similarly, it is obvious that if a high-temperature treatment is applied when the flower bud differentiation is already well-advanced, no beneficial effect on the earliness of flower bud differentiation will then be observed.

If flower bud and root differentiation are both promoted by a high-temperature treatment, it seems however that these two processes could perhaps have slightly different requirements, and that rapid rooting would require longer periods of high temperature than are necessary to obtain rapid flower bud differentiation (20). Thus, the results of Table 2 show that a treatment of several weeks at 30°C, which lead to a delayed flower bud differentiation nevertheless allows an earlier rooting even if it is applied fairly late. On the contrary, a treatment of the same duration at 9°C, applied at the beginning of storage, considerably delays rooting while it allows earlier flower differentiation than the 30°C treatment.

The active differentiation of the axillary buds only starts after completion of the flower bud. By allowing an earlier flower bud differentiation, a high-temperature treatment, applied immediately after lifting, also allows an earlier subsequent scale primordia differentiation in the axillary buds.

In summary, it appears that the high temperature induces the possibility of a more rapid organogenesis, which however will not take place unless the bulbs are subsequently placed in a temperature

Table 1. Effects of duration of a 30°C storage of tulip bulbs on subsequent flower bud differentiation and rooting; cv. "Paul Richter" (adapted from Le Nard, (19)).
Control : bulbs placed, immediately after harvest, at 20°C for flower bud differentiation or planted at 15°C.
Stage G : completion of flower differentiation (Beijer, (21)).

	Flower bud differentiation at 20°C		Rooting, at 15°C	
	Date of stade G	Days at 20°C	Date 75 % rooting	Days after planting
Control	6 Aug.	49	12 Oct.	116
1 week 30°C ...	26 Jul.	31	6 Oct.	103
2 weeks 30°C ..	26 Jul.	24	24 Sept.	84
3 weeks 30°C ..	2 Aug.	24	14 Sept.	67

which allows the actual organogenesis to proceed.

The results reported in Table 2, confirming other observations (20,22) show that there does not necessarily exist a correlation between the earliness of flower bud differentiation and the earliness of rooting. The observation of the degree of flower bud differentiation on its own, therefore does not make it possible to characterize the physiiological state of a bulb, in the sense that its subsequent behaviour cannot be accurately predicted.

When small-sized bulbs are stored, immediately after lifting, for several weeks at high temperature, there has sometimes been observed an increase in the number of bulbs which show a floral initiation (23). Therefore, the high temperature would not only affect the rate of flower bud differentiation but would also influence floral induction.

Storage Temperature and Induction of Growth Processes

Storage at high temperature, > 20°C. At 20°C, after its completion, the flower bud can immediately begin to grow. This growth however is very slow and limited, the tip of the bud only growing at maximum a few centimeters past the top of the bulb. When storage

Table 2. Effects of a 4-week storage at 30°C, 20°C or 9°C, applied at two different dates, on subsequent flower bud differentiation and rooting at 15°C; cv. "Paul Richter." Treatments applied on 2 July : bulbs previously stored at 20°C.
Stages of flower bud differentiation, see Beijer, (21).

		4 w. 30°C	4 w. 20°C	4 w. 9°C
Treatments applied on 14 Jun., i.e. one week after harvest	Stage of flower bud differentiation at the end of treatment	I	P_2	II
	Date stage G	5 Aug.	23 Jul.	2 Aug.
	Date 75 % rooting	27 Sept.	14 Oct.	10 Nov.
	Days for rooting	77	94	121
Treatments applied on 2 Jul. (Stage II of flower bud differentiation)	Stage of flower bud differentiation at the end of treatment	II	G^+	A_2
	Date stage G	19 Aug.	21 Jul.	5 Aug.
	Date 75 % rooting	28 Sept.	6 Oct.	16 Oct.
	Days for rooting	60	68	78

continues beyond 6-7 months, dissections make it possible to establish that the flower is dry, although the exterior aspect of the bulb is quite normal (See Fig. 3). The flower bud then progressively dries up entirely.

The root primordia also lengthen during storage at 20°C and the rooting plate becomes more and more protuberant. However, if the bulbs are planted at a temperature maintained above or at 20°C, rooting, i.e growth of roots out of the plate, does not take place and elongation of the flower bud remains restricted.

During storage at 20°C, the number of scale primordia increases in the axillary buds. When storage is of long duration, some of these scale primordia tend to change into leaves. Sometimes it is even possible to observe a floral differentiation from an axillary bud, usually the innermost one (24).

In the case of storage, even of very long duration, at a temperature above or at 20°C, the axillary buds do not change into daughter-bulbs, that is, bulbing does not take place.

If bulbs showing a completely differentiated flower bud are then exposed to a high temperature, 30°C - 35°C, drying of the flower can be rapidly observed (25,26).

Storage at low temperature. Blaauw et al.,(11) have clearly shown that a rapid extension growth of the flower bud can only be observed after a bulb has been exposed to low temperatures. Other results have shown that the low temperature also induces bulbing (24,27), and that its influence varies according to the level of the temperature, the length of its duration and the moment of its application.

(1) Influence of the level of the low temperature. When bulbs possessing a completely differentiated flower bud are stored at different temperatures, it can be observed, after planting, that plant growth is more rapid when the storage temperature is lower : rooting, elongation of the aerial organs and daughter-bulb enlargement are more rapid (Fig. 1). The maturity of the plants is also advanced and the duration of the growth period is shorter.

Thus, low temperature simultaneously induces a stimulation of the elongation and bulbing. The induction is all the stronger, i.e elongation and bulbing can subsequently proceed faster, when storage temperature is lower, at least as far as 2°C. Other observations suggest that a slightly negative temperature, - 1°C, would be as effective as 5°C, (28).

Earlier flowering corresponds with more rapid elongation and the production of longer flower shoots at anthesis and at the end of growth if the bulbs are planted at a mean temperature and if the environmental conditions then remain constant.

(2) Influence of the duration of application of low temperature. Extending the duration of low-temperature storage leads, first, to a more rapid growth, i.e. elongation plus bulbing, but shorter lasting (Fig. 2).

Later, when the duration of storage reaches a certain length which varies according to the level of the low temperature and the cultivar, the subsequent plant growth is very rapid but reduced (Fig. 2; sample 25 w. 10°C). The plants are then short and slender, and show blasted flowers. Finally, in the case of long duration storage at low temperature, dissections of bulbs make it possible to observe that bulbing has already begun during storage, that is the enlargement of the daughter-bulbs has started (Fig. 3). If such bulbs are then planted, neither rooting nor elongation of

Fig. 1. Tulip : effect of bulb storage on subsequent elongation and bulbing; cv. "Paul Richter" (from Le Nard and Cohat, (24)); left to right : bulbs stored for 10 weeks at 2°C - 3°C (S_1F_2), 10 weeks at 10°C (S_1M_2) or 15 weeks at 20°C (S_3) before planting at 14°-16°C.

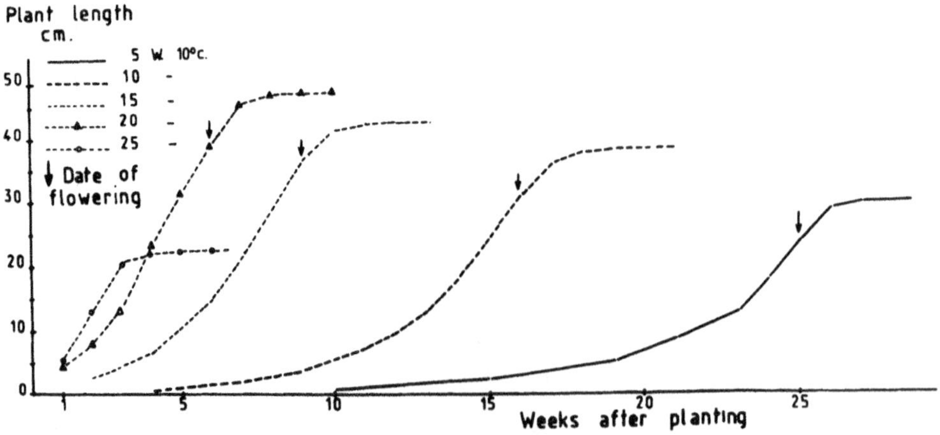

Fig. 2. Tulip : Growth of flower shoots produced by bulbs stored for 5,10,15,20 or 25 weeks at 10°C, before planting at 14°-16°C (from Le Nard and Cohat, (24)).

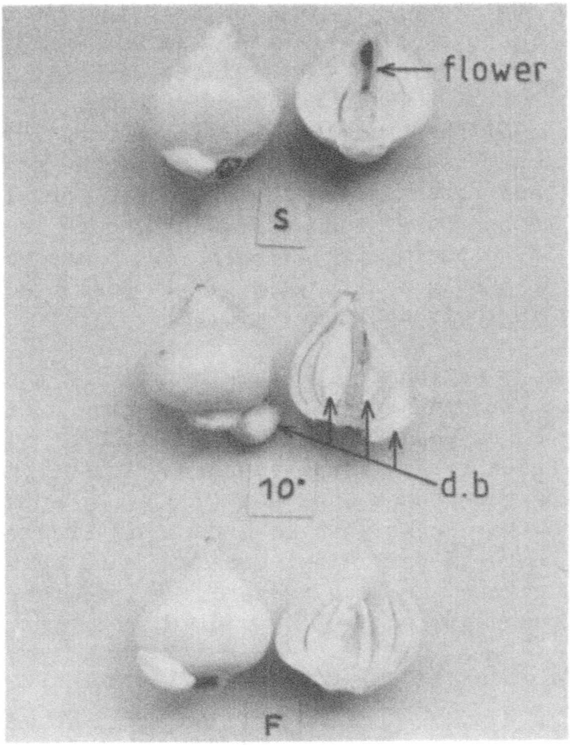

Fig. 3. Tulip : Aspect of bulbs stored, immediately after lifting, for 30 weeks at 18°C-20°C (S), 10°C or 2°-3°C (F); cv. "Pax" (from Le Nard and Cohat, (24)). d.b = daughter-bulbs

aerial organs are observed, and the only growth process to take place is bulbing.

Bulbing induction is all the stronger when the storage temperature is lower. However, bulbing during storage is noticeable earlier at 10°C than at 2°C (Fig. 3). This can be explained by the fact that a temperature of 10°C is more favorable than 2°C to the manifestation of growth processes.

These observations show that when storage at low temperature is extended beyond a certain period, bulbing becomes the predominant growth process and this results in a greater or lesser reduction in elongation (roots and aerial organs).

If bulbs formed during storage at low temperature, 4°-5°C, are in turn maintained at low temperature for a long period, they also produce daughter-bulbs during storage. In this case, the apical buds do not differentiate any aerial organs and directly give rise to daughter-bulbs which only possess one or two scales (Fig. 4). Thus, it appears that if the complete cycle of a tulip bulb proceeds at low temperature, organogenesis in the apical bud is very limited but, even in this case, bulbing takes place.

(3) Influence of the moment of application of low temperature. In natural conditions, because of the rhythm of the seasons, tulip bulbs possess well-differentiated buds and roots when they are subjected to low temperatures and subsequent growth is "normal." On the contrary, if, experimentally, bulbs are placed at low temperature at other times in their cycle, it is possible to show a certain diversity in their behaviour (19,29).

By lifting bulbs periodically, then by storing them immediately at low temperature, it can be observed that during a first phase, which practically corresponds to the whole period of bulb enlargement, the bulbs do not react to low temperature. A low-temperature treatment applied during this phase does not induce any stimulation of the subsequent growth : bulbs planted after this treatment give rise to plants whose growth is very slow, comparable to that of plants produced by bulbs planted at mean temperature immediately after harvest (Fig. 5; bulbs placed at 2°C, on 14 March, 11 April and 9 May).

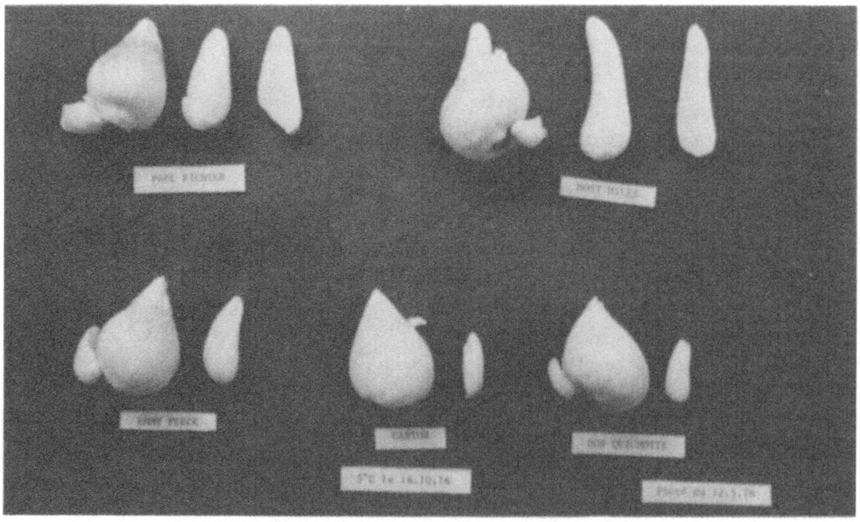

Fig. 4. Tulip : Apical bulbing in bulbs formed and maintained at low temperature, 4°-5°C : grandmother-bulbs were placed at 4°-5°C on 16 Oct. 1976; the mother-bulbs, shown on this photograph, were harvested and, in turn, kept at 4°-5°C; one or two bulbs per cultivar have been dissected to show the apical daughter-bulbs; photographed on 12 May 1978.

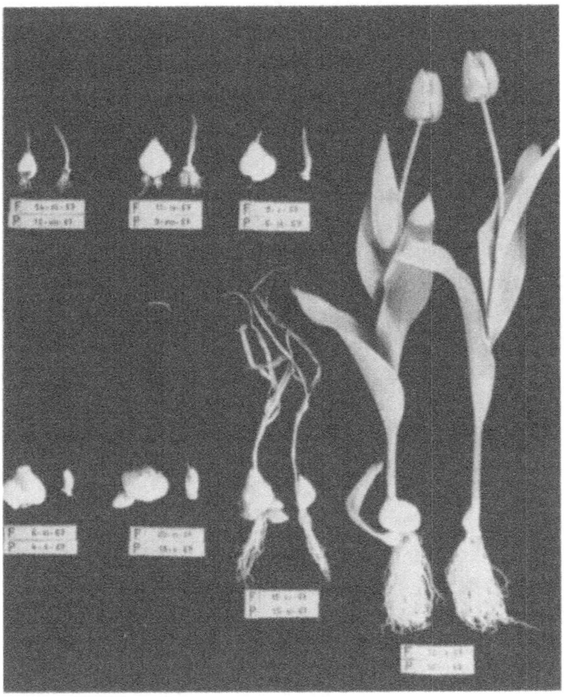

Fig. 5. Tulip : Effect of the date of treatment at 2°-3°C for 120 days on the subsequent behaviour of the bulb; cv, "Paul Richter" (from Le Nard, (29)); F : date placed in 2°-3°C; P : date planted at 14°-16°C; top : left to right; F : 14 March, 11 April, 9 May; bottom : left to right; F : 6 June, 20 June, 18 July, 12 September.

From a certain stage, which we have called "physiological maturity," (19), the bulb becomes sensitive to the action of low temperature which then induces growth of the daughter-organs. The growth process, or processes, which can then take place will depend on the degree of bud and root differentiation at the time the low-temperature treatment has been applied.

When the apical bud only possesses one or two leaf primordia, these primordia change into scales and give rise to an apical daughter-bulb (Fig. 5 : bulbs placed at 2°C one 6 June). When the low-temperature treatment is applied a little later, the axillary buds undergo bulbing, but the possibilities of elongation of the flower bud depend on its degree of differentiation. If the differentiation of the flower bud is incomplete, and although it can later take place, the flower bud cannot elongate and dries up (Fig. 5; bulbs placed on 2°C on 20 June) or it can show some elongation but the plants are slender and present blasted flowers (Fig. 5; bulbs placed at 2°C on 18 July). Finally, if the low-temperature treatment is applied even later to bulbs possessing a flower bud and roots which are well differentiated, the subsequent growth is "normal" and comprises elongation, roots and aerial organs, plus bulbing (Fig. 5; bulbs placed at 2°C on 12 September). All the transition stages exist between bulbing of the apical bud and "normal" growth.

All these observations together lead to the conclusion that the behaviour of a tulip bulb after a low-temperature storage depends on the one hand on its physiological state and on the other on the degree of bud and root differentiation. The physiological state determines the possibility of induction of growth processes by the low temperature, while the degree of bud and root differentiation determines the growth process, or processes, which can take place; bulbing alone or elongation and bulbing.

Bulb evolution towards "physiological maturity" is accelerated by high temperatures and is very slow at low temperature (19,29). Having reached this stage, the bulb becomes progressively more and more sensitive to the action of low temperatures, and a low-temperature treatment of fixed duration leads to a bulbing induction all the stronger when it is applied later to bulbs previously stored at high temperature (22) (Fig. 6)

The presence of buds which are sufficiently differentiated constitutes a preliminary to all growth possibilities and it could be supposed that before the "physiological maturity" stage a low-temperature treatment would not induce any growth process solely because the bulb does not possess sufficiently differentiated buds. It does not seem that this is the cause, for it has been possible to show that the physiological evolution of a bulb towards a greater sensitivity to low temperature can proceed in the absence of any visible bud differentiation and that bulbing induction due to a low-temperature treatment can be very strong even if the differentiation of the apical bud is not far advanced at the time the treatment is applied, (19,22). Besides, it has already been pointed out that in the case of bulbs formed at low temperature, (see (2)) bulbing can take place even if the apical bud has only produced one or two primordia. In fact, it appears, and this has been confirmed by other results, (20), that the degree of apical bud differentiation *per se* does not adequately define the physiological state of a tulip bulb.

When a bulb has reached its "physiological maturity" and becomes sensitive to the action of low temperature, it is sometimes possible to observe that bulbing of the apical bud stops shortly after beginning. In this case, which we have called "incomplete bulbing," the first primordium produced by the apical bud thickens

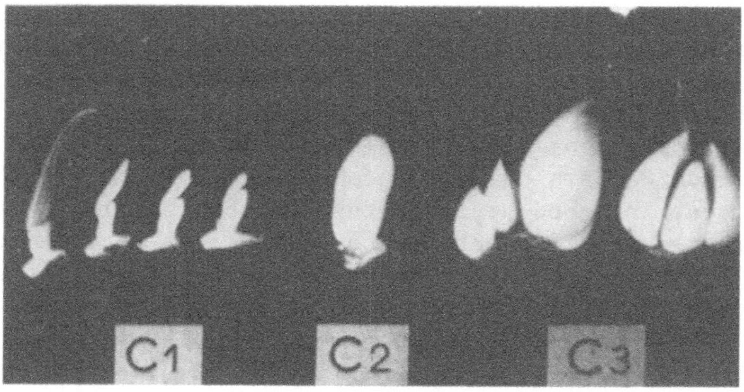

Fig. 6. Tulip : Importance of bulbing according to the treatment applied to the bulbs before a 15-week storage at 2°-3°C; cv. "Paul Richter" (from Le Nard,(19)).
Bulbs lifted on 30 April and then placed for 1 (C_1, 2 (C_2) or 3 weeks at 30°C (C_3) before cold treatment; C_1 = incomplete bulbing (only one scale primordia has started to thicken) and differentiation of aerial organs, a flower bud (left) or one leaf; C_2 = apical bulbing; C_3 = bulbing of different buds, apical and axillary (left), axillary only, the flower bud having dried up (right).

and turns into a fleshy scale, but this bulbing stops and the apical meristem of the bud produces again aerial organs, (19). This organogenesis can go as far as the production of a flower if the "incomplete bulbing" is not too great (Fig. 6).

Storage at High Temperature Following Storage at Low Temperature

The effect of a low-temperature storage, applied immediately after bulb lifting, can sometimes be nullified if the bulbs are then transferred at high temperature, (24,27,29). Recent observations, (22,30) have made it possible to determine more precisely the effects of a high temperature applied after low-temperature storage.

Studies on bulbing have shown that in fact, a high temperature has two opposed effects : it promotes the manifestation of bulbing induced by low temperature but at the same time it nullifies progressively this induction. Thus, the behaviour of bulbs first stored at low temperature then moved to high temperature, will depend on the one hand on the strength of bulbing induction and on the other on the duration of the high-temperature treatment. A weak induction can be easily nullified : this is the case of bulbs exposed to low temperature straight after lifting, that is, at a stage where they are not very sensitive to the action of low temperature. On the countrary, if bulbs are subjected later to low-temperature treatments of fairly long durations, bulbing induction is strong and it can no longer be completely nullified. Bulbing can then rapidly proceed at high temperature. Between the two extreme situations, weak induction completely nullified or rapid and complete bulbing, all the transition stages can be observed. In these intermediate situations, which we have called "incomplete bulbing," (19,29), the primordia present in a bud begin to thicken, but bulbing stops and new aerial organs are differentiated by the apical meristem of the bud. This differentiation can even go as far as flower differentiation in the case where "incomplete bulbing" is not very great and concerns the apical bud (27). When bulbing induction is completely nullified, the bulb behaves practically as if it had never previously been exposed to low temperature and organogenesis can proceed in the normal way.

The influence of low temperature on the stimulation of subsequent elongation can also be nullified by a high-temperature treatment, insofar as the previous low-temperature treatment is applied fairly soon after bulb lifting and it is of fairly short duration (24). On the contrary, if a low-temperature treatment is applied to bulbs possessing well-differentiated flower buds and roots, subsequent storage of short duration at high temperature, 25°-27°C, leads to a clear stimulation of elongation, that is, it leads to the same final result as that produced by extending the low-temperature treatment (30). However, looking at the plant growth curves suggests that the mechanisms of action of the two temperatures would not be the same.

When storage at 25°-27°C, or at a higher temperature, is extended, the flower is destroyed.

The effects of a high-temperature storage therefore vary considerably according to the moment it is applied and/or according to the previous storage conditions of bulbs.

Conclusions

The observations reported here, and summed up in Figure 7 show the importance of storage temperature on the behaviour of a tulip bulb. They indicate that:

A high temperature applied at the beginning of storage, or one which the bulbs have already been subjected to in the ground, induces a more rapid subsequent organogenesis when the bulbs are then exposed to a temperature which allows the manifestation of this induction, that is, a mean temperature. Simultaneously, high temperature speeds up the evolution of bulbs towards their "physiological maturity" and a greater sensitivity to low temperatures. The behaviour of bulbs formed at low-temperature and that of bulbs early-stored at low temperature suggests that bud differentiation would be restricted if the bulbs are not subjected to a high temperature for a sufficient duration.

The behaviour of a bulb following storage at low temperature depends on the one hand on its physiological state which determines the possibility of growth induction, and on the other it depends on the degree of organogenesis which determines the growth process, or processes, which can take place, bulbing only or elongation plus bulbing, these two processes being induced simultaneously by a low-temperature treatment.

It would seem necessary here to draw attention to the fact that only the use of relatively extreme temperatures, 30°C and 2°C, allows the induction of a process, organogenesis or growth, to separate from its manifestation. Indeed, at a mean temperature the induction of a process and its manifestation can both be expressed, as the cycle of the tulip can proceed slowly, but completely, at a temperature in the order of 15°-17°C for example.

A great deal of work has been devoted to the influence of low temperature on the extension growth of the flower but while neglecting its influence on bulbing. Now, the observations reported above show that bulbing appears to be a fundamental growth process.

"Incomplete bulbing" of the apical bud observed following an early storage at low temperature (see Fig. 6) and in the course of study of the nullification of the influence of a low-temperature treatment by a subsequent storage at high temperature, shows that in a bud, the start of bulbing corresponds to an interruption of organogenesis (19,29). Where the apical bud itself is concerned this can lead to an apical bulbing, while where the axillary buds are concerned this results in the production of daughter-bulbs which will possess fewer scales the earlier the bulbs have been exposed to low temperature and/or the more rapidly bulbing has been able to start (24). Inversely, if bulbs are stored at a temperature which does not allow the induction of growth processes, > 20°C, or after bulbing induction has been nullified, organogenesis can continue longer and the axillary buds can produce aerial organs and even sometimes a flower.

These results lead us to distinguish two main phases in the cycle of the tulip : a phase of organogenesis during which meristematic activity is great, and a phase of growth during which meristematic activity is considerably reduced. As far as the vegetative buds are concerned, we have seen that bulbing corresponds to an interruption of differentiation of new organs by their apical meristems. Organogensis resumes during the daughter-bulb formation but it only becomes active after the end of bulb enlargement. In the case of the flower bud, cell divisions are especially active before the extension growth, this chiefly resulting from the elongation of pre-existing cells (18,31).

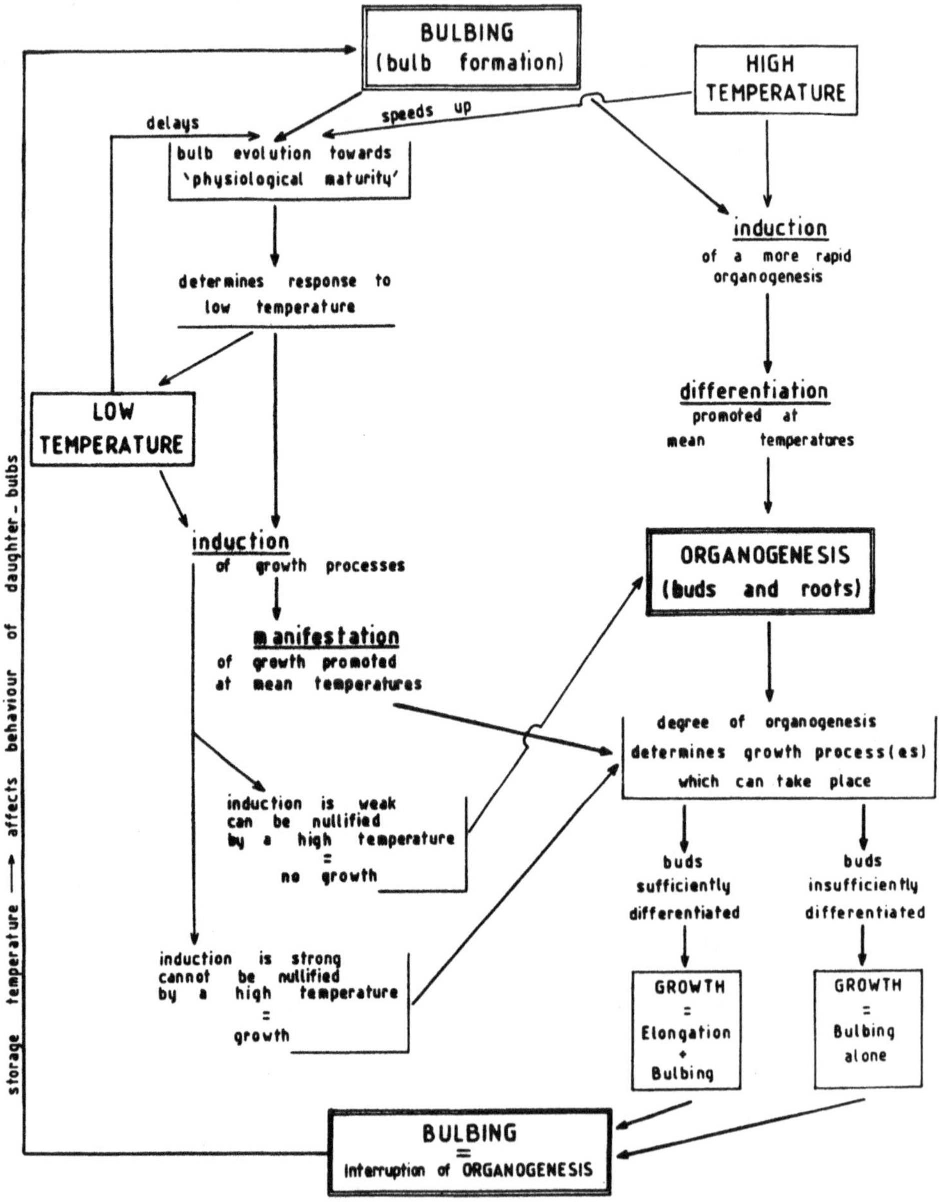

Fig. 7. Tulip: Role of storage temperature in the cycle of the tulip.

The importance of the physiological state of a bulb has been noted above, for that determines its reaction to low temperature. As long as the bulb has not reached its "physiological maturity," a low temperature only delays the physiological evolution of the bulb and does not induce any growth process. It can be supposed that the bulb is then in a state such that at the end of a low-temperature treatment its reserves could not be mobilized and, consequently, no growth would be possible. On the contrary, when the stage of "physiological maturity" is reached, low temperatures induce growth processes and the bulb reserve mobilization, necessary to ensure daughter-organ growth, is then possible. Mobilization of the reserves is all the more rapid the later low-temperature treatment is applied to bulbs previously stored at high temperature (20,22).

It has been shown repeatedly that the physiological state of a bulb cannot be characterized by the degree of flower bud differentiation, and that the physiological evolution of a bulb can even proceed in the absence of any visible organogenesis in the apical bud. This has an important consequence in practice where the degree of flower bud differentiation is used as morphological point of reference during the preparation of bulbs which will be forced. It appears in fact that bulbs showing flower buds at the same degree of differentiation will not necessarily be in the same physiological state and, consequently, their subsequent behaviour can be different (20).

On the practical level, the observations reported here, show that all storage of tulip bulbs at a positive temperature, and even slightly below 0°C (unpublished results), is limited in time. A fairly long storage at low temperature, 1°-2°C, can only be planned in the case of bulbs which have not reached their "physiological maturity," (29).

A great deal of work has been carried out to determine the temperature treatments which should be applied in order to obtain winter flowering, and a very rich literature exists on this subject (see de Hertogh (32)). It must always be kept in mind that flowering of good quality can only be obtained if the two growth processes, elongation and bulbing, can proceed harmoniously. If bulbing becomes predominant (the case of bulbs placed too early or stored for too long at low temperature, and bulbs whose planting or rooting is delayed after the end of low-temperature treatment), the plants are poor and, very often, show blasted flowers (33).

Bulbing in the absence of elongation, obtained in the case of a long storage at low temperature, cannot usually be observed in natural conditions where the environmental factors allow rooting and growth of the aerial organs.

BULBOUS IRIS

Temperature Requirements for Flower Differentiation

Bulbous iris of the dutch iris group (I. x *hollandica*) have a cycle very similar to that of the tulip. In temperate climates of Western Europe, bulbs planted in autumn root rapidly, the growth of the aerial organs begins in winter and continues into spring, and daughter-bulb enlargement ends at the beginning of summer.

However, there is a great difference between tulip and bulbous iris. In the case of the tulip, flower differentiation occurs in the weeks following bulb lifting, at rather high temperatures. In the case of bulbous iris, flower differentiation takes place after bulb planting, during autumn and winter, that is, at a rather low temperature (34).

At lifting time, the apical bud of an iris bulb, usually possesses 3 leaf primordia. The subsequent evolution of this bud depends on several factors. It depends first of all on the size of the bulb, that is, on its weight. There is a critical size, variable according to the cultivars, below which flower initiation cannot take place. The apical bud in that case, will only produce, besides the leaf primordia, scale primordia and will give rise to an apical daughter-bulb.

Temperature also greatly affects apical bud evolution. A high temperature, above 20°C, favors subsequent flower initiation. However, the actual flower differentiation will only proceed if the bulbs are subsequently moved to a mean or low temperature, the most favorable temperature for flower differentiation being between 9° and 15°C (11).

A high temperature therefore allows iris bulbs, of adequate size, to acquire the capacity to differentiate a flower, capacity which can only be expressed when temperature becomes favorable. The duration of high-temperature storage necessary for the acquisition of this capacity is all the longer the nearer the bulb is to the critical size.

Iris bulbs can be stored for long periods at high temperature, 25°C-30°C. In these conditions, flower differentiation cannot proceed, but the bulb retains its flowering capacity and it is possible to produce flowers all the year round (10,35,36). After planting, the development of the flower shoot will depend on the environmental factors (11,37,38,39,40,41).

The different temperature treatments applied to iris bulbs after lifting therefore affect greatly apical bud differentiation. However, storage temperatures do not only affect the possibilities of differentiation of this bud but also influence the whole bulb evolution.

Storage Temperature and Subsequent Bulb Behaviour

Influence of low temperature. A low temperature treatment, applied to bulbs having acquired the capacity to initiate a flower, induces simultaneously a stimulation of the elongation of the aerial organs and bulbing (42,43). However manifestation of these effects of low temperature depends on the one hand on the length of its application and on the other on the time it is applied.

(1) Influence of the duration of low-temperature treatment. In the case of bulbs possessing the capacity to initiate a flower, extending the low-temperature storage leads first to a more and more rapid elongation of the flower shoot (earlier flowering) and a more and more rapid daughter-bulb enlargement (Fig. 8; C_7,C_6M_1, C_5M_2, C_4M_3 and some plants of C_3M_4).

Later, when low-temperature storage is extended beyond a period which varies according to the cultivar and the temperature level, the extension growth of the flower bud is still very rapid but very reduced and the flower, although present, sometimes aborts and dries up (Fig. 8; C_3M_4). Bulbing proceeds still more and more rapidly.

Finally, if storage at low temperature lasts very long, bulbing can begin during storage. If the bulbs are then planted, elongation of the daughter-plants is reduced or nil (Fig. 8; C_2M_5).

In the case of the bulbous iris, as in that of the tulip, it therefore appears that elongation will be limited or nil if bulbing becomes the predominant growth process.

When elongation of the aerial organs is rapid, as a result of an extension of the duration of low-temperature storage, a decrease in the leaf length, and sometimes also in the flower shoot length, can be observed (37,42,44,45,46).

(2) Influence of the time at which low temperature is applied. Kamerbeek (37), has shown that early application of low temperature limits the differentiation of the apical bud which then produces only 2 or 3 leaves, and scales. By applying low-temperature treatments earlier, or of longer duration, it is even possible to obtain an apical daughter-bulb in the absence of production of any aerial organ (42,43). In this case the leaf primordia of the apical bud change into fleshy scales (Fig. 8; M_7). An early apical bulbing then results in a limited organogenesis.

The time at which low temperature treatment is applied does not

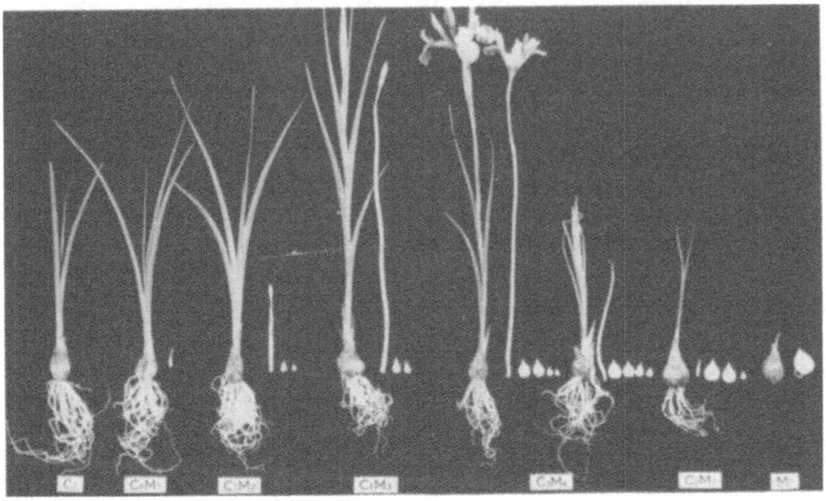

Fig. 8. Bulbous iris : Flower shoot elongation and bulbing following storage at 10°C applied after, or in the absence of, a previous treatment at 30°C; cv. "Wedgwood" (from Le Nard, (42)).
C = 30°C, M = 10°C; left to right : C_7 = 28 weeks at 30°C; C_6M_1 = 24 w. 30°C + 4 w. 10°C ; C_5M_2= 20 w. 30°C + 8 w. 10°C; C_4M_3= 16 w. 30°C + 12 w. 10°C ; C_3M_4= 12 w. 30°C + 16 w. 10°C (two types of plants) ; C_2M_5= 8 w. 30° C + 20 w. 20°C; M_7 = 28 w. 10°C.

only influence the possibilities of apical bud differentiation but has other consequences on the subsequent bulb behaviour.

Thus, bulbing begins earlier and is then completed more rapidly when the bulbs have been stored at higher temperature before exposure to low temperature (47). This result suggests that a high temperature allows a more rapid subsequent bulb reserve mobilization after a low-temperature treatment.

When a similar low-temperature treatment is applied to bulbs previously stored for fairly short periods at different temperatures, the plants issued from the bulbs stored for the longest time at high temperature produce fewer leaves, flower earlier and better, i.e. flowering percentage is higher (47). These observations, which agree with those of Kimura and Stuart (45) and Uhring (48) suggest that apical bud differentiation at low or mean temperature proceeds more quickly when previous storage at high temperature is extended. Corresponding on the one hand to an earlier flower differentiation would be a decrease in the leaf number and on the other an earlier shoot elongation. A better flowering could be explained, at least in part, by the favorable influence exerted by the high temperature on subsequent bulb reserve mobilization.

All these observations show that the behaviour of an iris bulb, after a low-temperature treatment depends on its physiological state which determines the bulb's response to low temperature, and on the possibilities of differentiation of the apical bud which remains vegetative or becomes floral.

Storage at high-temperature following storage at low temperature. In certain conditions a high-temperature treatment can nullify the influence of a previous storage at low temperature (42,43).

When a low temperature is applied immediately after bulb lifting and for a fairly short period, its influence can then be completely nullified by a subsequent storage at high temperature. In this case bulbing does not take place and organogenesis in the apical bud can then continue normally if the bulb is exposed at favorable temperatures. When low-temperature storage is of long duration, a subsequent high-temperature treatment can no longer nullify its influence, bulbing occurs and the mother-bulb scales shrivel.

Between these two situations, bulbing induction completely nullified or rapid bulbing at high temperature, there are intermediate situations. In these cases, bulbing induced by low temperature can start but stops before complete shrivelling of mother-bulb scales, and, later, a new differentiation of aerial organs is observed in the bud which had begun to form a bulb (42).

Conclusions

In an iris bulb, high temperatures and low temperatures induce processes which occur successively in the buds in the course of the bulb cycle. High temperatures induce an organogenesis and simultaneously promote the bulb's physiological evolution towards a greater sensitivity to low temperature. Organogenesis induced by a high temperature only proceeds rapidly if the bulbs are subsequently exposed to a medium to low temperature, 9° to 15°C.

Low temperatures, necessary to the stimulation of elongation, also induce bulbing. The start of bulbing seems to lead to an interruption of organogenesis in the buds. Thus, following the degree of apical bud differentiation at the time bulbing starts, so bulbing will be apical or axillary. The importance of the physiological state of a bulb on its behaviour at the end of a low-temperature treatment has been demonstrated. High temperatures, which allow a more rapid subsequent bulb reserve mobilization, accelerate bulb evolution towards a greater sensitivity to low temperature. This evolution, as in the case of a tulip bulb, can take place in the absence of any visible organogenesis. It has appeared besides, that a high-temperature treatment of sufficient duration to allow subsequent flower initiation, does not necessarily allow satisfactory growth of the flower shoot (47).

All these results indicate that the effects of storage temperature on the behaviour of an iris bulb appear to be very similar to those observed in the case of a tulip bulb.

Study of the effects of temperature on the behaviour of iris bulbs furnishes the basis for storage methods used in practice. Use of the sequence high temperature-low temperature makes it possible to control flower initiation and the choice of the length and level of temperatures to be applied, depends on what result is desired, early or late flowering, and on the type of plant looked for (33,36, 37,49,50,51).

The sequence high temperature-low temperature can also be used to prevent flower initiation. This objective, especially desired in the case of seed bulbs, can sometimes be attained if the low temperature applied is considerably lower than that which promotes flower differentiation (23,33). This result could be explained by the fact that a direct effect of low temperature is to slow down organogensis and, simultaneously, to induce bulbing. This could then start rapidly after planting and would limit the differentiation of the apical bud. However, the low-temperature treatment must not be applied too early and/or for a too long period, for if bulbing becomes predominant, plant growth will be poor.

One of the greatest impediments for a rational application of these different temperature treatments, is the absence of criteria enabling the bulb's physiological state to be defined. In the course of storage there is no morphological point of reference comparable to that which is used in the case of a tulip bulb, for in the iris bulb, flower differentiation generally occurs after planting. It is therefore not always easy to predict accurately the behaviour of iris bulbs nor to reproduce certain results.

GLADIOLUS

In the temperate climates, the cycle of the gladiolus differs from that of the tulip and the iris. Corms are planted in spring. Flower differentiation occurs after planting and growth of the aerial organs proceeds in spring and summer. The new corms are lifted in autumn and are at rest in winter.

The buds which will produce the aerial organs of the daughter-plant are initiated early, during corm formation, and at lifting time the apical bud has already produced several leaf primorida (52).

The rate of subsequent apical bud differentiation and the daughter-plant growth can be greatly affected by corm storage temperature.

Influence of Storage Temperature on Sprouting, Growth of the Aerial Organs and Formation of New Corms

Immediately after lifting, the corms and cormels cannot usually show any growth, even if they are placed in favorable conditions. Earliness of bud and root growth depends first of all on the cultivar and on the environmental conditions during the formation of the corms : corms produced at high temperature and/or at high light intensity sprout later than corms produced at mean temperature and/ or at weak light intensity (53,54,55,56,57).

Storage temperature, and sometimes moisture conditions, also greatly affect sprouting of corms and cormels (53,54,58,59,60,61, 62,63,64). These different studies indicate that sprouting results from the successive realization of two processes : first the acquisition of growth capacity (dormancy release), then the differentiation and growth of buds and roots.

The acquisition of growth capacity is favored by low temperature, 5° to 10°C. The length of the low-temperature treatment needed to acquire this capacity depends on the cultivar and on the physiological state of the corms at lifting. Cormels require longer low-temperature storage than corms.

At low temperature, organogenesis is practically stopped. It occurs later when the corms or cormels are placed at high temperature, 20°-25°C. To obtain homogeneous early sprouting it is therefore necessary to store the corms first at low temperature then at high temperature.

High temperature storage does not only promote organogenesis but also affects subsequent plant growth : leaf growth, flowering, cormel formation and new corm enlargement are speeded up (58,63,64, 65,66,67,68,69). Extending the high-temperature storage beyond a certain duration can result in the production of less vigorous plants : leaf and flower number per plant and the final plant height are decreased (67).

In the temperate climates of Western Europe, storage at high temperature has a favorable effect not only on the earliness of new corm formation but also on their final size. Daughter-corms produced by corms stored at high temperature have the peculiarity of possessing more tuberized internodes and, for a given size of corm, therefore have a more globular appearance (64,67). (Fig. 9).

As storage at high temperature leads to earlier sprouting, it could be supposed that this would explain a more rapid subsequent plant growth. It has been possible to show that this is not the case and that differences between growth in plants produced by corms stored at different temperatures could originate from the fact that the corms show different physiological states at planting.

Influence of Storage Temperature on the Physiological Evolution of Corms

The physiological evolution of corms has been studied by storing them for increasing periods at different temperatures, then by planting them in the dark, at 20°C (67,68). Following a high-temperature storage applied to corms which have already acquired growth capacity, the weight of sprouts produced in the darkness within a certain time increases, rapidly reaches a maximum, holds its own for a while, then progressively decreases (Fig. 10). Simultaneously, the new corms are formed more and more rapidly after planting. When high-temperature storage is extended beyond a certain period, daughter-corm formation can even begin during storage. In this case, after planting, the growth of leaves and roots is very poor or nil.

After storage at low temperature, the growth rate of the sprouts increases progressively to a maximum, then stays constant for a long period (Fig. 10). In the case of corms stored for a very long time at 2°-4°C, up to three years, daughter-corm formation has not been able to be observed during storage.

It seems therefore that high temperatures induce a possibility of a more rapid growth when the corms are subsequently placed in conditions favorable to the expression of this induction. When corms are stored for a longer period at a higher temperature, at least up to 30°C, the subsequent growth is more rapid (roots, sprouts, corm formation). This indicates that growth induction is stronger after a 30°C storage than after a 20°C storage for example. However, in the case of a long storage, daughter-corm formation occurs earlier at 20°C than at 30°C. This result can be explained by the fact that 20°C is more favorable than 30°C to corm formation (68,70). Manifestation of tuberization, in a given environment, is therefore the result of the double action of the temperature of this environment on the induction of tuberization and on the rate of its expression.

Conclusions

Having acquired the capacity for growth (emergence from dormancy), a gladiolus corm evolves constantly during storage, rapidly at high temperature and slowly at low temperature. The physiological state of a corm at planting, all things being equal everywhere else, influences the behaviour of the daughter-plant. Corms physiologically

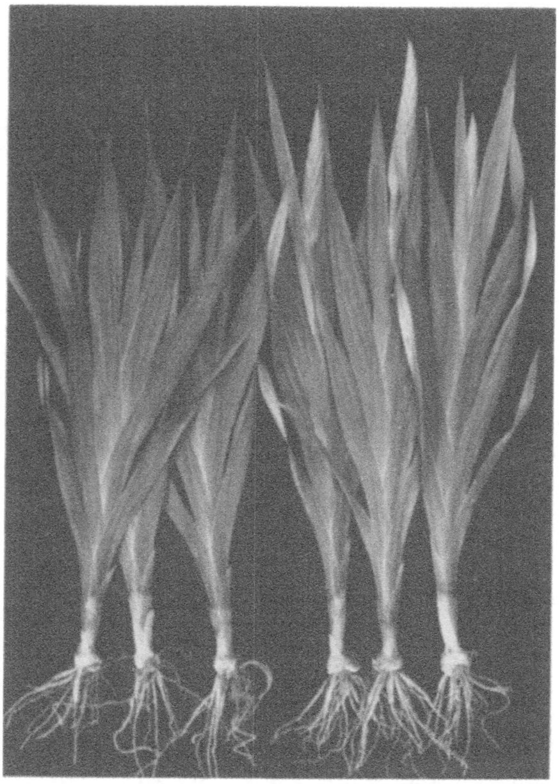

Fig. 9. Gladiolus : Effects of corm storage on the earliness of growth, flower shoot elongation and new corm formation; cv. "Happy End"; plants produced by corms stored at low temperature, 5°-12°C (left), or placed at 20°C, 79 days before planting (right).

little-evolved, that is, stored at low temperature, produce vigorous plants whose sprouting, flowering, tuberization and maturity are delayed. Corms physiologically evolved produce less vigorous plants whose growth is accelerated. Only a low temperature, which slows down the physiological evolution of corms, allows a long-term storage.

Daugher-corm production during storage at high temperature, > 20°C, can be compared to what has been observed in the case of tulip and iris bulbs stored for a long time at low temperature. In these last two species, it has been possible to show that organogenesis and bulbing are mutually exclusive in a bud. In the case of gladiolus corms, a similar relation does not seem to have

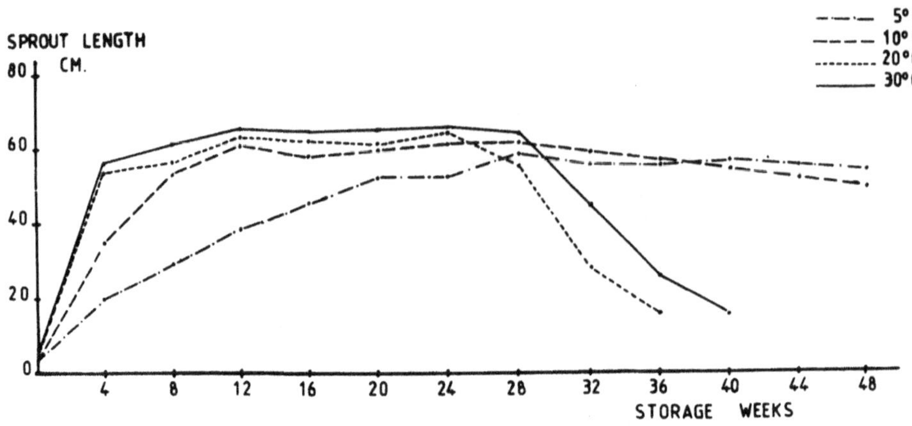

Fig. 10. Gladiolus : Evolution of the length of sprouts produced by corms stored for increasing periods at 5°, 10°, 20° or 30°C (adapted from Cohat (68)); sprouts measured 28 days after planting at 20°C, in darkness.

been observed. Some observations nevertheless show that when the formation of daughter-corms starts off rapidly, the formation of new organs by a bud would be limited. Thus, following a quite long storage at high temperature, large corms produce plants with fewer flowers (67), while small corms produce a higher percentage of vegetative plants, that is, organogenesis in their apical buds would be limited (64, 71). Finally, when daughter-corm formation begins during storage, after planting no leaf differentiation can be seen.

CONCEPTS AND NATURE OF DORMANCY IN BULBS

Concepts of Dormancy

The notion of dormancy in bulbs has not always been well-defined and is often rather subject to discussion. A review of dormancy in bulbs and corms has been made by Kamerbeek et al., (72). Referring to the definition of dormancy proposed by Amen, (73) these authors have classed the bulbous species in three groups.

The first group includes species like gladiolus, lilium, allium, etc..., which show a relatively long period during which differentiation of new organs as well as their elongation are totally interrupted. This period, which occurs after the bulb or corm has been formed, would correspond to the state of dormancy. Dormancy would disappear slowly, and would sometimes last for several months as in the case of gladiolus cormels (54).

The second group includes bulbs in which hardly any true physiological dormancy can be observed : tulipa, hyacinthus, narcissus, etc... In these species, dormancy would be very slight as can be noted in the case of the tulip where bulb formation is immediately followed by flower bud differentiation. The leaves and the flower can grow even if the bulb is not subjected to low temperature. Only the flower shoot does not show practically any growth in the absence of low-temperature treatment. Dormancy would therefore be restricted to the stem. In opposition to the dormancy of the first group (gladiolus, etc...), which begins before flower formation, dormancy in the tulip will become manifest after flower formation.

The third group includes species like bulbous iris, in which apparently no physiological dormancy can be observed. At bulb lifting, initiation of leaf primordia has already started, and their growth can be observed if bulbs are exposed to a mean temperature. On the other hand, if bulbs are stored at high temperature, > 25°C, the growth of leaf primordia is arrested. This would be a typical case of imposed dormancy.

The concept proposed by Kamerbeek et al., is based first of all on "the fact that many bulbs and corms only develop well after having been subjected to a period of cold. Therefore, it seem justifiable to suppose that during certain developmental periods, these plants are in dormancy that can be broken by cold." Besides, Kamerbeek et al. only take into account the ability of the floral bud to develop new organs and to elongate.

When comparing the basis of this classification with the behaviour of tulip bulbs, iris bulbs and gladiolus corms during and after storage, two remarks have to be made. First, it is not always low temperatures which accelerate the development of a bulb. Second, it has been shown that the development of buds depends very greatly on the physiological state of a bulb. The importance of the physiological state of the gladiolus corm on the subsequent bud development has also been demonstrated in in vitro culture (74).

It appears difficult therefore to dissociate a bud's behaviour from a bulb's or corm's general evolution, and it seems to us more logical to consider the bulb in its entirety. The results reported above for tulip bulbs, iris bulbs and gladiolus corms, show that it is possible to discern a certain unity in the behaviour of these species. In taking the period of bulb or corm formation as a starting point, two important phases in their cycle can be seen.

During a first phase, more or less long according to the species, no growth of daughter-organs is possible, even if the bulbs or corms are placed in conditions which are favorable to the induction and/or the manifestation of growth processes. At the end of enlargement, then in the course of storage or in the ground in natural conditions, a physiological evolution takes place which leads the bulbs or corms to acquire a growth capacity. According to whether the species grows in spring and summer, gladiolus, or in winter and spring, tulip and iris, this physiological evolution is accelerated, respectively, either by low temperatures or by high temperatures.

Once the growth capacity has been acquired, the second phase, which corresponds to growth, will only be able to come about rapidly if the temperatures change, high and mean temperatures for gladiolus, low and mean temperatures for tulip and iris.

In the case of the gladiolus corm, the first phase coincides with the period of dormancy proposed by Kamerbeek et al (72) a term used moreover by many other authors.

In the case of tulip, we have shown (19,29) that as long as the bulb has not reached its "physiological maturity," it cannot react to external factors favorable to induction and manifestation of growth processes. This situation seems close to a state of dormancy which for Chouard (9) is "a state of rest for which the actual cause resides in the dormant organ itself." This type of dormancy is also called "true dormancy," "internal dormancy" or "rest" (75).

The behaviour of iris bulbs is very similar to that of tulip bulbs, as has been shown above (see section BULBOUS IRIS).

If it can be accepted that we apply the term dormancy to a bulb in its entirety, it then appears that the period of bulb dormancy could be defined as being the period during which a bulb is not able to react to environmental factors which induce and/or allow the manifestation of growth processes. In the case of the tulip, iris and gladiolus, this period of dormancy would begin during bulb or corm formation and would end all the more rapidly when the bulbs or corms are then subjected to favorable temperatures in the ground or during storage : low temperatures for gladiolus, high temperatures for tulip and iris.

It can be noted here that temperatures which promote bulbs' emergence from dormancy are also those which, later, slow down their physiological evolution, in this sense they are unfavorable to the induction of growth processes. After emergence from dormancy, bulbs and corms would then stay quiescent if the temperatures do not change. Temperatures which accelerate dormancy release would also be those which allow long-term bulb storage.

Two major characteristics of the bulb dormancy period appear to be the following. First, the observations made in the case of tulip and iris bulbs show that during the period of dormancy, bulbs are in such a physiological state that their reserves would not be available for the daughter-organ growth (19,22). High temperatures, which lead to an earlier and a more rapid reserve mobilization when the bulbs are subsequently placed in environmental conditions which allow the actual mobilization, that is, low and mean temperatures, therefore accelerate bulbs' physiological evolution towards emergence from dormancy. Dormancy release would be gradual, the bulbs becoming more and more sensitive to the action of low temperatures.

The second characteristic of bulb dormancy period concerns organogenesis. It has been pointed out that organogenesis begins during bulb or corm formation, but it is not however very advanced at the end of bulb or corm enlargement. It can only become active after the bulbs or corms have been subjected either to high temperatures for tulip and iris, or to low temperatures for gladiolus, that is, temperatures which promote dormancy release. The period of bulb dormancy would correspond then also to a period of low meristematic activity. However, as already mentioned above, the physiological evolution of bulbs or corms towards emergence from dormancy, can occur in the absence of any visible organogenesis. Thus, it appears that the physiological evolution of storage tissues, and eventually that of the basal plate tissues, could proceed independently of that of the buds. These would express growth potentialities conditioned by the physiological state of other tissues (storage tissues, basal plate). The type of growth would also depend on the state of the bud, its degree of differentiation and the correlations it is subjected to.

The term of dormancy, although it is not well defined has already been applied to the iris bulb by different authors (48,76,77). Tsukamoto (77) considers that high temperature breaks dormancy in winter-growing bulbs and that low temperature breaks dormancy in summer-growing bulbs. This agrees with the concept proposed here.

The behaviour pattern which has just been presented for tulip and iris bulbs, and gladiolus corms, seems to be relevant to other species. For example, the results of Aoba, (78,79) show that freesia corms behave very similarly to tulip or iris bulbs. Moreover, it has long been admitted that high temperatures break dormancy in freesia corms (10). In the case of narcissus bulbs, whose temperature requirements for development are very similar to those of tulip bulbs (11,15), we have observed (unpublished result) that a low-temperature treatment applied before the end of bulb enlargement, did not lead to any stimulation of the subsequent daughter-organ growth, but, on the contrary, had a retarding effect as in the case of the tulip.

The concept of bulb dormancy as it has just been proposed for the tulip, iris and gladiolus, seems applicable to a larger number of bulbous species whose behaviour patterns appear then more homogeneous. The length of the period of bulb dormancy can however vary from one species to another : it is relatively short in iris while it can be long in gladiolus. For a species, according to the growing conditions (year, area, etc...), the length and the degree of dormancy would also vary.

In conclusion, this short discussion of the problem of dormancy in bulbs brings to the fore differences of conception in the approach of the phenomenon and shows, once again, the difficulties encountered on the level of terminology.

It can be noted here, that, apparently, a similar situation exists in the case of another storage organ, the potato tuber. In this plant, according to Burton (80,81) dormancy would begin at tuber initiation, but the concept of dormancy raises the same problems as in the case of bulbs, for if many authors (see, for ex. Rappaport and Wolf, (82) Burton, (83))talk of bud dormancy, some results (Madec and Perennec (84), Bailey et al.(85)) suggest that both the buds and the storage tissues exhibit dormancy, and that breaking of this during after ripening operates independently.

Finally, perhaps, it ought to be considered as with de Hertogh (32) that "the question of dormancy in bulbous plants is highly academic." while recognizing that "the fact is that bulb species have evolved mechanisms to survive adverse climatic conditions and man has attempted to describe these processes." What is sure is that our understanding of these processes is very limited, especially in some species.

Breaking of Dormancy and Biochemical Bases of Bulb Dormancy

In horticulture, the existence of bulb dormancy, in the sense we have defined it, constitutes an obstacle when early flowering is desired. For this reason a great deal of work has been carried out to determine the factors which can be used to reduce the length of the dormancy period.

The role of temperatures has already been pointed out, so only a short review of the chemical substances which break dormancy will be made. The major observations concerning the biochemical bases of dormancy will also be reported.

Extensive research on these subjects has especially been done in the case of a species like gladiolus whose corms and cormels sometimes show long periods of dormancy. In most other species knowledge is much more limited.

Gladiolus. Corm and cormel sprouting can be accelerated by a certain number of chemical substances among which are : ethylene chlorhydrin, methanol, 2-chloroethylphosphonic acid (2-CEPA), aqueous solution of calcium cyanamide, cytokinins (55,59,60,86,87,88,89,90,91 92). The action of these substances is more or less pronounced and can vary according to the conditions in which they are used. Thus, 2-CEPA which accelerates sprouting of dormant corms and cormels, tends to inhibit or delay sprouting of corms or cormels previously stored at low temperature, that is, which are no longer dormant.

Proof of the efficacity of cytokinins in breaking dormancy of corms (91) and cormels (55) is at the origin of work which has furnished some information on the mechanisms of dormancy regulation (55,56,74,77,92,93). These different authors have proved in the case of corms and cormels the presence of inhibitors, including abscisic acid, fatty acids (linoleic, linolenic, stearic, palmitic) and phenolic compounds. Tsukamoto (77) has shown synergism in the inhibitory activities of these substances.

The inhibitor content is related to the degree of dormancy and corms or cormels of slightly dormant cultivars contain less than those of very dormant cultivars. The abscisic acid level in dormant cormels is 5 to 10 times that of non-dormant ones. The inhibitor content decreases during a low-temperature storage which allows dormancy release. On the other hand, applications of abscisic acid, fatty acids and phenolic compounds inhibit sprouting of non-dormant cormels.

Abscisic acid and benzyladenine exert antagonist actions, but a treatment with benzyladenine which breaks dormancy, does not lead to a decrease in the content of inhibitors. It would prevent or counteract their activities.

Word done by Yazawa (74) shows that during a low-temperature storage, which eliminates dormancy, it is difficult to detect any cytokinin activity. However, if the corms are subsequently placed at high temperature, cytokinin activity rapidly increases. Low temperature would therefore induce cytokinin synthesis, and the actual synthesis takes place later, at high temperature. This synthesis would be necessary to allow sprouting, the decrease in inhibitor content not on its own being able to explain this. Antagonism between cytokinin and inhibitors in breaking dormancy would be related to RNA metabolism.

Gibberellin treatments do not break dormancy of gladiolus corms. However, Halevy et al. (94) have shown an evolution in the relative importance of three forms of gibberellin during dormancy release, an evolution which could affect synthesis of enzymes influencing the breakdown of reserves in the storage tissues.

All these results together show that dormancy regulation in gladiolus corms and cormels would result from a balance between inhibiting and promoting substances.

Other species. Most of the other bulbous species do not seem to have been the object of such detailed study as the gladiolus.

In the bulbous iris, sprouting can be accelerated by soaking bulbs in a solution of 2-CEPA (89). Our own observations (unpublished results) confirm this result and show that the influence of 2-CEPA is very comparable to that of high temperature in that it promotes simultaneously subsequent flower initiation and a more rapid growth of flower shoot. The effects of high temperature and 2-CEPA seem moreover to add to each other. These results have to be compared to the observations of Stuart et al. (3) and Uhring (48) concerning the influence of bulb exposure to ethylene, and those of Durieux and Kamerbeek (95) and Kamerbeek et al. (96) concerning the influence of plant spraying with 2-CEPA. The mechanisms of action of ethylene are not known however. Tsukamoto and Ando (76) and Tsukamoto (77) have shown the presence of inhibitors, abscisic acid and fatty acids (capric acid), in iris bulbs. Storage at high temperature decreases the inhibitor content and would thus allow dormancy release. It must also be pointed out that Rodrigues Pereira (97,98,99,100) has shown that iris bulbs contain a certain quantity of free gibberellin.

In the case of tulip bulbs, very few studies have borne on what we have defined as being the dormancy period, that is the period which precedes the "physiological maturity." During bulb formation, the presence of abscisic acid and the existence of a gibberellin activity have been proved (101,102,103,104,105). These substances also exist in the bulb after lifting (106,107). Ito et al. (108) have shown that during storage, inhibiting substances decrease while auxin content increases in scales. But, in the tulip, the influence of growth regulators and their eventual role have especially been studied in relation to the possibilities of flower shoot elongation.

In the case of lily bulbs, gibberellin treatments would promote sprouting (72). Wang and Roberts (109) and Tsukamoto (77,110) have proved the presence of inhibitors and auxin, A.I.A., in the bulb scales. The young scales contain more inhibitors than the old ones and their removal favors sprouting. Removing old scales can, on the contrary, delay sprouting. The influence of scale removing seems however to vary in time.

The observations made in the case of lily bulbs show some analogies with those made on the onion by Kato (111,112,113,114). This author distinguishes between a rest period and a dormant period in the onion bulb.

In the hyacinth, a study of the possible role of growth regulators in the development of the plant has been made and a conception of the hormonal control of growth and development has been suggested, (115). From this study, where dormancy applies to flower stem elongation, it appears that knowledge relative to bulb formation is still quite limited.

CONCLUSION

This review of the physiology and storage of bulbs, although limited to three species, has nevertheless made it possible to prove on the one hand the importance of storage temperature in bulbs' physiological evolution, and on the other a certain unity in behaviour of species which at first glance seem quite different.

Certainly, according to the species, a same temperature does not have the same effects, but this must be related to the cycle of the species in its original environment. Thus, tulips which come mainly from central Asia and Asia Minor, grow in winter and spring, and are apparently at rest during the hot dry summer. Gladiolus, which come from the more or less mountainous areas of tropical Africa, grow during the hot damp summer and are, also, apparently at rest during the dry and cool season. It must be kept in mind that during the apparent rest period or during storage, bulbs and corms are continuously evolving, more or less rapidly according to the temperature. In natural conditions, the temperatures encountered during this period slow down the development of bulbs or corms in that they are unfavorable to induction and/or manifestation of the growth processes.

The influence of storage temperature on the behaviour of a fairly large number or bulbs is at present well known and used in horticulture to manipulate the bulbs in order to act on their flowering. However, it must be recognized that our knowledge of the fundamental processes which determine the behaviour of bulbs is very limited, especially in some species. This is clearly shown when it is a question of tackling a problem like dormancy. In this field, and abstracting problem of terminology, it appears that in species like tulip and bulbous iris very little is known about the mechanisms which intervene in the formation of the storage organ which enables these plants to survive.

This situation results perhaps in part, from the fact that most research work, often done for practical purposes, has been aimed at studying flowering while relatively little work has borne on bulb formation and its physiological evolution. Yet, it appears obvious that basic research on this subject is necessary if we want to understand how a bulb works. Such research could have practical applications in two areas: the characterization of the physiological state of a bulb, and the use of growth regulators to replace, at least partially, some long and costly temperature treatments. The possibility of characterizing the physiological state of bulbs would also allow a more rational application of temperature treatments at present in practice.

Fundamental research would also furnish some information on floral induction and perhaps would make it possible to determine why small bulbs do not initiate flowers.

REFERENCES

1. E. S. Haber, A preliminary report on the stimulation of growth of bulbs and seeds with ethylene. Proc. Amer. Soc. Hort. Sci., 23, 201 (1926).
2. F. E. Denny, Shortening the rest period of Gladiolus by treatment with chemicals. Contr. Boyce Thompson Inst., 2, 523 (1930).
3. N. W. Stuart, S. Asen and C. J. Gould, Accelerated flowering of bulbous iris after exposure to ethylene. Hort. Science, 1, 19 (1966).
4. W. J. de Munk, "Some disorders in tulips due to storage conditions and ethylene." Thesis L.B.O. Lisse, Netherlands (1973).
5. M. Masuda and T. Asahira, Effect of ethylene on breaking dormancy of freesia corms. Scientia Hortic., 13, 85 (1980).
6. O. N. Purvis, Recent Dutch research on the growth and flowering of bulbs. I. The temperature requirements of hyacinths. Scient. Hortic., V, 127 (1937).
7. O. N. Purvis, Recent Dutch research on the growth and flowering of bulbs. II. The temperature requirements of tulips and daffodils. Scient. Hortic., VI, 160 (1938).
8. F. W. Went, Thermoperiodicity, in "Vernalization and photoperiodism," Murneek, A.E. and White R.O., 145, Chronica Botanica Co., Waltham, Massachusetts, U.S.A. (1948).
9. P. Chouard, "Dormances et inhibitions des graines et des bourgeons, Préparation au forcage. Thermopériodisme." Cours du C.A.M., C.D.U., Paris (1951).
10. A. H. Hartsema, Storage of bulbs. Bull. Inst. Intern. du Froid, Annexe I, 115 (1954).
11. A. H. Hartsema, Influence of temperatures on flower formation and flowering of bulbous and tuberous plants, in : "Handbuch der Pflanzenphysiologie," 16, 123 W. Ruhland (ed.) Springer Verlag, Berlin (1961).
12. J. P. Nitsch, Physiology of flower and fruit development, in : "Handbuch der Pflanzenphysiologie," 15, 1 1537; W. Ruhland (ed.), Springer Verlag, Berlin (1965).
13. A. Vegis, Die Bedeutung von physikalischen und chemischen Aussenfaktoren bei der Induktion and Beeindigung von Ruhezuständen bei Organen und Geweben höherer Pflanzen, in :

Handbuch der Pflanzenphysiologie," 15, 2, 534; W. Ruhland (ed.), Springer Verlag, Berlin (1965).
14. A. R. Rees, The physiology of ornamental bulbous plants. Bot. Rev., 32, 1 (1966)
15. A. R. Rees, "The growth of bulbs." Academic Press, London (1972).
16. R. Mulder and I. Luyten, De periodieke ontwikkeling van de Darwin-tulp. Verh. Kon. Akad. Wet. (Sectie II), 26, 3, 1 (1928)
17. A. R. Rees, The initiation and growth of tulip bulbs. Ann. Bot., 32, 125, 69(1968).
18. J. Schoub and A. A. de Hertogh, Growth and development of the shoot, roots and central bulblet of Tulipa gesneriana L. cv. Paul Richter during standard forcing. J. Amer. Soc. Hort. Sci., 100, 1, 32 (1975).
19. M. Le Nard, Incidence de séquences de hautes et basses températures sur la différenciation des bourgeons, l'enracinement et la bulbification de la tulipe. Ann Amélior. Plantes, 22, 1, 39 (1972).
20. M. Le Nard, Effects of duration of high temperature treatment on subseuqent flower differentiation, rooting and flowering of tulip bulbs. Acta Hortic., 109, 65 (1980).
21. J. J. Beijer, De terminologie van de bloemaanleg der bloembolgewassen. Meded. Landbouwhogesch. Wageningen, 46, 5, 1 (1942)
22. M. Le Nard, Influence de différents traitements thermiques sur l'intensité de l'induction de la bulbification et sa suppression chez le bulbe de tulipe. Ann Amelior. Plantes, 30, 4, 455 (1980).
23. M. Le Nard, La différenciation florale chez la tulipe et l'iris bulbeux ; Relations entre floraison et bulbification. Sélectionneur fr., 23, 33 (1977).
24. M. Le Nard and J. Cohat, Influence des températures de conservation des bulbes sur l'élongation, la floraison et la bulbification de la tulipe (Tulipa gesneriana L.) Ann. Amelior. Plantes, 18, 2, 181 (1968).
25. A. H. Hartsema and I. Luyten, Over het blindstoken van tulpen. Meded. Landbouwhogesch., Wageningen, 50, 1 (1950).
26. A. R. Rees, The effect of high-temperature treatment of tulip bulbs ("blindstoken") on flowering and bulb yield. Rep. Glasshouse Crops Res. Inst. 61 (1967).
27. T. Aoba, Effect of temperature on bulb and tuber formation in bulbous and tuberous crops. IX. On the bulb formation in tulip. Bull. Yamagata Univ. Agr. Sci., 7, 3, 387 (1976).
28. R. Moe and A. Wickstrom, Effect of precolling at 5 or - 1°C on shoot growth, flowering and carbohydrate metabolism in tulip bulbs. Scientia Hortic., 10, 187 (1979)
29. M Le Nard, Studies on the possibility of delaying flowering of tulip. Acta Hortic., 47, 251 (1975).
30. M. Le Nard, Influence d'un traitement à température élevée appliquée après une conservation des bulbes à température basse sur l'élongation de la hampe florale de la tulipe. Ann Amelior. Plantes, 30, 4, 469 (1980).

31. J. McD. Gilford and A. R. Rees, Growth of the tulip shoot. Scientia Hortic., 1, 143 (1973)
32. A. A. de Hertogh, Principles for forcing tulips, hyacinths, daffodils, easter lilies and dutch irises. Scientia Hortic., 2, 313 (1974).
33. M. Le Nard, Travaux sur tulipe, iris bulbeux et narcisse, in : "Recherches sur les plantes florales à bulbes en France," 10, I.T.I.H. (ed.) Paris (1978).
34. A. H. Blaauw, De periodieke ontwikkeling van een bol-iris (Iris xiphium praecox "Imperator"). Verh. Kon. Akad. Wet. (Sectie 2), 34, 3, 1 (1935).
35. J. J. Beijer, Experiments on the retardation of Dutch Irises. Acta bot. neerl., 1, 268 (1952).
36. W. J. de Munk, All year round culture of dutch irises. Ann. Rept. Jersey Bulb and Flower Grower Assoc., St. John, Jersey (1966).
37. G. A. Kamerbeek, Temperature treatment of dutch iris bulbs in relation to the development. Rep. 1st Intern. Symposium on iris, Florence, May 14-18, 1963,459 (1965).
38. G. A. Kamerbeek, Influence of light, temperature and other factors on bud blast of dutch irises. Proc. 17th Int. Hort. Congress, vol. 1, 233 (1966).
39. G. A. Kamerbeek, Influence of light and temperature on flower-bud development in bulbous irises (iris cv. "Wedgwood") and lilies (lilium cv. "Enchantement"). Acta Hortic., 14, 175 (1969).
40. E. J. Fortanier and A. Zevenbergen, Analysis of the effects of temperature and light after planting on bud blasting in iris hollandica. Neth. J. agr. Sci., 21, 145 (1973).
41. T. Mae and C. R. Vonk, Effect of light and growth substances on flowering of iris x hollandica cv."Wedgwood." Acta Bot. Neerl., 23, 3, 321 (1974).
42. M. Le Nard, Influence de la température de conservation de bulbes sur la différenciation d'organes aériens, leur élongation et la bulbification chez l'iris bulbeux hollandais, var. "Wedgwood." Ann. Amélior. Plantes, 23 3, 265 (1973).
43. T. Aoba, Effect of temperature on bulb-and tuber formation in bulbous and tuberous crops. VI. On the bulb formation in bulbous iris. J. Japan. Soc. Horti. Sci., 43, 3, 273 (1974)
44. G. A. Kamerbeek and J. J. Beijer, Vroege bloei van iris "Wedgwood." Meded. Dir. Tuiub., 27, 598 (1964).
45. Y. Kimura and N. W. Stuart, Exponential nature of heat exposure duration relative to temperature change in the curing and flowering of bulbous iris. J. Amer. Soc. Hort. Sci. 97, 3, 424 (1972).
46. I. Walla and T. Kristoffersen, Some factors affecting the result of early forcing of iris x hoolandica "Wedgwood." Acta Hortic., 14, 187 (1969).
47. M. Le Nard, Bulbing and flowering of iris bulbs stored at different temperatures before a cold treatment. Acta Hortic., 109 141 (1980).

48. J. Uhring, Morphological studies of flower bud initiation and development in bulbous iris stored at various temperatures. J. Amer. Soc. Hort. Sci., 98, 1, 54 (1973)
49. N. W. Stuart, C. J. Gould, W. E. Vassey and M. Hickman, Heat-curing, cool storing speed flowering of Wedgwood iris. Florists' Review 132, 3425 (1966).
50. C. J. Gould, N. W. Stuart and T. Sabelis, Suggestions and comments on forcing Washington-grown iris in greenhouse in the U.S. Florists' Review, 154, 3996, 33 (1974).
51. A. J. B. Durieux, Bloei van hollandse irissen. Praktijkmededeling, n° 44, L.B.O. Lisse, Netherlands (1975).
52. A. H. Hartsema, De Periodieke ontwikkeling van gladiolus hybridus, var. "Vesuvius," Verh. Kon. ned. Akad. Wet. Natuurk., Sectie II, 36, 3, 1 (1937)
53. G. F. Ryan, Effects of temperature on rest in gladiolus corms. Proc. Amer. Soc. Hortic. Sci., 65, 463 (1955).
54. S. S. Apte, Dormancy and sprouting in gladiolus. Meded. Landbouwhogesch. Wageningen, 62, 5, 1 (1962).
55. C. Ginzburg, Hormonal regulation of cormel dormancy in Gladiolus grandiflorus. J. Exp. Bot. 24, 80, 558 (1973).
56. Y. Tsukamoto and H. Konoshima, Changes in endogenous growth regulators in the gladiolus corm during dormancy. Physiol. Plant., 26, 244 (1972)
57. H. Konoshima and Y. Tsukamoto, Effect of light conditions on the dormancy of gladiolus corms. J. Jap. Soc. Hort. Sci., 45, 4, 414 (1977).
58. D. C. Fairburn, Growth responses of the gladiolus as influenced by storage temperatures. Iowa Agric. Exp. Stn. Res. Bull., 170, 95 (1934).
59. F. E. Denny and L. P. Miller, Storage temperatures and chemical treatments for shortening the rest period of small corms and cormels of gladiolus. Contr. Boyce Thompson Inst., 7, 257 (1935).
60. F. E. Denny, Storage temperatures for shortening the rest period of gladiolus corms. Contr. Boyce Thompson Inst., 8, 137 (1936).
61. F. E. Denny, Prolonging, then breaking the rest period of gladiolus corms. Contr. Boyce Thompson Inst., 9, 403 (1938)
62. F. E. Denny, Effects of a few hours chilling upon germination of gladiolus corms subjected to an artificially prolonged rest period. Contr. Boyce Thompson Inst., 12, 375 (1942).
63. A. Paulin, Etude de l'action de traitements thermiques ou chimiques sur les cormus de glaïeuls en vue de hatêr leur floraison. C.R. Acad. Agric. Fr., 47, 953 (1961).
64. J. Cohat, La production des bulbes chez le glaïeul. Pep. Hort. Mar., 122, 23 (1971).
65. A. Paulin, Etude de l'action de températures élevées sur les cormus de glaïeuls en vue de hâter la floraison. Proc 15th Intern. Hortic. Cong., II, 422 (1958)
66. A. Paulin, Etude de l'action de températures élevées sur les

cormus de glaïeuls en vue de hâter la floraison. Rev. hortic., 3, 1921 (1958).

67. J. Cohat and M. Le Nard, Influence des températures de conservation des cormus de glaïeul sur la précocité de floraison et le grossissement du nouveau cormus. Ann. Amélior. Plantes, 18, 3 287 (1968).

68. J. Cohat, Contribution a l'étude de la tubérisation du glaïeul d'été (Gladiolus grandiflorus Hort.). Ann. Amélior. Plantes, 24, 1, 13 (1924).

69. T. Aoba, Effect of temperature on bulb-and tuber formation in bulbous and tuberous crops. VIII. On the corm formation in gladiolus. J. Yamagata Agr. For. Soc., 32, 48 (1975).

70. T. Asahira, H. Imanishi and Y. Tsukamoto, Studies on the cormel formation in gladiolus. Mem. Coll. Agric. Kyoto Univ., 93, 21 (1968).

71. J. Cohat, Influence de la densité de plantation et de la température de conservation des cormus de semence sur le résultat de la culture de grossissement chez le glaïeul. Documentation Bulbes, n° 5, 27 (1977).

72. G. A. Kamerbeek, J. C. M. Beijersbergen and P. K. Schenk, Dormancy in bulbs and corms. Proc. 18th Intern. Hortic. Congr., vol. 5, 233 (1972).

73. R. D. Amen, A model of seed dormancy. Bot Rev.,, 34, 1 (1968).

74. S. Yazawa, The action mechanism of cytokinin in breaking dormancy of gladiolus corm. Sci Rep. Kyoto Prefect. Univ. Agric., 28, 40 (1976).

75. T. C. Moore, "Biochemistry and physiology of plant hormone" (p. 181-182), Springer Verlag, New York (1979).

76. Y. Tsukamoto and T. Ando, The change of amount of inhibitors inducing dormancy in dutch iris bulb. Proc. Japan. Acad., 49, 627 (1973).

77. Y. Tsukamoto, Changes in endogenous regulators and dormancy in bulbous plants. Proc. 19th Intern. Hortic. Congr. 293 (1974).

78. T. Aoba, Effect of temperature on bulb-and tuber formation in bulbous and tuberous crops. III. Effect of the heat treatment on the pupa formation in freesia. J. Japan. Soc. Hort. Sci., 42, 4, 341 (1974).

79. T. Aoba, Effect of temperature on bulb-and tuber formation in bulbous and tuberous crops. IV. On the pupa formation in freesia corms at various states of dormancy. J. Japan Soc. Hort. Sci., 43, 1, 69 (1974).

80. W. G. Burton, The dormancy and sprouting of potatoes. Fd. Sci. Abstracts, 29, 1 (1957).

81. W. G. Burton, Concepts and mechanism of dormancy in : "The growth of the potato," Butterworth, London, 17 (1963).

82. L. Rappaport and N. Wolf, The problem of dormancy in potato tubers and related structures. Symp Soc. Exp. Biol., 23 219 (1969).

83. W. G. Burton, Post-harvest behaviour and storage of potatoes, in: "Applied Biology," Vol. III, Academic Press, London, 86 (1978).

84. P. Madex and P. Pérennec, Levée de dormance de tubercules de pomme de terre d'âge différent : action de la rindite, de la gibberelline et de l'oeilletonnage. Eur. Potato J., 12 96 (1969.
85. K. M. Bailey, I. D. J. Philips and D. Pitt, The role of buds and gibberellin in dormancy and the mobilization of reserve materials in potato tubers. Ann. Bot., 42, 649 (1978).
86. F. E. Denny and L. P. Miller, Hastening the germination of dormant gladiolus cormels with vapors of ethylene chlorhydrin. Contr. Boyce Thompson Inst., 6, 31 (1934).
87. O. D. K. Bautista and T. G. Cadiz, Breaking the dormancy of gladiolus. Philippine Agric., 48, 288 (1964)
88. A. H. Halevy, R. Shilo and S. Simchon, Effect of 2-chloroethanephosphonic acid (Ethrel) on health, dormancy, and flower and corm yield in gladioli. J. Hort. Sci., 45, 427 (1970).
89. S. Nakamura and S. Yoshida, Studies on the dormancy of bulbs and corms. I. Freesia, iris and gladiolus. Bull. Fac. Agric. Yamagata Univ., 24, 591 (1973).
90. C. Ginzburg, Studies on the role of ethylene in gladiolus cormel germination. Plant Sci. Lett., 2, 133 (1974).
91. Y. Tsukamoto, Breaking dormancy in the gladiolus corm with cytokinins. Proc. Jap. Acad., 48 34 (1972).
92. Y. Tsukamoto, Breaking dormancy of gladiolus corms and inhibitors inducing dormancy. Acta Hortic., 43, 393 (1974)
93. H. Konoshima, S. Yazawa and Y. Tsukamoto, Inhibitors concerned in the dormancy of the gladiolus corm. J. Jap. Soc. Hort. Sci., 42, 1, 35 (1973).
94. A. H. Halevy, S. Simchon and R. Shilo, Changes in "free" and two forms of "bound" gibberellins in the various stages of dormancy of gladiolus corms, in : "Plant growth substances," Hirokawa Pub. Co. Inc., Tokyo, 64 (1973).
95. A. J. B. Durieux and G. A. Kamerbeek, The prevention of "three leaved" plants in the forcing of iris x hollandica by early heat treatment of stored bulbs by ethephon field spraying. Scientia Hortic., 2, 101 (1974).
96. G. A. Kamerbeek, A. J. B. Durieux and J. A. Schipper, An analysis of the influence of ethrel on flowering of iris "Ideal"; an associated morphogenetic physiological approach. Acta Hortic., 109, 235 (1980).
97. A. S. Rodrigues Pereira, Flower initiation in excised stem disk of Wedgwood iris. Science, 134, 2044 (1961).
98. A. S. Rodrigues Pereira, Physiological experiments in connection with flower formation in Wedgwood iris (Iris, cv. "Wedgwood"). Acta Bot. neerl., 11, 97 (1962).
99. A. S. Rodrigues Pereira, Endogenous growth factors and flower formation in Wedgwood iris bulbs. Acta Bot. neerl., 13, 303 (1964).
100. A. S. Rodrigues Pereira, Physiological analysis of flower formation in Wedgwood iris. J. Exp. Bot., 16, 48, 405 (1965).

101. L. H. Aung and A. A. de Hertogh, The occurrence of gibberellin-like substances in tulip bulbs (Tulipa sp.). Pl. Cell. Physiol. Tokyo, 8, 201 (1967)
102. L. H. Aung and A. A. de Hertogh, Temperature regulation of growth and endogenous abcissic acid-like content of Tulipa gesneriana L. Plant Physiol., 63, 1111 (1979).
103. L. H. Aung and A. R. Rees, Changes in endogenous gibberellin levels in Tulipa bulblets during ontogeny. J. Exp. Bot., 25, 87, 745 (1974).
104. A. E. Einert, G. L. Staby and de Hertogh A. A., Gibberellin-like activity from organs of Tulipa gesneriana. Can. J. Bot., 50, 5, 909 (1972)
105. G. R. Hanks and A. R. Rees, Growth substances of tulip ; the activity of gibberellin-like substances in field-grown tulips from planting until flowering. Z. Pflanzenphysiol. 98, 3, 213 (1980).
106. G. Syrtanova and I. Rakhimbaev, Endogenous gibberellins during dormancy and growth of tulip bulbs. Fiziologiya Rastenii, 20, 4, 721 (1973).
107. G. Syrtanova, I. Rakhimbaev and V. I. Kefeli, A natural growth inhibitor of tulip. Fiziologiya Rastenii, 22, 1, 165 (1975).
108. H. Ito, T. Kato and T. Toyoda, Metabolism of tulip bulbs as related to thermoperiodicity. J. Jap. Soc. Hort. Sci., 29, 323 (1960).
109. S. Y. Wang and A. N. Roberts, Physiology of dormancy in Lilium longiflorum "Ace" Thunb. J. Amer. Soc. Hort. Sci., 95, 5, 554 (1970).
110. Y. Tsukamoto, Changes of endogenous growth substances in Easter lily as affected by cooling. Acta Hortic., 23, 75 (1971).
111. T. Kato, Physiological studies on the bulbing and dormancy of onion plants. VII. Effects of some environmental factors and chemicals on the dormant process of bulbs. J. Jap. Soc. Hort. Sci., 35, 49 (1966).
112. T. Kato, Physiological studies on the bulbing and dormancy of onion plants. VIII. Relations between dormancy and organic constituents of bulbs. J. Jap. Soc. Hort. Sci., 35, 142 (1966).
113. T, Kato, Physiological studies on the bulbing and dormancy of onion plants. IX. The relationship between bulb dormancy and the components of the juice. J. Jap. Soc. Hort. Sci., 35, 297 (1966).
114. T. Kato, Physiological studies on the bulbing and dormancy of onion plants. X. A germination inhibitor in onion juice. J. Jap. Soc. Hort. Sci., 35, 395 (1966).
115. R. M. Rudnicki, Hormonal control of growth and development of hyacinth. Acta Hortic., 91, 185 (1979).

THE FORMATION OF ENZYMATIC PRODUCTS IN THE FRUITS DURING GROWTH AND STORAGE

George Karaoulanis

Food Technology Institute
Lykovrissi-Amaroussion
Athens, Greece

INTRODUCTION

It is well-known that research on fruits is important not only for their commercial and nutritive value, but also for the realization that biological processes that are common to all living organisms can be studied in fruits.

It is also known that there are several reasons for using fruits for the study of the general biological problems of ageing.

The fruits are capable of prolonged independent existence even after removal from the parent plant and as long as they are kept under suitable conditions (humidity and temperature) they do not require external supply of water to maintain the cells in a turgid condition.

Most fruits are amply supplied with respiratory substrates, in the ordinary sense, although the depletion of certain specific substances required in small amounts and not synthesized in detached fruits may play a role in the metabolic pattern. On the other hand, the study of fruits avoid complexities resulting from the interplay of several processes, such as happens with photosynthesis and respiration in leaves.

It is well-known, that the central reactions in ripening are biological oxidation. This term is used advisedly to include glycolic as well as respiratory activities.

The former refers to the conversion of carbohydrates to pyruvic acid in the presence or absence of oxygen. The pyruvic acid is converted in the absence of oxygen to fermentative end-products such as alcohol, lactic acid, and propionic acid. The term respiration is reserved then for the reactions requiring oxygen for the enzymatic oxidation of carbon compounds to CO_2, water and energy. Anaerobic respiration of higher plants was first noted (as evolution of CO_2 in absence of air) by William Cruikshann in 1797. In 1872 Pasteur and also Lechartier and Balley, reported formation of alcohol by apples and pears in absence of oxygen and Pasteur said this was due to the "chemical and physical life of the cells" and linked it to "that of the cells of a ferment" (yeast).

Argument followed amongst plant physiologists, some (including de Saussure) holding that aerobic and anaerobic respiration were distinct and separate, while others (Pfeffer, Palladin 1900-1903) thought they were basically similar, and that fermentation was the first stage in respiration, the second stage being oxidation of alcohol.

In 1913 Kostytschew, on the basis of experiments in which he measured the CO_2: Alcohol ratios, concluded that anaerobic respiration was not identical with respiration. Boysen-Jensen (1932) was rather more cautious, concluding that there is not necessarily any connection.

Then work on the intermediate metabolites notably by Newberg, Meyrhof, Embden and Parnas, built up the modern theory, in which aerobic and anaerobic respiration are linked by common intermediates.

In subsequent works Thomas (1925) and Fidler (1933) observed that the anaerobic respiration of carbohydrates by apple fruits is qualitatively and quantitatively similar to alcohol fermentation of yeast.

This presentation is an attempt to review the research works on the formation of the anaerobic enzymatic products (ethyl alcohol and acetaldehyde) in the fruits, the causes of their concentration, and to discuss whether the concentration of ethyl alcohol can be used as an additional criterion for the determination of a) the stage of maturity of the fruits, b) the duration of their preservation under different storage conditions,and c) the optimum storage conditions.

Formation of Enzymatic Products in the Fruits During Their Growth

G. Karaoulanis (1968,1978 and 1979) and P. Davis (1970) observed that during the growing of citrus fruits, mainly oranges, the increase of alcohol (ethanol) concentration is higher than the ratio Brix/acidity, Index which is used for the determination of their maturity.

In experiments with oranges of W. Mariel, Valencia and Marsh and Ruby Red variety, picked at different steps of maturity (during their ripening) the concentration of ethyl alcohol in the peel and in the flesh during ripening was observed. (Table 1.). Similar observations were reported for apples (several years ago) by other researchers (Fidler, 1951).

Formation of Enzymatic Products in the Fruits During Their Storage in Air and Under Different Temperatures

Apples of the varieties Newton Wonder and Bramley's Seedlings stored at 3°C showed low concentrations of alcohol and acetaldehyde. However, when exposed to high temperatures even for a short period of time, i.e. at 18°C, the concentration of the above substances increased in general and more markedly during their senescence (Fidler, 1933) (Table 2.). The same was observed when the oranges W. Navel were kept at 2.5°C, 5°C and 10°C in air (Karaoulanis, 1979).

Formation of Enzymatic Products in the Fruits During Storage Under Low O_2 Concentration

High quantities of ethyl alcohol and acetaldehyde accumulate in apples of different varieties (Thomas and Fidler, 1933) and oranges W. Navel and Valencia (Karaoulanis, 1969) and in most vegetable tissues, when preserved under low oxygen concentrations. This was verified by the work of Furlong (1961) for apples of the Coxes variety which showed very high concentrations of enzymatic products when preserved under anerobic conditions, compared to those developed under normal conditions. He also observed that the Coxes apples produce more alcohol than the Bramley's Edward or Cox's Orange Pippin apples in a low oxygen concentration environment. Work at Ditton Laboratory has shown that ethanol accumulation in Cox's Orange Pippin, Barnack and "Lambourne" apples stored at O_2 levels below 2% at 3.5°C increases. Further experiments with Cox's Orange Pippin and Bramley's Seedling held for 7 days in 0% O_2 at 3.5°C and then in 3% O_2 at the same temperature yielded the following results: After 7 days at 0% O_2, the ethanol content of Cox's Orange Pippin was 250 mg/100 gr. and Bramley's Seedling 140 mg/100 gr. of flesh (Furlong, 1961b).

Fidler (1951) also studied the effect of air and N_2 atmospheres at various temperatures on the loss of carbohydrate and acid and the production of CO_2 and ethanol in Sturmer Pippins and Bramley's Seedling apples. The experimental results indicated that the rate of acid loss was not affected by the presence or absence of O_2; the presence of O_2 had a converse effect on the loss of carbohydrate and acid, accounted quantitatively for the production of CO_2 and ethanol in air and in N_2 atmospheres.

It was also observed that during anaerobiosis the apples

Table 1. Monthly averages of ethanol, solids, acid and pH of juice of Hamlin and Valencia oranges

Variety	Season	Month	Ethanol (mg/100 ml)	Solids (%)	Acid (%)	Solids/ acids	pH
Hamlin on Cleopatra mandarin rootstocks	1968-69	October	4.0[1]	9.3	1.05	8.9	-
Do	1968-69	November	20.0	9.5	0.89	10.6	-
Do	1968-69	December	25.4	10.4	0.87	12.0	-
Do	1969-70	September	0.2	8.9	1.46	6.3	3.044
Do	1969-70	October	1.7	9.4	1.04	9.1	3.243
Do	1969-70	November	9.3	10.1	0.94	10.7	3.371
Do	1969-70	December	24.5	10.8	0.91	12.0	3.407
Hamlin on sour orange rootstock	1969-70	September	0.2	9.2	1.53	6.1	3.057
Do	1969-70	October	0.8	9.4	1.21	7.9	3.043
Do	1969-70	November	9.1	10.0	0.97	10.3	3.333
Do	1969-70	December	38.1	10.6	0.94	11.2	3.417
Valencia on rough lemon rootstock	1968-69	October	0.5	8.5	2.83	3.0	-
Do	1968-69	November	4.7	9.1	2.28	4.0	-
Do	1968-69	December	9.5	10.6	1.88	5.8	-
Do	1968-69	January	28.9	11.3	1.67	6.7	-
Do	1968-69	February	40.0	11.7	1.46	8.0	-
Do	1968-69	March	45.7	12.4	1.29	9.7	-
Do	1968-69	April	48.0	11.8	1.04	11.5	-

[1]Figures represent an average of four weekly samplings of 10 fruits each. From Davis 1970

showed an increase in the rate of ethyl alcohol accumulation, with a subsequent increase in acetaldehyde. When apples are fed with ethanol, the ratio acetaldehyde to ethanol remained the same as in the absence of ethanol feeding.

Formation of Enzymatic Products in the Fruits During Storage under High CO_2 Concentrations

The high concentrations of ethanol and aldehyde observed in fruit tissues, mainly apples, oranges etc. during their preservation

Table 2. Influence of the harvest, the temperature and the duration of the storage on the concentration of ethyl alcohol in the flesh of oranges of the W. Navel variety (mg. of ethyl alcohol per 100 g of flesh weight)

Days in storage	CONDITIONS OF PRESERVATION 1st Harvest 0°C	2.5°C	5.0°C	10°C	Mean	2nd Harvest 0°C	2.5°C	5°C	10°C	Mean	3rd Harvest 0°C	2.5°C	5°C	10°C	Mean
0	8.60	8.60	8.60	8.60	8.60	8.80	8.80	8.80	8.80	8.80	17.57	17.57	17.57	17.57	17.57
60	27.20	13.44	26.50	23.50	22.66	19.60	8.97	21.50	22.50	18.13	11.12	15.52	33.95	39.26	24.96
120	29.19	28.80	29.92	50.12	34.51	27.15	14.84	24.94	27.80	23.64	37.42	20.83	38.36	81.12	44.43
160	77.40	41.46	48.70	156.61	81.09	38.45	33.64	36.94	37.72	36.68	53.00	35.24	50.29	84.41	55.73
Mean	34.09	23.07	28.43	59.68		23.51	16.50	22.83	24.20		29.78	22.29	35.04	55.59	
L.S.D.		5%		1%		5%		1%			5%		1%		
Btn Cond.Pr.*		±1.620		±2.240		±0.250		±0.344			±0.215		±0.296		
" Days "		±1.620		±2.240		±0.250		±0.344			±0.215		±0.296		
Interaction		±3.240		±2.240		±0.501		±0.690			±0.430		±0.593		

*Between Conditions of Preservation

From Karaoulanis 1979

under high concentrations of CO_2 are due either to the anaerobic conditions or to damage in the fruit tissues or to both reasons.

Thomas (1925) attributed the severe damages of the fruit cells to the "CO_2 zymosis" and to the high concentration of aldehyde (poisoning of their internal atmosphere).

Also Fidler (1951) studying fruits preserved under high CO_2 concentration observed the following:

a) The rate of increase of acetaldehyde concentration is higher than that of ethanol.
b) The appearance of injuries is speeded as the CO_2 concentration increases.
c) When the CO_2 concentration was higher than 30% the fruits showed damages similar to Brown Heart.
d) The different varieties of apples have different sensitivity to the high CO_2 concentration, i.e. the Newton Wonder variety is more sensitive to injuries than other commercial varieties.
e) The CO_2 zymosis can be induced in a wide variety of plant tissues. Already in 1932 it was observed that when the CO_2 concentration in the storage room is higher than 80% the ratio of ethyl alcohol to acetaldehyde concentration produced in the plant tissues can have the following values:

    a. 30:1 to 100:1 in swede, beet, horseradish, potato, barley seedlings, pea seedlings.
    b. 3:1 to 10:1 in pears, oranges, bananas.

<u>Formation of Enzymatic Products in the Fruits with Internal or External Physiological Disorders</u>

As to whether zymosis, precedes injury, or injury leads to zymosis, Thomas (1929) came down firmly on the side of disorganization of state of the tissue being the precursor of zymosis. Part of the evidence on which this conclusion was based is given in effect of injury of surface tissue of Newton and Wonder apples by tapping with a pestle, in which the external tissue was mechanically damaged. This damage leads to increase of alcohol number (i.e. aldehyde+ alcohol, no separate figures are given).

Effect of injury of surface tissue by tapping with a pestle gave the following results:

Newton)	apple varieties	(Whole apple	100 mg Alcohol/100 gr.
Wonder)		(Untreated apple	6 mg " / "

Further, Thomas (1929) showed that subjecting apples to -5°C for 3 days led to an increased "alcohol number" when kept 3 days subsequently at ordinary temperatures. Complete killing of the tissues at -10°C for 7 days has less effect. A similar autolytic effect was found in work by Fidler when apples were treated with chloroform vapour.

Anaerobic zymosis is characterized by a high proportion of alcohol related to aldehyde; in low temperature injury, senescent breakdown and brownheart, the quotient alcohol/aldehyde is lower and in freshly damaged "brownheart" tissue can be as low as unity.

a. Low temperature injuries, freezing injuries etc. It is known that the injury becomes more severe at ordinary temperatures, subsequent to storage at very low temperature. It is observed that in Coxes apples the enzymatic products increase with increased injuries.

On the other hand, Wills and McGlasson (1968, 1971) observed that the appearance of low temperature breakdown is related to the presence in the apple tissue of esters of Oxalic Acid and Alcohols, compounds which appear in high concentrations at low temperatures. Also, the hardening of the skin, because of low temperatures, seems to prevent the diffusion of gases leading to accumulation of CO_2 and the appearance of anaerobic environment in the fruits.

b. Brownheart: It has been stated previously that a feature of brownheart, when the injuries are of recent occurrence, is the high concentration of acetaldehyde. If, however, the injury took place some considerable time before the trouble was evident, and the injured zones dried out to form the characteristic cork cavities then it has been found that both acetaldehyde and ethyl alcohol have to some extent disappeared.

c. Breakdown: Neal and Hulme (1958) showed that the peel of Bramley's Seedlings apples decarboxylated malic acid to acetaldehyde. The NADP-specific malic enzyme was isolated from McIntosh and Cox's Orange Pippin apples (Dilley et al. 1964, Hulme et al. 1962). Also Clijsters (1965) has found that a) The malic acid content of diseased "Jonathan apples" decreased and that several volatile products accumulated, especially ethanol (1963 a and b and 1965 b). "Jonathan" apple tissue decarboxylates malic acid to acetaldehyde and ethanol. Malic acid decarboxylation occurs as well in healthy apples as in diseased fruits affected by "Jonathan" Internal Breakdown.

Formation of Enzymatic Products in the Fruits During their Storage

Work with apples, pears (Furlong 1961 and Karaoulanis 1968) and oranges (Karaoulanis 1978) showed that the concentration of anaerobic zymosis products increases with time of preservation (Table 2) even at the suitable conditions.

In the case where the concentration of ethanol was higher than

150-1500 mg/100 gr. of fresh weight, the fruits were unsuitable for consumption.

Changes in the Internal Atmosphere of Fruits During Senescence and Their Influence on Metabolism

Changes in the internal atmosphere of the fruit during ripening. The relationship between respiratory activity, its course and the composition of the internal atmosphere has been studied in several kinds and varieties of fruits.

In the case of banana (Musa Sapientum) and papaya (Carica papaya) Wardlaw and Leonard (1939) observed that the climacteric phase coincides with the maximum concentration of CO_2 inside these fruits. This concentration drops as the respiratory activity increases and reaches the lowest value in the last stages of senescence. These remarkable changes in the composition of gases inside the fruits led the above reserarchers to conclude that the concentration of O_2 in their internal atmosphere regulates the mechanism of the climacteric phase.

Trout and his colleagues (1960) reported also that the lack of O_2 depresses the respiratory activity and delays the ripening of mature apples. They also showed that Granny Smith apples stored at 7°C and 29°C in air had 17% O_2 and 2% respectively in their internal atmosphere, while the CO_2 concentrations were 2% and 17% respectively.

Ulrich (1958) and Ulrich and Marcellin (1955) showed that the highest concentration of CO_2 in the intercellular space occurs during the climacteric peak. They also investigated the percentages of gases in the internal atmosphere and made observations on the epidermal layers as the tissue that offers the major resistance to gaseous diffusion for the apples.

Smith (1954) reported that the maximum concentration of O_2 is found near the epidermis and drops as we proceed to the center of the fruit, while the CO_2 concentration increases. Biale et al. (1962) observed that after the climacteric phase the concentration of CO_2 in the intercellular spaces increases while that of O_2 decreases due probably to their filling with fluids from the surrounding cells. The rate of fluid leakage from the cells increases with ripening and this is a factor leading to the senescence of the fruits.

Henze(1969) in a series of experiments showed that the decrease of the percentage of the intercellular space combined with the high respiratory activity leads quicker to the maximum tolerable CO_2 and O_2 concentration level, above which anomalous metabolism takes place in the fruit.

Ben-Yehoshua et al. (1963) observed for the Avocado fruits that there is a correlation between internal atmosphere and respiration rate for the lag period and respiratory rise; upon softening this correlation changed. Despite the decline in respiratory activity the oxygen percentage dropped to as low as 1-2% and CO_2 percentage rose to 15%.

The cross-sectional area of air-space of firm fruit was calculated to be at least four times larger than the cross-section of air space of ripe fruit.

Lyons et al. (1961) experimenting with Cantaloupe fruits found similarity to the changes in the internal concentrations of CO_2 and O_2 in cantaloupe to those reported earlier by Wardlaw and Leonard (1939) for bananas.

To conclude from all the above observations one can say that during senescence the atmosphere within the fruits changes, resulting in an increase of CO_2 and decrease of O_2 concentration. The new environmental composition depresses the oxidative processes and favours the accumulation of enzymatic products.

<u>Changes in the internal atmosphere of oranges and apples</u>. Henze (1969) experimenting with apples of the Alexander Lucas and Coxes Orange varieties observed that as the temperature increases the concentration of CO_2 increases and that of O_2 decreases within the fruits with an increase of ethyl alcohol accumulation in the same time.

Eaks and Ludi (1960) have shown that the internal atmosphere of citrus fruits is affected by temperature, washing and waxing.

Vines and Oberlacher (1961) found that certain packing house treatment increased CO_2 concentrations within citrus fruits.

Also Davis et al. (1967) found for the Temple oranges, that the internal O_2 and CO_2 of fruits held at 0°C were usually within 1% of those of the external concentrations. Fruits held at 4.5°C varied somewhat more in their internal O_2 and CO_2 levels. This is probably due to the greater metabolic activity at 4.5°C, wherein O_2 concentration tended to be lower and CO_2 higher. Waxed fruits consistently had lower O_2 and higher CO_2 in the internal gases than in the atmosphere.

According to Ben-Yehoshua (1969) the commercial degradation of oranges is connected with transpiration, the respiratory activity and the concentrations of CO_2, O_2 and ethylene within oranges. He also observed that during storage of oranges their respiratory activity decreases, the internal concentration of CO_2 increases from 2-4% to 5-10%, and that of O_2 decreases from 17-19% to 10% and even lower.

The epidermis mainly the flavedo hardens resulting in a reduction of gaseous diffusion and a change in the composition of the internal atmosphere.

Acetaldehyde and Ethanol Metabolism in Plant (fruit) Tissues

As it is already shown many plant tissues are known to produce ethanol when they are subjected to experimental stresses such as low oxygen supply or a variety of respiratory inhibitors. Ethanol production has also been observed under natural conditions in such tissues as germinating seeds, fruits and root tips (Cossins et al. 1963). There is one question whether ethanol, particularly when it has accumulated in periods of natural or imposed anaerobiosis, can be metabolized when the tissues subsequently gain better access to O_2.

The question has two directions: a) Whether plant tissues which have become alcoholic as a result of accidental deprivation of oxygen during storage, can be rendered palatable and b) what is the fate of any alcohol which may be lost. On the first point, that of disappearance of alcohol formed within the tissues in the absence of oxygen, Furlong (1961) gives some figures in the D. L. Annual Reports for 1960 and 1961. Typical data are given in Table 3.

Two further sets of data exist, one relating to a sample from Norfolk fruit growers, from a store which accidentally went anaerobic, and the other obtained by Karaoulanis (Table 4).

From Tables 4 and 5 it will be seen that alcohol disappears from Coxes apples in air at 38°F only very slowly, the rate of loss in 3% O_2 at 38°F is of the same order as in air. At 65F the rate of loss is more rapid, but if the initial concentration is of the order of 150 mg/100 gr. or higher, the apples lose quality before they are acceptable to the taste. A low oxygen store can probably be anaerobic for a maximum of about 6 days, after which the apples would slowly recover in 3% O_2 at 38°F or could be marketed after 10-14 days at 65°F.

Turning now to the question of the fate of alcohol which is lost, Fidler (1951), gives data on the disappearance of alcohol fed as vapour to apples in air (Table 5). Analyses of the data show that in the first periods of aeration, for both Sturmers and Dunns, the alcohol could have been oxidized. But in periods (ii) and (iii) for Sturmers and (ii) for Dunns, the CO_2 and oxygen

Table 3. Changes in concentration of alcohol and aldehyde in Coxes Apples

Sample	mg/100 g ald.	alc.	Alc./ ald.	Loss of alcohol	Loss per day
Initial	--	152			
After 7 days, AIR, 65°F	--	62		90	13
Initial	--	302			
After 26 days, AIR, 65°F	--	52		250	10
Initial	4	207	70		
After 9 days, AIR, 65°F	3	123	41	84	9
" 11 "	--	59	-	148	13
Initial	3	169	57		
After 9 days, AIR, 65°F	--	98	-	71	8
Initial	4	207	70		
After 28 days, 3% O_2, 38°F	3	134	41	73	2.5

From Furlong 1961b

equivalents of the alcohol lost exceed the observed total gaseous exchange, and therefore the alcohol must have been lost by some process other than oxidation.

Cossins and Beevers (1963) experimented with pears, beans and corn on metabolizing of ethanol, observed clearly that all the tissues investigated were able to convert the supplied ethanol labelled with C^{14} (ethanol-1-C^{14}) into organic acids, aminoacids, lipids and sugars.

These observations are consistent with the conversion of ethanol to acetyl coenzyme A, which is then involved in the reactions of established pathways. Based on what has been already discussed, one can say the following:

a) The presence of ethyl alcohol in the fruits is connected to their maturity and accompanies them until their death.
b) The storage of fruits under different conditions leads to increase in the concentration of ethyl alcohol. The presence of ethanol in some cases results in such a change of their quality that they become unsuitable for consumption.
c) Since there is a relation of concentration of ethanol to length and temperature of storage and to the degree of maturity at the time of storage, we would say that the content of ethanol, may be used as an additional criterion for determination of 1) The maturity of the fruits; 2) the suitable storage conditions; and, 3) the length of their storage life.

Table 4. Changes in concentration of aldehyde and alcohol in Cox's Orange Pippins, in air

Sample	Subsequent history T°F	days	mg/100g. ald.	alc.	Loss of alc.	Notes
N.F.G. recd.22/11/65)	initial samples		4.5	312	--	
N.F.G. " 24/11/65)						
	38	8	6.4	296	18	
	38	15	5.4	296	18	
	38	22	--	--	--	
	38	36	5.2	242	70	
	65	8	6.0	250	62	
	65	15	5.1	172	140	
	65	22	3.7	69	243	Alcoholic taste
	65	36	2.2	23	289	Senescent
Karaoulanis						
After 6 days N_2,38°F	Initial		3.2	125	--	
	38	7	2.7	109	16	
	"	14	2.5	111	14	
	65	7	2.7	51	74	
	"	14	1.8	21	104	Yellowing,normal
After 12 days N_2, 38°F	Initial		5.4	215	--	
	38	7	3.7	210	5	
	"	14	3.2	187	28	
	65	7	4.4	131	84	
	"	14	2.2	64	151	Alcoholic
After 18 days, N_2,38°F	Initial		5.9	274	--	
	38	7	4.0	256	18	
	"	14	3.8	238	36	
	65	7	4.3	163	111	
	"	14	3.1	113	161	

From Karaoulanis 1968

Table 5. Changes in concentration of aldehyde and alcohol following alcohol feeding (as vapour) in air

Date	Variety	T°F	Days in Air	Aldehyde	Alcohol	Change in concen.of Alcohol	mg/100 g Alcohol Transpired into Air Steam	Alcohol Lost by Processes Other Than Transpiration	
X1.33	Bramley	68	Initial	3	148	--	--	--	
			2	4	156	+8	14	None, gain of	22
			9	4	143	-5	52	" , " "	47
			15*	4	230	+80	84	" , " "	164
			22	5	290	+140	115	" , " "	255
*on this date, an untreated sample in air had formed 3 aldehyde + 100 alcohol									
II.34	Sturmer	54							
		(i) After 1st period of feeding:							
			Initial	0.7	64	--	--	--	
			6	0.8	51	13	8	5	
		(ii) After 2nd period of feeding:							
			Initial	1.1	200	--	--	--	
			6	0.8	136	64	13	51	
		(iii) After 3rd period of feeding:							
			Initial	2	525	--	--	--	
			4	2	413	110	15	95	
		The R.Q. fell progressively from 1.35 initially to 1.00 at the end.							
IV.34	Dunn's Seedling	54							
		(i) After 1st period of feeding:							
			Initial	1.6	163	--	--	--	
			4	2	144	19	9	10	
		(ii) After 2nd period of feeding:							6
			Initial	4	602	--	--	--	6
			4	4	430	170	17	153	6
		The R.Q. fell progressively from 1.8 initially, to 0.95 at end.							

From Fidler (unpublished).

REFERENCES

Ben-Yehoshua, S., Robertson, R. N. and Biale, J. 1963. Respiration internal atmosphere of avocado fruits. Plant Physiol. 38: 194.

Ben-Yehoshua, S., 1969. Gas exchange transpiration and the commercial deterioration in storage of orange fruit. J. Amer. Soc. Hort. Sci. 94:521.

Boysen-Jensen P. 1932. The gentic connection between normal and intramolecular respiration of plants. Kgl. Danske. Vidensk. Selsk. Biol. Meddel. 4,1.

Clijsters, H. 1963a. Omzetting van appelzuurinetanol bijverschillende appelvarieteiten. Arch. Int. Physiol. Bioch. 71:127

Clijsters, H., 1963b, Quelques aspects biochimiques de la pourriture de la pomme Jonathan, en frigo. Rept. XVI, 4th Int. Hort. Congress, Brussels 1962 (Ed. Dueulot S.A. Gemblous, 3:449.

Clijsters, H. 1965. Malic acid metabolism and initiation of the internal breakdown in "Jonathan" apples. Physiol. Plant.18:85.

Cossins, E. A. and Beevers, H. 1963. Ethanol metabolism in plant tissues. Plant Physiol. 38:375

Davis, P. L. and Cubbedge, P. H., 1967. Factors affecting internal O_2 and CO_2 concentration of citrus fruits. Hortscience. 2:168.

Davis, P. 1970. Relation of ethanol content of citrus fruits to maturity and storage conditions. Proc. Florida State Hort. Soc. 23:27.

Dilley, D.R., McLean, D.C. and Dedolph, R.R. 1964. Aerobic and anaerobic CO_2 production by apples fruits following air and controlled atmosphere storage. Proc. Amer. Soc. Hort. Sci. 84:59.

Eaks, I. L., and Ludi, W. A. 1960. Effects of temperature, washing and waxing on the composition of the internal atmosphere of orange fruits. Proc. Amer. Soc. Hort. Sci. 76:220.

Fidler, J. C., 1951. A comparison of aerobic and anaerobic respiration of apples. J. Exp. Bot. 2:41.

Fidler, J. C., 1933. Studies in zymosis. IV. The accumulation of zymosis products in apples during senescence. Biochem. J., 27: 1614.

Furlong, C. R., 1961a. Accumulation of alcohol in anaerobic tissues. Agr. Res. Council Rept. Ditton and Covert Garden Lab. 1959 and 1960:44.

Furlong, C. R., 1961b. Accumulation of alcohol in anaerobic tissues. Agr. Res. Council Rept. Ditton and Covert Garden Lab. 1961:24.

Henze J. 1969. Beziehungen zwischen respiration and innerer atmosphäre bei CO_2-empfindlichen Kernobstorten Qual. Plant Mater. Veg. XIX, 1-3:229.

Hulme, A. C., Smith, W. H., and Wooltorton, L. S. 1964. Biochemical changes associated with the development of low temperature breakdown in apples. J. Sci. Food Agric. 15:303.

Karaoulanis, D. G. 1968. The effects of storage under controlled atmosphere conditions on the aldehyde and alcohol of oranges and grapes. Agr. Res. Council Rept. Ditton Lab. 1968:36.

Karaoulanis, D. G. 1979. Biochemical changes in washing on Navel oranges during their growing and under different storage conditions. XVth Inter. Congr. of Refrig. Venezia III:837.

Karaoulanis, D. G. 1978. The changes of ethanol and other chemical characteristics of Washington Navel oranges during their growing and under different storage conditions. Abst. XXth Int. Hort. Cong., Sydney.

Kostytschew, S., 1913. The character of anaerobic respiration in various seed plants. Ber.d.d.Bot.Ges. 21:125.

Lyons, M. J., McGlasson, W. B., and Pratt, H. K. 1962. Ethylene production, respiration and internal gas concentrations. Plant Physiol. 36:31.

Neal, G. E., and Hulme, A. 1958. The organic acid metabolism of Bramley's seedling apple peel. J. Exp. Biol. 9:142.

Pfeffer, W. 1900-1903. The physiology of plants Vol. I, II and III. Translated by A. J. Ewart, Osford.

Smith, W. H. 1954. Structure of the mature apple fruit in relation to gaseous exchange. Int. Bot. Congr. Paris, 8: Secs. 11/12, pp. 405.

Thomas, M. 1925. The controlling influence of CO_2 V. A qualitative study of the production of ethyl alcohol and acetaldehyde by cells of the higher plants in relation to concentration of O_2 and CO_2. Biochem. J. 19:921.

Thomas, M. 1929. The production of ethyl alcohol and acetaldehyde by apples in relation to the injuries occurring in storage. Part I. Ann. Applied Biol. 16:444.

Trout, S. A., Huelin, F. E. and Tindale, E. B. 1960. The respiration of W. Navel and Valencia oranges. Con. Sci. Industr. Res. Org. Aust. Bull. No. 14.

Ulrich R., and Marcellin, P. 1955. Voies et modalités des échanges de gaz carbonique et d'oxygène des fruits avec l'atmosphere ambiante. J. Recherches Centre nat Recherche Sci. Labs. Bellevue (Paris) 6:241.

Ulrich, R. 1958. Postharvest physiology of fruits. Ann. Rev. Plant Phys. 9:385.

Vines, H. M. and Oberbacher, M. F. 1961. Changes in CO_2 concentrations within fruit and containers during storage. Proc. Fla. State Hort. Soc. 74:243.

Wardlaw, C. W., and Leonard E. R. 1939. Studies in tropical fruits. IV. Methods in the Investigations of respiration with special reference to the banana. Ann. Bot. N.S. 3:27.

Wills, R. B. H. and Skott, K. J. 1968. Relation between low temperature breakdown and the volatiles given off by individual apples. Aust. J. Biol. Sci. 21:1307.

Wills, R. B. H. and McGlasson, W. B. 1971. Effect of storage temperature on apple volatiles associated with low temperature breakdown. J. Hort. Sci. 46:119.

Young, R. E. and Biale, J. B. 1962. Carbon dioxide effects on fruit respiration. Endeavour. 21:164.

HOST-PATHOGEN INTERACTIONS IN POSTHARVEST DISEASES

Joseph W. Eckert and Malinie Ratnayake

Department of Plant Pathology
University of California
Riverside, CA 92521

INTRODUCTION

Every fruit, vegetable and root crop is exposed to hundreds of species of microorganisms during its lifetime yet the vast majority of these are incapable of extensively invading the plant tissues as long as the cells are living and the tissues are coherent. Approximately 25 species of fungi and bacteria are responsible for the major decays of plant products after harvest. Most of these microorganisms have the potential for attacking only a few different products from related species of plants. For example, Penicillium digitatum causes green mold on citrus fruits but does not cause disease in apples and pears. P. expansum on the other hand, attacks apples and pears, but not citrus fruits. Successful pathogens must have the capability to overcome the host defenses, must be able to initiate growth under conditions of water potential, pH and nutrients present in the host tissues, and must elaborate enzymes that macerate host tissue and cause a release of nutrients required to sustain the indeterminate parasitic development of the pathogen in the host tissues.

As living entities, the host and pathogen possess the potential to interact with each other as well as with the physicochemical environment. Their individual reaction to a stimulus may be vastly different, however, because of physiological differences of the cells of higher plants and microorganisms. For example, ethylene accelerates senescence of the tissues of many fruits and increases their susceptibility to attack by pathogenic microorganisms. Yet, this plant hormone has no measurable effect upon most pathogenic microorganisms. Fungal infection of certain fruits is stimulated by certain chemical constituents of that fruit, whereas other chemicals inhibit the growth of the pathogen in the host. The attacking pathogen may elicit the production of fungitoxic substances by the host to ward off invasion. Finally, constituents of the host often induce the pathogen to elaborate enzymes that destroy host tissues, whereas other plant constituents inhibit the action of these enzymes.

Some host-pathogen interactions promote the development of pathogenesis, whereas other interactions retard or prevent the process. Since living fruits and vegetables are usually inhospitable substrates for extensive invasion by most microorganisms, the unfavorable interactions evidently outweight the favorable interactions in most plant-microorganism combination. The nature of these interactions and their influence upon the development of postharvest diseases is the subject of this chapter.

DEVELOPMENT OF POSTHARVEST DISEASES

Diseases of fruits and vegetables observed in the marketplace and in the home may be traced chronologically to: 1) infection or infestation of the product during the period of its development in the field, or 2) infection through wounds created during the harvesting operation and through physiological injuries resulting from an unfavorable storage environment (Table 1). The subsequent progress of an infection depends upon the growth requirements and enzymatic capabilities of the pathogen and upon the physiological resistance of the plant tissue to invasion.

Table 1. Time and Site of Infection in some Representative Postharvest Diseases

Disease	Pathogen	Infection Site
I. Field Infection		
Apple lenticel rot	*Gloeosporium*	lenticel
Banana anthracnose	*Colletotrichum*	epidermis
Peach brown rot	*Monilinia*	epidermis
Orange stem-end rot	*Diplodia*	calyx
Strawberry gray mold	*Botrytis*	flower
II. Postharvest Infection		
Banana crown rot	*Colletotrichum*	crown cushion
Pineapple black rot	*Thielaviopsis*	fruit stem
Peach *Rhizopus* rot	*Rhizopus*	surface injuries
Citrus blue-green mold	*Penicillium*	surface injuries
Tomato sour rot	*Geotrichum*	surface injuries
Potato soft rot	*Erwinia*	surface injuries

Infections of several species of fungi, especially if initiated early in the growing season, enter a latent or quiescent state soon after the initial stages are completed because the host tissues resist further invasion. After harvest the fruit gradually loses this resistance as it ripens, and the pathogen resumes growth, giving rise to an active decay lesion on the host. Diplodia, Phomopsis, and Alternaria establish latent infections in the stem ends of citrus fruits because this part is vulnerable to attack when the fruits are very small. Alternaria may also cause a latent infection in the stylar end of the navel orange. Gloeosporium rots of apples arise from quiescent infections in the lenticels of the fruits which are initiated months before the rot is apparent in storage. Anthracnose diseases of bananas, papayas, and mangoes arise from latent infections initiated randomly over the entire surfaces of these fruits.

Gray mold (Botrytis) on grapes and brown rot (Phytophthora) on citrus fruits come from infections of fruits in the field that begin shortly before harvest. These infections may not be truly quiescent, but grow rather slowly in mature fruits, depending upon the disease resistance of the host tissues and the ambient temperatures.

The second major class of postharvest diseases originates by infection through wounds which occur during the harvest and preparation of the product for market. These infection sites may be localized--the severed stem of the pineapple, the avocado, and the pear--or more or less randomly distributed on the entire surface as abrasions or cuts caused by rough handling. Several examples of disease caused by preharvest and postharvest (wound) infection are given in Table 1. The etiology of the postharvest diseases of tropical and temperate crops has been considered in detail in other reviews (24,25,26).

ROLE OF PATHOGENS' ENZYMES IN DISEASE DEVELOPMENT

Plant cell walls are held together by intercellular material composed mainly of pectic polysaccharides. Development of a postharvest disease depends upon the capability of the pathogen to secrete enzymes that depolymerize these insoluble pectic polymers, leading to loss of tissue coherence and separation of the individual cells, a process referred to as tissue "maceration." The cells of the macerated tissues increase in permeability and die, allowing outward diffusion of host metabolites which may be used as substrates for growth by the pathogen (52).

Pathogens involved in postharvest diseases produce endopolygalacturonases, endo pectin lyases, and endo pectate lyases (7). Although the random hydrolysis of α-1,4 glycosidic linkages in polygalacturonans is sufficient to cause tissue maceration and cell death, cellulases and hemicellulases may also be involved in pathogenesis. Recently, a proteinaceous factor has been isolated which stimulates tissue maceration by fungal enzymes, but the factor is essentially devoid of activity alone or in *in vitro* assays with pure enzymes (42).

The endopolygalacturonases are enzymes which hydrolyze the α-1, 4 galacturonide linkages of pectic acid with different degrees of randomness to produce a series of oligogalacturonides. These enzymes have been associated with tissue maceration by several microorganisms that cause postharvest disease, e.g., *Penicillium*, *Rhizopus*, *Geotrichum*, *Sclerotinia* and the bacterium *Erwinia carotovora*, as well as many non-pathogenic microorganisms growing in culture (61). Maceration of citrus peel by *Geotrichum* (73), sweet potatoes by *Rhizopus* (65), and persimmons by *Gloeosporium* (71) can be accounted for entirely by the action of extracellular endopolygalacturonases produced by the pathogens.

The endopectin lyases split the α-1, 4 galacturonide linkages of pectin to different degrees of randomness by a transelimination mechanism so that a series of C_4-C_5 unsaturated methylated oligogalacturonides are produced. The pectin lyases which have been characterized are all of fungal origin. Pectin lyases have been implicated in the maceration of citrus fruit peel attacked by *Penicillium digitatum* and *Penicillium italicum* (19), but this has been questioned recently by Barmore and Brown (3,4) who identified endopolygalacturonase and exopolygalacturonase as the principal enzymes responsible for the maceration of orange peel by *Penicillium italicum* and *P. digitatum*, respectively. These investigators did not find pectin lyases in extracts of fruit invaded by either of these pathogens.

The third group of enzymes associated with postharvest diseases, the endopectate lyases, are similar to the pectin lyases except that their substrate is pectic acid and their product, oligogalacturonides. This group of enzymes is commonly produced by pectolytic bacteria and the pectate lyases are the major cause of softening of potatoes attached by soft rot bacteria (53), although endopolygalacturonases may also be involved (8).

The production of extracellular endopolygalacturonases and pectin lyases by fungi is generally induced by pectate or pectin in their environment. Cooper and Wood (21) reported that these enzymes in the fungi *Verticillium* and *Fusarium* were induced by low concentrations of galacturonic acid rather than pectin or pectic acid and that the production of the enzymes was repressed when the inducer was in slight excess over that required for growth of the fungi. The extracellular pectate lyases produced by *Erwinia caratovora* are usually induced by pectin and, may or may not, be subject to catabolite repression, depending upon the individual isolate of the bacterium and the environment (56, 79).

Although microorganisms which cause postharvest disease must produce extracellular pectolytic enzymes, it does not follow that microorganisms that produce pectolytic enzymes, can cause postharvest diseases. A number of nonpathogenic fungi and bacteria are capable of producing large amounts of enzymes which macerate slices of potatoes and other plant products (42). Both virulent and avirulent isolates of Penicillium spp. and Erwinia carotovora produce pectolytic enzymes in culture, indicating that the ability to attack native pectin in the host is not measured by pectolytic activity in vitro. However, avirulent strains of Erwinia carotovora showed lower activity of pectate lyase and endopolygalacturonase than did virulent strains (8,31).

SUSCEPTIBILITY OF HOST TISSUE TO DISEASE DEVELOPMENT

The susceptibility of fruits and vegetables to postharvest decay is influenced dramatically by the maturity of the crop at the time of harvest and by subsequent physiological changes, collectively termed "ripening." The susceptibility of apples to blue mold (Penicillium expansum) increases with both maturity and ripeness, and appears to result from an increase in susceptibility of the flesh to bruise damage. Likewise, navel oranges become more susceptible to invasion by Penicillium molds with advance in maturity during the season. The resumption in activity of quiescent infections during ripening of fruits is a clear example of increasing disease susceptibility, rather than increased susceptibility of the host tissue to mechanical injury. Lenticel rotting of apples is virtually unknown at the time of harvest but develops in storage as the fruit ripens. Different pathogens appear at different stages of ripening of apples in storage. A striking increase in Alternaria-stem-end-rot occurs when lemons pass a certain threshold of ripeness during storage. The lemon peel is highly resistant to invasion by Alternaria citri prior to this stage of development. Latent infections of Colletotrichum on banana, papaya and mango fruits seldom become a serious problem until the fruit approaches ripeness. A number of investigators have observed that potatoes become more susceptible to Fusarium dry rot during storage; the susceptibility of carrots to Botrytis and Centrospora at 5°C also increases with duration of the storage period (22, 23, 32).

A few examples of disease susceptibility decreasing with maturity or storage of the crop are known. Several investigators have recorded that green tomatoes are more susceptible than red tomatoes to bacterial soft rot. White potatoes become less susceptible to bacterial soft rot during the first few weeks of storage because the periderm layer matures during this period.

Since susceptibility to disease increases with ripeness, it is not surprising that the application of ethylene gas to accelerate ripening has been observed to increase Alternaria rot on tomatoes (63), and stem end rot and anthracnose on citrus fruits (17,50). McCornack (50) demonstrated that ethylene degreening of oranges increased Diplodia stem end rot, but not Penicillium green mold, and that there was a direct relationship between the ethylene concentration (4-50 ppm) and stem end rot (Table 2). Five to ten parts per million ethylene is the minimum required for satisfactory coloring of the fruit. The ethylene treatment accelerates senescence and abscission of the fruit button, thereby stimulating the development of quiescent infections of Diplodia in this organ and providing an avenue for the fungus to enter the fruit (18).

Ethylene treatment of green tangerines stimulates appressoria of Colletotrichum present on the surface of the fruit to form infection hyphae which then penetrate the epidermal cells of the fruit (14, 17). Apparently the ethylene treatment simultaneously triggers an unknown resistance mechanism since the fruit become resistant to infection if they are treated with ethylene three days before inoculation with Colletotrichum spores.

The intrinsic and variable resistance of fruits and vegetables to postharvest disease, especially with aging, might be associated with one or a combination of several properties of the host (Table 3).

STIMULATION OF PATHOGEN BY HOST CONSTITUENTS

The development of a pathogen on the surface of its host may be influenced by chemicals exuded from the interior of the host plant. There are many reports that simple sugars and nitrogenous compounds are released to the surface of fruits and leaves where they stimulate spore germination and germ tube growth of plant pathogens. However, these substances often delay the formation of infection structures such as appressoria.

Anthranilic acid, which has been identified in the leachate from banana fruit, is metabolized to 2,3-dihydroxybenzoic acid by the fungus Colletotrichum musae, the pathogen of banana anthracnose (36, 70). The latter compound stimulates conidial germination and appressorium formation by the pathogen. Similarly, leachates of apple fruits (cv. Bramley's Seedling) contained chlorogenic acid and p-coumarylquinic acid, both of which stimulated spore germination and formation of appressoria of the apple-rotting pathogen Diaporthe perniciosa (11). The leachates showed most stimulatory activity in mid-August when the concentration of p-coumarylquinic acid was highest in the leachate.

Table 2. Effect of Ethylene on Anthracnose, Stem End Rot, and Green Mold of Citrus Fruits

Ethylene (µl/liter air)	Percent fruit decayed by Anthracnose[1]	Stem end rot[2]	Green mold[2]
0	0	0	6
5	-	13	3
10	17	-	-
20	57	-	-
50	86	23	2

[1]Data of Brown and Barmore 1977. Anthracnose incited by Colletotrichum gloeosporioides. Robinson tangerines exposed to ethylene for 2 days and then stored for 1 wk at 26°C and 100% relative humidity.

[2]Data of McCornack 1972. Stem end rot incited mostly by Diplodia natalensis and green mold by Penicillium digitatum. Valencia oranges exposed to ethylene for 2 days and then stored at 21°C for 4 wk.

Table 3. Intrinsic Factors Influencing Resistance and Susceptibility of Plant Products to Attack by Microorganisms

1. Stimulators of pathogen growth and infection
2. Inhibitors of pathogen growth or enzyme action
   a. Preformed
   b. Elicited by pathogen attack
3. Hypersensitivity of invaded tissue
4. Water status
5. pH and nutritional level
6. Morphological barriers to pathogen spread
7. Resistance of host tissue to pathogen's enzymes

INHIBITION OF PATHOGEN GROWTH AND ENZYME ACTIVITY BY HOST CONSTITUENTS

Two types of microbial growth inhibitors have been recognized in plant tissues: 1) preformed compounds and 2) inhibitors that are synthesized by the host in response to attempted infection by a pathogen. The latter are referred to as "phytoalexins." Examples of preformed inhibitors that may offer resistance in fruits to infection by pathogens are tannins and 3,4-dihydroxybenzaldehyde in bananas (5,34,54) and phenolic compounds in apples (20,41,55).

The biosynthesis of antimicrobial compounds by plant tissue in response to injury is a rather common phenomenon (45). Several investigators have reported that the production of 6-methoxymellein by the carrot is stimulated by the application of spores of *Rhizopus stolonifer*, *Botrytis cinerea* and other storage pathogens to the injured surface of the roots (33,35,37,38,51). The loss of biosynthetic ability of the carrot to produce this compound during long-term storage is believed to be an important factor in the increase in susceptibility of carrot roots during this period (32,37). The carrot also synthesizes *p*-hydroxybenzoic acid and the polyacetylene falcarinol which are fungitoxic and may play a role in the resistance of the carrot to postharvest disease (35).

The pulp of green tomatoes contains substantial amounts of tomatin (an alkaloid saponin) but the concentration decreases with fruit maturity until only traces are detectable in yellow and red fruit. Tomatin is a weak base and possesses maximum fungitoxic activity in its non-ionized form which exists in neutral to alkaline solutions. Although the tissue of green tomatoes is acidic, Schlosser (62) reported that the fungus *Gloeosporium fructigena* increased the pH of the tomato tissue at the infection site to pH 6.4 at which pH the amount of tomatin present was inhibitory to the pathogen. Hence, *G. fructigena* is capable of colonizing mature tomato fruits only and not green fruits. Verhoeff and Liem (75, 76) believed that latency of the pathogen *Botrytis* in tomato fruit could also be explained in part by tomatin.

The antimicrobial terpenoid, rishitin, formed in potato tubers inoculated with *Erwinia carotovora* and stored in air may be a factor in restricting the expansion of lesions of bacterial soft rot which develops under these conditions (48).

Nectria galligena infects apple fruits in the late summer but resistance of the fruit tissue prevents further growth of the fungus. Attack by the pathogen causes the host tissue to synthesize benzoic acid around the site of infection. At the pH of the immature apple flesh (pH 3), benzoic acid is in its most fungitoxic form (undissociated) and prevents extensive growth of the fungus in the apple (9,10). During storage, the pH of the fruit tissue increases, benzoic acid dissociates to the anionic form and the fungus resumes growth producing disease lesions in the fruit. Synthesis of benzoic acid is elicited by an extracellular protease produced by *Nectria galligena*; other apple pathogens such as *Penicillium expansum* that do not elaborate the protease do not elicit the production of benzoic acid by the apple tissue (68, 69).

Brown and Swinburne (12) demonstrated recently that phytoalexins probably play a role in the latency of *Colletotrichum musae* on banana fruit. Drops of water containing spores of this fungus placed on green banana fruit cause necrosis of cells in the tissues directly beneath the drops. Five compounds that inhibited the growth of C. musae were isolated from the necrotic tissue on green fruit, but the fungitoxic compounds disappeared as the fruit ripened. Brown and Swinburne (12) isolated a glycan containing polysaccharide from the cell wall of C. musae which elicited necrotic lesions, as well as the two major fungitoxic compounds elicited by living conidia of C. musae. It seems likely that the fungitoxic compounds elicited by this pathogen play a role in the latent behavior of the fungus on green fruit and the fungus resumes growth when the compound disappear as the fruit ripens.

The distribution of substances in plant tissues that may inhibit enzymes of the pathogen is probably more general than is usually appreciated. General enzyme inhibitors such as polyphenols are known to be present in the flesh of apples (20, 41, 55) and the level of these compounds generally decreases as the fruit mature. Dicotyledonous plants have proteins associated with their cell walls which are capable of inhibiting the action of endopolygalacturonase secreted by plant pathogens. Albersheim and Anderson-Prouty (2) suggest that a pathogen is incapable of attacking a plant unless the pathogen can secrete sufficient endopolygalacturonase to overcome the inhibitor present in the wall of the plant.

HYPERSENSITIVITY OF HOST TISSUE CHALLENGED BY THE PATHOGEN

Appressoria of Colletotrichum gloeosporioides are commonly present on the surface of Robinson tangerines during the period of fruit development in the grove but they do not germinate and infect green fruit. Treatment of the fruit after harvest with ethylene caused a loss of chlorophyll and an increase in carotenoids in the fruit, as well as stimulating the appressoria to send out infection hyphae into the fruit (14,15,16,17). However, if the appressoria are removed from the surface of green fruit before treatment with ethylene, the ethylene-treated fruit become orange in color and simultaneously develop resistance to invasion by Colletotrichum spores applied subsequently. Brown (16) has shown that penetration of an epidermal cell of orange-colored fruit by an infection hypha from an appressorium of C. gloeosporioides results in the death of four layers of cells surrounding the infection hyphae. The necrotic cells become filled with phenolic compounds. Apparently, the dead phenol-filled cells of the host form a barrier to further spread of the fungus hyphae, localizing the pathogen to the area penetrated by the infection hyphae.

WATER STATUS OF THE HOST

Many fruits and vegetables are more susceptible to invasion by pathogens when their tissues are turgid. This is usually attributed to the presence of a water film associated with injuries which supports the growth of the pathogen during the infection process. Thus, slight desiccation reduces the susceptibility of citrus fruits to Penicillium digitatum (27), potatoes to bacterial soft rot (58) and carrots to Centrospora acerini (22,23). In contrast, both Botrytis and Rhizopus preferentially infect wilted carrots (23,32). The susceptibility of carrots to infection by these fungi increases when the tissues exceed 8% water loss; more turgid roots are resistant. The increase in susceptibility of slightly flaccid carrots has been attributed to an increase in the intercellular spaces as the cells begin to separate at approximately 7% water (72). Cabbage also seems to be more resistant to Botrytis and other pathogens when stored at very high humidity which delays the onset of senescence of the outer leaves (74,77,78). These effects of host water status upon the disease reflect the physiological status of the host rather than a direct effect upon the pathogen, since most microorganisms involved in postharvest diseases grow well at water potential deficits larger than those that are likely to be encountered in the tissues of palatable fruits and vegetables.

pH AND NUTRITIONAL LEVEL OF HOST TISSUES

Acidity of the tissue of many fruits may be one of the most important reasons for their general resistance to bacteria that cause soft rot of many vegetable crops (47). Tomatoes, peppers cucumbers, and pears are the few fruits known to be seriously affected by bacterial soft rot. Fruit tissues are usually below pH 5 which inhibits the growth of most bacteria capable of degrading plant tissues; vegetable tissues are generally less acid (23, 47). A decrease in the acidity of the flesh of Bramley's Seedling apple during storage may be responsible for the decrease in resistance to Nectria galligena (68). The increase in pH of the flesh associated with ripening in storage favors the dissociation of benzoic acid in the apple to the less fungitoxic dissociated form.

Several examples of plant constituents that stimulate preinfection development of the pathogen on the host surface have been described. Clear-cut examples of unique nutrients or growth stimulants that determine susceptibility in fruits and vegetables are rare, but the possibility of their existence must none-the-less be considered. Penicillium digitatum, a unique pathogen on citrus fruits, requires a complex medium for rapid germination and vigorous hyphal growth. Citrus fruits contain both ascorbic acid and certain terpenes that are known to stimulate germination of P. digitatum spores (30, 57) as well as a complement of organic nutrients that sustain vigorous growth during the infection process. Otazu and Secor (56) found a highly significant correlation between reducing sugar content and susceptibility to bacterial soft rot in Norgold Russet potato tubers of various ages held in different storage temperatures, and Bartz et al. (6) reported that bacterial soft rot susceptibility of tomato fruit was influenced by nitrogen fertilization of the plant in the field.

RESTRICTION OF PATHOGEN GROWTH BY MORPHOLOGICAL BARRIERS OF THE HOST

In response to injury, plant tissue may form protective barriers of tightly packed cells, provided that the tissue is still capable of cell division, or the cells surrounding the injury can deposit lignin and suberin in their walls to protect against the action of macerating enzymes of microorganisms. Lenticels of potatoes are normally blocked with a layer of suberized periderm which prevents the entry of soft rot bacteria into the cortex of the tuber (1,29). Rupture of this periderm layer, either mechanically or by proliferation of underlying cells of the tuber in response to free water, makes the lenticels susceptible to invasion by soft rot bacteria (29,58). A similar situation exists in the case of lenticels of immature apples (28).

If potato tubers are superficially wounded and then placed at high humidity and moderate temperatures, suberin begins to form in the walls and intercellular spaces of the living cells surrounding the injury and, within several days, a suberized periderm forms beneath these suberized surface cells (29). These structure offer substantial resistance to the invasion of pathogens at wounded areas on the tuber. The surface cells of carrots also become suberized in response to wounding and this suberized surface layer is a barrier to infection by Botrytis (32, 37, 39).

Suberization of cells around injuries is probably the most important defense of the potato against infection by microorganisms, since this reaction takes place within 24 hours under optimal environmental conditions (29). The suberized periderm which follows in about four days is highly resistant to infection by microorganisms because it is composed of tightly packed cells, the walls of which are impregnated with suberin and have only a small amount of pectin (29). Periderm formation is most rapid at 21-27°C at high humidity, but storage temperatures of 16-21°C are usually recommended to reduce the rate of growth of Erwinia carotovora and Fusarium. Smith and Smart (64) found that the periderm formed on potato slices at 10°C in four days was adequate to protect against bacterial soft rot whereas slices held at 4.4°C formed only a slight amount of suberin, no periderm, and were highly susceptible to decay after transfer to 21°C. Low temperatures favor the development of gangrene (Phoma) on potatoes in storage because the pathogen continues to develop while the low temperature prevents the formation of barriers by the host. Carrots held at 22-26°C and high humidity for two days to permit wound healing before storage had less decay after long-term storage at 5°C than an uncured lot (22). Curing of sweet potatoes and yams for 4-7 days at 26-32°C and 85-90% relative humidity allows for the tubers to form a suberized periderm to protect harvest injuries against invasion by Rhizopus and Endoconidiophora during subsequent storage of the crop (67). Wound healing proceeds most rapidly at 30°C whereas Rhizopus grows most rapidly at 24°C (49).

The ability of tissues of some plants to form a barrier of lignin around an injury is apparently an important mechanism of defense against invasion by a pathogen. Holding oranges at 30°C and 90-96° RH for several days after harvest reduces Penicillium decay (13,43) because this environment is highly favorable to the synthesis of lignin and the temperature is too high for the development of Penicillium (Table 2). The lignin barrier is not attacked by enzymes of the pathogen and probably prevents pectic enzymes of the pathogen from depolymerizing the middle lamella of the host cells (60). Resistance of immature papaya fruit to Colletotrichum gloeosporioides has been correlated with the ability of the wounded immature fruit to ward off the pathogen by the formation of a lignified periderm (66).

RESISTANCE OF HOST TISSUE TO PATHOGEN'S ENZYMES

The coherence of the flesh of fresh fruits and vegetables is dependent, to a large degree, upon the pectic substances in the middle lamella which functions as an adhesive, binding the cells together. The pectic materials are made up of linear chains of D-galacturonic acid, methylated to varying degrees (pectin), and with side chains of neutral sugars. As initially laid down by the cell, the pectic substances in the middle lamella are in an insoluble form known as protopectin. The insolubility of protopectin is a consequence of high molecular weight and an anchoring of the linear chains of pectin and pectic acid to the cellulose of the cell wall through bridges of neutral polysaccharide and perhaps proteins (59). In this insoluble state, the middle lamella is resistant to hydrolysis by pectolytic enzymes of plant pathogens(7). Kuc et al. (44) concluded that this cross linking of pectin to the cell walls was a major reason for the resistance of immature apple fruits to the rot caused by Botryosphaeria ribis.

As fruits and vegetables ripen, the bonds anchoring the pectic materials to the cell wall are broken with the result that the pectic materials become soluble and the tissue begins to soften. The increased solubility of the pectic substances and perhaps, increased openness of the middle lamella makes the tissue more vulnerable to maceration by the pectolytic enzymes of pathogens. The susceptibility of strawberries to attack by Botrytis cinerea is highly correlated with the soluble pectin content of the fruit (40). Barmore and Brown (3) found that pectin was demethylated by pectin methylesterase that leaked from the cells of orange peel attacked by Penicillium digitatum and that the resulting polygalacturonic acid was more readily depolymerized by the exopolygalacturonse of the pathogen than was native citrus pectin.

CONCLUSION

The fact that relatively few microorganisms are capable of parasitizing fresh plant products, i.e., living fruits, tubers and leafy parts, reveals that a unique relationship exists between host and pathogen in postharvest pathology. This situation suggests that opportunities may exist to reduce postharvest deterioration by increasing the resistance of the host to infection or at least by delaying the onset of senescence that is invariably accompanied by an increase in disease susceptibility.

Harvested fruits and vegetables possess the ability, to varying degrees, to reduce the vulnerability of surface injuries to pathogen invasion by the deposition of biologically-resistant barriers of lignin, suberin, callose and layers of tightly packed cells (wound periderm). These anabolic reactions are favored by high relative humidity and usually the temperature-rate relationship is distinct from that of the pathogen growth curve. Thus environmental conditions can often be discovered that encourage the development of a barrier in wounds before the pathogen can invade the tissue.

Another interesting possibility for disease control involves the intensification of the natural defense mechanisms of the plant, pathogen-inhibiting compounds or structural barriers, by the application of chemical or physical agents to the plant product after harvest. Langcake (46) recently reviewed the status of research in this field as applied to the treatment of growing plants.

Numerous investigations on the optimum environment for storage of plant products have clearly demonstrated the benefits of delaying senescence upon resistance to disease arising from pathogens that are suppressed in the immature host (latent infections) and those that invade through wounds. Temperature management, relative humidity, inert atmospheres, ethylene removal and hypobaric conditions can all have a beneficial role in reducing disease lossess. Senescence-inhibiting compounds such as 2,4-D and gibberellin are used in practice to reduce susceptibility of citrus fruits and this approach may develop greater applicability as more versatile agents of this type are discovered.

REFERENCES

1. Adams, M. J. (1975). Potato tuber lenticels: susceptibility to

infection by Erwinia carotovora var. atroseptica and Phytophthora infestans. Ann. Appl. Biol. 79: 275.
2. Albersheim, P. and Anderson-Prouty, A. J. (1975). Carbohydrates, proteins, cell surfaces and the biochemistry of pathogenesis. Ann. Rev. Plant Physiol. 26: 31.
3. Barmore, C. R. and Brown, G. E. (1979). Role of pectolytic enzymes and galacturonic acid in citrus fruit decay caused by Penicillium digitatum. Phytopathology 69: 675.
4. Barmore, C. R. and Brown, G. E. (1980). Polygalacturonase from citrus fruit infected with Penicillium italicum. Phytopathology 71: 328.
5. Barnell, H. R. and Barnell, R. (1945). Studies in tropical fruits. 16. The distribution of tannins within the banana and the changes in their condition and amount during ripening. Ann. Botany N. S. 5: 608.
6. Bartz, J. A., Geraldson, G. M. and Grill, J. P. (1979). Nitrogen nutrition of tomato plants and susceptibility of the fruit to bacterial soft rot. Phytopathology 69(2): 163.
7. Bateman, D. F. and Millar, R. L. (1966). Pectic enzymes in tissue degradation. Ann. Rev. Phytopath. 4: 119.
8. Beraha, L., Garber, E. D. and Billeter, B. A. (1974). Enzymes profiles and virulence in mutants of Erwinia carotovora. Phytopath. Z. 81: 15.
9. Brown, A. E. and Swinburne, T. R. (1971). Benzoic acid: an antifungal compound formed in Bramley's Seedling apple fruits following infection by Nectria galligena Bres. Physiol. Plant Pathol. 1: 469.
10. Brown, A. E. and Swinburne, T. R. (1973). Factors affecting the production of benzoic acid in Bramley's Seedling apples infected with Nectria galligena. Physiol. Plant Pathol. 3: 91.
11. Brown, A. E. and Swinburne, T. R. (1978). Stimulants of germination and appressorial formation by Diaporthe perniciosa in apple leachate. Trans. Br. Mycol. Soc. 71(3):405.
12. Brown, A. E. and Swinburne, T. R. (1980). The resistance of immature banana fruits to anthracnose (Colletotrichum musae). Phytopath. Z. 99: 70.
13. Brown, G. E. (1973). Development of green mold in degreened oranges. Phytopathology 63: 1104.
14. Brown, G. E. (1975). Factors affecting postharvest development of Colletotrichum gloeosporioides in citrus fruits. Phytopathology 65: 404.
15. Brown, G. E. (1977). Ultrastructure of penetration of ethylene-degreened Robinson tangerines by Colletotrichum gloeosporioides. Phytopathology 67(3): 315.
16. Brown, G. E. (1978). Hypersensitive response of orange-colored Robinson tangerines to Colletotrichum gloeosporioides after ethylene treatment. Phytopathology 68: 700.
17. Brown, G. E. and Barmore, C. R. (1977). The effect of ethylene on susceptibility of Robinson tangerines to anthracnose.

Phytopathology 67: 120.
18. Brown, G. E. and Wilson, W. C. (1968). Mode of entry of Diplodia natalensis and Phomopsis citri into Florida oranges. Phytopathology 58: 736.
19. Bush, D. A. and Codner, R. C. (1970). Comparison of the properties of the pectin transeliminases of Penicillium digitatum and Penicillium italicum. Phytochemistry 9: 87.
20. Byrde, R. J. W. (1963). Natural inhibitors of fungal enzymes and toxins in disease resistance. In Perspectives of Biochemical Plant Pathology, Ed. by S. Rich. Conn. Agr. Expt. Sta. Bull 663.
21. Cooper, R. M. and Wood, R. K. S. (1975). Regulation of synthesis of cell wall degrading enzymes by Verticillium albo-atrum and Fusarium oxysporum f. sp. lycopersici. Physiol. Plant Pathol. 5: 135.
22. Davies, W. P. (1977). Infection of carrot roots in cool storage by Centrospora acerini. Ann. Appl. Biol. 85: 163.
23. Dennis, C. (1977). Susceptibility of stored crops of microbial infection. Ann. Appl. Biol. 85: 430.
24. Eckert, J. W. (1977). Control of postharvest diseases, pp. 269-352, In Antifungal Compounds. Vol. 1, Ed. by M. R. Siegel and H. D. Sisler, Marcel Dekker Inc., New York.
25. Eckert, J. W. (1978). Pathological diseases of fresh fruits and vegetables, pp. 161-209. In Postharvest Biology and Biotechnology, Ed. by H. O. Hultin and N.Milner, Food and Nutrition Press, Westport, Conn.
26. Eckert, J. W. (1978). Postharvest diseases of citrus fruit. Outlook on Agric. 9: 225.
27. Eckert, J. W. and Kolbezen, M. J. (1963). Control of Penicillium decay of oranges with certain volatile aliphatic amines. Phytopathology 53: 1053.
28. Edney, K. L. (1958). Observations on the infection Cox's Orange Pippin apples by Gloeosporium perennans Zeller and Childs. Ann. Appl. Biol. 46: 622.
29. Fox, R. T. V., Manners, J. G. and Myers, A. (1971). Ultrastructure of entry and spread of Erwinia carotovora var. atroseptica into potato tubers. Potato Res. 14: 61.
30. French, R. C., Long, R. K. Latterell, F. M. Graham, C. L., Smoot, J. J. and Shaw, P. E. (1978). Effect of n-nonanal citral, and citrus oils on germination of conidia of Penicillium digitatum and P. italicum. Phytopathology 68: 877.
31. Friedman, B. A. (1962). Physiological differences between a virulent and weakly virulent radiation-induced strain of Erwinia carotovora. Phytopathology 52: 328.
32. Goodliffe, J. P. and Heale, J. B. (1977). Factors affecting the resistance of cold-stored carrots to Botrytis cinerea. Ann. Appl. Biol. 87: 17.
33. Goodliffe, J. P. and Heale, J. B. (1978). The role of 6-methoxymellein in the resistance and susceptibility of

carrot root tissue to the cold storage pathogen Botrytis cinerea. Physiol. Plant Pathol. 12: 27.
34. Greene, G. L. and Morales, C. (1967). Tannins as the cause of latency in anthracnose infections of tropical fruits. (Gloeosporium musarum in bananas). Turrialba 17:447.
35. Harding, V. K. and Heale, J. B. (1980). Isolation and identification of the antifungal compounds accumulating in the induced resistance response of carrot root slices to Botrytis cinerea. Physiol. Plant Pathol. 17: 277.
36. Harper, D. B. and Swinburne, T. R. (1979). 2,3-Dihydroxybenzoic acid and related compounds as stimulants of germinations of conidia of Colletotrichum musae (Berk. & Curt) Arx. Physiol. Plant Pathol. 14: 363.
37. Heale, J. B., Harding, V., Dodd, K. and Gahan, P. B. (1977). Botrytis infection of carrot in relation to the length of the cold storage period. Ann. Appl. Biol. 85: 453.
38. Heale, J. B., Dodd, K. and Gahan, P. B. (1978). Cytochemical changes in carrot root tissue treated with heat-killed and live conidia of Botrytis cinerea. Ann. Appl. Biol. 89: 310.
39. Heale, J. B. and Sharman, S. (1977). Induced resistance to Botrytis cinerea in root slices and tissue cultures of carrots (Daucus carota L.) Physiol. Plant Pathol. 10: 51.
40. Hondelmann, W. R. E. and Richter, E. (1973). On the susceptibility of strawberry clones to Botrytis cinerea Pers. in relation to pectin quantity and quality in the fruit. Gartenbauwissenschaft 38: 311.
41. Hulme, A. C. and Edney, K. L. (1960). Phenolic substances in the peel of Cox's orange pippin apples with reference to infection by G. perennans. pp. 87-99. In Phenolics in Plants in Health and Disease, Ed. by J. B. Pridham, Pergamon Press, New York.
42. Ishii, S. (1977). Purification and characterization of a factor that stimulates tissue maceration by pectolytic enzymes. Phytopathology 67: 994.
43. Ismail, M. A. and Brown, G. E. (1979). Postharvest wound healing in citrus fruit: induction of phenylalanine ammonia-lyase in injured 'Valencia' orange flavedo. J. Amer. Soc. Hort. 104: 126.
44. Kuc, J., Williams, E. B., Maconkin, M. A., Ginzel, J., Ross, A. F. and Freedman, L. J. (1966). Factors in the resistance of apple to Botryosphaeria ribis. Phytopathol. 57: 38.
45. Kuc, J. and Shain, L. (1977). Antifungal compounds associated with disease resistance in plants, pp. 497-528. In Antifungal Compounds, Vol. 2. Ed. by M. R. Siegel and H. D. Sisler, Marcel Dekker, Inc., New York.
46. Langcake, P. (1981). Alternate chemical agents for controlling plant disease. Proc. Royal Soc. (in press).

47. Lund, B. M. (1971). Bacterial spoilage of vegetables and certain fruits. J. Appl. Bacter. 34: 9.
48. Lyon, G. D., Lund, B. M., Bayliss, C. E. and Wyatt, G. M. (1975). Resistance of potato tubers to Erwinia carotovora and formation of rishitin and phytuberin in infected tissue. Physiol. Plant Pathol. 6: 43.
49. McClure, T. T. (1960). Chlorogenic acid accumulation and wound healing in sweet potato roots. Amer. J. Bot. 47: 277.
50. McCornack, A. A. (1972). Effect of ethylene degreening on decay of Florida citrus fruit. Proc. Florida State Hort. Soc. 84: 270.
51. Menke, G. H., Patel, P. N. and Walker, J. C. (1964). Physiology of Rhizopus stolonifer infection on carrot. Z. Pflanzenkrank. 71: 128.
52. Mount, M. S. (1978). Tissue is disintegrated, pp. 279-297. In Plant Disease, Vol. III, Ed. by J. G. Horsfall and E. B. Cowling, Academic Press, New York.
53. Mount, M. S., Bateman, D. F. and Basham, H. G. (1970). Induction of electrolyte loss, tissue maceration and cellular death of potato tissue by an endopolygalacturonate trans-eliminase. Phytopathology 60: 924.
54. Mulvena, D., Webb, E. C. and Zerner, B. (1969). 3,4-Dihydroxybenazaldehyde, a fungistatic substance from green Cavendish bananas. Phytochemistry 8: 393.
55. Ndubizu, T. O. C. (1976). Relations of phenolic inhibitors to resistance of immature apple fruits to rot. J. Hort. Sci. 51:311.
56. Otazu, V. and Secor, G. A. (1981). Soft rot susceptibility of potatoes with high reducing sugar content. Phytopathology 71: 290.
57. Pelser, P. du T. and Eckert, J. W. (1977). Constituents of orange juice that stimulate the germination of conidia of Penicillium digitatum. Phytopathology 67: 747.
58. Perombelon, M. C. M. and Lowe, R. (1975). Studies on the initiation of bacterial soft rot in potato tubers. Potato Res. 18: 64.
59. Pilnik, W. and Voragen, A. G. J. (1970). Pectic substances and other uronides. In The Biochemistry of Fruits and their Products, Vol. 1, Ed. by A. C. Hulme, Academic Press, New York.
60. Ride, J. P. (1978). The role of cell wall alterations in resistance to fungi. Ann. Appl. Biol. 89: 302.
61. Rombouts, F. M. and Pilnik, W. (1972). Research on pectin depolymerases in the sixties - a literature review. Critical Rev. Food Technol. 3: 1.
62. Schlosser, E. (1975). Role of saponins in antifungal resistance. III. Tomatin dependent development of fruit rot organisms on tomato fruits. Z. Pflkrankh. Pflschutz. 82: 476.
63. Segall, R. H. (1967). Bacterial soft rot, bacterial necrosis,

and Alternaria rot of tomatoes as influenced by field washing and postharvest chilling. Plant Disease Reptr. 51: 151.
64. Smith, W. L., Jr. and Smart, H. F. (1955). Relation of soft rot development to protective barriers in Irish potato slices. Phytopathology 45: 649.
65. Spalding, D. H. (1969). Toxic effect of macerating action of extracts of sweet-potatoes rotted by Rhizopus stolonifer and its inhibition by ions. Phytopathology 59: 685.
66. Stanghellini, M. E. and Aragaki, M. (1966). Relation of periderm formation and callous deposition to anthracnose resistance in papaya fruit. Phytopathology 56: 444.
67. Steinbauer, C. E. and Kushman, L. J. (1971). Sweet potato culture and disease. U. S. Dept. Agr. Handbook 388.
68. Swinburne, T. R. (1973). The resistance of immature Bramley's Seedling apples to rotting by Nectria galligena Bres. In Fungal Pathogenicity and the Plant's Response, Ed. by J. W. Byrde and C. V. Cutting, Academic Press. New York.
69. Swinburne, T. R. and Brown, A. E. (1975). The biosynthesis of benzoic acid in Bramley's Seedling apples infected by Nectria galligena Bres. Physiol. Plant Pathol. 6: 259.
70. Swinburne, T. R. (1976). Stimulants of germination and appressoria formation by Colletotrichum musae (Berk. & Curt) Arx. in banana leachate. Phytopath. Z. 87:74.
71. Tani, T. (1967). The relation of soft rot caused by pathogenic fungi to pectic enzyme production by the host, pp. 40-57. In The Dynamic Role of Molecular Constituents in the Plant-Parasite Interaction, Ed. by C. J. Mirocha and I. Uritani, Am. Phytopathol. Soc.
72. Thorne, S. M. (1972). Studies on the behavior of stored carrots with respect to their invasion by Rhizopus stolonifer Lind. J. Food Technol. 7: 139.
73. Toibin, M. (1974). Pectic enzymes in sour rot of orange caused by Geotrichum candidum var. citri-aurantii (Ferr.) M. Sc. Thesis National University of Ireland, Dublin.
74. Van Den Berg, L. and Lentz, C. P. (1973). High humidity storage of carrots, parsnips, rutabagas and cabbage. J. Am. Soc. Hort. Sci. 98: 129.
75. Verhoeff, K. (1974). Latent infections by fungi. Ann. Rev. Phytopath. 12: 99.
76. Verhoeff, K. and Liem. J. (1975). Toxicity of tomatine to Botrytis cinerea in relation to latency. Phytopath. Z. 82: 333.
77. Yoder, O. C. and Whalen, M. L. (1975). Factors affecting post-harvest infection of stored cabbage tissues by Botrytis cinerea. Can. J. Botany 53: 691.
78. Yoder, O. C. and Whalen, M. L. (1975). Variations in susceptibility of stored cabbage tissue to infection by Botrytis cinerea. Can. J. Botany 53: 1972.
79. Zucker, M., Hankin, L. and Sands, D. (1972). Factors governing pectate lyase synthesis in soft rot and non-soft rot bacteria. Physiol. Plant Pathol. 2:59.

CONTROL OF POSTHARVEST DISEASES WITH ANTIMICROBIAL AGENTS

Joseph W. Eckert

Department of Plant Pathology
University of California
Riverside, CA 92521

INTRODUCTION

Pathological diseases of fruits and vegetables can be reduced to a certain extent by maintaining the natural resistance of the host, low temperature storage, low oxygen atmospheres, and treatment with growth regulators that delay senescence. However, these beneficial practices may not adequately protect the crop from microbial attack, especially during prolonged storage or movement of the crop through export market channels. This is particularly true for crops of tropical origin such as bananas, sweet potatoes and lemons, that suffer physiological injury at the near-freezing temperatures that are required to inhibit growth of fungi over an extended period of time. The maximum storage life of many fresh fruits and vegetables can be realized only by treating the product with an antifungal agent before storage in an optimal environment. The antimicrobial treatment is not a substitute for a satisfactory storage environment, since these agents rarely influence the rate of physiological deterioration of the fresh product. However, the antimicrobial agent is most efficient when the host possesses intrinsic resistance to infection and the environmental conditions are least favorable for the growth of the pathogen.

STRATEGIES AND METHODS OF DISEASE CONTROL

Several strategies for controlling postharvest diseases are outlined in Table 1. The application of a protective fungicide to the developing crop in the field was formerly the principal means for controlling postharvest diseases initiated through preharvest infection. This approach has been used extensively in the tropics

Table 1. Strategies for the Control of Pathogenic Microorganisms on Fresh Fruits and Vegetables

Strategy	Agents
1. Spray developing crop with fungicides to prevent preharvest infection.	Protective fungicides-captan, maneb, etc.
2. Inactivate pathogens in postharvest environment.	Broad spectrum sanitizing agents - formaldehyde, hypochlorite, quaternary ammonium compounds.
3. Inactivate latent infections in harvested crops.	Systemic fungicides - benomyl, thiabendazole.
4. Eradicate spores in harvest-related injuries.	_o_-phenylphenate, sec-butylamine.
5. Protect surface of product against infection through wounds created after application of fungicide.	Benomyl
6. Inhibit fungus sporulation and the spread of hyphae to adjacent fruits.	Biphenyl, benomyl, SO_2 dicloran, metalaxyl.

to control anthracnose on mangoes and papayas and in Europe to control lenticel rots of apples (1,2,3,4). The necessity for repeated applications of fungicides in the field has been reduced considerably in recent years due to the development of systemic fungicides that can be applied after harvest to inactivate the pathogen situated within the host. However, even today preharvest fungicide treatment is the only satisfactory means for control of brown rot (_Phytophthora_) of citrus fruits, a disease that is initiated by zoospore infection of fruit on the tree (5). The destruction of spores in packing houses is an essential practice to reduce the inoculum that may find its way into harvest-related injuries and to suppress the build-up of fungicide-resistant mutants that accumulate under the selection pressure created by the fungicide treatment. Formaldehyde and quaternary ammonium compounds are used as broad spectrum sanitizing agents for applications that do not involve contact with the food product. Solutions of hypochlorite and sodium _o_-phenylphenate are used to surface sterilize the product itself or to sterilize water that is recycled to wash, transport or cool the product.

Over the past thirty years, about twelve postharvest fungicides have been developed that are highly effective against latent infections, wound-pathogens, and the spread of pathogens between adjacent fruits. Table 2 lists the antimicrobial compounds that are in practical use today, the crops treated and the diseases controlled. The chemical structures of these compounds are shown in Figure 1.

The selection of an antimicrobial compound for a specific postharvest application depends upon: 1) the sensitivity of the pathogens to the chemical agent; 2) the ability of the agent to penetrate through the boundary layer of the host to the infection site; 3) the tolerance of the plant product to the chemical agent, both from

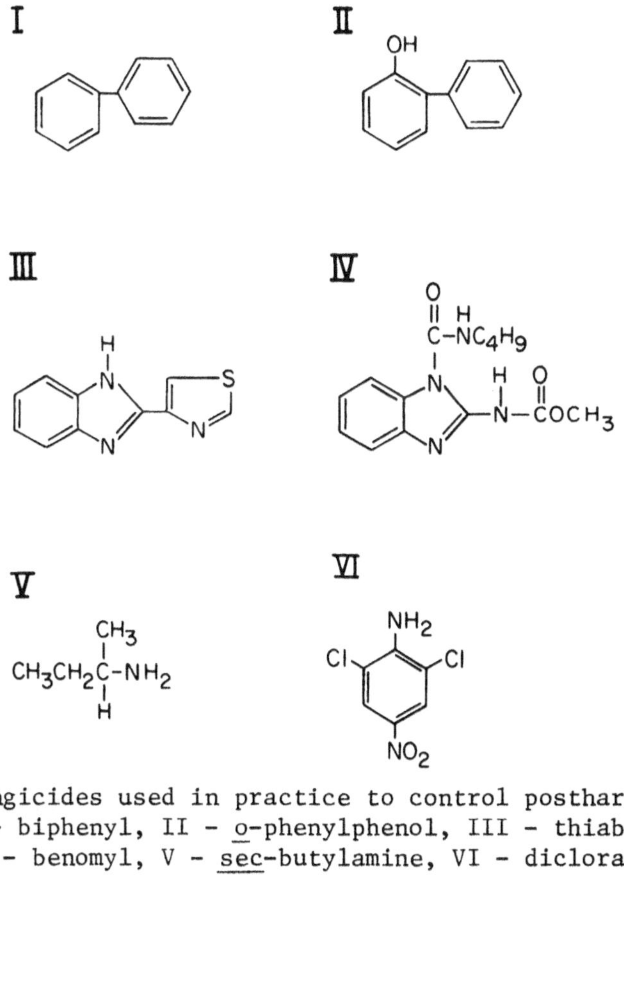

Fig. 1. Fungicides used in practice to control postharvest diseases. I - biphenyl, II - o-phenylphenol, III - thiabendazole, IV - benomyl, V - sec-butylamine, VI - dicloran.

Table 2. Practical Treatments for Control of Postharvest Diseases of Fruits and Vegetables

Crop	Disease pathogens	Antimicrobial agents
Apple Pear	Penicillium Botrytis Pezicula	*o*-Phenylphenate Thiabendazole Benomyl
Banana	Colletotrichum Fusarium Ceratocystis	Thiabendazole Benomyl Thiophanate methyl Imazalil
Citrus	Penicillium Diplodia Phomopsis	*o*-Phenylphenate Thiabendazole Benomyl *sec*-Butylamine Biphenyl 2,4-D Imazalil
Grape	Botrytis	Sulfur dioxide
Mango Papaya	Colletotrichum Diplodia	Thiabendazole Benomyl
Melons	Alternaria Cladosporium	Dimethyldithio-carbamate
Peach Nectarine Cherry	Monilinia Rhizopus	Dicloran Benomyl *o*-Phenylphenate
Pineapple	Ceratocystis	*o*-Phenylphenate Benomyl
Potato	Fusarium Phoma Oospora Helminthosporium	Thiabendazole Benomyl *sec*-Butylamine
Sweet potato	Ceratocystis Rhizopus	Dicloran *o*-Phenylphenate
Tomato	Botrytis Rhizopus Geotrichum	Discloran *o*-Phenylphenate

the standpoint of phytotoxicity and the effect of the treatment upon the sensory properties of the product. No specific bactericidal compounds have been developed for postharvest use and, until recently, there were few compounds available that were effective against phycomycetes, *Alternaria* or *Geotrichum* and diseases incited by these pathogens were very difficult to control. Even those compounds that exhibited *in vitro* activity against these pathogens did not prevent their development in the host, apparently because these compounds were unable to penetrate to the sites of infection, e.g. latent infections. Postharvest fungicides of the first generation, e.g. sodium *o*-phenylphenate, dicloran and *sec*-butylamine are effective in preventing decay by the wound-invading pathogens, e.g. *Penicillium* and *Rhizopus*, but have little effect on the development of latent and other deepseated infections (2). Several fungicides developed since 1965 have shown a high degree of efficiency against such latent infections as well as protective and antisporulant action in controlling certain post-harvest diseases (4,5).

Fungicides may be applied to harvested products either in a gaseous state or mixed with water or wax formulations, depending upon the physical and chemical properties of the fungicide and the convenience of each method of application. A fumigation treatment is advantageous for treatment of strawberries and other highly perishable commodities that are not normally washed after harvest because they might be adversely affected by hydration. Sulfur dioxide is extensively used to fumigate grapes for control of *Botrytis* and seems to be effective on raspberries as well. Ethylene oxide is an effective fumigant for control of yeast on dried dates, but was discontinued in the late 1960's when residues of toxic ethylene chlorohydrin were found in treated fruit (4). Acetaldehyde has received considerable attention as a fumigant for apples and strawberries in recent years because it is effective in controlling mold growth and also is a natural metabolite in fruits.

Today, most fungicide treatments are applied in water-base formulations that are sprayed or flooded onto the plant product as it is conveyed mechanically through the packing houses. Also, formulations of fungicides in water-base waxes and in hydrocarbon-solvent-base waxes are in popular use. All of the organic fungicides listed in Table 2 have been applied in these ways. Fungicides applied in water-base formulations to prevent infection at injury sites may be divided into two categories according to their distribution on the surface of the treated product: 1) Non-ionic water insoluble compounds (e.g., thiabendazole) that are applied as a uniform deposit over the entire surface of the product; 2) water soluble salts such as sodium *o*-phenylphenate, *sec*-butylamine, and sodium carbonate that are applied in relatively concentrated solutions (0.5-3.0%) in water. Aqueous solutions of these salts are absorbed by fruit injuries but the intact cuticle is not permeable to such ionic compounds. After treatment, the product is rinsed lightly with fresh water to remove most of the fungicide from the surface of the fruit except for a significant residue that remains in the injury site to inhibit growth of the pathogen there (Fig. 2). The pH of injured tissue of most fruits is considerably below the pKa of o-phenylphenol (10.01), resulting in the precipitation of o-phenylphenol from the solution of sodium o-phenylphenate (6). An advantage of uniform non-selective coverage of the product with a fungicide is that the entire surface is protected to some degree against infections at injuries incurred after the treatment is applied. For example, benomyl-treated citrus fruits are difficult to inoculate through superficial injuries made after the benomyl treatment is applied. Apparently, benomyl enters the surface cells of the host, making the tissue an unsatisfactory substrate for the growth of the pathogen (7). Fungicides that carry some risk of phytotoxicity are applied selectively to the potential infection sites and the excess compound must be removed by rinsing the fruit with water to prevent injury to the fruit surface. sec-Butylamine and thiabendazole are not phytotoxic; therefore, substantial deposits may be left on the fruit without causing injury. Rinsing with water after treatment significantly reduces the fungicidal residue on the product, but also reduces the antimicrobial effectiveness of the treatment as well.

CONTEMPORARY FUNGICIDE TREATMENTS (Fig. 1)

o-Phenylphenol (II) is a broad spectrum antimicrobial agent that is toxic to fungi, bacteria and plants. The undissociated phenol is lethal to microorganisms and is injurious to most plant products at concentrations of 200-400 mg/L, depending upon the temperature of the solution and the period of contact. The dissociated form (o-phenylphenate anion) is not phytotoxic and a solution of sodium o-phenylphenate (SOPP), containing excess alkali to suppress hydrolysis of the salt, is quite safe for treatment of several fruits and vegetables. The intact waxy surface of citrus fruits, apples and other fruits is impermeable to the o-phenylphenate anion as indicated by the low residue levels of fruit treated at pH 12 (6). However, the o-phenylphenate anion diffuses selectively into injuries where the surface cells of the fruit have been ruptured and is hydrolyzed to o-phenylphenol in the presence of fruit acids and metabolic carbon dioxide. Rinsing the treated fruit with water removes most of the sodium o-phenylphenate residues from the surface of the fruit, but a substantial amount remains associated with injuries and prevents the development of pathogens at these sites. The selective accumulation of o-phenylphenol in injuries and its role in the prevention of fruit decay is shown in Fig. 2.

The benefits of treating fruits and vegetables with sodium o-phenylphenate are two-fold: 1) Fungus spores and bacteria on the

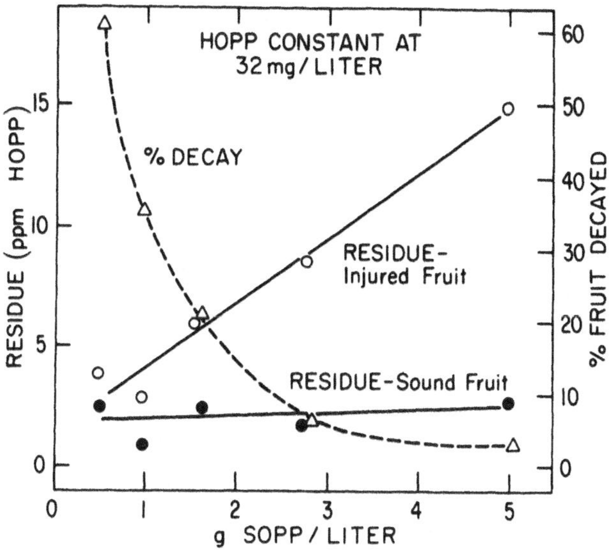

Fig. 2. Effect of concentration of SOPP (sodium o-phenylphenate tetrahydrate) upon decay of oranges inoculated with Penicillium digitatum and upon the residues of HOPP (o-phenylphenol) in sound and injured tissues of the fruits peel(6).

surface of the fruit or in cleaning solutions are killed; 2) a residue of o-phenylphenol is deposited in wounds on the fruit surface, preventing infections at these sites during marketing and storage of the crop. Solutions of sodium o-phenylphenate have been used extensively for the control of postharvest diseases of citrus fruits, apples, pears, peaches, sweet potatoes and other perishable fruits and vegetables. The treatment is applied in one of several ways. The fruit may be soaked in a bath of SOPP for 1-3 minutes; the fruit may be flooded with a recirculated solution of SOPP; the fruit may be brushed in a foam of SOPP for 15 seconds. In all cases, the fruit are rinsed with a spray of fresh water after treatment. Krochta et al. (8) demonstrated that a foam prepared from a solution of 0.5% SOPP and 0.3% sodium lauryl sulfate was highly effective as a cushion for mechanically harvested tomatoes as they dropped into a bin and also prevented decay of fruit caused by Geotrichum and Rhizopus.

Biphenyl (I) has been added to the packaging material for citrus fruits for the past forty years to provide vapor phase inhibition of sporulation of Penicillium digitatum and P. italicum during long distance shipment and marketing of the crop. Despite these many years of successful use, biphenyl has three recognized problems that limit its value as a postharvest treatment. 1) Treated fruit may have temporary chemical odor. Although this problem is minimal with citrus fruits, biphenyl has never been used on other fruits and vegetables because of this odor problem. 2) Residues of biphenyl on treated fruits may exceed the 70 mg/kg tolerance in Europe and Japan. This problem is greatest with fruit that tend to absorb substantial quantities of biphenyl, especially if they are not adequately cooled during shipment. 3) Strains of *Penicillium* that are resistant to biphenyl and cross-resistant to *o*-phenylphenol are commonplace in citrus packing houses.

sec-Butylamine (V) may be applied to harvested crops either as a fumigation treatment or as a solution of a salt of *sec*-butylamine. The fumigation treatment controls *Penicillium* on citrus fruits. The injuries on the fruit surface become alkaline after absorption of the amine vapor and fungistatic residues of neutral *sec*-butylamine salts persist in injuries after the fumigation treatment is terminated (9). The *sec*-butylammonium cation is uniquely fungistatic to several species of fungi (10,11). The optical isomer (-) *sec*-butylamine is substantially more active than the (+) isomer, although the racemic mixture is used in practice to control postharvest diseases. Fumigation of seed and warehouse potatoes in the U.K. with *sec*-butylamine controls *Oospora* and *Phoma* during storage (12). Gaseous *sec*-butylamine penetrates deeply into the potatoes so that the fumigation can be delayed for 14 days after harvest without diminishing the effectiveness of the treatment. However, *sec*-butylamine is not effective against Fusarium dry rot.

Oranges in California are drenched with a solution of 1% *sec*-butylamine (phosphate or HCl salt, pH 9) to protect superficial injuries against infection by *Penicillium* species during the ethylene degreening period and during storage (10). The degreening operation is highly conducive to the development of decay since it involves exposure of the fruit to a low concentration of ethylene gas (5 ppm) for several days in a warm humid environment. *sec*-Butylamine is also added to wax formulations applied to lemons before storage to control *Penicillium*, but the treatment has no effect upon diseases caused by *Alternaria* or *Geotrichum* which are resistant to *sec*-butylamine. *Penicillium expansum* and *Pezicula* spp. have been controlled on apples treated with solutions of *sec*-butylamine salts (13,14). The effectiveness against *Penicillium* was diminished, however, when treated fruit were stored for several months at low temperatures. The major problems that have beset the use of *sec*-butylamine are injury to Navel oranges if the peel absorbs the solution, and the build-up of *sec*-butylamine resistant strains of *Penicillium*. In California, *sec*-butylamine is used mostly on lemons in a schedule with other fungicides which do not select strains of *Penicillium* that are cross-resistant to *sec*-butylamine.

Dicloram (VI) is the most effective fungicide for control of Rhizopus stolonifer, a wound invading pathogen of many fruits and vegetables (15). Cherries, peaches and other stone fruits are usually treated with this fungicide before shipment to distant markets. Dicloran is much less active against Monilinia fructicola, another major pathogen of stone fruits, and is only weakly active against other Rhizopus species and Mucor species (16). Nectarines, peaches and cherries are usually treated with a combination of dicloran and benomyl after harvest to control the major postharvest pathogens, Monilinia fructicola and Rhizopus stolonifer (17).

Dicloran is the standard postharvest treatment for control of Rhizopus rot of sweet potatoes, but the treatment is ineffective against Ceratocystis black rot or Penicillium rots (18). Recent trials in California have demonstrated that dicloran applied to tomatoes after harvest is highly effective in controlling Botrytis and Rhizopus rots (19). However, the dicloran treatment has no influence upon Geotrichum rot.

The principle postharvest fungicides prior to the mid 1960's, borax, sodium carbonate, SOPP and sec-butylamine controlled decay by preventing the invasion of pathogenic microorganisms through harvest-related injuries on the surface of the fruit. Biphenyl and dicloran function in this fashion; but in addition, inhibit the sporulation of the pathogen and prevent its spread to adjacent fruits. The introduction of the benzimidazole fungicides (Fig. 3) thiabendazole, benomyl, carbendazim, and thiophanate-methyl in the late 1960's made it possible for the first time to prevent the development of latent infections of such fungi as Diplodia, Phomopsis and Colletotrichum on certain tropical fruits. These fungicides also are highly effective in preventing infection of fruit injuries incurred during harvest and in handling after application of the fungicide treatment. Finally, at high dosage rates, thiabendazole (4-6 g/L) and benomyl (2-3 g/L provide a protective barrier on the fruit surface that inhibits the sporulation of Penicillium and other fungi on diseased fruits. The significance of this latter property is that the benzimidazole fungicides potentially offer an acceptable alternative to the biphenyl treatment for the control of spoilage of citrus fruits.

All of the benzimidazole fungicides and their derivatives share a common antifungal spectrum although they differ quantitatively in activity (20). In general, the benzimidazole fungicides have been used interchangeably for certain postharvest applications while, in other cases, a specific compound is preferred because of its biological or chemical properties. The benzimidazole fungicides have been used world-wide to control Penicillium decay and stem end rots of citrus fruits, Penicillium blue mold and Gloeosporium (Pezicula) lenticel rots of apples (1,21), Colletotrichum and Ceratocystis on several tropical fruits (22,23,24,25,26) and

Monilinia rot of stone fruits (peaches, cherries, plums). Benomyl is usually combined with dicloran for treatment of stone fruits in order to control both Monilinia and Rhizopus, the major postharvest diseases of these fruits (17). The success of the benzimidazole fungicides in the control of postharvest diseases can be attributed to two factors: 1) high intrinsic activity of these compounds against many of the fungi responsible for postharvest diseases and 2) the ability of the benzimidazole fungicides to penetrate at least some degree through the waxy surface of the host to reach the site of infection (7,27,28). Benomyl appears to penetrate the host's surface more efficiently than thiabendazole, carbendazim or thiophanate methyl and may be more effective in disease situations in which the pathogen is situated below the surface of the host.

Numerous trials conducted in the U.S. and in Europe have demonstrated that the benzimidazole fungicides can control several diseases of potato tubers during storage--dry rot (Fusarium), gangrene (Phoma), skin spot (Oospora) and silver scurf (Helminthosporium (29). Thiabendazole appears to be somewhat more effective than benomyl for the control of Fusarium and Phoma (30,31). Thiabendazole is applied at 42 g/metric ton by means of an ultra low volume mist applicator as the potatoes are transported into storage (31). High volume application may be more effective, but the residual water on the potatoes in storage may increase the risk of bacterial soft rot. A similar treatment of sugar beets has been effective in reducing decay by Penicillium and other fungi that cause a sugar loss in the beets during storage in outdoor piles.

Thiabendazole and carbendazim are stable under most conditions that might be encountered in the treatment of fruits after harvest. Thiabendazole can even be applied as "smoke" in fruit and vegetable storage rooms. Benomyl and thiophanate methyl, on the other hand, are unstable in aqueous solution or in organic solvents. Benomyl suspensions in pure water appear to be stable because of the very low solubility of benomyl in water (ca. 4 mg/L); however, benomyl is not stable in very dilute aqueous solutions or in formulations of wax coating applied to fruits and vegetables after harvest. Under neutral conditions, benomyl decomposes significantly within a few hours to methyl 2-benzimidazolecarbamate (carbendazim); under alkaline conditions, benomyl is converted to 1,2,3,4-tetrahydro-3-butyl-2, 4-dioxo-s-triazino (alpha)-benzimidazole(STB) within a few hours (7,32,33; Fig. 3A). Carbendazim is almost as active as benomyl in preventing infection of citrus fruits by Penicillium, but STB is essentially devoid of antifungal properties. Carbendazim is more polar than benomyl and is significantly less effective in controlling sporulation of Penicillium on citrus fruits because it does not penetrate into the peel of the fruit as efficiently as benomyl (7). Carbendazim and thiophanate methyl, a progenitor of carbendazim (Fig. 3B), have been used as substitutes for benomyl (24,34). For some applications these compounds are equivalent

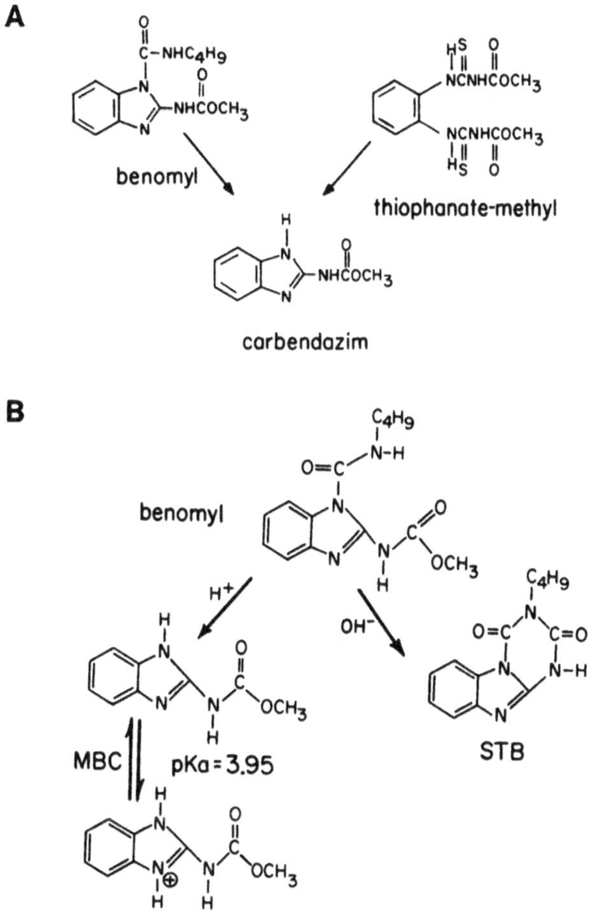

Fig. 3A. Reactions of benomyl in aqueous formulations resulting in the formation of carbendazim (MBC) and STB.

B. Spontaneous conversion of benomyl and thiophanate-methyl to carbendazim (MBC).

in effectiveness to benomyl, but for uses that require significant penetration into the host, MBC and thiophanate methyl appear to be inferior.

All benzimidazole fungicides have two serious biological limitations: 1) All compounds in this group are inactive against several microorganisms responsible for important postharvest diseases (20)--*Rhizopus* and other mucors, *Alternaria*, *Geotrichum*, *Phytophthora* and all bacteria, including *Erwinia* spp. which causes soft rot of potatoes and vegetable crops. 2) Pathogens that are usually controlled by treatment with benzimidazole compounds have built up benzimidazole-resistant strains following intensive use of these fungicides.

Diseases of secondary importance caused by *Alternaria*, *Geotrichum* and other pathogens of citrus fruits that are resistant to benzimidazole fungicides have become a limiting factor in the storage and handling of this crop after treatment with a benzimidazole fungicide. The resistance of strains of *Penicillium* to benzimidazole fungicides has seriously diminished the usefulness of these compounds as postharvest treatments for citrus fruits and it now appears that this problem may also arise on other crops.

RESISTANCE OF PATHOGENS TO POSTHARVEST FUNGICIDES

Strains of *Penicillium* resistant to biphenyl were recognized in Israel in 1940. These strains were not regarded as a practical problem until the late 1960's when several shipments of California lemons packed with biphenyl-impregnated papers arrived in distant markets with high levels of Penicillium decay and sporulation. Harding (35) observed that many of these problem shipments originated in packing houses where the lemons were treated with sodium *o*-phenylphenate before they were stored for several months prior to shipment. Harding found that strains of *Penicillium digitatum* that were resistant to biphenyl were also resistant to SOPP, a relationship that would be anticipated from the structural similarity of the two fungicides (Fig. 1). He demonstrated that residues of *o*-phenylphenol on SOPP-treated lemons were sufficient to suppress the growth of biphenyl sensitive strains, thereby allowing the biphenyl resistant mutants in the population to build up on the fruit. In a survey of California citrus growing areas, Harding observed biphenyl resistant variants of *P. digitatum* at a frequency of about 0.15% of the isolates collected in citrus groves and packing houses that had never been exposed to SOPP. Packing houses that had a history of SOPP use had a resistant strain frequency of 14-18% of the *Penicillium digitatum* population and 60% of the *Penicillium* isolates were resistant to biphenyl in packing houses that had used SOPP continuously over a period of years.

The origin of the SOPP/biphenyl resistance problem can be explained by the preexistence of mutants of P. digitatum that are cross-resistant to both of these fungicides (36). These strains are apparently less well adapted in the natural population of Penicillium digitatum in the citrus grove, but under selection pressure of SOPP/biphenyl treatment they selectively proliferate in the packing house to become the dominant component of the P. digitatum population. Mutants of several fungi that are cross-resistant to biphenyl, SOPP and dicloran have been reported, but dicloran resistance does not appear to be a real problem at this time in the crops that are treated with this fungicide after harvest. Dicloran is not used on citrus fruits.

sec-Butylamine may be applied to oranges and lemons before they are place in storage. These applications create a situation that is highly favorable for the selection and proliferation of sec-butylamine resistant strains of P. digitatum and P. italicum in the natural population. Thus, the continuous use of sec-butylamine over a period of months may lead to a failure of the treatment to control Penicillium decay of citrus fruits.

The most serious problem of fungicide resistance arose in connection with the continuous use of benomyl and thiabendazole on citrus fruits (36,37). Mutants resistant to one benzimidazole fungicide are also resistant to every other member of this group. Benzimidazole resistance in Penicillium spp. was observed in California citrus packing houses about fifteen months after thiabendzole was incorporated into the wax formulation applied to lemons before storage. These resistant isolates occurred at a high frequency in the atmosphere of packing houses that used thiabendazole intensively as a prestorage treatment for lemons and they were not controlled by treating the fruit with the highest dosage of thiabendazole recommended for commercial treatment. Benzimidazole resistant strains of P. digitatum and P. italicum have been isolated also in Florida, Israel, Australia, and Japan. A survey of citrus fruits arriving in the port of Rotterdam revealed that approximately one-half of the isolates of Penicillium digitatum and P. italicum recovered from fruit from twenty countries were resistant to benzimidazole fungicides (38). Benzimidazole resistant strains of P. expansum have been reported on harvested apples and pears in Oregon, New York and Australia (14, 39, 40). Most observations of benzimidazole resistant strains of Penicillium spp. have followed intensive use of thiabendazole or benomyl over a period of some months but Kuramoto (41) reported a severe resistance problem on satsuma mandarins in Japan following application of a single annual spray of thiophanate methyl immediately before harvest. Several investigators have recorded the isolation of benzimidazole resistance strains from citrus groves and packing houses where benzimidazole fungicides had never been used (36).

CURRENT DEVELOPMENTS IN POSTHARVEST FUNGICIDES

Several new fungicides (Fig. 4) now at various stages of commercial development could correct most of the deficiencies of postharvest treatments in commercial use today. Guazatine (VII), a diguanadine, is a broad spectrum fungicide that is effective against wound infection of citrus fruit by Penicillium spp., including strains that are resistant to the benzimidazoles and other fungicides now in popular use. Guazatine also controls Geotrichum sour rot on citrus fruits, but not the sporulation of Penicillium since it is not a systemic fungicide (42). Guazatine is used on a commercial basis in Australia, but the treated fruit are not acceptable in most importing countries at this time.

The imidazole fungicides (Fig. 4), imazalil (VIII), phenapronil (IX) and prochloraz (X) are more effective than benomyl and thiabendazole in controlling Penicillium decay and sporulation on citrus fruits, including strains of this pathogen that are highly resistant to the benzimidazole fungicides (43,44). On the other hand, the imidazole fungicides are less effective than benomyl and thiabendazole against Diplodia stem end rot, the most important cause of early-season decay in citrus fruit produced in the humid subtropics. Imazalil, phenapronil and prochloraz (45) in common with the benzimidazole fungicides, do not control Geotrichum sour rot or Phytophthora brown rot diseases. Imazalil does control Alternaria rot on tomatoes (46) and preliminary observations indicate some benefit against this disease on citrus fruits as well.

The effectiveness of imazalil in controlling Penicillium decay has been demonstrated in large scale trials conducted throughout the world (44). Imazalil offers outstanding potential for the control of Penicillium decay in citrus fruits due to its effectiveness against strains of Penicillium that are resistant to thiabendazole and benomyl (47). Despite the obvious advantages, imazalil is not widely used at present because several countries, including the U.S. and Japan, have not established permanent tolerances for this fungicide on citrus fruits. Imazalil was very effective in controlling benzimidazole-resistant strains of Penicillium expansum on apples during long-term storage in a modified atmosphere (14, 48). Our tests showed that imazalil was more effective than benomyl in controlling Monilinia fructicola (benzimidazole sensitive strains) on sweet cherries. Others have reported good control of this pathogen on peaches (49). The triazole fungicides, CGA 64251 and its propyl homolog, CGA 64250 (50) are equivalent to the imidazole fungicides in controlling Penicillium decay and sporulation. Remarkably, these fungicides are highly effective against Geotrichum sour rot (51), but like all other organic fungicides currently in use, the triazole fungicides have no effect against Phytophthora

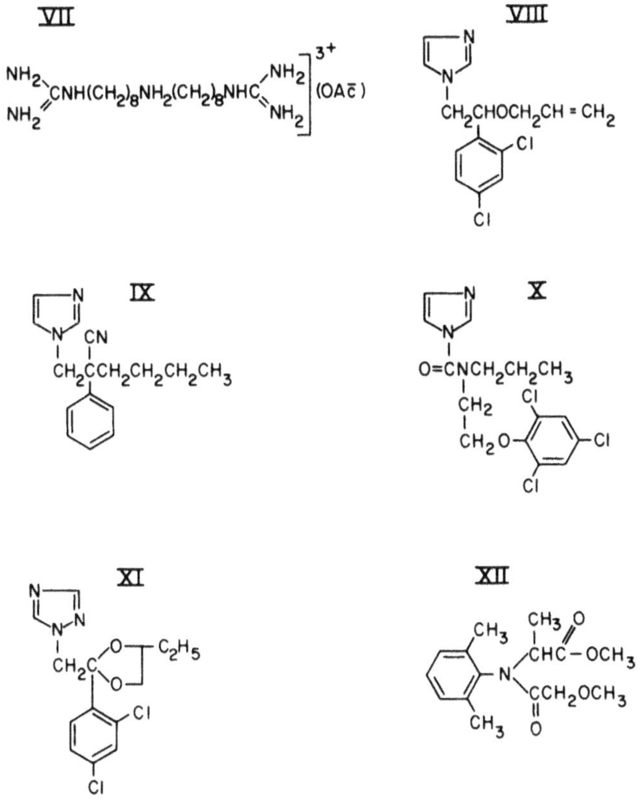

Fig. 4. New fungicides currently being evaluated for control of postharvest diseases. VII-guazatine, VIII-imazalil, IX-phenapronil, X-prochloraz, XI-CGA 64251, XII-metalaxyl.

brown rot. Experiments conducted in California, Greece and Israel have demonstrated the effectiveness of metalaxyl (Fig. 4, XII) in controlling Phytophthora rot on citrus fruits (52) and apples (53). If applied early in the incubation period, metalaxyl can penetrate the waxy surface layer of the fruit to eradicate incipient infections of Phytophthora beneath. Short range systemic action is essential for control of these diseases because Phytophthora infection is initiated in the orchard, but a postharvest fungicide treatment cannot be applied until several days later. Current practical control measures for Phytophthora rots consist of protective fungicides (copper or captan) applied to the citrus grove before the onset of the rainy season.

SUCCESSES AND CHALLENGES IN POSTHARVEST DISEASE CONTROL

Substantial progress has been made over the past decade in the chemical control of postharvest diseases. The benzimidazole fungicides provided a conceptual breakthrough in that they demonstrated, for the first time, that latent or incipient infections located within host tissues could be inactivated by chemical treatment applied to the surface of the product after harvest. Stem-end rots of citrus fruits, lenticel rots of apples, latent infections of Colletotrichum on bananas and other tropical fruits, and late season infections of Monilinia on peaches have all been controlled by postharvest fungicide treatment to an extent thought impossible in earlier times. Surface deposits of thiabendazole, benomyl and imazalil on citrus fruits, within legal residue tolerances, can suppress the sporulation of Penicillium mold on the surface of citrus fruits more effectively than biphenyl, the standard fungicide for export shipments of citrus fruits for four decades. Thiabendazole has equalled or surpassed the organomercurial fungicides in controlling certain important diseases of seed potatoes and this technology has been extended to the reduction of postharvest losses due to fungal pathogens in warehouse and fresh-packed potatoes as well. The outstanding successes of the benzimidazole fungicides in reducing storage losses in these crops has emphasized, by comparison, the inadequate control of diseases incited by microorganism species that are tolerant to this class of fungicides, i.e., Alternaria, Geotrichum, Phytophthora, Mucors, bacteria and strains of fungal pathogens that have been selected for benzimidazole resistance.

The new imidazole and triazole fungicides discovered over the past few years potentially can correct the deficiencies of the benzimidazole fungicides. As a group, they are able to control Penicillium mutants that are resistant to the benzimidazoles and several compounds can control Alternaria and Geotrichum sour rot as well. Furthermore, genetic studies with this class of compounds suggest that they are less vulnerable to the problem of fungicide resistance than the benzimidazole fungicides. The benzimidazole compounds bind to fungal tubulin causing mitotic arrest (54, 55), whereas the imidazole compounds are inhibitors of sterol biosynthesis (56). Large increases in resistance to benzimidazole fungicides can result from a simple mutation involving the benzimidazole binding sites, whereas laboratory-induced resistance to the imidazole fungicides involves only a small increase in resistance to this class of compounds. Speculation regarding the ease of resistance development has not been tested since these new fungicides have not yet been subjected to the enormous selection pressure that prevails in citrus packing houses. However, strains of Penicillium expansum (an apple fruit pathogen) that are resistant to imazalil were isolated recently in Israel (57). The fact that imazalil and related compounds are less effective against Diplodia stem end rot than the benzimidazole compounds could result in a use pattern of the two fungicides that could delay the emergence of strains of Penicillium that are resistant to both classes of fungicides. Benomyl or thiabendazole are preferred early in the season when Diplodia is the major problem, but they should be replaced by an imidazole fungicide at mid-season when Penicillium decay and sour rot become the major problems. Benzimidazole resistant strains of Penicillium that are selected by the early season use of benomyl and thiabendzole could be controlled by an imidazole compound later in the season.

Recent tests with metalaxyl have shown that incipient infections of Phytophthora can be eradicated by postharvest treatments dictated when environmental conditions before harvest forecast the probability of an impending disease problem in storage. Although a few diseases of fresh fruits and vegetables (e.g. bacterial diseases) cannot be satisfactorily controlled by existing postharvest treatments, currently available fungicides have demonstrated the potential for controlling the major fungal pathogens responsible for postharvest losses. The major thrust of current research in postharvest pathology should be directed towards the development of existing fungicides into practical and economical treatments for the reduction of crop losses during storage and distribution to the consumers.

REFERENCES:

1. Burchill, R. T. and Edney, K. L. (1972). As assessment of some new treatments for the control of rotting of stored apples. Ann. Appl. Biol. 72: 249.
2. Eckert, J. W. (1967). Application and use of postharvest fungicides in Fungicides - An Advanced Treatise, Vol. 1, D. C. Torgeson, ed., Academic Press, New York, 287.
3. Eckert, J. W. (1975). Postharvest pathology - general principles in Postharvest Physiology, Handling and Utilization of Tropical and Subtropical Fruits and Vegetables, E. B. Pantastico, ed., Avi Publ. Co., Westport, CT, 393.
4. Eckert, J. W. (1977). Control of postharvest diseases, In Antifungal Compounds. Vol. 1, M. R. Siegel and H. D. Sisler, ed., Marcel Dekker Inc., New York. pp. 269.
5. Eckert, J. W. (1978). Post-harvest diseases of citrus fruits. Outlook on Agric. 9: 225.
6. Eckert, J. W., Kolbezen, M. J. and Kramer, B. A. (1969). Accumulation of o-phenylphenol by citrus fruits and pathogenic fungi in relation to decay control and residues. Proc. 1st Intern. Citrus Symp., Univ. Calif. (Riverside) 2: 1097.
7. Eckert, J. W., Kolbezen, M. J., Rahm, M. L. and Eckard, K. J. (1979). Influence of benomyl and methyl 2-benzimidazole carbamate on the development of Penicillium digitatum in the pericarp of orange fruit. Phytopathology 69: 934.
8. Krochta, J. M., Carlson, R. A., Ogawa, J. M. and Manji, B. T. (1977). Harvesting into foam reduces tomato losses.

Food Technol. 313: 42.
9. Eckert, J. W., and Kolbezen, M. J. (1970). Fumigation of fruits with 2-aminobutane to control certain postharvest diseases. Phytopathology 60: 545.
10. Eckert, J. W., and Kolbezen, M. J. (1964). 2-Aminobutane salts for control of postharvest decay of citrus, apple, pear, peach and banana fruits. Phytopathology 54: 978.
11. Eckert, J. W., Rahm, M. L. and Kolbezen, M. J. (1972). Fungistatic activity of cations of non-aromatic amines. J. Agric. Food Chem. 20: 104.
12. Graham, D., Hamilton, G. A., Quinn, C. E. and Ruthven, A. D. (1973). Use of 2-aminobutane as a fumigant for control of gangrene, skin spot and silver scurf diseases of potato tubers. Potato Res. 16: 109.
13. Bompeix, G. and Morgat, F. (1977). Cires, anti-èchaudures, fongicides et conservation des pommes. Fruits 32: 189.
14. Koffman, W., Penrose, L. J., Menzies, A. R., Davis, K. C. and Kaldor, J. (1978). Control of benzimidazole tolerant Penicillium expansum in pome fruit. Scientia Hortic. 9: 31.
15. Luepschen, N. W., Rohrbach, K. G., Jones, A. C., and Peters, C. L. (1971). Methods of controlling Rhizopus decay and maintaining Colorado peach quality. Colorado State Univ. Expt. Sta. Bull. 547S.
16. Ogawa, J. M., Mathre, J. H. Weber, D. J., and Lyda, S. D. (1963). Effects of 2,6-dichloro-4-nitroaniline on Rhizopus species and comparison with other fungicides on control of Rhizopus rot of peaches. Phytopathology 53: 950.
17. Wade, N. L. and Gipps, P. G. (1973). Postharvest control of brown rot and Rhizopus in peaches with benomyl and dicloran. Austrl. J. Exptl. Agr. Animal Husbandry 13: 600.
18. Steinbauer, C. E. and Kushman, L. J. (1971). Sweetpotato culture and diseases. U. S. Dept. Agr. Handbook 388: 35.
19. Chastagner, G. A. and Ogawa, J. M. (1979). A fungicide-wax treatment to suppress Botrytis cinerea and protect fresh-market tomatoes. Phytopathology 69: 59.
20. Eckert, J. W. and Rahm, M. L. (1979). The antifungal activity of alkyl benzimidazole-2-yl-carbamates and related compounds. Pestic. Sci. 10: 473.
21. Edney, K. L. (1970). Some experiments with thiabendazole and benomyl as postharvest treatments for the control of storage rots of apples. Plant Pathol. 19: 189.
22. Cho, J. J., Rohrbach, K. G. and Apt, W. G. (1977). Induction and chemical control of rot caused by Ceratocystis paradoxa on pineapples. Phytopathology 67: 700.
23. Couey, H. M. and Farias, G. (1979). Control of postharvest decay of papaya. Hort. Sci. 14: 719.
24. Löcher, P. and Hampel, M. (1973). Control of postharvest diseases of bananas, pineapples and citrus with carbendazim. Proc. 7th British Insecticide Fungicide Conf. 1:301.

25. Meredith, D. S. (1977). Recent advances in control of post-harvest deterioration using thiabendazole with special reference to tropical and subtropical crops. Proc. 1977 British Crop Protection Conf.
26. Muirhead, I. F. (1976). Postharvest control of mango anthracnose with benomyl and not water. Austral. J. Exptl. Agr. Animal Husbandry 16: 600.
27. Ben-Aire, R. (1975). Benzimidazole penetration, distribution and persistence in postharvest treated pears. Phytopathology 65: 1187.
28. Phillips, D. J. (1975). Detection and translocation of benomyl in postharvest-treated peaches and nectarines. Phytopathology 65: 255.
29. Meredith, D. S. (1975). Control of fungal diseases of seed potatoes with thiabendazole. Proc. 8th British Insectic. Fungic. Conf. 2: 581.
30. Boyd, A. E. W. (1975). Fungicides for potato tubers. Proc. 8th British Insectic. Fungic. Conf. 3: 1035.
31. Logan, C., Copeland, R. B. and Little, G. (1975). Potato gangrene control by ultra low volume sprays of thiabendazole. Ann. Appl. Biol. 80: 199.
32. Chiba, M. and Doornbos, F. (1974). Instability of benomyl in various conditions. Bull. Envir. Contam. Toxicol. 11: 273.
33. White, E. R., Bose, E. A., Ogawa, J. M., Manji, B. T. and Kilgore, W. W. (1973). Thermal and base-catalyzed hydrolysis products of the systemic fungicide, benomyl. J. Agr. Food Chem. 21: 616.
34. Vonk, J. W. and Kaars-Sijpesteijn, A. (1971). Methyl benzimidazole-2-yl-carbamate, the fungitoxic principle of thiophanate methyl. Pestic. Sci. 2:160.
35. Harding, P. R. Jr. (1962). Differential sensitivity to sodium o-phenylphenate by biphenyl-sensitive and biphenyl-resistant strains of Penicillium digitatum. Plant Disease Reptr. 46: 100.
36. Eckert, J. W. and Wild, B. L. (1981). Problems of fungicide resistance in Penicillium rot of citrus fruits. in Pest Resistance to Pesticides: Challenges and Prospects, C. P. Georghiou and T. Saito, eds., Plenum Press (in Press).
37. Harding, P. R. Jr. (1972). Differential sensitivity to thiabendazole by strains of Penicillium italicum and P. digitatum. Plant Disease Reptr. 56: 256.
38. McDonald, R. E., Risse, L. A., and Hillebrand, B. M. (1979). Resistance to thiabendazole and benomyl of Penicillium digitatum and P. italicum isolated from citrus fruit from several countries. J. Amer. Soc. Hort. Sci. 104:333.
39. Bertrand, P. F. and Saulie-Carter, J. L. (1978). The occurrence of benomyl-tolerant strains of Penicillium expansum and Botrytis cinerea in the mid-Columbia region of Oregon and Washington. Plant Disease Reptr. 62:302.

40. Rosenberger, D. A. and Meyer, F. W. (1979). Benomyl-tolerant Penicillium expansum in apple packing houses in Eastern New York. Plant Disease Reptr. 63: 37.
41. Kuramoto, T. (1976). Resistance to benomyl and thiophanate-methyl in strains of Penicillium digitatum and P. italicum in Japan. Plant Disease Reptr. 60: 168.
42. Kuramoto, T. and Yamada, S. (1976). DF-125, a new experimental fungicide for control of satsuma mandarin postharvest decays. Plant Disease Reptr. 60: 809.
43. Kaplan, H. J. and Dave, B. A. (1979). The current status of imazalil: a postharvest fungicide for citrus. Proc. Fla. State Hort. Soc. 92: 37.
44. Laville, E. Y., Harding, P. R. Dagan, Y. Rahat, M., Kraght, A. J., and Rippon, L. E. (1977). Studies on imazalil as a potential treatment for control of citrus fruit decay. Proc. Int. Soc. Citriculture 1: 259.
45. Birchmore, R. J., Brookes, R. F., Copping, L. G., and Wells, W. H. (1977). BTS 40 542 - A new broad spectrum fungicide. Ninth British Insect. Fungicide Conf. 2: 593.
46. Spalding, D. H. (1980). Control of Alternaria rot of tomatoes by postharvest application of imazalil. Plant Disease Reptr. 64: 169.
47. Anonymous (1980). Imazalil: a new weapon in the fruit decay battle. Citrograph 65: 95.
48. Rosenberger, D. A., Meyer, F. A. and Cecilia, C. V. (1979). Fungicide strategies for control of benomyl-tolerant Pencillium expansum in apple storages. Plant Disease Reptr. 63: 1033.
49. Bompeix, G., Coeffic, M., and Greiffier, P. (1979). Lutte contre les pourritures des peches à Monilia sp., Botrytis sp. et Rhizopus sp. Fruits 34: 423.
50. Van Gestel, J., Heeres, J., Janssen, M. and Van Reet, G. (1980). Synthesis and screening of a new group of fungicides: 1-(2-phenyl-1, 3-dioxolan-2-yl-methyl)-1,2,4-triazoles. Pestic. Sci. 11: 95.
51. Brown, G. E. (1979). Biology and control of Geotrichum candidum, the cause of citrus sour rot. Proc. Fla. State Hort. Soc. 92: 186.
52. Chitzanidis, A. and Laskaris, D. (1980). Effects of postharvest dips of oranges in metalaxyl against brown rot. Proc. 5th Congress of the Mediterranean Phytopathological Union. Patras, Greece, Sept. 1980. pp 151.
53. Edney, K. L. and Chambers, D. A. (1981). Postharvest treatments for the control of Phytophthora syringae storage rot of apples. Ann. Appl. Biol. 97: 237.
54. Davidse, L. C. and Flach, W. (1977). Differential binding of methyl benzimidazole-2-yl-carbamate to fungal tubulin as a mechanism of resistance to this antimitotic agent in mutant strains of Aspergillus nidulans. J. Cell Biol. 72: 174.

55. Davidse, L. C. and Flach, W. (1978). Interaction of thiabendazole with fungal tubulin. Biochem. Biophys. Acta 543: 82.
56. Leroux, P. and Gredt, M. (1978). Effets de quelques fongicides systemiques sur la biosyntheses de l'Ergostérol chez Botrytis cinerea Pers., Penicillium expansum Link et Ustilago maydis (DC) Cda. Ann. Phytopathology 10: 45.
57. Prusky, D., Bazak, M., and Bendrie, R. (1980). Tolerance of Penicillium expansum in Israel to postharvest fungicide treatments. Proc. 5th Congress Mediterranean Phytopath. Union, Patras, Greece, Sept. 21-27. Published by the Hellenic Phytopath Soc.

HYDROXYPROLINE-RICH GLYCOPROTEINS IN THE CELL WALL OF DISEASED PLANTS AS A DEFENSE MECHANISM

Marie Thérèse Esquerré-Tugayé, Dominique Mazau and Alain Toppan

Université Paul Sabatier
Centre de Physiologie Végétale - L.A. 241 CNRS
118, route de Narbonne
31062 Toulouse, France

INTRODUCTION

The plant cell wall may be considered as a barrier which separates a protoplast from the surrounding medium. Yet it is not a passive barrier, as indicated by active modifications under natural and stress physiology processes such as growth, maturation, ripening, abscission, senescence, and plant-microorganism interactions. This activity results, in part, from the presence of proteins and glycoproteins inside the cell wall. Although already mentioned sixty years ago(1), cell wall proteins were mainly studied after the discovery in the 1960's (2), that some of these proteins contain high amounts of hydroxyproline (Hyp). This paper is concerned with the study of these Hyp-containing proteins that we will abbreviate, according to their chemical nature, i.e. HRGP (Hydroxyproline Rich Glycoproteins). The following points will be successively considered:

1 - Hydroxyproline-containing components in plants: their occurrence and structural features;
2 - Hydroxyproline-rich glycoproteins in plant-pathogen interactions;
3 - Significance of the accumulation of hydroxyproline-rich glycoproteins in diseased plants;
4 - Role of cell surface interaction in the elicitation, via ethylene, of hydroxyproline-rich glycoprotein biosynthesis.

1- HYDROXYPROLINE-CONTAINING COMPONENTS IN PLANTS

Hyp-containing components are present in varying amounts in all plant parts: roots (3), stems and leaves (4), tubers (5), seeds (6,7), sap, and in the growth medium of cells in culture (8,9). They are either located within the cell wall, or found inside or outside the cell. The main features of the chemical composition of these molecules are summarized in Table I. They all have in common a high level of Hyp, of the β-hydroxyaminoacids serine and threonine, of alanine or glycine, and they all contain sugars, mainly arabinose and galactose. Accordingly these molecules are glycoproteins. In addition to their sugar to protein ratio which varies from about 50 to 90%, they differ by their solubility. Intra- and extracellular Hyp-glycoproteins are quite soluble, whereas cell wall Hyp-glycoproteins (HRGP) are highly insoluble in the native state once they are incorporated into the wall. Despite this limitation, cell wall HRGP are better characterized because chemical and radiochemical techniques were originally devised at the aim of understanding the structure and biosynthesis of these glycoproteins.

Chemical (HCl(10), $Ba(OH)_2$(11), N_2H_4OH (12), NaOCl-CH_3COOH (13)) and enzymic (14,15) hydrolyses and oxidation (Table II) allow recovery of either Hyp alone, or glycoaminoacids and glycopeptides rich in hydroxyproline. The results indicate that most Hyp residues in isolated peptides are found in a series of consecutive or distant pentapeptides in which 4 Hyp residues are linked to serine at the N-terminus (14) : -Ser-Hyp_4-. In these pentapeptides, the 1 D-galactopyranosyl residue is O-glycosidically linked to serine(14), and 1 to 4 L-arabinofuranosyl residues are O-glycosidically linked to Hyp (11,16), although varying levels of non-glycosylated serine and hydroxyproline are also found, depending on the group (Dicots as compared to Monocots) (17,18), or on the species and on the age of a particular plant (19). Altogether, the results allow prediction of a left-handed helix with 3 Hyp residues per turn for the secondary structure of the Hyp-rich portions of the wall HRGP (20).

In addition to structural investigations, one can also measure the biosynthesis of cell wall HRGP using radiolabeled techniques (21, 22,23). The rate of ^{14}C-Hyp and thus of HRGP deposition into the cell wall can be measured in tissues or cells incubated in the presence of ^{14}C-proline, since peptidyl Hyp originates from peptidyl proline (21,24).

A combination of these techniques was retained to study cell wall HRGP in plant microorganism interactions.

Table I. The major amino acids and sugars of Hydroxyproline-Rich Glycoproteins (HRGP) in plants

Amino acid[a], sugar	MELON stems cell walls (10)	TOMATO leaves intracell. (4)[b]	POTATO tubers intracell. (5)	CORN pericarp (6)	TOBACCO cells extracell. (8)	WHEAT endosperm (7)	LOLIUM multiflorum cells intracell. (9)	LOLIUM multiflorum cells extracell. (9)
HYP	11.7	23.4	20.4	12.2	23.8	16.7	9.9	14.8
ASP	10.9	S	S	S	S	6.3	6.3	S
THR	6.2	9.1	5.7	11.4	9.3	6.1	6.0	6.5
SER	8.6	14.3	12.6	7.7	14.8	9.6	9.5	9.8
GLU	7.5	S	6.9	S	S	12.2	6.6	-
GLY	11.5	19.6	12.2	15.5	S	S	7.8	6.7
ALA	S[c]	S	S	S	21.3	24.3	17.3	22.5
PRO	S	S	6.9	10.3	S	S	S	S
ARA	+	+	++	+	++	+	+	+
GAL	+	+	+	+	++	+	++	++
Other HEXOSES	-	-	-	++	+	-	-	-
GlcN	-	-	-	+	-	+	+	+

[a]Molar % of total amino acids residues.

[b](4) Recalculated from Lamport 1970.A.R.Pl.Phys 21:235 (1970) ; (5) Allen et al. Biochem J. 171:665 (1978) ; (6) Boundy et al. J. Biol. Chem. 242:2410 (1967) ; (7) Fincher et al. Biochem J. 139:535 (1974) ; (8) Hori et al. Phytochem. 16:1485 (1977) ; (9) Anderson et al. Austr. J. Pl. Physiol. 4:143 (1977) ;(10) Esquerré-Tugayé C.R. Acad. Sci. Paris Ser. D 276:525 (1973). Numbers correspond to reference list.

[c]Small amounts.

Table II. A typical hydroxyproline-rich glycopeptide of the wall HRGP and a summary of A, chemical and B, radiochemical methods used to study the major structural features and the biosynthesis of this glycoprotein.

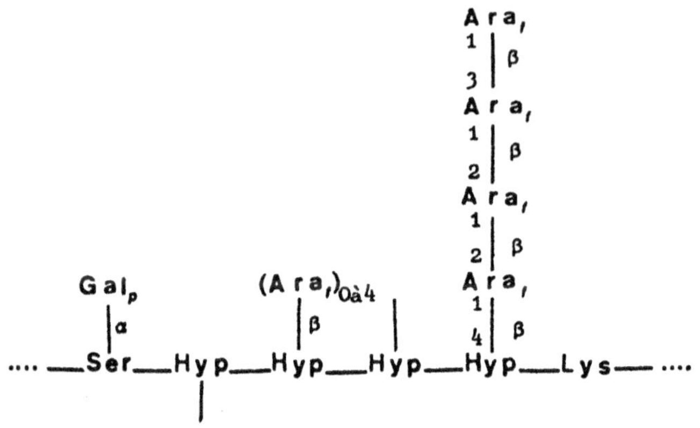

Method	Result	Recovery of
A - CHEMICAL and BIOCHEMICAL METHODS[(a)]		
Acid hydrolysis[15] 6 N HCl, 100°C, 18 h	Cleavage of peptide and glycosidic bonds in glycoproteins	Hyp, other amino acids
Partial acid hydrolysis[14] 0.1 N HCl, 100° C, 1 h	Cleavage of furanosidic linkages	Deglycosylated Hyp-containing material
Partial alkaline hydrolysis[26] 0.44 N $Ba(OH_2)$, 90° C, 20 h	Cleavage of peptide linkages	Hyp-arabinosides
Hydrazinolysis[15] N_2H_4OH, 110° C, 18 h	Elimination of glycosylated -Ser and-Thr in HRGP	Non glycosylated -Ser and-Thr
Oxidation by 1 % NaOCl-[13] 1 % CH_3COOH, 75° C, 30 min	Cleavage of phenolic crosslinks	Hyp-rich glycopeptides
Enzymic hydrolysis[14] Trypsine pH 8.0, 23° C, 18 h	Cleavage of peptide bonds involving basic amino acids residues	Hyp-rich glycopeptides
B - RADIOCHEMICAL METHOD		
^{14}C-Proline incorporation[21]	Synthesis of ^{14}C-peptidyl proline, leading to ^{14}C-peptidyl Hyp after hydroxylation	Cell wall ^{14}C-peptidyl Hyp

[a]- Numbers correspond to reference list.

2 - HYDROXYPROLINE-RICH GLYCOPROTEINS IN PLANT-PATHOGEN INTERACTIONS

Working on the hypothesis that host proteins could serve as substrates for proteolytic degradation by fungal pathogens we discovered, in 1973, that, instead of being lowered, the level of cell wall protein and of Hyp is higher upon infection (10). This was originally found in melon seedlings (cv. Cantaloup charentais) inoculated with Colletotrichum lagenarium, a fungal pathogen which develops on the aerial part of this plant, notably on the fruits, but the roots are left symptomless. Under greenhouse conditions, melon seedlings are killed 7 to 9 days after inoculation.

Infected seedlings exhibit a general increase in both cell wall protein and Hyp, whatever the organ considered, the roots included. As shown in the stems, this increase begins 3 days after inoculation (19), at the time of symptom appearance, and then proceeds throughout the disease to a level of Hyp which can be ten times as high as that in healthy plants.

The structural features which normally characterize cell wall HRGP, namely : a, their glycosylation pattern and, b, the presence of extractable Hyp-rich glycopeptides were studied upon infection.
a. Hyp-arabinosides were isolated after hydrolysis of the wall with $Ba(OH)_2$ and ion exchange chromatography (26). Galactosyl serine was measured in the presence of hydrazine (N_2H_4OH) which cleaves this O-glycosidic linkage by a β-elimination process (12,22). It was found that the two glycosylated amino acids were present in much higher amounts in the cell wall of diseased plants but that the ratios of glycosylated serine to Hyp were very similar in healthy and infected plants.
b. Serine and Hyp were also found together in glycopeptides obtained by trypsinolysis of the wall and, upon sequencing, were recovered under the form of the pentapeptide -Ser-Hyp_4-which typifies the wall HRGP (15).

From these data, it was concluded that infected melon seedlings have an increased ability to accumulate HRGP of the kind already present in the wall of healthy plants, but in minor quantities.

This response was later shown to occur in several plants challenged by different microorganisms, as indicated by Hyp and Hyp-arabinoside analyses of the cell walls of plants inoculated with either fungi (37) or bacteria (unpublished data). However, it was noticed that increases in Monocots were very low as compared to increases found in Dicots.

Because of its widespread occurrence, the significance of cell wall enrichment in HRGP upon infection was investigated.

3 - SIGNIFICANCE OF HYDROXYPROLINE-RICH GLYCOPROTEIN ACCUMULATION IN DISEASED PLANTS

Melon seedlings were retained throughout this study because of their ability to accumulate large amounts of cell wall HRGP in a relatively short time upon infection by C. lagenarium. Experiments were designed in order to control the extent at which this accumulation proceeds, and to measure the amount of pathogen present in the seedlings under these conditions.

Plants with higher than normal amounts of HRGP were obtained by treatment with ethylene, 6 days prior to inoculation. As formerly reported (27), this hormone promotes a Hyp enrichment of the cell wall, and this increase corresponds to the deposition of HRGP, as demonstrated by Hyp-arabinosides analysis (22). Amounts of this glycoprotein, lower than in control infected plants, were also obtained by feeding plants with L-trans Hyp for 6 days during the course of the disease. This amino acid specifically inhibits the synthesis of the wall HRGP, which results in lower amounts of hyp-arabinosides (22).

The amount of pathogen present in both situations was assessed by measuring glucosamine, a component which characterizes the cell wall of many fungi, but is present in only trace amounts in plants (28). An inverse relationship was found between the amount of HRGP and the extent of pathogen spread inside the plants (22). Plants pretreated with ethylene contained much less glucosamine, i.e. pathogen, and showed restricted symptoms as compared to untreated, infected controls. As a consequence, a large number of them recovered, instead of being killed as usual. However, plants whose HRGP accumulation was inhibited by free Hyp showed an accelerated susceptibility to the disease : they contained more pathogen and, as a result, were killed earlier (Fig. 1a, 1b).

These experiments allowed us to conclude that the accumulation of HRGP in melon seedlings is closely associated to their defense against C. lagenarium. This defense becomes efficient provided it starts early, before appearance of symptoms. The fact that ethylene or 1 mM Hyp bring about a two to three fold increase, or a 30% inhibition in HRGP deposition, only, suggests that even moderate changes can be significantly associated to resistance. Such moderate differences were later found in cultivar specific resistance and also in Legume-Rhizobium symbiosis under conditions where the roots were resistant to nodulation either constitutively or in the presence of inhibitory concentrations of nitrates (unpublished data).

Because of the high insolubility of HRGP once they are incorporated into the wall, one cannot explain yet the biochemistry of their association with resistance. One may hypothesize that they are structural proteins which strengthen the cell wall, or that they

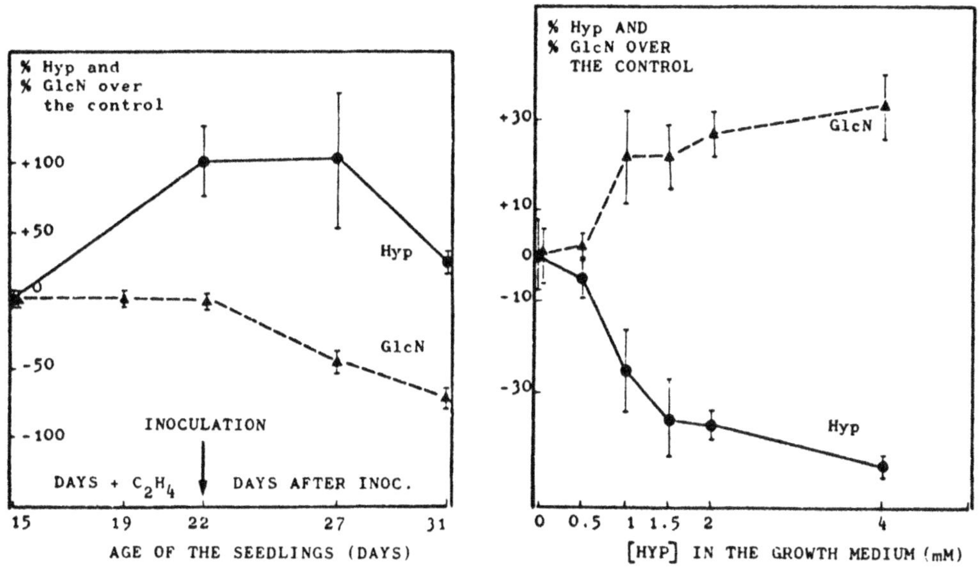

Fig.1a. Cell wall hydroxyproline (●) and glucosamine (▲) in plants treated with ethylene before inoculation in a % of the same components in control plants. Melon seedlings treated with ethylene received this hormone for 7 days. These plants and non-ethylene treated controls were then inoculated with Colletotrichum lagenarium and grown in the air of the growth room without added ethylene. The stems of both kinds of plants were harvested from day 0, before starting ethylene treatment, to day 9 after inoculation, and their cell wall extracted. Data represent the mean ± SE of three replicates (From M.T. Esquerré-Tugayé et al. 1979 Plant Physiol. 64:320).

Fig.1b. Cell wall hydroxyproline (●) and glucosamine (▲) in diseased plants grown in the presence of L-trans hydroxyproline, as a % of the same components in control plants. The addition of hydroxyproline at different concentrations to the culture medium started one day after inoculation of melon seedlings with Colletotrichum lagenarium, and lasted 6 days. Control plants were simultaneously inoculated and grown without hydroxyproline. The stems of both kinds of plant were harvested 7 days after inoculation and their cell wall extracted. Data represent the mean ± SE of three replicates (From M,T Esquerré-Tugayé et al. 1979 Plant Physiol. 64:320).

have a lectin-like activity. A Hyp-containing lectin which bears structural similarities with cell wall HRGP has been isolated from potato tubers (5,29). This Hyp-lectin is thought to be associated with the resistance of potatoes against Pseudomonas solanacearum through its agglutinating activity towards avirulent, but not towards virulent, bacteria (30). This agglutination involves the LPS fraction of the bacterial cell wall and is inhibited by oligosaccharides that contain N-acetyl glucosamine residues.

Probably the most recent evidence that HRGP accumulation is a novel parameter, which must be taken into account in plant-microorganism interactions, comes from the study of its elicitation.

4 - ROLE OF CELL SURFACE INTERACTION IN THE ELICITATION, VIA ETHYLENE, OF HYDROXYPROLINE-RICH GLYCOPROTEIN BIOSYNTHESIS

The fact that HRGP begins to accumulate in infected seedlings at a time when the pathogen is still restricted in the host (31) suggests that cell surface interaction only, between both partners, could be sufficient to elicit this accumulation. The role of ethylene as a mediator in the elicitation of this phenomenon was investigated since ethylene alone can promote an enrichment of the cell wall in HRGP (22), and that infected seedlings accumulate large amounts of this hormone early on (32). Using several inhibitors of the ethylene pathway in plants (33), namely L-Canaline, Benzyl isothiocyanate and Aminoethoxyvinylglycine (AVG), as well as 1-Aminocyclopropane-1-carboxylic acid (ACC) as a natural precursor of this hormone (34), it was found that HRGP biosynthesis in infected plants is ethylene-dependent (35). The deposition of protein-bound ^{14}C-Hyp in the cell wall closely parallels the biosynthesis of ethylene,both being simultaneously inhibited or stimulated.

The role of cell surface interaction in promoting, via ethylene, HRGP biosynthesis is evidenced by the presence of ethylene elicitors on fungal cell surfaces (Table III). These elicitors are glycoconjugates that we reported for the first time in plant-microorganism interactions three years ago (32). They elicit the production of ethylene by healthy tissues 60 to 90 minutes after incubation in their presence (36) (Fig. 2). Their activity is such that even microgram amounts induce both the synthesis of ethylene and of ^{14}C-Hyp rich glycoproteins in otherwise healthy tissues. Ethylene is a necessary, intermediate step, since inhibitors of the ethylene pathway also inhibits this elicitation process (37).

This approach provides a means of eliciting in healthy plants the accumulation of HRGP which normally occurs in diseased plants, so we can study the host genome activity responsible for this accumulation, without the interfering presence of the pathogen genome itself. It also suggests that triggering of wall HRGP biosynthesis

Table III. Production of ethylene by healthy petioles incubated in the presence of elicitors extracted from Colletotrichum lagenarium. Elicitors correspond to a crude extract containing sugars and nitrogen.

Elicitor source	Activity nl C_2H_4/g petiole/h[a]	Crude extract composition mg Glc eq.	mg N_2
Mycelium	0.27	3.81	1.24
Cell walls	0.14	1.49	0.78
Culture filtrate	0.29	-	-
Control	0.06	0	0

[a]Petioles (2g) are incubated in 5 ml of phosphate buffer 10 mM pH 6.0, with or without elicitors (elicitors from 1 g mycelium).

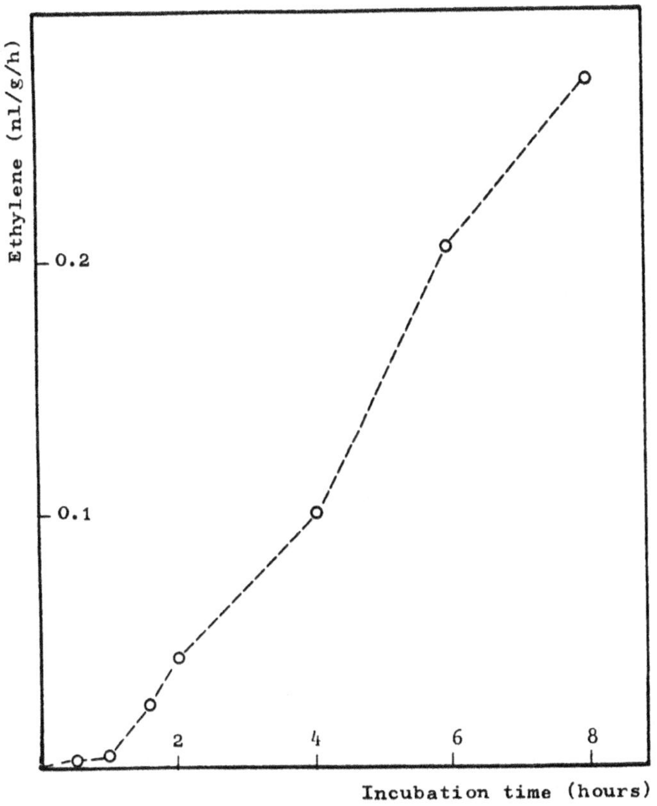

Fig. 2. Production of ethylene by healthy petioles treated with fungal elicitors. The assay was performed under the conditions given in Table III.

in diseased plants is mediated by cell to cell recognition between host and pathogen.

CONCLUSION

An understanding of the nature of defense reactions occurring in plants is of importance for the practice of crop productivity and storage. Due to their location as the most outer layer of the cell, plant cell walls are particularly exposed to unfavorable environmental conditions. It is the purpose of this paper to demonstrate that the cell wall actively participates in the defense of plants through the accumulation of a Hyp-rich glycoprotein. Because this glycoprotein is very insoluble, it is, as yet, not clear as to whether it acts as a structural protein or whether it fulfills other functions.

Ethylene plays a major role in the elicitation of this defense response. The fact that this hormone is also involved in aging processes might explain, in part, why mature plants are often more resistant to disease.

REFERENCES

1. R.M. Tupper-Carey and J.H. Priestley, The composition of the cell-wall at the apical meristem of stem and root, Proc. Roy. Soc. B 95:109 (1923).
2. D.T.A. Lamport and D.H. Nortcote, Hydroxyproline in primary cell walls of higher plants, Nature 188: 665 (1960).
3. M.C. Jotterand-Dolivo and P.E. Pilet, Hydroxyproline in cell wall and cytoplasmic fractions from maize roots, Plant. Sci. Lett. 4:183 (1975).
4. D.T.A. Lamport, Cell wall metabolism, Ann. Rev. Plant Physiol. 21:235 (1970).
5. A.K. Allen, N.N. Desai, A. Neuberger and J.M. Creeth, Properties of Potato lectin and the nature of its glycoprotein linkages, Biochem. J. 171: 665 (1978).
6. J.A. Boundy, J.S. Wall, J.E. Turner, J.H. Woychik and R.J. Dimler, A mucopolysaccharide containing hydroxyproline from corn pericarp, J. Biol. Chem. 242:2410 (1967).
7. G.B. Fincher, W.H. Sawyer and B.A. Stone, Chemical and physical properties of an arabinogalactanpeptide from wheat endosperm, Biochem. J. 139:535 (1974).
8. H. Hori and S. Sato, Extracellular hydroxyproline-rich glycoprotein of suspension-cultured tobacco cells, Phytochemistry 16:1485 (1977).
9. R.L. Anderson, A.E. Clarke, M.A. Jermyn, R.B. Knox and B.A.Stone, A carbohydrate-binding arabinogalactan-protein from liquid suspension cultures of endosperm from <u>Lolium multiflorum</u>, Aust. J. Plant Physiol. 4:143 (1977).

10. M.T. Esquerré-Tugayé, Influence d'une maladie parasitaire sur la teneur en hydroxyproline des parois cellulaires d'epicotyles et de pétioles de plantes de melon, C.R. Acad. Sci. Paris Serie D 276:525 (1973).
11. D.T.A. Lamport, Hydroxyproline-O-glycosidic linkage of the plant cell wall glycoprotein extensin, Nature 216:1322 (1967).
12. D.T.A. Lamport, The glycopeptide linkages of extensin : O-D-galactosyl serine and O-L-arabinosyl hydroxyproline, in: "Biogenesis of plant cell wall polysaccharides," F. Loewus, ed., Acad. Press, New York and London 149 (1973).
13. A.J. Mort and D.T.A. Lamport, Evidence for polysaccharide attachment of extensin in cell walls obtained from tomato cell suspension cultures, Plant Physiol. 56, Suppl. Abstract 80:16 (1975).
14. D.T.A. Lamport, L. Katona and S. Roerig, Galactosyl serine in extensin, Biochem. J. 133:125 (1973).
15. M.T. Esquerré-Tugayé and D.T.A. Lamport, Cell surfaces in Plant-Microorganism interactions. I. A structural investigation of cell wall hydroxyproline-rich glycoproteins which accumulate in fungus-infected plants, Plant Physiol. 64:314 (1979).
16. Y. Akiyama and K. Kato, Structure of hydroxyproline-arabinoside from tobacco cells, Agric. Biol. Chem. 41:79 (1977).
17. D.T.A. Lamport and D.H. Miller, Hydroxyproline arabinosides in the Plant kingdom, Plant Physiol. 48:454 (1971).
18. M.T. Esquierré-Tugayé and D. Mazau, Les glycoproteines à hydroxyproline de la paroi végétale, Physiol. Vég. In Press (1981).
19. D. Mazau and M.T. Esquerré-Tugayé, Hydroxyproline et peroxydases dans les parois cellulaires et le cytoplasme de tiges de Melons atteints d'anthracnose, Plant Sci. Lett. 7:119 (1976).
20. D.T.A. Lamport, Structure and function of Plant Glycoproteins, in "The Biochemistry of plants." J. Preiss, ed., Academic Press, New York, Vol. 3:501 (1980).
21. M.J. Chrispeels, Synthesis and secretion of hydroxyproline-containing macromolecules in carrots. II. In vivo conversion of peptidyl proline to peptidy hydroxyproline, Plant Physiol. 45:223 (1970).
22. M.T. Esquerré-Tugayé, C. Lafitte, D. Mazau, A. Toppan and A. Touzé, Cell surfaces in plant-microorganism interactions. II. Evidence for the accumulation of hydroxyproline-rich glycoproteins in the cell wall of diseased plants as a defence mechanism, Plant Physiol. 64:320 (1979).
23. J.E. Barner and J.E. Burton, An in vivo assay for the synthesis of hydroxyproline-rich proteins, Plant Physiol. 66:1044 (1980).
24. D. Ashford and A. Neuberger, 4-Hydroxyl-L-proline in plant glycoproteins, Trends in Biochemical Sciences 5:245 (1980).
25. M.T. Esquerré-Tugayé, Les glycoproteines des surfaces cellulaires végétales : étude particulière d'une glycoprotéine à hydroxyproline dans les plantes de melon au cours d'une

maladie parasitaire, Thèse Doct. Etat, Université de Toulouse (1977).

26. M.T. Esquerré-Tugayé and D. Mazau, Effect of fungal disease on extensin, the plant cell wall glycoprotein, J. Exp. Bot. 25: 509 (1974).
27. I. Ridge and D.J. Osborne, Hydroxyproline and peroxydases in cell walls of Pisum sativum : regulation by ethylene, J. Exp. Bot. 21:843 (1970).
28. A. Toppan, M.T. Esquerré-Tugayé and A. Touzé, An improved approach for the accurate determination of fungal pathogens in diseased plants, Physiol. Plant Path. 9:241 (1976).
29. N.N. Desai and A.K. Allen, The purification of potato lectin by affinity chromatography on an N, N', N"-triacetylchitotriose-sepharose matrix, Anal. Biochem. 93:88 (1979).
30. L. Sequeira, Lectins and their role in host-pathogen specificity, Annu. Rev. Phytopathol. 16:453 (1978).
31. R. Dargent and A. Touzé, Etude cinétique, en microscopie électronique, des interactions entre Colletotrichum lagenarium et les cellules foliaires de Cucumis melo, Can. J. Bot. 52: 1319 (1974).
32. M.T. Esquerré-Tugayé, D. Mazau and A. Toppan, Triggering by fungal components of a defence mechanism involving cell wall glycoproteins in plants, 3rd Int. Congress of Plant Path. München, Section VI:229 (1978).
33. M. Lieberman, A.T. Kunishi, L.D. Owens, Specific inhibitors of ethylene production as retardants of the ripening process in fruits, in: "Facteurs et régulation de la maturation des fruits." Coll. Int. CNRS n° 238, Paris (1975).
34. D.O. Adams, S.F. Yang, Ethylene biosynthesis : identification of 1-aminocyclopropane-1-carboxylic acid as an intermediate in the conversion of methionine to ethylene, Proc. Natl. Acad. Sci. USA 76:170 (1979).
35. A. Toppan, D. Roby and M.T. Esquerré-Tugayé, Cell surfaces in plant-microorganism interactions. III. In vivo regulation by ethylene of hydroxyproline-rich accumulation in the cell wall of diseased plants, Plant Physiol. (Submitted for publication).
36. A. Toppan, D. Mazau, A. Touzé and M.T. Esquerré-Tugayé, Production of ethylene as a result of recognition between plants and fungal pathogens, in: "Proc. Vth Int. Symp. on Glycoconjugates." R. Schauer, P. Boer, E. Buddecke, J.F. Kramer, J. Vliegenthart and G. Wiegandt, eds., Georg Thieme Publishers, Stuttgart, Kiel (1979).
37. M.T. Esquerré-Tugayé, D. Mazau, A. Toppan and D. Roby, Elicitation, via l'éthylène, de la synthese de glycoprotéines pariétales associées à la défense des plantes, in: "Actes du Symposium Phytoalexines et phenomenes d'élicitation chez les plantes", Ann. Phytopathol. (sous presse) (1980).

STRESS METABOLITES

Norman F. Haard

Department of Biochemistry
Memorial University of Newfoundland
St. John's, Newfoundland, Canada A1B 3X9

INTRODUCTION

A phenomenon, comparable to induced immunity in animals, was reported for the late blight disease of potatoes over 40 years ago (73). The phytoalexin theory of plant immunity evolved from these early observations. According to this theory, a specific metabolic interaction between a plant host and a challenging microorganism may give rise to the formation of phytoalexins (from Greek: to ward off). The phytoalexins formed in potato acted to inhibit the growth of the virulent race 1 of Phytophthora infestans. Originally, the theory suggested that phytoalexins are absent in healthy plant tissues and that the phytoalexin response is highly specific with regard to pathogen and host cultivar. In the example of late blight disease, potato tubers carrying the gene R_1 and an avirulent race 0 to P. infestans represented the specific metabolic interaction. Recent studies have demonstrated that phytoalexins may occur in trace quantities in apparently healthy plant tissues and that a variety of biological agents can trigger the accumulation of phytoalexins in a given plant tissue. In view of these findings, the phytoalexin concept has been redefined, e.g. "the term phytoalexin should serve as an umbrella under which chemical compounds contributing to disease resistance can be classified whether they are formed in response to injury, physiological stimuli, and the presence of infectious agents or are the products of such agents" (58). It is now clear that these secondary metabolites may accumulate in response to a wide array of chemical and physical stresses in addition to biological agents. As a result of these findings the term "stress metabolite" has been used to describe secondary plant metabolites which accumulate in response to varied trauma to plant tissue. While the terms "stress metabolite" and "phytoalexin" have been used interchangeably, we shall consider phytoalexins as a special class of stress metabolites, i.e. substances which have an implicit role in disease resistance. It should be noted that additional factors, such as naturally occurring constituents and mechanical barriers also play a role in the resistance of plants to microbial challenge (114). There are numerous reviews of work on phytoalexins and their role in plant immunity to disease (29,55,58,59,67,72,107).

Generally stress metabolites possess some degree of toxicity to living matter and given that this toxicity can be extended to humans, one must consider their importance in the context of human and livestock safety (113). Consideration of stress metabolites with respect to mammalian toxicity is of continued concern because of the reality of on-going plant breeding programs and because of the increased complexity of food processing and handling operations. The focus of this lecture will therefore be on the occurrence of stress metabolites in edible plants, the nature of the triggering mechanism by which these secondary metabolites accumulate in plants, and our present understanding of their significance with regard to food safety.

CHEMICAL NATURE AND DISTRIBUTION OF STRESS METABOLITES

Present evidence indicates that stress metabolites are widespread in the plant kingdom. Serious attempts to identify involvement of phytoalexins in host-parasite interactions invariably demonstrate the formation of secondary metabolites possessing antibiotic activity. The number of new stress metabolites identified in the past few years is indicative of the considerable activity in this field of study (3,16,19,20,25,43,45,47,48,50,56,62,67,76,87,89,98). There appears to be a taxonomic basis to the distribution of phytoalexins in the plant kingdom (49). Members of the *Leguminosae* tend to accumulate isoflavanoids and according to a recent report (88), there is a dichotomy in phytoalexin distribution amongst the Viceae which has taxonomic relevance. When the different members of the genera *Lathyrus*, *Lens*, *Pisum* and *Vicia* were challenged with *Helminthosporium carbonum* or *Botrytis cinerea* it was observed that *Pisum* and *Lathyrus* spp. produced the isoflavanoid pisatin and *Vicia* and *Lens* species produced furanoacetylenes. Stress metabolites have been identified in most plant families of agronomic importance.

Green beans (*Phaseolus vulgaris*) produce a number of isoflavanoid stress metabolites including phaseollin, phaseollidin, phaseollinisoflavan, 2'-methoxyphaseollinisofalvan, kievitone, coumesterol, genistein and 2-hydroxygenistein (8). Garden pea (*Pisum sativum*) also forms various isoflavanoids in response to stress including pisatin, inermin, 3,4-hydroxy-2,3,9-trimethoxypterocarpan, 3-hydroxy-2,9-dimethoxypterocarpan and 2,3,9-trimethoxypterocarpan (17). Lima bean roots (Phaseolus lunatus) produce coumesterol and other coumestans as a result of nematode infection (86). Broad beans (Vicia faba) produce the isoflavanoid medicarpin and, in addition, the furanoacetylene compounds wyerone, wyerol, wyerone acid and wyerone epoxide (38). Lentil seeds (Lens culinaris) yield a similar family of compounds as broad bean including wyerone, wyerone epoxide and the isoflavanoid variabilin (87). Stress compounds identified in soybean (Glycine max) include glyceollin and its derivatives diadzein, coumesterol and sojagol (64). Alfalfa (Medicago sativa) produces at least eleven isoflavanoid derivatives including sativan, sativol, trifoliol, medicarpin and diadzein (78). Peanuts (Arachis hypogoea) produce stilbene-like compounds (46) and cowpeas (Vigna sinensis)produce isoflavonoids such as phaseollin and a 2-arylbenzofuran compound called vignafuran (82).

The Solanaceae family has also been extensively studied with respect to stress metabolites. The principle class of compounds formed in this family are diterpenes although sterols, steroidal alkaloids and phenylpropanoid compounds are also formed in this family. White potato (Solanum tuberosum), the tissue examined by Müller and Börger in their classic study, is now known to form more than twenty sesquiterpene stress metabolites including rishitin, katidinone, dehydrokatadinone, isolubimin, lubimin, oxylubimin, phytuberin and rishitinol (93,102). Other examples of sesquiterpene stress metabolites in this family include capsidiol in green pepper (Capsicum brutesceus); 9-oxonerolidol and aubergenone in eggplant (S. melongena); and rishitin in tomato (L. esculentum) (31).

The Solanaceae also contain steroidal alkaloids as natural constituents (e.g. α-solanine, α-chaconine in potato or tomatine in tomato) which increase in amounts as a result of stress conditions (19,35). Other alkaloids, such as α- and β-solamarine are absent from normal potato tissue but accumulate as a result of physical stress (94). Phenolic substances (e.g. chlorogenic acid and scopoletin and sterols (cholesterol, stigmasterol, β-sitosterol) are also recognized as stress metabolites in Solanaeceous plants (79).

The Umbelliferae family also produce a variety of stress metabolites. Parsnips may contain the coumarin compound xanthotoxin as a normal constituent which increases as a result of stress (53). Sweet potatoes (Ipomea batatas) produce coumarin derivatives, (e.g. umbelliferone, esculentin), phenylpropanoids (e.g. chlorogenic acid) and furanoterpenes. Two families of furanoterpenes are the 15 carbon group, noteably ipomeamarone, and 9-carbon terpenes such as 4-ipomeanol (32). Carrot roots produce isocoumarins, noteably 8-hydroxy-3-methyl-6-methoxy-3,

4-dihydroisocoumarin and other phenolics which were originally recognized due to their contribution to the bitterness resulting from improper vegetable storage (92,101).

Other examples of stress metabolites in edible plants include betavulgarin and betagarin in sugar beets (Beta vulgaris) (27); viniferins in grapevines (62); safynol and at least eleven other polyacetylene compounds in safflower (106); moracins A and B in mulberry (104); momilactones A and B in rice, wheat and rye, avenacin in oats and unidentified components in corn (67).

In general, stress metabolites most often occur as homologous families of structurally related compounds. The ratio of one compound to another may vary with the part of the plant and the nature of the stress. In addition, different chemical families of stress metabolites may arise simultaneously or in response to different stress conditions. A competitive relationship, with respect to biosynthesis, may occur as is the case for steroidal alkaloids and terpenes in white potato (11,95). It is important to recognize the multiplicity of stress metabolites because it undoubtedly complicates our understanding of their role as phytoalexins and as potential mammalian toxicants. Furthermore, it is often assumed (perhaps wrongly) that the most abundant stress metabolite(s) in a given tissue is of most importance from the toxicological viewpoint.

MECHANISM OF ELICITOR RESPONSE - BIOTIC ELICITORS

Considerable effort has been made towards elucidating the molecular basis for elicitation of stress metabolite(s) formation in plant tissues. When sweet potato root is challenged by Ceratocystis fimbriata the further invasion of the host (black rot) is checked by a layer of necrotic tissue and by a layer of furanoterpenoid phytoalexins in the adjacent healthy tissue. In general, fungal or insect extracts can also elicit stress metabolite formation (18,57,60). Most often the race specificity of the pathogen is lost as a result of cell disruption of the pathogen and much lower levels of metabolite are often observed. Specific cellular fractions or pathogens have also been shown to possess the capability of switching on stress metabolite synthesis in the host. Numerous studies indicate that cell wall components are active in eliciting stress metabolites (1,3,14,34,40,60). Chitosan, a polymer of B-(1 $\rightarrow$ 4) glucosamine isolated from Fusarium solani cell walls is an effective phytoalexin elicitor (33). The elicitation by chitosan can be potentiated by host chitosanase (75). The cell walls of Clasdosporium fulvum contain a glycoprotein which elicits rishitin accumulation in tomato fruit (22). A β-glucan, isolated from the cell walls of Phytophthora megasperma and yeast, served to stimulate isoflavanoid accumulation in kidney bean and soybean and terpene accumulation in potato (14).

Other studies have related fungal metabolites or other intracellular components of the pathogen to the elicitor response (18, 55, 97). There has also been an attempt to relate limited host cell necrosis to the elicitor response. While most biotic or abiotic stress conditions generally cause a hypersensitive response, this is not always the case (11). However, it was recently demonstrated that phaseollin and phaseollidin were formed in live bean hypocotyl tissue when incubated in contact with dead bean hypocotyl tissue (37). The authors provide evidence that a water soluble (elicitor) component is released during host cell damage.

In most cases the specificity (race or cultivar) associated with the native systems is lost with molecular elicitors. This has prompted interest in suppressors of phytoalexin as the basis of specificity. Suppressors may be molecular components contained in the pathogen. For example, glucans (β-(1 $\rightarrow$ 3), β-(1 $\rightarrow$ 6)) isolated from compatible races of P. infestans are more active in suppressing the hypersensitive reaction than those from incompatible races (60). Similarly, Mycophaerella pinodes, a pathogen of pea, contains a low molecular weight peptide which acts to suppress phytoalexin formation by the host (97). These studies indicate that virulence of the pathogen may relate more to its ability to provide suppressors of phytoalexin formation rather than to a lack of an elicitor. Other factors such as ageing time after cut injury appear to influence the elicitor response. Potato slices preincubated for 5 hours are no longer responsive to abiotic elicitors (11) or to P.infestans (95). This may be related to the spontaneous formation of glycoalkaloids accumulation resulting from cut injury in the absence of a terpene elicitor. Other factors such as membrane barriers (76), excessive necrosis and soil components can suppress the elicitor response in certain systems (11,36).

Another factor which may influence the specificity of the phytoalexin response is the finding that the pathogen can metabolize phytoalexins (26,42,61,63,65,99,110). Generally, phytoalexins appear to be metabolized to more polar compounds. This would appear to be another means of achieving race specificity for the pathogen. Moreover, it is known that the host may also metabolize phytoalexins (9,51,112). Often, the formation of specific metabolites is transient indicating further metabolism (11). Many of these stress metabolites are phytotoxic so their further metabolism by the host may serve an adaptive function.

ABIOTIC ELICITORS

It is now clear that a variety of physical and chemical stresses can evoke the accumulation of phytoalexins in plant tissues.

While it was earlier argued that substances which accumulate in response to abiotic stress are not true phytoalexins, it is now clear that most, if not all, phytoalexins may accumulate as a result of such non-specific elicitors.

Physical

Mechanical injury in the form of bruising or slicing is generally necessary to elicit stress metabolite formation. In most instances mechanical damage alone is insufficient to trigger stress metabolite formation although this is the only factor involved with glycoalkaloid accumulation in potato (94); phaseollin accumulation in bean pod (24); or coumarin synthesis in carrot (92). Other physical stresses, in combination with cut injury, can evoke stress metabolic accumulation. For example, exposure of tissue slices to ultraviolet light for a brief time triggers furanoterpene accumulation in sweet potato (32), diterpene accumulation in potato slices (11), isoflavanoid accumulation in lucerne (21), and bean pod (24), soybean (84) and cowpea (69). Other physical factors known to evoke stress metabolite accumulation are chilling injury, visible light, anoxia and reduced atmospheric pressure (32). In all examples other than chilling injury, preparation of tissue slices appear to be a prerequisite for elicitation.

Chemical

There is now abundant evidence that a wide array of chemical agents can evoke accumulation of stress metabolites in plant tissues. In many instances the chemical treatment evokes a reaction causing localized tissue necrosis, although this hypersensitive response is not always observed (11). Certain chemical agents appear to elicit stress metabolite accumulation in all tissue systems when applied under controlled conditions of time, temperature and concentration. For example, heavy metal salts of copper or mercury trigger ipomeamorone accumulation in sweet potato (32), rishitin accumulation in potato (11), vestitol accumulation in lucerne (21) and phaseollin in bean (36). Actinomycin D and other transcriptional and translational inhibitors also appear to be effective elicitors of stress metabolites when applied under carefully controlled conditions (12,13). Metabolic inhibitors such as sodium flouride, detergents such as sodium lauryl sulfate, various sulfyhydryl reagents and salts such as 5 mM sodium acetate are additional examples of chemical agents known to evoke stress metabolite biosynthesis (13). Moreover, recent studies indicate various pesticides act to elicit phytoalexin accumulation (74, 108) indicating that the efficacy of certain fungicides and insecticides may be partly dependent on switching on a native biological defense mechanism of the plant.

The findings that relatively innocuous conditions can give

rise to stress metabolite formation is further supported by reports that produce obtained from wholesale and retail markets may contain significant quantities of toxic stress metabolites (10,15,25, 70). There is a need for more information on the occurrence of stress resulting from postharvest handling and storage of produce on metabolite accumulation.

Role of Ethylene

We have discussed three factors which influence the accumulation of stress metabolites in plant tissues - elicitors, suppressors and catabolism of the induced metabolite by the host or invading organism. Various lines of evidence point to the involvement of ethylene gas in the accumulation of stress metabolites by plant tissues. The subject has recently been reviewed (32). It has long been recognized that ethylene of endogenous or exogenous origin is necessary for the accumulation of isocoumarin in carrot root (101). Ethylene gas can potentiate the accumulation of terpenes resulting from heavy metal treatment (32) and influence the balance of different terpenes resulting from the fungal induced reaction in sweet potato (32) and white potato (2). Phenylpropenoid accumulation in carrot slices is also potentiated by exogenous ethylene (92) and the isoflavanoid pisatin accumulates in garden pea as a result of ethylene treatment.

It has been suggested that ethylene acts to alter stress metabolite accumulation via its suggested role (100) in cyanide resistant respiration (2,32). Whether this apparent relationship is a direct one or is incidentally related will have to await further study.

There is relatively little information on the role of other plant hormones in stress metabolite accumulation. Synthetic auxins can influence phaseollin accumulation (23) and abscisic acid acts to prevent rishitin and lubimin accumulation in *P. infestans* infected potato but acts to increase the levels of these terpenes in *C. cucumerimun* challenged potatoes (41). It has also been reported that the highly unsaturated eicosapentaenoic and arachidonic acids elicit antimicrobial stress metabolite accumulation and cause necrosis in potato tuber slices (5) and this relate to the wound hormone traumatin (115). There would, accordingly, appear to be a need for additional information on the interplay of hormones in the elicitation of stress metabolites.

TOXICITY OF STRESS METABOLITES

Most, if not all, stress metabolites possess antibiotic activity which is more or less specific. For example, the isoflavanoid stress metabolites from French bean are selectively toxic to gram positive but not to gram negative bacteria (28). The sensitivity of microorganisms to the isoflavanoids pisatin and phaseollin is dependent on the growth conditions employed (111).

Members of the various chemical classes of stress metabolites are also phytotoxic at microgram concentrations. Examples where phytotoxicity of stress metabolite has been characterized include pisatin (96), phaseollin (83), rishitin (66) and wyerone (39). The phytotoxicity of stress metabolites is probably related to some lesion formations in postharvest fruits and vegetables.

The toxic characteristics of stress metabolites in mammalian systems is not very well understood although there is ample evidence of toxicity for certain compounds. The general concern about stress metabolites with regard to food safety has been reviewed by Wood (113). Concern was heightened following Renwick's hypothesis that blighted potatoes containing diterpene stress metabolites such as rishitin are linked to the severe birth defects spina bifida and anencephaly (85). While some experimental evidence has supported this hypothesis (80), later studies did not confirm the involvement of diterpenes in this response (81). Later studies have implicated glycoalkaloid stress metabolites in teratogenic effects (54). The current status of the Renwick hypothesis is complex and was recently reviewed (52).

The glycoalkaloids α-solanine and α-chaconine have been responsible for isolated cases of human death and innumerable instances of food poisoning (52). In many instances, minor, unidentifield alkaloids, occurring in potatoes have been implicated in food poisoning (71).

The sweet potato furanoterpenoids have been responsible for several instances of livestock death and one suspected case of human death (32). Ipomeamarone, the major stress metabolite in black rot infected roots is a hepatotoxin and 4-ipomeanol, the major component in chill injured and Fusarium infected roots is a lung toxin which has been extensively studied as to toxicology in recent years (6). It has also been suggested that excessive dietary consumption of sweet potato may be linked to lung cancer.

Other studies on the toxicology of stress metabolites are preliminary in nature and their significance to mammalian organisms is not clear. Pisatin and related isoflavanoid compounds like phaseollin and medicarpin are reported to lyse human and animal erythrocytes (77). Certain steroidal alkaloids, such as α-tomatine are known to complex with membrane sterols and cause lysis (90). Phytoalexins formed in rye grass infected with C. rathayi were found to be acutely toxic to mice at relatively low doses (103). Accordingly, there is sufficient evidence to justify concern about the toxicological influence of stress metabolites on livestock and people. Virtually nothing is known about possible toxicity of these substances resulting from long-term consumption of sub-acute doses. It may well be that the body has adequate detoxification systems to handle the low levels of these agents which may be associated with marketable vegetables

CONCLUSION

Recent evidence indicates that stress metabolites, a family of noxious compounds, may accumulate in edible plant tissues as a result of relatively innocuous circumstances of stress and, in the few vegetable systems studied, may accumulate in blemished areas of marketable produce. Current toxicological data does not justify alarm but there is presently sufficient information to justify caution in the use of new storage, handling and processing techniques since they may comprise a sufficient trauma to the tissue to evoke accumulation of these noxious agents. Similarly, plant breeders should recognize the benefits of disease resistance achieved in new cultivars may be accompanied by a greater propensity of the tissue to form stress metabolites or to the formation of new substances.

REFERENCES

1. Albersheim P. and Valente, B.S. (1978). Host pathogen interactions in plants: When exposed to oligosaccharides of fungal origin, defend themselves by accumulating antibiotics. J. Cell Biology, 78(3), 627.
2. Alves, L. M., Heisler, E. G. Kissinger, J. C., Patterson, J. M. and Kalan, E. B. (1979). Effects of controlled atmospheres on production of sesquiterpenoid stress metabolites by white potato tuber. Possible involvement of cyanide resistant respiration. Plant Physiol., 63 (2), 359.
3. Anderson, A. J. and Albersheim, P. (1975). Isolation of a pathogen-synthesized fraction rich in glucan that elicits a defense response in the pathogens host. Plant Physiol. 56(2), 286.
4. Baker, R., Hanchey, P. and Dottarar, S. D. (1978). Protection of carnation against Fusarium roseum stem rot by fungi. Phytopathology, 68 (10), 1495.
5. Bostoc, R.M., Kuc, J.A. and Laine, R.A. (1981). Eicosapentaenoic and arachidonic acids from Phytophthora infestans elicit fungitoxic sesquiterpenes in the potato. Science, 212, 67.
6. Boyd, M.R., Burka, L.T., Wilson, B.J., and Sasume, H.A. (1978) In vitro studies on the metabolic activation of the pulmonary toxin 4-ipomeanol by rat lung and liver microsomes. J. Pharmacol. Exp. Ther., 207, 677.
7. Buregger, B.B. and Keen, N.T. (1979). Specific elicitors of glyceollin accumulation in Pseudomonas glycinea soybean

host parasite system. Physiol. Plant Pathol., 15(1), 43.
8. Burden, R.S., Bailey, J.A. and Dawson, G.W. (1972). Structure of the isoflavanoids from Phaseolus vulgaris infected with tobacco necrosis virus. Tetrahedron Lett., 41,4175.
9. Burka, L.T. and Kuhnert, L. (1977). Biosynthesis of furano-terpenoid stress metabolites in sweet potatoes. Oxidation of ipomeamarone to 4-hydroxymyoporone. Phytochemistry, 16 (12), 2022.
10. Catalano, E.A., Hasline, V.C., Dupuy, H.P. and Constatin, R.J. (1976). Ipomeamarone in blemished and diseased sweet potatoes. J. Agric. Food Chem., 25, 94.
11. Cheema, A.S. and Haard, N.F. (1978). Induction of rishitin and lubimin in potato tuber slices by nonspecific elicitors and the influence of storage conditions. Physiol. Plant Pathol., 13, 233.
12. Cheema, A.S. and Haard, N.F. (1979). Induction of rishitin and lubimin synthesis in potato tuber slices by non-specific elicitors-role of gene derepression. Journal of Food Protection, 42 (6), 512.
13. Cheema, A.S. and Haard, N.F. (1980). Induction of rishitin and lubimin synthesis in potato tuber slices by non-specific elicitors-role of gene derepression. Journal Indian Potato Association, 7 (2), 48.
14. Cline, K., Wade, M. and Albersheim, P. (1978). Host-parasite interactions: Fungal glucans which elicit phytoalexin accumulation in soybean also elicit the accumulation of phytoalexins in other plants. Plant Physiol., 62(6), 918.
15. Coxon, D.T., Curtis, R.F. and Howard, B. (1975). Ipomeamarone, a toxic furanterpenoid in sweet potatoes (Ipomea batatas) in the United Kingdom. Fd. Cosmet. Toxicol., 13, 87.
16. Coxon, D.T., Oneill, T.M., Mansflied, J.W. and Porter, A.E. (1980). Identification of 3 hydroxyflavan phytoalexins from daffodil Narcissus pseudonaricissus bulbs. Phyto chem., 19 (5), 889.
17. Cruickshank, I.A.M. and Perrin, D.R. (1960). Isolation of the phytoalexin from Pisum sativum. Nature, 187, 799.
18. Daniels, D.L. and Hadwiger, L.A. (1976). Pisatin-inducing components in filtrates of virulent and avirulent Fusar ium solani cultures. Physiol. Plant Pathol., 8(1), 9.
19. Debnovetskii, G. and Basova, S.V. (1979). Phytoalexin activity of poplars as an index of their resistance to rust. Mikol. Fitopatol., 13 (5), 428.
20. Dewick, P.M. and Ingham, J.L. (1980). Isopterofuran a new 2 aryl benzofurna phytoalexin from Coronilla emerus Phytochem., 19 (2), 289.
21. Dewick, P.M. and Martin, M. (1979). Biosynthesis of pterocarpan isoflavan and coumestan metabolites of Medicago sativa. Chalcone isoflavone and isoflavanone precursors. Phytochemistry, 18 (4), 587.

22. Dewit, P.J. and Roseboom, P.H. (1980). Isolation and partial characterization and specificity of glycoprotein elicitors from culture filtrates, mycelium, cell walls of Cladosporium fulvum. Physiol. Plant. Pathol., 16 (3), 391.
23. Dixon, R.A. and Fuller, K.W. (1976). Effect of synthetic auxin levels on Phaseollin production and phenylalanine amminia-lyase activity in tissue cultures of Phaseolus vulgaris. Physiol. Plant Pathol., 9 (3), 299.
24. Doherty, J. and Beusher, R. (1978). Occurrence of the phyto-alexin phaseollin in pods of Phaseolus vulgaris. J. Sci. Food Agric., 29 (10), 853
25. Flores, G. and Hubbes, M. (1979). Phytoalexin production by aspen Populus tremuloides in response to infection by Hypoxylon mammatum and Alternaria spp. Eur. J. For. Pathol., 9 (5), 280.
26. Fuchs, A., Devries, F.W., Lantheer, C.A., and Vanveldhuizen, A. (1980). 3-hydroxymaackiain isoflavan a pisatin metabolite produced by Fusarium oxysporum. Phytochemistry, 19 (5), 917.
27. Geigert, J., Stermitz, F.R., Johnson, D., Maag, D. and Johnson, D. (1973). Two phytoalexins from sugarbeet Beta vulgaris leaves. Tetrahedron, 29, 2703.
28. Gnanamanickam, S.S. and Smith, D.A. (1980). Selective toxicity of isoflavanoid phytoalexins to gram positive bacteria. Phytopathology, 70 (9), 894.
29. Grisebach, H. and Ebel, J. (1978). Phytoalexins chemical defense substances of higher plants. Agnew. Chem. Int. Engl., 17 (9), 635.
30. Grzelinska, A. (1976). Fitoalesksyny. Post. Biochem., 22, 53.
31. Grzelinska, A. and Sierakowska, A. (1978). Isolation of rish-itin from tomato plants. Phytopathol. Z., 91 (4), 320.
32. Haard, N.F. and Cody, M. (1978). Stress metabolites in post-harvest fruits and vegetables-role of ethylene. In: Post-harvest Biology and Biotechnology (Hultin, H.O. and Milner, M., Eds.) Food and Nutrition Press, Westport, 111.
33. Hadwiger, L.A. and Beckman, J.M. (1980). Chitosan as a com-ponent of pea and Fusarium solani interactions. Plant Physiol., 66 (2), 205
34. Hahn, M.B., Cline, K., Wade, M., Valent, B.S. and Alberschiem, P. (1977). Isolation of a glucan elicitor from yeast and the role of such glucans as general stimulants of phyto-alexin accumulation. Plant Physiol., 59 (6), 67.
35. Hampson, M. and Haard, N.F. (1980). Pathogenesis of Synchy-trium endobioticum: 1. Infection responses in potato and tomato. Canadian J. Plant Pathology, 2, 143.
36. Hargreaves, J.A. (1979). Investigations into the mechanisms of mercuric chloride stimulated phytoalexin accumulation in Physeolus vulgaris cultivar prince and Pisum sativum cultivar alaska. Physiol. Plant Pathol., 15 (3), 279.

37. Hargreaves, J.A. and Bailey, J.A. (1978). Phytoalexin production by hypocotyls of Phaseolus vulgaris in response to constitutive metabolites released by damaged bean cells Physiol. Plant Pathol., 13 (1), 89.
38. Hargreaves, J.A., Mansfield, J.W. and Rosall, S. (1977). Changes in phytoalexin concentrations in tissues of the broad bean plant following innoculation with species of Botrytis. Physiol. Plant Pathol., 11, 227.
39. Harrison, J.G. (1980). Phytotoxicity of wyerone acid, wyerone and wyerone epoxide to field bean leaves. Phytopathol. Z., 97 (1), 14.
40. Henfling, J.W., Bostock, R.M. and Kuc, J. (1980). Cell walls of Phytophthora infestans contain an elicitor of terpene accumulation in potato tuber slices. Phytopathology, 70 (8), 772.
41. Henfling, J.W.D.M., Bostock, R. and Kuc, J. (1980). Effect of abscisic acid on rishitin and lubimin accumulation and resistance to Phytophthora infestans and Cladosporium cucumerinum in potato tuber tissue slices. Phytopathology, 70 (11), 1074.
42. Ingham, J.L. (1976). Fungal modification of pterocarpin phytoalexins from Melilotus alba and Trifolium pratense. Phytochemistry, 14 (10), 1489.
43. Ingham, J.L. (1980). Induced isoflavanoids of Erythrinia sandwicensis. Z. Naturforsch. Sect. C. Biosci., 35(5), 384
44. Ingham, J.L. (1972). Phytoalexins and other natural products as factors in plant disease resistance. Bot. Rev., 38, 343.
45. Ingham, J.L. (1978). Phytoalexin production by species of the genus Caragana. Z. Naturforsch. Sec. C. Biosci., 34(3), 293
46. Ingham, J.L. (1976). 3,5,4'-trihydroxystilbene as a phytoalexin from groundnuts, Arachis hypogaea. Phytochemistry, 15, 1791.
47. Ingham, J.L. and Dewick, P.M. (1979). A new isoflavan phytoalexin from leaflets of Lotus hispidus. Phytochem., 18 (10) 1711.
48. Ingham, J.L. and Dewick, P.M. (1980). Sparticarpin: A pterocarpin phytoalexin from Spartium junceum. A. Naturforsch. Teil. C. Biochem. Biophys. Biol. Virol., 35 (3), 197.
49. Ingham, J.L. and Harborne, J.B. (1976). Phytoalexin induction as a new dynamic approach to the study of systematic relationships among higher plants. Nature, 260(5548),241.
50. Ingham, J.L. and Marksham, K.R. (1980). Identification of the Erythina crista galli phytoalexin cristacarpin and a note on the chirality of other 6 hydroxpterocarpans. Phytochem., 19 (6), 1203.
51. Iniguri, Y., Tomiyama, K., Doke, N., Murai, A., Katsui, N., Yagihash, F. and Matsume, T. (1978). Induction of rishitin-metabolizing activity in potato tuber discs by wounding and identification of rishitin metabolites. Phytopathology, 68 (5), 720.

52. Jadhav, S.J., Sharma, R.P. and Salunkhe, D.K. (1981). Naturally occurring toxic alkaloids in foods. CRC Critical Reviews in Tolicology, 9 (1), 21.
53. Johnson, C. and Brannon, D.R. (1973). Xanthotoxin: A phytoalexin of Pastinaca sativa root. Phytochemistry, 12, 2961.
54. Keeler, F.R., Brown, D.R., Douglas, D.R., Staliknecht, G.F. and Young, S. (1976). Teratogenicity of the solanum alkaloid solasodine and of kennebec potato sprouts in hamster. Bull. Environ. Contam. Toxicol., 15, 522.
55. Keen, N.T. and Breugger, B. (1977). Phytoalexins and chemicals that elicit their production in plants. In: Host plant resistance to pests (Hedin, P.A., Ed.). American Chemical Society, Washington, D.C.
56. Keen, N.T. and Littlefield, L.J. (1979). The possible association of phytoalexins with resistance gene expression in flax to Melampora Lini. Physiol. Plant Pathol., 14(3), 265.
57. Kim, W.K. and Uritani, I. (1974). Fungal extracts that induce phytoalexins in sweet potato roots. Plant Cell Physiology, 14 (6), 1093.
58. Kuc, J. (1972). Phytoalexins. Ann. Rev. Phytopath., 9, 207.
59. Kuc, J. (1976). Phytoalexins. In: Physiological Plant Pathology (Heitefuss, R. and Williams, P.H., Eds.), Springer Verlag,N.Y.
60. Kuc, J., Henfling, J., Garas, N. and Doke, N. (1979). Control of terpenoid metabolism in the potato Phytophthora infestans interaction. Journal of Food Protection, 42 (6), 508.
61. Kuhn, P.J. and Smith, D.A. (1979). Isolation from Fusarium solani of an enzymatic system responsible for kievitone and phaseolidin detoxification. Physiol. Plant Pathol., 14 (2), 179.
62. Langcake, P., and Pyrce, R.J. (1977). The production of reseratrol and the viniferins by grapevines in response to ultraviolet irradiation. Phytochemistry, 16, 1193.
63. Lappe, V. and Barz, W. (1978). Degradation of pisatin by fungi of the genus Fusarium. X. Naturforsch. 33(3), 301.
64. Lyne, R.L., Mulheirn, L.J. and Leworthy, D.P. (1976). New pterocarpenoid phytoalexins of soybean. J. Chem. Soc. Sec.D., 497.
65. Lyon, G.D. (1976). Metabolism of the phytoalexin rishitin by Botrytis spp. J. General Micro., 96, 225.
66. Lyon, G.D. and Mayo, M.A. (1978). The phytoalexins rishitin effects the viability of isolated plant protoplasts. Phytopath. Z., 92 (4), 298.
67. Manibhushushanrao, K. and Zuber, M. (1979). Disease resistance in cereals. Acta Phytopath. Acad. Sci. Hung., 13, 3.
68. Martin, M. and Dewick, P.M. (1980). Biosynthesis of pterocarpan isoflavan and coumestan metabolites of Medicago sativa. The role of an isofla-3-ene. Phytochemistry, 19 (11), 2341.
69. Martin, M. and Dewick, P.M. (1979). Biosynthesis of the 2-aryl benzofuran phytoalexin Vignafuran in Figna unguiculata. Phytochemistry, 18 (8), 1309.

70. Martin, W.J., Hasling, V.C. and Catalano, E.A. (1976). Ipomeamarone content in disease and nondiseased tissues of sweet potato infected with different pathogens. Phytopathology 66, 678.
71. McMillan, M. and Thompson, J.C. (1979). An outbreak of suspected solanine poisoning in school boys. Examination of criteria of solanine poisoning. Q. J. Med., 48, 227.
72. Melitskii, L.V. and Ozeretskovskaya, O.L. (1978). Induced plant immunity to parasitic fungi. Izu. Akad, Nauk. SSR Ser. Biol., 5, 700.
73. Müller, K. and Börger, H. (1940). Experimentelle untersuchungen uber die Phytophthora resistenz der kartoffel. Arb. Biol. Rerchsanstalt. Lander. Forstu. Berlin, 23, 189.
74. Mustava, M. and Dyakov, Y.T. (1978). The connection between supersensitive cell death and phytoalexin production in protective reactions of potato against Phytophthora infestans. Biol. Nauki., 11, 90.
75. Nichols, E.J., Beckman, J.M. and Hadwiger, L.A. (1980). Glycosidic enzyme activity in pea tissue and pea-Fusarium solani interactions. Plant. Physiol., 66 (2), 199.
76. Obi, I.M. (1979). Additional phytoalexin-like compounds in HT gene resistance of corn to Helminthosporium turicum. Ann. Appl. Biol., 92 (3), 377.
77. Oku, H., Ouchi, S., Shiraishi, T., Utsumi, K. and Seno, S. (1976). Toxicity of the phytoalexin pisatin to mammalian cells. Proc. Jap. Acad., 52, 33.
78. Olah, A.F. and Sherwood, R.T. (1971). Flavones, Isoflavones, and coumestans in alfalfa infected by Ascochyto imperfecta. Phytopathology, 61, 65.
79. Osman, S.F., Zacharius, R.M., Kalan, E.B., Fitzpatrick, T.J. and Krulick, S. (1979). Stress metabolites of the potato and other Solanaceous plants. Journal of Food Protection, 42 (6), 502.
80. Poswillo, D.E., Soper, D. and Mitchell, S. (1972). Experimental induction of foetal malformation with blighted potato: A preliminary report. Nature, 239, 462.
81. Poswillo, D.E., Soper, D., Mitchell, S.D., Coxon, R.F. and Price, K.R. (1973). Further investigations into the teratogenic potential of imperfect potatoes. Nature, 244, 387.
82. Preston, N.W., Chamberlain, K. and Skipp, R.A. (1975). A 2-arylbenzofuran phytoalexin from cowpea. Phytochemistry 14, 1843.
83. Rahe, J.E. and Arnold, R.M. (1975). Injury related phaseollin accumulation in Phaseolus vulgaris and its implications with regards to specificity of host-parasite interaction. Canadian Journal Botan., 53 (9), 921.
84. Reilly, J.J. and Klarman, W.L. (1980). Thymine dimer and glyceollin accumulation in U.V. irradiated soybean suspension cultures. Environ. Exp. Bot., 20 (2), 131.

85. Renwick, J.H. (1972). Hypothesis-anencephaly and spina bifida are usually preventable by avoidance of a specific but unidentified substance present in certain potato tubers. Brit. J. Prev. Soc. Med., 26, 67.
86. Rich, J.R., Keen, N.T. and Thomas, I.J. (1977). Association of coumestans with the hypersensitivity of lima bean roots to Pratylenchus schribneri. Physiol. Plant Pathol., 10, 105.
87. Robeson, D.J. (1978). Furanoacetylene and isoflavanoid phytoalexins in Lens culinaris. Phytochenistry, 17, 807.
88. Robeson, D.J. and Harborne, J.B. (1980). A chemical dichotomy in phytoalexin induction within the tribe Viceae of the Leguminosae. Phytochemistry, 19 (11), 2359.
89. Robeson, D.J. and Ingham, J.L. (1979). New pterocarpan phytoalexins from Lathyrus nissolia. Phytochem., 18 (10), 1715.
90. Roddick, J.G. (1979). Complex formation between solanaceous steroidal glycoalkaloids and free sterol in vitro. Phytochemistry, 18 (9), 1467.
91. Salunkhe, D.K. and Wu, M.T. (1979). Control of postharvest glycoalkaloid formation in potato tubers. Journal of Food Protection, 42 (6), 519.
92. Sarkar, S.K. and Phan, T. (1979). Naturally occurring and ethylene induced phenolic compounds in carrot root. Journal of Food Protection, 42 (6), 526.
93. Sato, N., Tomiyama, K., Katsui, N. and Masamune, T. (1968). Isolation of rishitin from tubers of interspecific potato varieties containing late blight resistance genes. Ann. Phytopathol. Soc. Japan, 34, 140.
94. Shih, M.J. and Kuc, J. (1974). Alpha-solanine in kennebec leaves and aged tuber slices. Phytochemistry, 13, 997.
95. Shih, M., Kuc, J. and Williams, E.B. (1973). Suppression of steroids in potato tubers. Phytopathology, 63, 821.
96. Shiraishi, T., Oku, H., Isono, H. and Ouchi, S. (1975). The injurious effect of pisatin on the plasma membrane of pea. Plant Cell Physiol., 16 (5), 939.
97. Shiraishi, T., Oku, H., Yamishita, M. and Ouchi, S. (1978). Elicitor and suppressor of pisatin induction in spore germination fluid of pea pathogen Mycophaerella pinodes. Ann. Phytopathol. Soc. Jpn., 44 (5), 659.
98. Shirata, A. (1978). Production of phytoalexin in cortex tissue of mulberry shoot. Ann. Phytopath. Soc. Jpn., 44(4), 485.
99. Smith, D.A., Kuhn, P.J., Bailey, J.A. and Burden, R.S. (1980). Detoxification of phaseolidin by Fusarium solani. Phytochemistry, 19 (8), 1673.
100. Solomos, T. and Laties, G. (1976). Induction of cyanide resistant respiration by ethylene. Biochem. Biophys. Res. Comm., 70, 663.
101. Sondheimer, E. (1957). Bitter flavor in carrots. 3. Isolation of a compound with spectral characteristics similar to hydrocarbon extracts of bitter carrots. Food Res., 22,296.

102. Stoessel, A., Stothers, J.B. and Ward, E.W.B. (1978). Biosynthetic studies of streee metabolites from potatoes. Incorporation of sodium acetate carbon-13 into 10 sesquiterpenes. Can. J. Chem. 56 (5), 645.
103. Stynes, B.A., Petterson, D.C., Lloyd, J., Payne, A.L. and Lanigan, G.W. (1979). The production of toxin in annual rye grass Lolium regidum infected with a nematode and Corneybacterium rathayi. Aust. J. Agric. Res., 30(1), 201.
104. Takasugi, M., Nagao, S. and Masamune, T. (1978). Structure of moracin A and B, new phytoalexins from diseased mulberry. Tetrahedron Lett., 9, 797.
105. Terekhova, V.A., Chalova, L.I., Pijakov, Y.T. and Ozertskovskayaw, L. (1980). Effect of phytoalexin inducing metabolites from Phytophthora infestans mycelium on potato protoplasts. Mikol. Fitopatol., 14 (2), 111.
106. Thomas, C.A. and Allen, E.H. (1970). An antifungal polyacetylene compound from Phytophthora infected safflower. Phytopathology, 60, 361.
107. Tibor, E. (1975). Fitoalexinek. Novenztermeles, 24, 359.
108. Tyusterev, S.L., Tarlakovskii, S.A. and Meloyan, V.V. (1979). Effect of some fungicides and biologically active substances on the biosynthesis of phytoalexins in potato tuber induced by Phytophthora infestans. Nauk, Im. V. Lenina, 9, 18.
109. Uritani, I. (1967). Abnormal substances produced in fungus contaminated foodstuffs. J. Assoc. Off. Anal. Chem., 50, 105.
110. Vanetten, H.D., Mathews, P.S., Tegtmeier, K.J., Dietert, M.F. and Stein, J.I. (1980). The association of pisatin tolerance and demethylation with virulence on pea in Nectria haematococca. Physiol. Plant Pathol., 16(2), 257.
111. Vanetten, H.D. and Stein, J.I. (1978). Differential response of Fusarium solani isolates to pisatin and phaseollin. Phytopathology, 68 (9), 1276.
112. Ward, E.W.B., Stoessl, A. and Stothers, J.B. (1977). Metabolism of the sesquiterpenoid phytoalexins capsidiol and rishitin to their 13-hydroxy derivatives by plant cells. Phytochemistry, 16 (12), 2024.
113. Wood, G.E. (1979). Stress metabolites of plants - a growing concern. J. Food Protection, 42 (6), 496.
114. Wood, R.K.S. and Graniti, A. (1976). Specificity in Plant Diseases. Plenum Press, N.Y., 353 pp.
115. Zimmerman, D.C. and Coudron, C.A. (1979). Identification of traumatin, a wound hormone, as 12-oxo-trans-10-dodecenoic acid. Plant Physiol., 63, 536.

MYCOTOXINS AS A DETERIORATING FACTOR IN STORED CROPS

Palle Krogh

Department of Veterinary Microbiology
Purdue University, West Lafayette, IN 47907, USA
and
Department of Microbiology
Royal Dental College, Copenhagen, Denmark

INTRODUCTION

Mycotoxins are secondary fungal metabolities which by ingestion in man and animals cause intoxication, termed mycotoxicosis. Already in 1968 more than 100 toxin-producing mold species were known (1), mainly belonging to the fungal genera *Aspergillus*, *Penicillium* and *Fusarium*, and this list has since been considerably extended. Information on occurrence of mycotoxins in foodstuff is however limited, and survey data are only available for a few mycotoxins. Similarly the number of recognized mycotoxicoses in domestic animals is limited and amounts to less than ten, and in man even fewer mycotoxicoses are known. Thus four mycotoxins or groups of mycotoxins were considered to be causally associated with disease in humans, proven or hypothetically, according to a recent publication by the World Health Organization (2). These mycotoxins have been selected for detailed description in this chapter.

MYCOLOGICAL AND CHEMICAL NATURE OF SELECTED MYCOTOXINS

The aflatoxins are produced by *Aspergillus flavus* and *Aspergillus parasiticus*. They are derivatives of furano-coumarin, and although 17 compounds have been identified, aflatoxin B_1, B_2, G_1 and G_2 are the most commonly occurring in natural products. Aflatoxins, as most of the mycotoxins, are low-molecular weight compounds (2) (Table 1), and they are heat stable so that they will survive most cooking conditions. Lactating animals ingesting aflatoxin contaiminated feed excrete metabolites in the milk called aflatoxin M.

Table 1. Chemical Characteristics of Aflatoxins

Aflatoxin	Molecular formula	Molecular weight	Melting point
B_1	$C_{17}H_{12}O_6$	312	268-269
B_2	$C_{17}H_{14}O_6$	314	286-289
G_1	$C_{17}H_{12}O_7$	328	244-246
G_2	$C_{17}H_{14}O_7$	330	237-240
M_1	$C_{17}H_{12}O_7$	328	299

Ochratoxins are produced by a number of species within the genera _Aspergillus_ and _Penicillium_, with _Penicillium viridicatum_ being the dominating producer. The ochratoxins constitute a group of closely related derivatives of isocoumarin linked to an amino-acid, L-β-phenylalanine. Of the nine toxins only ochratoxin A is occurring in natural products. Zearalenone is a phenolic resorcylic acid lactone, produced by a number of _Fusarium_ species, e.g. _Fusarium graminearum_ and _Fusarium moniliforme_. The trichothecenes possess the tetracyclic 12,13-epoxy-trichothec-O-ene skeleton, and more than 30 trichothecenes have been identified, with T-2 toxin, nivalenol and deoxynivalenol being most commonly found. They are produced by a number of species belonging to the fungal genera: _Fusarium_, _Cephalosporium_, _Myrothecium_, _Trichoderma_ and _Stachybotrys_.

FORMATION OF MYCOTOXINS

Mycotoxin production can occur in most plant products, but cereal and oil seeds constitute the crops most likely to be contaminated. Apart from the toxin-producing fungal organism mycotoxin formation is influenced by a number of factors, the major ones being temperature and water activity; other factors are : mechanical damage, nature of substrate and invertebrate vectors (3). The water activity and temperature conditions for ochratoxin A production is illustrated in Fig. 1 (4). A water activity of 0.83 appears to be the minimal value also for aflatoxin production, but the minimal temperature for aflatoxin production is much higher (10°C), compared to ochratoxin A (5,6), pointing to the fact that whereas aflatoxin is a contaminant of crops from warmer areas, ochratoxin has been encountered in products from temperate climatic zones. The water activity and temperature conditions for production of zearalenone and trichothecenes are inadequately known.

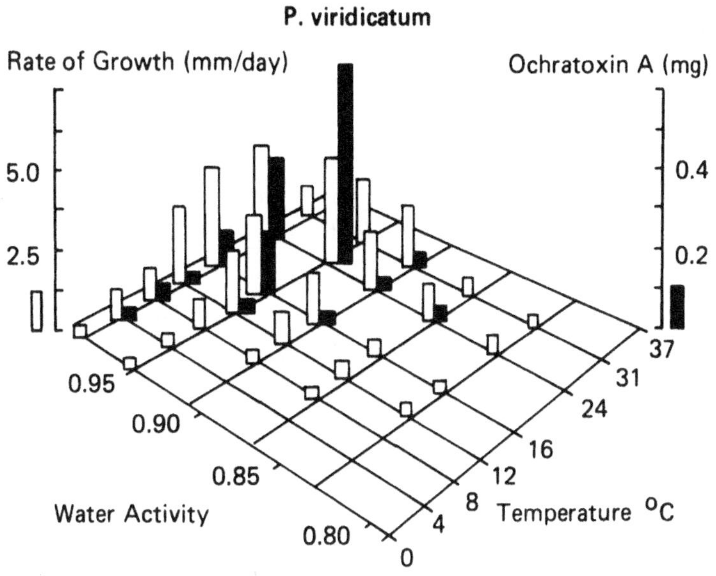

Fig. 1. Growth and ochratoxin A production by a strain Penicillium viridicatum on maize sucrose agar under various conditions of temperature and water activity.

OCCURRENCE OF SELECTED MYCOTOXINS IN FOODSTUFF

In reporting on contamination of foodstuffs with mycotoxins two parameters are used, namely, concentration and frequency. Monitoring foodstuffs for mycotoxins is therefore dependent on the availability of chemical methods of analysis, which can identify and quantify individual mycotoxins, whereas biological procedures are unreliable for that purpose. Chemical methods of analysis are available only for a limited number of mycotoxins. In the following the occurrence of four mycotoxins is reported, namely, aflatoxins, ochratoxins, zearalenone and the trichothecenes, because these toxins have been associated with disease in man or animals. Other mycotoxins for which causal associations in animal diseases have been documented, such as sporidesmin causing facial eczema in ruminants (7), are not included because no survey data have been reported.

Aflatoxin

By far the most information on mycotoxin contamination is available for aflatoxin. The occurrence of aflatoxin (aflatoxin $B_1 + B_2 + G_1 + G_2$) in a number of plant products is illustrated in the following Tables 2-4. Aflatoxin was first detected in the 1960's as a contaminant in groundnuts, and this is a commodity likely to be contaminated, with frequencies up to 49% (Table 4). Also maize is often contaminated, with frequencies up to 97%, and concentrations up to 15.6 mg/kg have been reported (Table 2). In addition to the commodities mentioned here, aflatoxin has been found in treenuts (often heavily contaminated), copra and spices. It is evident from Tables 2-4 that aflatoxin contamination occurs in crops from tropical-subtropical areas. This reflects the fact that aflatoxin-producing *Aspergillus flavus* strains are found mainly in tropical-subtropical areas, where also the optimal temperature conditions for aflatoxin production are present. Contaminated crops can then be shipped to other areas, and thus practically all countries are faced with the problem of aflatoxin contaminated foodstuffs. As a result, many countries, in the aflatoxin producing areas and outside, have established tolerances for aflatoxin in foods and feeds (9). The use of aflatoxin-contaminated feed for lactating animals results in carry-over into the milk of hydroxy-metabolized compounds, aflatoxin M_1 and M_2, with almost unchanged properties. As indicated in Table 5 marketed milk samples have been found contaminated up to 80%, when very sensitive methods able to detect 0.01 μg aflatoxin/l have been used. It is also evident from Table 5 that aflatoxin-contamination of milk occurs in a number of North-European countries (Belgium, East- and West Germany, the Netherlands, U.K.) where aflatoxin is not usually produced, but merely imported, in contaminated crops used for animal feed. In addition, aflatoxin residues may occur in the carcass and organs of pigs, as indicated by experimental data as well as findings in marketed products.

Ochratoxin

As indicated in Table 6 ochratoxin A is, like aflatoxin, a contaminant of grain in particular, but has also been found in coffee beans and peanuts. Feed samples are more frequently contaminated than foods. The high frequency (above 50%) of ochratoxin contamination in feeds in Canada and in Denmark is due to biased sampling plans being used in the Canadian study, whereas only samples associated with porcine nephropathy were screened in the Danish investigation. It is evident from Table 6 that ochratoxin A is found in the temperate climatic areas of Europe and North America, the leading producer being *Penicillum viridicatum*. Whether ochratoxin A occurs in tropical-subtropical crops, produced by the more thermophilicfungal producers in the Aspergillus ochraceus group, cannot be established at present, as no survey data are available from such areas, but the observation that coffee beans have been found

Table 2. Occurrence of Aflatoxin in Maize (8)

Country	No. of Samples	% Contaminated	Range of Concentration (μg/kg)
USA	1594	2.5	3 - 37
USA	2866	8.2	
USA(South-East)	60	35.0	6 - 348
Thailand		35.0	average: 400
Uganda		40.0	average: 133
Philippines		97.0	average: 213
India	5		250 - 15600
France	461	0.2	52

Table 3. Occurrence of Aflatoxin in other Cereals (2,8)

Country	No. of Samples	% Contaminated	Range of Concentration (μg/kg
USA	66(sorghum)	3.0	13 - 50
USA	157 (rice)	0.6	5
Uganda	69(sorghum)	23.0	average: 152
Viet Nam	139(rice)	31.0	
USSR	169(wheat)	0.6	100
USSR	138(wheat)	17.4	10 - 333
USSR(Kazakstan)	50(wheat)	4.0	5 - 10

Table 4. Occurrence of Aflatoxin in Groundnuts (8)

Country	No. of Samples	% Contaminated	Range of Concentration (μg.kg)
USA	361	15.0	5 - 50
Sudan	173	41.0	5 - 1000
Thailand		49.0	average:1530
Uganda		17.0	average:363

Table 5. Occurrence of Aflatoxin M_1 in Milk (2)

Country	No. of Samples	% Contaminated	Range of Concentration (μg/kg)
Belgium	68	61.8	0.02 - 0.2
DDR	36	11.1	1.7 - 6.5
FRG	260	45.4	0.05 - 0.33
FRG	419	18.9	0.05 - 0.54
India	21	14.3	up to 13.3
Holland	95	77.9	0.09 - 0.5
UK	278	30.6	0.03 - 0.5
South Africa	56	0	
USA	320	7.5	0.1 - 0.4
USA	302	63.6	up to 3.9

contaminated with *Aspergillus ochraceus*-produced ochratoxin A may support this assumption. Residues of ochratoxin A may occur in food of animal origin particularly from pigs and poultry (10).

Zearalenone

Fusarium-elaborated zearalenone is found predominantly in maize as indicated in Table 7. Only a few surveys of foodstuffs have been carried out, in a restricted number of countries. However, the compound is surely occurring in many other areas, and in other cereals, as indicated by reports on the associated disease in pigs, the estrogenic syndrome, which have been reported from many countries.

Trichothecenes

Very few reports on the trichothecenes have been published so far (Table 8), so no frequency of contamination can be calculated, and the understanding of the trichothecenes as contaminants of foodstuffs is very inadequate. However, recent improvements in methodology for the detection of trichothecenes in foodstuffs may soon change the present concept that the trichothecenes are only rarely occurring.

ADVERSE EFFECTS OF MYCOTOXINS

The adverse effect of mycotoxins in man and animals has been elucidated through observations of field cases (for man exclusively) and of experiments. Field cases are of major interest in the con-

Table 6. Natural Occurrence of Ochratoxin A in Foods and Feeds of Plant Origin (10)

Commodity	Country	No. of Samples Analyzed	Percentage Contaiminated	Range of Ochratoxin A Level (μg/kg)
Foods				
maize	USA	293	1.0	83 - 166
maize (1973)	France	463	2.6	15 - 200
maize (1974)	France	461	1.3	20 - 200
wheat (red winter)	USA	291	1.0	5 - 115
wheat (red spring)	USA	286	2.8	5 - 115
barley (malt)		50	6.0	9 - 189
barley	USA	127	14.2	10 - 40
coffee beans	USA	267	7.1	20 - 360
maize	Yugoslavia	542	8.3	6 - 140
wheat	Yugoslavia	130	8.5	14 - 135
wheat bread	Yugoslavia	32	18.8	
barley	Yugoslavia	64	12.5	14 - 27
barley	Czechoslovakia	48	2.1	3800
bread	U.K.*	50	2.0	210
flour	U.K.	7	28.5	490 - 2900
rice	Japan	2	(100)	230 - 430
beans	Sweden	71	8.5	10 - 442
peas	Sweden	72	2.8	10

*All samples suspicious of containing mycotoxins

Table 6 (Continued)

Commodity	Country	No. of Samples Analyzed	Percentage Contaiminated	Range of Ochratoxin A Level (μg/kg)
<u>Feeds</u>				
barley, wheat, oats, rye, maize	Poland	150	5.3	50 - 200
mixed feed	Poland	203	4.9	10 - 50
maize	Yugoslavia	191	25.7	45 - 5125
barley, oats	Sweden	84	8.3	16 - 409
wheat, hay	Canada	95	7.4	30 - 6000
wheat, oats barley, rye	Canada	32	56.3	30 - 27000
barley, oats	Denmark	33	57.6	28 - 27500

text of this paper, because the reporting of field cases is the first step in the epidemiologic calculations providing the basis for estimation of economic loss. The adverse effects of the four mycotoxins are listed in Table 9 - 11. Outbreaks of liver damage, consisting of necrosis and subsequent cirrhosis, have been reported for man and domestic animals, associated with levels of aflatoxin in foodstuffs in the range 1 - 10 mg/kg (Table 9). Only for primary liver carcinoma in man are data on incidence rates available. Aflatoxin B_1 particularly is a potent liver carcinogen in experimental animals, and in areas in Africa and Southeast Asia with high incidence rates of primary liver carcinoma a strong correlation between aflatoxin intake and liver carcinoma incidence has been observed. This does not, however, exclude other factors in the etiology of primary liver carcinoma in these areas (2). At levels in foodstuffs below 10 mg/kg ochratoxin A is exclusively a nephrotoxin, causing nephropathy in single-stomach animals, particularly pigs (Table 10) and most feed lots analyzed contained less than 10 mg/kg (Table 6). Ochratoxin A is not a proven nephrotoxin in man, but preliminary data seem to indicate that this food contaminant may play a causal role in endemic nephropathy occurring in rural populations on the Balkan peninsula (10). Outbreaks of zearalenone-induced estrogenic syndrome in pigs has been reported from many countries (Table 11), and a few outbreaks have been observed of trichothecene-induced intoxications in farm animals, characterized by necrosis of the alimentary tract. No cases of zearalenone-associated disease in man have been reported, and the hypothesis that trichothecenes were causally involved in Alimentary Toxic Aleukia (ATA), predominant in areas of USSR in the 1930's and 1940's, cannot be confirmed, because ATA is no longer encountered.

ASSESSMENT OF THE LOSS DUE TO MYCOTOXINS

Although information on concentrations in foodstuffs of mycotoxins associated with disease in man and animals is available as listed in Table 9 - 11, it is not possible to predict disease occurrence and distribution in human populations and economic loss in farm animals on the basis of that kind of information. The main reason for this is that the existing toxicologic concept for mycotoxicoses cannot explain the effect of variation in mycotoxin exposure, and it is evident from Table 2 - 8, that there is a pronounced variation in the frequency of mycotoxin contamination. Assessment of the economic loss due to mycotoxins in farm animals can however be made, when data on disease occurrence and distribution are available. This is unfortunately not generally the case for mycotoxicoses, but as evident from Table 10 data on prevalence rates are available for ochratoxin A-induced nephropathy in pigs. The temporal and spatial distribution of this disease has been documented for several countries in Northern and Central Europe (Table 12) using data obtained from the veterinary meat inspection

Table 7. Occurrence of Zearalenone in Foodstuffs (2)

Country	No. of Samples	% Contamination	Range of Concentration (μg/kg)
USA	576(maize)	1.0	450 - 800
USA	223(maize)	17.0	100 - 5000
USA	11(maize-meal)	81.8	12 - 69
Canada	1(maize-cornflakes)	100	13
Zambia	maize-beer		10 - 4600
Lesotho	140(beer)	12.0	300 - 2000

Table 8. Occurrence of Trichothecenes in Foodstuffs (2)

Country	Commodity	Compound	Concentration (μg/kg)
USA	maize	T-2	2000
USA	barley	T-2	25000
USA	barley	Nivalenol	
USA	maize (3 samples)	Deoxynivalenol	100 - 1800

at pig slaughterhouses. Although prevalence rates for the disease, during endemic and epidemic periods, are available only for Denmark at present, preliminary data indicate a similar distribution pattern in other countries listed in the table. In addition, ochratoxin A-induced porcine nephropathy is likely to occur in many other countries, when the worldwide occurrence is considered (Table 6). Associated with the renal damage induced in pigs by ochratoxin A is a growth depression (observed in many other mycotoxicoses as well), the extent of which has been experimentally determined. Both the decrease in daily gain and increase in feed consumption contribute to the economic loss, which amounts to approximately 8% if the exposure is 1 mg ochratoxin A per kg feed, a moderately low level, which does induce a kidney damage that can be detected at the meat inspection. An economic loss of 8% based upon the price level in Denmark 1980, corresponds roughly to US $ 10 per pig, affected by ochratoxin A-induced nephropathy (Table 13). Employing the data from Tables 12 & 13 the economic loss due to ochratoxin A-induced porcine nephropathy in the husbandry of

Table 9. Adverse Effects due to Aflatoxins (2)

Animal	Organ/Tissue Alteration	Concentration in Food (mg/kg)	Epidemiologic Data
Turkey	Liver necrosis	1 - 10	outbreaks
	" cirrhosis		
Cattle	" "	1 - 2	outbreaks
Pig	" "		
Dog	" "	1 - 2	outbreaks
	(hepatitis X)		
Man	Liver carcinogenesis	3.5-222 ng/kg bw/day	incidence rates 2.0-13.0/10(5)
Man	Liver fibrosis	0.25 - 15.6	outbreaks
Enfants	Reye's syndrome	?	outbreaks

Table 10. Adverse Effects due to Ochratoxin A

Animal	Organ/Tissue Alteration	Concentration in Food (mg/kg)	Epidemiologic Data
Pig (and other single stomach animals)	a) nephropathy	0.2 - 10	prevalence
	b) nephropathy + extrarenal damage	above 10	no data
Man?	Balkan endemic nephropathy	0.005-0.140	prevalence rates

Table 11. Effects of Zearalenone and Trichothecenes (2)

Animal	Organ/Tissue Alteration	Concentration in Food (mg/kg)	Epidemiologic Data
		<u>Zearalenone</u>	
Pig	estrogenic syndrome	0.1 - 6.8	outbreaks
Cattle	infertility	14	one outbreak
		<u>Trichothecenes</u>	
Cattle	hemorrhages	2(T-2 toxin)	one outbreak
Poultry	alimentary tract necrosis	25(T-2 toxin)	one outbreak
Man	ATA	?	epidemics

Table 12. Occurrence of Mycotoxic Porcine Nephropathy (11,12,13)

Country	Prevalence Rate (Cases per 10^4 Pigs) Endemic Level	Epidemic Level
Denmark	1 - 10	50 - 80
Sweden	?	?
Czechoslovakia	?	?
Hungary	Ca. 1	?

Table 13. Economic Loss of Experimentally OA-Induced (14) Porcine Nephropathy

Feed Level of OA (mg/kg) during 3 months	Decrease in Daily Gain %	Increase in Feed Consumption %	Economic in %	Loss in US$ per Pig
0.2	6.4	5.6		
1.0	4.2	4.9	8.0	10
4.0	17.2	19.3		

several areas of the world has been estimated (Table 14). The estimates are based upon the assumption that the ochratoxin A-induced nephropathy does occur in these areas, and that the prevalence rates and agricultural economy in these areas are similar to those in Denmark. Thus the combined annual economic loss in Europe (EEC), U.S.S.R. and U.S.A. is estimated to be US $ 1.4 mill. This is a conservative estimate, as higher feed levels of ochratoxin A than 1 mg/kg will increase the economic loss considerably (Table 13), and epidemic prevalence rates would result in an economic loss 10 - 50 times greater. If similar estimates of economic loss in animal husbandry were made for other mycotoxicoses it is evident that the combined loss would be considerable. The above estimates include only the loss due to growth depression. If the disease results in death of the animals, the loss would be much larger. Assessment of the occurrence and distribution of mycotoxin-associated diseases in human populations has so far been inhibited by the lack of chemical procedures for mycotoxin detection applicable in clinical chemistry. Recent improvements in methodology for aflatoxin and ochratoxin quantitation for clinical use may provide the necessary tool for assessment of the role of mycotoxins in human populations.

Table 14. Estimated Economic Loss of OA-induced Porcine Nephropathy in Various Countries

Country	Annual No. of Pigs(15,16) Slaughtered in 1979 (x1000)	Predicted No. of Pigs with Nephropathy	Estimated Annual Economic Loss Loss (in US $)
Denmark	13 389	6 690	66 900
EEC	119 850	59 925	599 250
USSR	75 000	37 500	375 000
USA	89 000	44 500	445 000
Total loss:	EEC + USSR + USA		1 419 250

ACKNOWLEDGEMENTS

The use of Fig. 1 from M.D. Northolt and coworkers (4) is gratefully acknowledged. Published as paper No.8572 of the Agricultural and Experiment Station, Purdue University, West Lafayette, Indiana, USA.

REFERENCES

1. D. E. Wright, Toxin produced by fungi, Ann. Rev. Microbiol. 22:269 (1968).
2. World Health Organization, "Environmental Health Criteria 11:Mycotoxins" WHO, Geneva (1969).
3. C. W. Hesseltine, Conditions leading to mycotoxin contamination of foods and feeds, in "Mycotoxins and other Fungal Related Food Problems," J. V. Rodricks, ed., American Chemical Society, Washington DC (1976).
4. M. D. Northolt, H.P. van Egmond, and W.E. Paulsch, Ochratoxin A production by some fungal species in relation to water activity and temperature, J. Food Prot. 42:485 (1979).
5. M. D. Northolt, H.P. van Egmond, and W. E. Paulsch, Differences between Aspergillus flavus strains in growth and aflatoxin B_1 production in relation to water activity and temperature, J. Food Prot. 40:778 (1977)
6. M. D. Northolt, C.A.H. Verhulsdonk, P.S.S. Soentoro, and W.E. Paulsch, Effect of water activity and temperature on aflatoxin production by Aspergillus parasiticus, J. Milk Food Tech. 39:170 (1976).
7. L. G. Atherton, D. Brewer, and A. Taylor, Pithomyces chartarum, A fungal parameter in the aetiology of some diseases of domestic animals, in: Mycotoxins, I.F.H. Purchase, ed., Elsevier, Amsterdam (1974).
8. L. Stoloff, Occurrence of mycotoxins in foods and feeds, in: Mycotoxins and other Fungal Related Food Problems, J. V. Rodricks, ed., American Chemical Society, Washington, D.C. (1976).
9. P. Krogh, Mycotoxin tolerances in foodstuff, Pure appl. Chem. 49:1719 (1977).
10. P. Krogh, Ochratoxins, occurrence, biological effects and causal role in diseases, in: "Natural Toxins," D. Baker and T. Wadström, ed., Pergamon Press, Oxford (1980).
11. P. Krogh, Epidemiology of mycotoxic porcine nephropathy, Nord. Vet.-Med.28:452 (1976).
12. L. Rutqvist, N.-E, Björklund, K. Hult, and S. Gatenbeck, Spontaneous occurrence of ochratoxin residues in kidneys of fattening pigs, Zbl. Vet. Med. A. 24:402 (1977).
13. A. Piskac, R. Halouzka, J. Drabek, J. Mala, and C. Smrcek, Mykotoxicka nefropatie prasat ve velkochovech, Veterinarstvi 29:253 (1979).
14. P. Krogh, N.H. Axelsen, F. Elling, N. Gyrd-Hansen, B. Hald,

J. Hyldgaard-Jensen, A.E. Larsen, A. Madsen, H.P. Mortensen, T. Møller, O.K. Petersen, U. Ravnskov, M. Rostgaard, and O. Aalund: Experimental porcine nephropathy. Changes of renal function and structure induced by ochratoxin A-contaminated feed. Acta path. microbiol. scan., Section A, Supplementum No. 246, 21 pp., (1974).

15. The Danish Pig Meat Sector, "Statistics 1979," Copenhagen (1980).
16. Food and Agriculture Organization, "Production Yearbook," Rome (1975).

HORMONAL AND CHEMICAL PREHARVEST TREATMENTS WHICH INFLUENCE POSTHARVEST QUALITY, MATURITY AND STOREABILITY OF FRUIT

F. Bangerth

Institut für Obst-, Gemüse- and Weinbau
Universität Hohenheim
Stuttgart, Germany

INTRODUCTION

The work of a postharvest physiologist is usually concerned with the quality and durability of a product at harvest and in storage. However, this quality, or potential for quality, is often affected or determined long before that time, and storage may in many respects be regarded as the preservation of predetermined quality. In addition, the way a fruit has to be stored greatly depends on its preharvest life, e.g. the extent of its accumulation of calcium, sugars, acids, on cell size and cell number etc. Growth regulators and phytohormones affect most of these processes and the postharvest physiologist should be aware of the potential as well as the problems related to their use.

Although only a few of the growth regulators and phytohormones applied in modern orchard management are used to affect the quality and post harvest behaviour of fruits directly, many of them do have side effects which influence the post harvest life of the treated fruit. Growth regulators are able to affect quality as well as ripening of fruits when applied as early as the time of blossoming or shortly thereafter. Because of the considerable period of time between most of these treatments and the physiological event, it seems unlikely that the effect of these growth regulators is a direct one. More likely, they act via an alteration of physiological processes occurring during or shortly after treatment (e.g. cell division, Ca^{2+} uptake, assimilate partitioning etc.). Quite often this may be mediated by an interaction between the growth regulator applied and the synthesis, metabolism, transport etc. of endogenous hormones. Whilst we have at least some ideas about the participation of endogenous hormones in ripening and senescence of fruit (Sacher

1973, McGlasson et al. 1978, Lieberman 1979, Bruinsma, this Vol.) our knowledge about the interactions of growth regulators and endogenous hormones is still rather restricted. It is this field, however, which has to be explored if we wish to make progress in our understanding of how growth regulators act. In addition, the use of growth regulators may be an excellent tool to modify endogenous hormone levels, metabolism, etc., thus enabling us to investigate better their role in growth and development.

Because many details of the practical use of growth regulators in fruit production in general and their effects on the postharvest behaviour of fruits in particular can be found in literature surveys of Faust (1973), Sharples (1973), and Monselise (1979), this review is restricted largely to more recent and less recognized developments. Due to the more limited availability of solid data and our even greater ignorance of the relevant physiological processes, pitfalls are likely and guesswork is more necessary. It is hoped, however, that in this way it is possible to draw the attention of investigators to those less developed fields and to stimulate activity in these areas.

INDUCED PARTHENOCARPIC FRUIT SET AND ITS EFFECTS ON POSTHARVEST BEHAVIOUR

A number of growth regulators and plant hormones are used in practice to induce or stimulate parthenocarpic fruit set. Gibberellins are the most prominent group among them but auxins also have an effect especially on tomatoes, cucurbits and some other multiseeded fruits (Dennis 1973). Recently a combination of GA_3, an auxin and a cytokinin has been widely used in experiments in the UK on a number of different pome and stone fruits in order to stimulate fruit set (Schwabe 1978, Webster and Goldwin 1981). Several of these treatments clearly affect postharvest quality physiology and marketability. With regard to effects on market standards, it is well known that some growth regulators particularly gibberellins can change the shape of parthenocarpic pear and apple fruits to such an extent that they are not further acceptable for the fresh market and therefore have to be processed (Stembridge 1973). Puffiness and malformation of parthenocarpic tomatoes induced by some synthetic auxins are other examples of this type of problem (Rylski 1979, Bangerth, unpublished). The hormonal and/or physiological reasons for these deviations from cultivar specific fruit shapes are barely understood.

As far as nutritional and storage characteristics are concerned some induced parthenocarpic fruits show clear disadvantages which are more serious than the alterations in fruit shape. Parthenocarpic pears e.g. but also some apple, tomato and sour cherry cultivars ripen faster, have a higher respiration rate, less acidity, firmness and Vitamin C and are more susceptible to some physiological disorders (Modlibowska 1966, Kotob and Schwabe 1975, Gil et al. 1972, Bangerth unpublished). These alterations (see e.g. Table 1.) are not simply a GA effect because they can also be observed in tomato fruits induced with some auxins. These effects seem to be related to parthenocarpic fruit development, and it would be most interesting to know whether natural parthenocarpic apple, pear or tomato fruits show a similar behavior. Most of the disadvantages mentioned are probably related to a reduced Ca^{2+} content in parthenocarpic fruits, as shown in Table 1. Other mineral constituents are obviously not affected. Ca is one of the most crucial factors in determining postharvest quality and physiology and has therefore been investigated in some detail for a number of fruits (Shear 1975, Sharples and Johnson 1977, Bangerth and Link 1972 and many others). Calcium affects respiration (Fig. 1), softening, acidity, vitamin C content, rate of ripening and a number of physiological disorders including chilling injury (Chaplin and Scott 1980) in a number of fruits. To increase its concentration in the fruit flesh is one of the most efficient preharvest methods to improve storage life. Due to these remarkable effects the transport of Ca^{2+} into storage organs has been the subject of many investigations and it seems that one component of this rather complex transport pathway is hormonally regulated (see Bangerth 1979 for lit.). A mutual interaction between auxin and Ca^{2+} transport seems to exist as can be concluded from results with auxin transport inhibitors, like TIBA or morphactins (Bangerth and Firuzeh 1971, Stahly and Benson 1976). To relate this to parthenocarpic fruits we have measured diffusible auxins from seeded and induced parthenocarpic fruits and found in most cases a considerably lower auxin concentration in diffusates from the latter, which may explain their lower Ca^{2+} content (Bangerth and Sjut 1978, Sjut and Bangerth in prep.). Thus although parthenocarpy is not an important issue in fruit production it demonstrates a role for growth regulators and endogenous hormones in determining the concentration of this very important element in fruits, and suggests a future possibility for the correction for low Ca^{2+} concentration and deficiencies by the application of appropriate growth regulators. In fact one new growth regulator from BASF can increase the Ca^{2+} concentration in fruits even when applied at low concentration to young fruitlets. Preliminary experiments in our laboratory seem to indicate a stimulated auxin diffusion out of tomato fruits treated with this growth regulator. If this can be verified it will be a good example of how more fundamental research into growth regulator-hormone interactions can eventually lead not only to a better understanding of the physiological processes involved but also to an important practical result.

FRUIT THINNING AND ITS EFFECT ON POSTHARVEST QUALITY

Selection and breeding for highly productive cultivars in combination with intensive orchard management methods have created

Table 1. Some Characteristics of Seeded (control) and Induced Parthenocarpic Fruits at Harvest Time

Fruit	Treatment	Fruit weight	Calcium content mg/100g	Fruit firmness kg/cm	Ascorbic acid mg/100g	Titratable acidity meq/g
Apple	Control	165	27	7.8	9.8	0.82
(Golden Del.)	$GA_{4/7}$	143	20	6.6	7.2	0.91
Pear	Control	128	49	10.4	7.0	-
(Alex. Lucas)	GA_3	204	23	7.7	4.4	-
Tomato	Control	48	82	-	35.0	73.0
	GA_3	18	63	-	28.0	65.0

trees which after normal fruit set produce more fruits than they can support without loss of size and quality. Fruit- thinning is therefore a common practice in modern fruit cultivation as far as apples, peaches, mandarines etc. are concerned. It would be an even more useful and desirable technique for many fruits if more effective and reliable growth regulators could be found for this purpose. Thinning - mostly achieved by growth regulators like NAA, 4 -CPA, carbaryl, ethephon etc. - reduces alternate bearing and increases fruit size. In addition, however, and this is seldom recognized, it improves colour development, soluble solids, titratable acidity and therefore taste and quality in apples (see Table 2), and also in pears and peaches. These are not specific effects of a particular growth regulator but are obviously caused by any substance which reduces a heavy fruit load.

On the other hand it is often claimed that thinning, mainly by increasing fruit size, reduces the Ca^{2+} content of fruits and therefore impairs storage life and increases postharvest physiological disorders. Link (1973), however, has shown that when fruits from thinned trees are compared with those from unthinned ones over a period of several years, physiological disorders like bitter pit in fact are reduced due to the more regular bearing of these trees. So in total, thinning produces an overall increase in fruit quality. This may change, however, if C_2H_4-releasing chemicals are used for this purpose more frequently. Even applied at early stages ethrel accelerates ripening of some fruits.

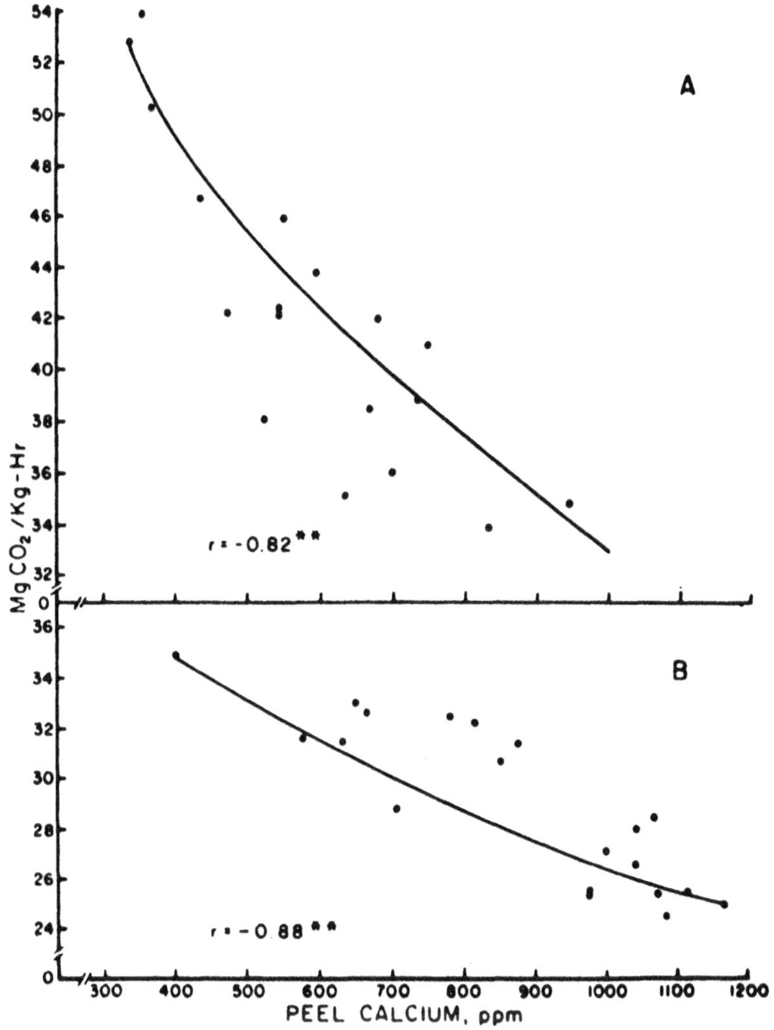

Fig. 1. Relationship of average respiration rate at 21°C to peel Ca^{2+} content of 'Baldwin' apples. A.1971-72. B. 1972-73 (Bramlage et al. 1974).

ASSIMILATE PARTITIONING BETWEEN FRUITS AND VEGETATIVE TISSUE

Fruit are almost entirely dependent on long distance transport of assimilates from the leaves and therefore subject to competition with vegetative growth. This competition may not only restrict yield but also quality. Growth regulators, especially the so-called growth retardants like alar, CCC or ethrel should be useful tools to achieve that goal. In reality, however, most of these substances not only check extension growth but also favour lateral thickening of twigs and branches (Barden 1968, Firuzeh 1971), thus a significant

Table 2. Effect of fruit thinning on fruit size and color, soluble solids and titratable acidity of 'Golden Delicious' apples (Link 1967)

Treatment	Character				
	% fruits with a diameter larger than 70mm*	%fruits with yellow green color		% soluble solids*	meq. titr. acidity per 100 ml juice*
Control	20.2	2	19	11.8	4.8
Hand thinned	32.5	8	13	12.7	5.8
Chemically thinned	31.0	10	4	12.5	5.3

*mean values of 4 year results

redistribution of assimilates in favour of fruits is doubtful. This is also emphasized by the fact that these growth retardants do not greatly affect carbohydrate content of treated fruits (Drake et al. 1980, Lüdders and Fischer-Bölukbasi 1980). In addition substances like alar and ethrel, when applied early, inhibit fruit growth and are therefore not very popular. especially with cultivars where fruit size is a problem.

Other substances like some synthetic auxins - especially NAA - when painted on to pruning cuts in late winter or early spring retard or even prevent the outgrowth of lateral buds around these cuts and therefore reduce considerably vegetative growth (Jackson et al. 1978 and Fig. 2). In our experiments (Bangerth, unpublished) these treatments improved colour, soluble solids content, firmness and storage behaviour of two apple cultivars in two years. Presently it is not clear to us whether this was the result of a change in partioning of assimilates and probably Ca^{2+}, or of the higher light intensities reaching the fruits because of the reduced leaf area especially on the top of the trees.

Chemical pinching agents which kill or temporarily check the growth of shoot apices are another class of growth regulators which may not only enhance fruit set and hence yield but also diminish

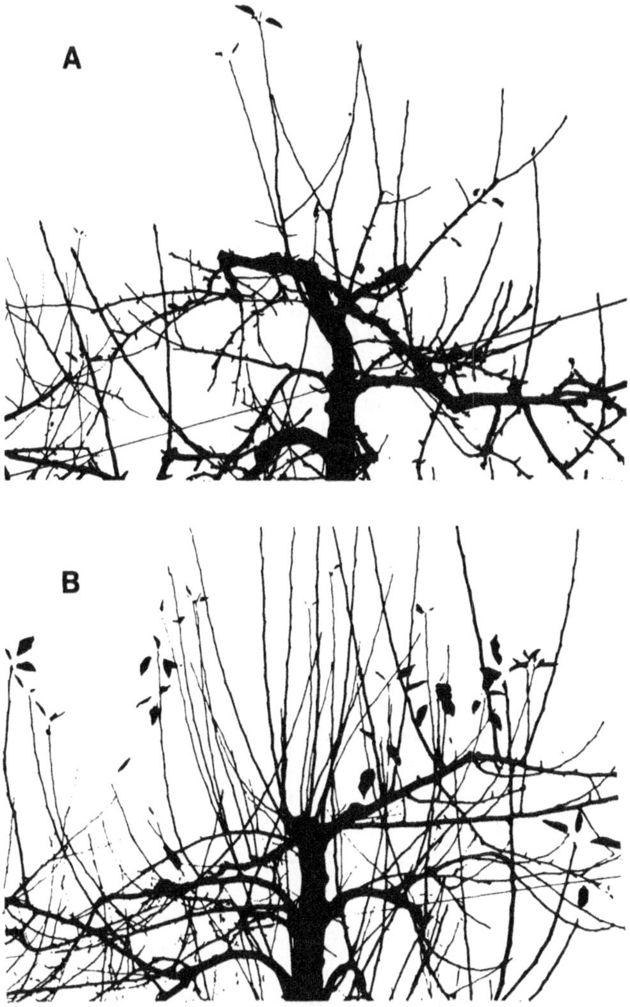

Fig. 2. "Golden Delicious" apple trees one year after treatment of pruning cuts with (A) or without (B) NAA

vegetative competition for Ca^{2+} and therefore improve fruit Ca^{2+} and probably storage life (Jackson et al. 1978).

These results, when confirmed by additional experiments, may open a new and promising field for further improvement of fruit quality and storeability without losses on yield. Special attention should be paid to a better understanding of the basic mechanisms of competition and nutrient partioning and how hormones and growth regulators affect them (Patrick 1976, Wareing 1978).

APPLICATIONS OF GA_3 AND THEIR EFFECTS ON FRUIT QUALITY

GA_s have been widely used in fruit production for quite different purposes and because they affect market -, nutritional -, and storage quality criteria it seems justified to treat them separately and more extensively. It has been already mentioned that GA_s are applied to induce especially parthenocarpic pome fruits. In addition they are frequently used to increase fruit set by reducing June - drop e.g. in apples (Wertheim 1971), citrus (Monselise 1979), stone fruits (Webster and Goldwin 1981) etc. sometimes in combination with a cytokinin and/or an auxin. These treatments clearly change market standards with a number of fruits. In some apple cultivars e.g. GA_3 and Benzyladenine applications after fruit shape as well as size in a way that may have some practical significance for some countries like the USA or Canada (Unrath 1974, Looney 1979). Positive effects on size have also been reported from peaches (Zucconi and Bukovac 1978), seedless grapes (Sachs and Weaver 1968) 'Alexander Lucas' pears (Buchloh and Kayali 1972, and Table 1) and a few other fruits. More often, however, GA applications to fruit at this early developmental stage reduce the number of seeds and, probably as a result of this, fruit size (Wertheim 1971, Krezdorn 1973, Bangerth and Götz 1975).

In seeded wine grapes this GA-induced reduction in the number of seeds per berry (20-50%) advances maturity and stimulates early sugar accumulation and acid catabolism, which may be of advantage in countries with short autumn season (Peynaud and Ribereau-Gayon 1971, Bangerth and Götz 1975 and Table 3). Currants obviously behave similarly (Hartmann, pers. communication).

Apart from these GA-effects on grapes and probably currants it has been shown that gibberellins can retard ripening of several climacteric as well as non-climacteric fruits, even when applied shortly after fruit set (see e.g. Webster and Goldwin 1981). A more pronounced effect can be observed with later applications, which may be related to the general decline of endogenous GA_s to very low levels during maturation (e.g. Jackson and Coombe 1966, Ebert and Bangerth in press). When applied during later stages of ripening GA retards the development of a great number of different fruits such as apples and pears (Sharples 1973), prunes(Proebsting and Mills 1966), pineapple (Norman 1978), citrus (Monselise 1979), apricots (Abdel-Gawad and Romani 1974), kaki fruits (Kitagawa et al. 1966) etc.

However, it is surprising that with the exception of citrus no practical applications have yet been developed from these esperimental results. In citrus preharvest GA treatments are used quite extensively to prevent premature peel senescence and physiological disorders related to this process (Coggins et al. 1960, Coggins 1973). Here again a sharp drop in endogenous gibberellins in the peel is probably the reason for the good response of these fruits to gibberellin applications (Goldschmidt and Galili (1974)). It is noticeable that this effect of GA_s on clorophyll degradation is counteracted by ethylene and vice-versa (Dostal and Leopold 1967, Goldschmidt et al. 1977 and Fig. 3).

Another example of GA/C_2H_4 interaction is the premature ripening of pears caused by low night temperatures before harvest (Wang et al. 1971). A rapid outburst of C_2H_4 is the result of these low night temperatures followed by an advanced respiratory climacteric. Similar observations have been made by Streif (1976) with apples. Wang et al. have been able to prevent this early autocataltic C_2H_4 production as well as the premature ripening by GA_3 (and alar) applications (Fig. 4)

Dostal and Leopold (1967) found the effect of dipping tomatoes into GA_3 solutions to be restricted to chlorophyll degradation and lycopene synthesis with almost no change in respiration pattern. Similar observations for citrus have been made by Goldschmidt et al. (1977) who found much more pronounced effects of GA on chloroplast reactions than on other ripening parameters. On the other hand in

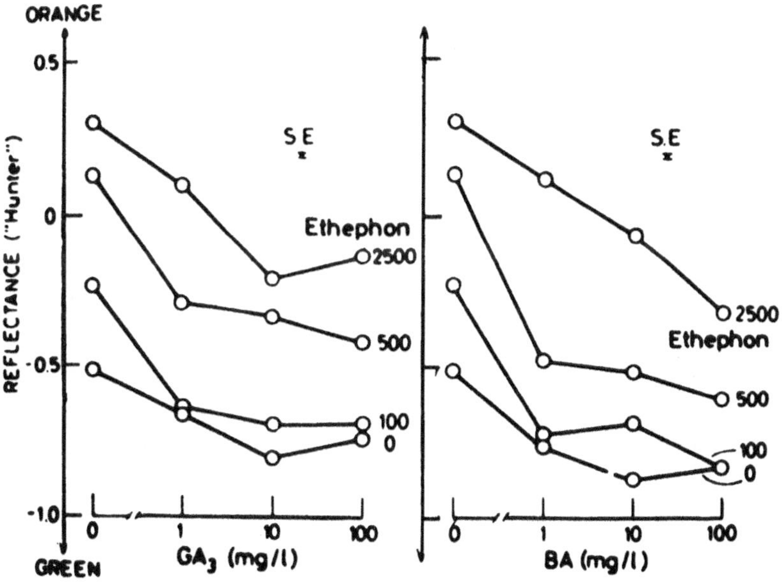

Fig. 3. Effect of combined GA_3-ethephon and BA-ethephon treatments on colour change of mature green harvested orange fruit. Fruits were stored after treatment for 7 days at 25°C prior to colour evaluation (Goldschmidt et al. 1977).

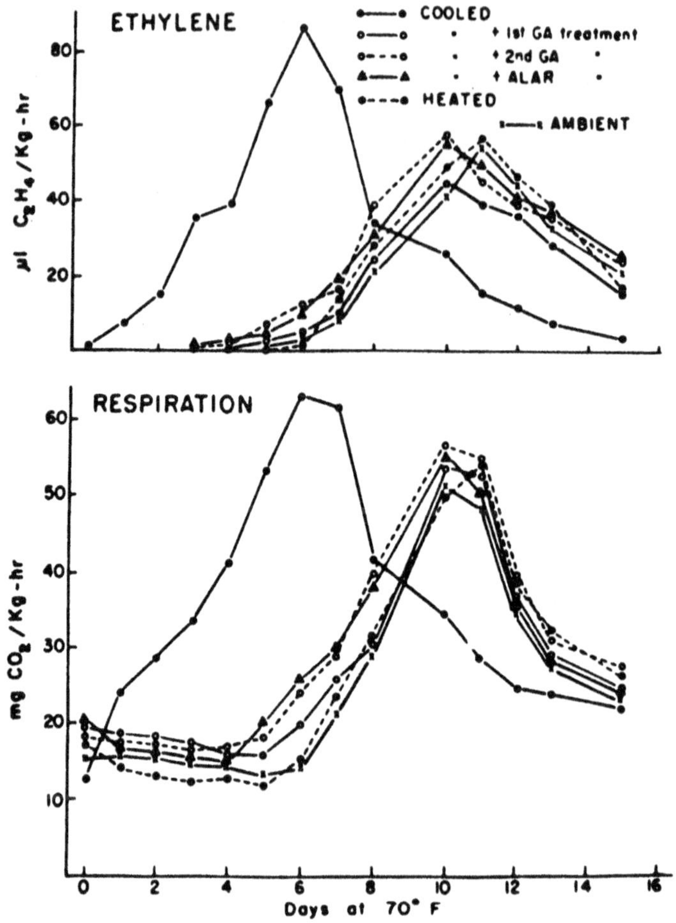

Fig. 4. Effect of preharvest temperature, GA and alar on respiration and ethylene production of 'Bartlett' pears after picking. (Wang et al. 1971).

pears and apricot GA applications affect respiration as effectively as chlorophyll degradation. Species - different reactions, and more likely, the different application methods used (dipping of detached tomato and citrus fruits but preharvest spraying of apricots and pears) may be responsible for the observed variations in GA effects, emphasizing once more the difficulties in comparing the effects of growth regulator/hormone treatments when different application methods have been employed.

Citrus rind disorders can be a serious problem especially when the fruits have to be left for longer periods on the tree or in storage. Under these conditions physiological as well as pathological rind disorders can be very efficiently reduced by preharvest GA applications (Coggins 1973, Chalutz pers. communication). Similarly some physiological apple and pear disorders related to senescence can be diminished by preharvest GA applications (see Sharples 1973 for lit.). Although not a postharvest disease russeting is a serious problem with a number of important apple varieties and considerably affects postharvest market standards. Early $GA_{4/7}$ applications effectively reduces this abnormal development of the fruits epidermis (Taylor 1975, Eccher and Boffelli 1981). Some disadvantages of GA applications like delayed degreening of citrus (Monseliese 1979) or inhibition of flower initiation in apples have to be mentioned.

ALAR AND ITS EFFECTS ON POSTHARVEST QUALITY

Alar is a good example of a growth regulator not primarily used to influence postharvest physiology and quality which has, however, prnounced side effects on these processes. A vast amount of literature exists about the use of alar in fruit production (Faust 1973, Lüdders 1973, Sharples 1973), although in retrospect it is difficult to justify the wealth of scientific attention that has been paid to this chemical. For this reason only the most fundamental effects of alar on the postharvest behaviour of fruits will be mentioned here.

From an applied point of view the retardation of ripening of pome fruits (especially apples) and the acceleration of ripening of stone fruits (even when climacteric, like peach) is the most interesting effect of alar. This allows the grower to extend the harvest period by spraying only part of the orchard and more importantly the ripening period can be concentrated, especially with stone fruits, allowing the number of pickings to be reduced considerably (see Fig. 5).

This may be particularly important for fruits harvested mechanically, e.g. for processing. Another desirable effect of alar is its inhibition of preharvest drop of apples which may be related to the delayed ripening of treated fruits. Retardation of chlorophyll loss and the better firmness of treated apple fruits also may be related to this effect on ripening.

Alar treated fruits frequently show a higher colouration due to a higher accumulation of anthocyanins. This is clearly unrelated to effects of the chemical on ripening because it is observed in fruits whose ripening is accelerated (cherries, peaches etc.) as well as in fruits with an inhibited ripening (apples e.g.). It seems possible that this is in some way linked to an alar induced shift in carbohydrate metabolism (Faust 1973), an hypothesis probably supported by the observation that alar aggravates the occurrence of two physiological disorders in some apple cultivars

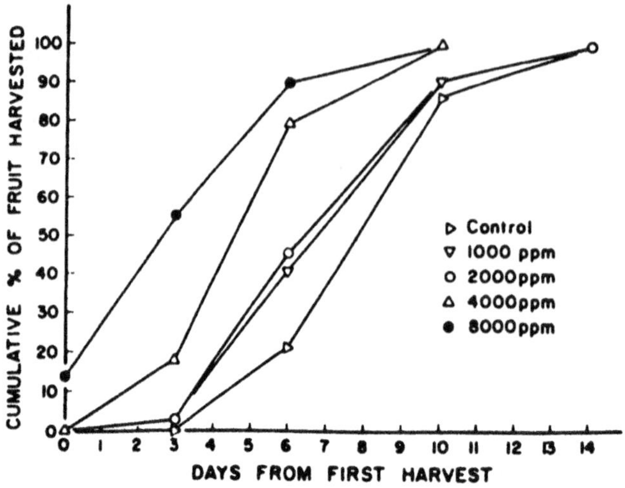

Fig. 5. Effect of alar on 'Redskin' peach fruit maturation when applied during stage I. (Byers et al. 1972)

(internal breakdown and core flush) which are also related to an alteration in carbohydrate metabolism. A clearly undesirable effect of early alar applications is their diminuation of fruit size. It is still not known whether this is due to a reduction in cell division or in cell expansion.

More interesting than its practical application seems to be the use of alar as a tool to explore in more detail the mechanisms controlling fruit ripening. Comparative studies on pome versus stone fruits are particularly interesting since they can show opposite responses to applications of alar. Surprisingly this opportunity has so far been used only to a limited extent. Looney (1969) and Rhodes et al. (1969) have shown that alar treated apple fruits have a delayed and reduced capacity for C_2H_4 biosynthesis during ripening and that exogenous ethylene can overcome this effect of alar on ripening (Fig. 6). In peaches on the other hand, where alar accelerates ripening, the production of ethylene during the final rapid growth phase is increased (Looney et al. 1974). Nevertheless the authors cast doubt on the opinion that this increased C_2H_4 production is the cause of the earlier ripening and favour instead a role for ABA in this fruit. Thus it seems unlikely that alar is acting directly on ethylene biosynthesis in apples and peaches.

Other hypotheses of how alar may exert its effects on fruit ripening include that of Williams and Stahly (1970) who suggested an interference with auxin metabolism, and Martin et al. (1964) who proposed an effect on gibberellin metabolism. Both lines, however,

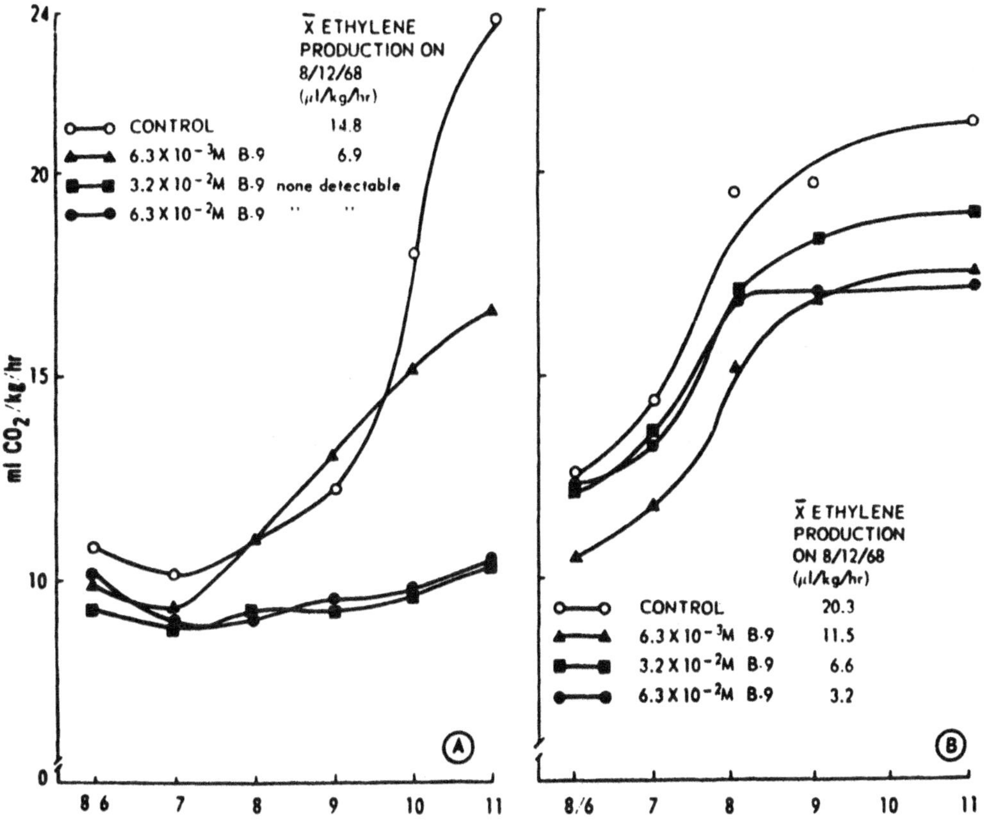

Fig. 6. Respiration patterns of and ethylene production by 'Tydeman's Early' apples. (A), treated with B-9; B-9 plus supplemental ethylene. (Looney 1969).

have not been followed up in recent years, although with the availability of modern analytical methods it seems worthwhile to look again into such growth regulator-hormone interactions. Whether the opposite effects of alar on fruit ripening of pome - and stone fruits are mediated through an interaction with endogenous phytohormones or with other physiological processes, the unravelling of such an interaction would be an important contribution to our understanding of the initiation of ripening in climacteric fruits in general.

REGULATION OF ETHYLENE LEVELS IN FRUITS BY PREHARVEST APPLICATIONS OF GROWTH REGULATORS

From a theoretical as well as from an applied point of view developments in ethylene physiology have been among the most promising ones in the field of hormones and growth regulators. Because of the dominating role of this hormone in the ripening behaviour of all climacteric fruits (Biale 1968, Burg and Burg 1965, Sacher 1973, McGlasson et al. 1978, Lieberman 1979) effects on quality are clearly expected. The more recent developments in this area started with the discovery of ethylene - releasing compounds (Abeles 1973) and have culminated with the almost complete elucidation of the biosynthetic pathway of ethylene (Adams and Yang 1979). More important, this biosynthesis can be effectively controlled from both sides: Endogenous ethylene production can be inhibited by applications of the rhizobitoxin analogue aminoethoxyvinylglycine (AVG) (Lieberman et al. 1974, Lieberman 1979) or of the aminooxyacetic acid (AOA) Amrhein and Wenker (1979), whereas the ethylene precursor 1-aminocyclopropane-1-carboxylic acid (ACC) (Adams and Yang 1979, Lürssen et al. 1979) or derivatives of this compound (Lürssen 1981) stimulate ethylene production considerably.

As far as the ripening of fruits is concerned a whole catalogue of more or less desirable effects can be achieved by applying these chemicals. Presently still dominating is the enhancement and concentration of ripening of climacteric fruits by ethylene-releasing compounds like ethrel. For example, when ethrel is sprayed onto indeterminate tomato plants less pickings are required (Fig. 7) whereas with determinate cultivars more fruits in the ripe stage can be harvested, especially when "once-over" mechanical harvesting machines are employed.

Applied together with a "stop-drop" agent such as NAA or 2,4,5 TP, ethrel advances remarkably the harvest date of apples or peaches, allowing the grower to take advantage of the higher prices of early marketed fruits. Furthermore, red colouration of treated fruits is often dramatically improved and their flavour is indistinguishable from control fruits harvested 1-2 weeks later at their normal harvest date (Luckwill and Child 1972 and Table 4).

In fact in our experiments (Bangerth, unpubl.) the aroma and flavour of ethrel (CEPHA) treated early ripening apples was always superior to control fruits, even when these were panel-tasted at their optimal eating time. More experiments involving analytical methods in addition to tasting panels are necessary to confirm this observation.

Ethylene clearly plays a decisive role in the production of volatile flavour substances in fruits and higher concentrations of this hormone are probably necessary for this purpose than for many other ripening parameters. This may indicate the existence of different ethylene threshold and saturating values for different physiological processes related to ripening and there are some indications for this in the literature. Ethylene stimulated anthocyanin or lycopene production e.g. is more sensitive to low C_2H_4 concentrations

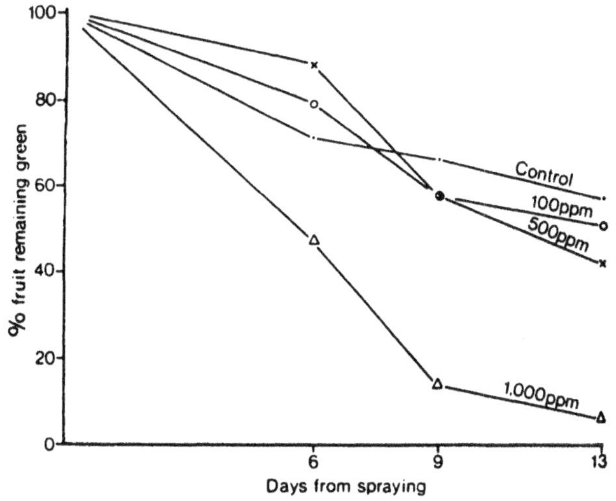

Fig. 7. The effect of ethephon applications on the ripening of 'Asic Cross' tomato fruits (Ashby 1972).

(Chalmers and Faragher 1977, Streif and Bangerth 1976) than is chlorophyll degradation (Knee 1976, Streif and Bangerth 1976) or aroma production (Blanpied and Blak 1976). Alternatively processes which apparently require higher C_2H_4 concentrations to be saturated may be nearer the end of an C_2H_4 initiated chain reaction and thus need a longer time to be fully operative.

Commercial applications of ethrel plus auxin sprays to accelerate ripening frequently lead to shorter ripening and harvest periods which is not solely of advantage. Efficient harvest as well as marketing methods are therefore necessary to get the fruits to the consumer at the best time. Treatments with ACC or derivatives of it probably do not have the disadvantage of a too fast softening (Lürssen 1981) and should therefore be tested more extensively for this purpose.

Ethrel or other ethylene releasing compounds not only promote the ripening of climacteric fruits but also of non-climacteric ones like currants, blueberries, cherries, raspberries, olives, pineapples, and even nuts and coffee (Pecheur and Ribaillier 1975). Desirable effects concerning these fruits are often not so much the acceleration of ripening but more a decrease in "fruit removal force" by the induction of an abscission layer, which facilitates mechanical harvest (Cooper and Henry 1973). Some drawbacks of ethrel sprayings have also to be mentioned. First of all, as with almost all other growth regulators, the responsiveness of the plant or plant organ to ethrel is not only species - but very much cultivar - dependent.

Table 4. Effect of 3 growth regulators on fruit quality of "Worcester Pearmain" apples 10 days after CEPHA application. (Child 1973)

Treatment	Character			
	Fruit Firmness (kg)	Starch Content	Anthocyanin Content	Weight of 100 Fruits (kg)
Control	8.5	1.4	170	7.3
CEPHA, 750 ppm + 2,4,5 TP 15 ppm	7.7	4.4	220	7.2
CEPHA 750 ppm + Alar, 2000ppm	8.5	1.9	270	7.2

In addition environmental effects during and after application are considerable. Together these factors make the selection of the appropriate time and concentration of ethrel sprays difficult. Even more serious is the ethrel - induced gummosis and dieback to twigs and branches in many stone fruits (Cooper and Henry 1973, Stösser 1978). It will be interesting to see whether the use of ACC or its derivatives can overcome some of these problems.

Procedures reducing the level of ethylene in fruits are as important as those which increase it. Again this is more important for climacteric fruits. This field has been widely opened by the recent discovery that rhizobitoxin and its analogue AVG greatly diminish C_2H_4 biosynthesis in apple fruits (Fig. 8). When sprayed onto apple trees AVG prevented autocatalytic C_2H_4 formation, the respiratory climacteric, softening and other ripening parameters in at least 4 apple cultivars (Fig. 9). Chlorophyll breakdown was not prevented to the same extent and very surprisingly red colour formation was almost unimpaired. The AVG-induced block in ripening could be completely overcome by subsequent applications of C_2H_4 at low concentrations (Halder-Doll, pers. Communication). Cold storage experiments with AVG treated apple fruits have shown that C_2H_4 production and respiration are greatly reduced for at least five months whereas differences in firmness have been found to be much smaller. Stored AVG treated apple fruits behave, therefore, somewhere in between cold- and CA - stored fruits.

Like some other growth regulators and storage methods AVG

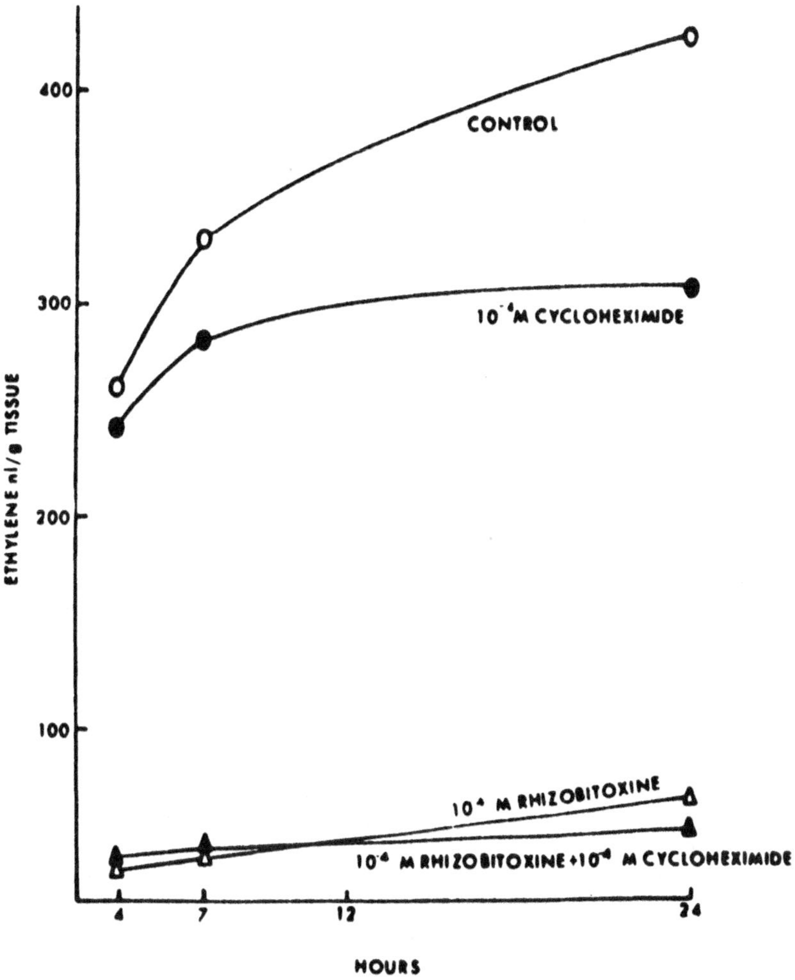

Fig. 8. Inhibition of ethylene production in apple tissue slices by Rhizobitoxin, Cycloheximide, and a combination of both. (Lieberman et al 1974)

treatment leads to a distintegration of the various ripening characteristics, e.g. as with hypobaric storage or CA, AVG strongly reduces aroma development whereas other ripening processes are not affected to the same extent. Another example of this kind is the enhanced and accelerated colour development after alar application (see above) without affecting sugar or acid concentration normally associated with this process. Such disintegrations of the ripening process by modern technology need further investigation so that losses in quality can be prevented.

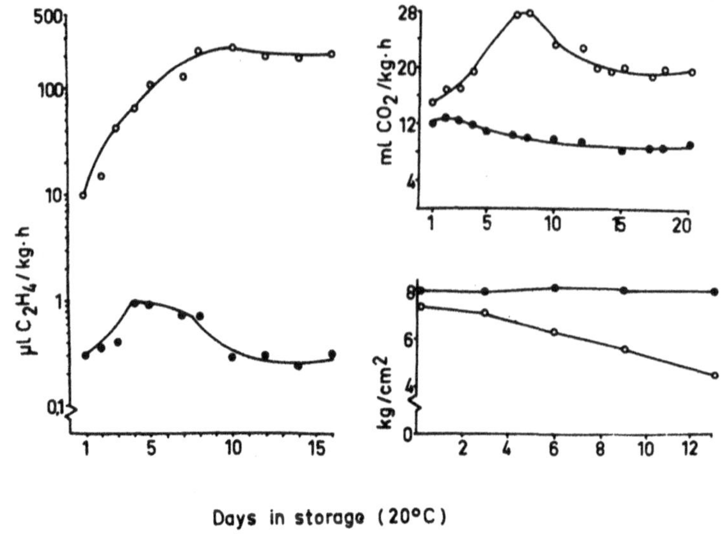

Fig. 9. Effect of preharvest AVG treatments on C_2H_4 production, respiration, and fruit firmness of 'Golden Delicious' fruits. o - o Control; ● - ● AVG treatment. (Bangerth 1978)

In contrast to apples, some other fruits, like tomatoes, apparently do not respond to AVG applications to the whole plant very much because ripening, as assessed by lycopene synthesis and chlorophyll breakdown, is not delayed. Bartlett pears do respond to AVG but not to the same degree as most apple cultivars (Bufler, pers. communication). Because apple fruits seem to obtain most of the applied AVG via the leaves (Bangerth 1978) it is possible that this leaf-to-fruit transport takes place not so readily in pears and tomatoes.

CONLUSIONS

Most review articles concerning the effects of preharvest hormone and growth regulator applications on postharvest quality and storeability of fruits have concentrated almost entirely on alar and ethrel. It seems, however, that by selective use of other perhaps more appropriate growth regulators or phytohormones more improvements in fruit quality and maturation can be achieve. Although most of these improvements may not be as spectacular as alar-induced intensified colouration, or the dramatically enhanced ripening caused by ethrel, they may in the long run be even more beneficial. Much more fundamental research is necessary in order to understand better the sometimes bewildering complexity of the mechanism, and regulation of processes such as cell division/expansion in fruits, Ca transport into fruits, assimilate partioning between vegetative and reproductive organs, initiation and coordination of ripening etc. Such an improved understanding of the basic physiological processes involved and how they are affected by growth regulators and phytohormones is necessary if we want to overcome the hitherto almost purely empirical methods used in this field. The progress made in the field of ethylene biosynthesis and physiology during recent years and the advances in growth regulator research related to these developments are encouraging, and leave us with the hope that similar progress in research into other such key processes will improve our ability to control the growth and development of fruits both before and after harvest by chemical treatments in the field.

ACKNOWLEDGMENT

The author expresses his appreciation to Dr. M. V. Palmer for reading the English manuscript.

REFERENCES

Abdel-Gawad, H. and Romani, R. J., 1974, Hormone induced reversal of color change and related respiratory effects in ripening apricot fruits, Physiol. Plant., 32: 161.

Abeles, F. B., 1973, Ethylene in plant biology, Academic Press, New York, 302 pp.

Adams, D. O., and Yang, S. F., 1979, Ethylene biosynthesis: identification of 1-aminocyclopropane-1-carboxylic acid as an intermediate in the conversion of methionine to ethylene, Proc. Natl. Acad. Sci. USA, 76: 170.

Amrhein, N., and Wenker, D., 1979, Novel inhibitors of the ethylene production in higher plants, Plant and Cell Physiol., 20: 1635.

Ashby, J. P. R., 1972, Ethephon stimulation of tomato ripening, Hort. Res., 12: 145.

Bangerth, F., 1976, A role for auxin and auxin transport inhibitors on the Ca content of artificially induced parthenocarpic fruits, Physiol. Plant, 37: 191.

Bangerth, F., 1979, Calcium related physiological disorders of plants, Ann. Rev. Phytopathol., 17: 97.

Bangerth, F., 1978, The effect of a substituted amino acid on ethylene biosynthesis, respiration, ripening and preharvest drop of apple fruits, J. Amer. Soc. Hort. Sci., 103: 401.

Bangerth, F., and Firuzeh, P., 1971, Der Einfluß von 2,3,5-Trijodbenzoesäure (TIBA) auf den Mineralstoffgehalt und die Stippigkeit von 'Boskoop' Früchten, Z. Pflanzenkr., 78:93.

Bangerth, F., und Götz, G., 1975, Zur Wirkung von Gibberellinsäure- und Kinetinapplikationen auf Qualität und Ertrag verschiedener Sorten von Vitis vinifera L., Weinwissenschaft, 30: 121.

Bangerth, F., und Link H. 1972, Möglichkeiten der Entstehung und Bekämpfung von Stippigkeit und Lentizellenflecken, Der Erwerbsobstbau, 14: 113 and 138.

Bangerth, F., and Sjut, V., 1978, Induced parthenocarpy- a tool for investigating hormone regulated physiological processes in fruits, Acta Horticulturae 80: 169.

Bardon, J. A., 1968, Effects of alar on the growth and distribution of the growth increment in one year old apple trees. Proc. Amer. Soc. Hort. Sci. 93: 33.

Biale, J. B., 1968, Respiration of fruits, in: "Handbuch fur Pflanzenphysiology," Vol. 12/2: 536. W. Ruhland, ed., Springer-Verlag Berlin.

Blanpied, G. D., and Blak, V. A., 1976, Relationship between ethylene level and varietal flavor in 'Delicious' apples, Hort Science, 11: 596.

Bramlage, W. J., Drake, M., and Baker, J. H., 1974, Relationships of calcium content to respiration and postharvest condition of apples, J. Amer. Soc. Hort. Sci., 99: 376.

Buchloh, G., and Kayali, A. Z., 1972, Die Wirkung von Gibberellinsäure und Kinetin auf die Entwicklung parthenokarper Birnenfrüchte, Erwerbsobstbau, 14: 22.

Burg, S. P., and Burg, E. A. 1965, Ethylene action and the ripening of fruits, Science 148: 1190.

Byers, R. E., Emerson, F. H., and Dostal, H. C., 1972, The effect of succinic acid-2,2-dimethylhydrazide (SADH) and other growth regulating chemicals on peach fruit maturation, J. Amer. Soc. Hort. Sci., 97: 420.

Chalmers, D. J., and Faragher, J. D., 1977, Regulation of anthocyanin synthesis in apple skin. II. Involvement of ethylene, Aust. J. Plant Physiol. 4: 123.

Chaplin, G. R., and Scott, K. J., 1980, Association of calcium in chilling injury susceptibility of stored avocados, Hort Sci., 15: 514.

Child, R.D., 1973, The interaction of SADH, CEPHA and 2,4,5-TP in improving the quality of early-harvested apples, Acta Horticulturae, 34: 441.

Coggins, Ch. W., 1973, Use of growth regulators to delay maturity and prolong shelf life of citrus, Acta Horticulturae, 34: 469.

Coggins, C. W., Hield, H. Z., and Boswell, S. B., 1960, The influence of potassium gibberellate on Lisbon lemon trees and fruit, Proc. Amer. Soc. Hort. Sci., 76: 199.

Cooper, W. C., and Henry, W. H., 1973, Chemical control of fruit abscission, in: "Shedding of plant parts," T.T. Kozlowski, ed., Academic Press, New York.

Dennis, F. G., 1973, Physiological control of fruit set and development with growth regulators, Acta Horticulturae 34: 251.

Dostal, H. C., and Leopold, A. C., 1967, Gibberellin delays ripening of tomatoes, Science 158: 1579.

Drake, S. R., Proebsting, E. L., Thompson, J. B., and Nelson, J. W., 1980, Effects of Daminozide, maturity, and cultivar on the color grade and character of sweet cherries, J. Amer. Soc. Hort. Sci., 105: 668.

Eccher, T., and Boffelli, G., 1981, Effects of dose and time of application of GA_{4+7} on russeting, fruit set and shape of 'Golden Delicious,' Scientia Hortic. 14: 307.

Faust, M., 1973, Effect of growth regulators on firmness and red color of fruit, Acta Horticulturae 34: 407.

Firuzeh, P., 1971, Die Auswirkungen von Alar-Spritzungen auf das vegetative und generative Verhalten zweier Sauerkirschensorten, Erwerbsobstbau, 10: 168.

Gil, G. F., Griggs, W. H., and Martin, G. C., 1972, Gibberellin - induced parthenocarpy in 'Winter Nelis' pear, Hort Science, 7: 559.

Goldschmidt, E. E., Aharoni, Y., Eilati, S. K., Riov, J. W., and Monselise, S. P., 1977, Differential counteraction of ethylene effects by gibberellin A_3 and N_6-benzyladenine in senescing citrus peel, Plant Physiol. 59: 193.

Goldschmidt, E. E., and Galili, D., 1974. The fate of endogenous gibberellins and applied radioactive gibberellin A_3 during natural and ethylene - induced senescence in citrus peel, Plant and Cell Physiol., 15: 485.

Jackson, J. E., Quinlan, J. D., and Preston, A. P. 1978, Chemical pruning and pruning aids, Acta Horticulturae, 65: 199.

Jackson, D. J. and Coombe, B. G., 1966, Gibberellin like substances in the developing apricot fruit, Science, 1954: 277.

Kitagawa, H., Suguiura, A., and Sngiyama, M., 1966. Effects of gibberellin spray on storage quality of 'Kaki,' Hort Science 7: 59.

Knee, M., 1976, Influence of ethylene on the ripening of stored apples, J. Sci. Food Agric., 27: 383.

Kotob, M. A., and Schwabe, W. W., 1975. Respiration rate and acidity in parthenocarpic and seeded 'Conference' pears, J. hort. Sci., 50: 435.

Krezdorn, A. H., 1973. Effect of growth regulators on set and development of citrus fruit, Acta Horticulturae, 34: 283.

Lieberman, M., 1979. Biosynthesis and action of ethylene, Ann. Rev. Plant Physiol., 30: 533.

Lieberman, M., Kunishi, A. T., and Owens, L. D., 1974. Specific inhibitors of ethylene production as retardants of the ripening process in fruits, Coll. Inter. C.N.R.S. Paris, No. 238: 161.

Link, H., 1967. Der Einfluß der Ausdünnung auf Fruchtqualität und Erntemenge bei der Apfelsorte 'Golden Delicious,' Gartenbauwiss., 32:423.

Link, H., 1973, Effect of fruit thinning on some components of fruit quality in apples, Acta Horticulturae, 34: 445.

Looney, N. E., 1969, Control of apple ripening by succinic acid 2,2 dimethyl hydrazide, 2-chloroethyltrimethylammonium chloride, and ethylene, Plant Physiol., 44: 1127.

Looney, N. E., 1979. Some effects of gibberellins A_{4+7} plus benzyladenine on fruit weight, shape, quality, Ca content, and storage behavior of 'Spartan' apple, J. Amer. Soc. Hort. Sci., 104: 389.

Looney, N. E., McGlasson, W. B., and Coombe, B., 1974. Control of fruit ripening in peach, Prunus persica: action of succinic acid -2,2 dimethylhydrazide and (2-chloroethyl) phosphonic acid, Austral. J. Plant Physiol., 1: 77.

Lüdders, P., 1973, Möglichkeiten der Ertragssteigerung durch Einsatz von triebwachstumshemmenden Mitteln, Mitt. Klosterneuburg, 23: 45.

Lüdders, P., und Fischer-Bölukbasi, T., 1980. Einfluß von Alar und TIBA auf den Säure- und Kohlenhydratgehalt von Äpfeln bei unterschiedlichem Fruchtbehang, Gartenbauwiss., 45: 102.

Lürssen, K., (1981), Manipulation of crop growth by ethylene and some implications of the mode of generation. 33rd Easter School in Agricultural Science. Chemical Manipulation of Crop Growth and Development. Butterworth, London (in press).

Lürssen, K., Naumann, K. and Schroder R., 1979, 1-Aminocyclopropane - 1-carboxylic acid - an intermediate of the ethylene biosynthesis in higher plants, Z. Pflanzenphysiol. 92: 285.

Luckwill, L. C., and Child, R. D. 1972, Growth regulator effects on quality and pre-harvest drop of 'Worcester Pearmain' apples, J. Hort. Sci., 47: 249.

Martin, G. C., William M. W., and Batjer, L. P., 1964, Movement and fate of labeled N-dimethyl amino succinamic acid (B-Nine) a size - controlling compound, in apple seedlings, Proc. Amer. Soc. Hort. Sci., 84:7.

McGlasson, W. B., Wad, N. L., and Adato., J. 1978, Phytohormones and fruit ripening, in: "Phytohormones and related compounds - A comprehensive treatise," Vol. 2, D.S. Letham, P. B., Goodwin, and T.J.V. Higgins eds., Elsevier/North Holland, Amsterdam.

Modlibowska, J., 1966, Inducing precocious cropping on young 'Dr. Jules Guyot' pear trees with gibberellic acid, J. Hort. Sci. 41:137.

Monselise, S. P., 1979, The use of growth regulators in citriculture, a review, Scientia Hortic., 11: 151.

Norman, J.C., 1978, Response of developing ethephon-induced fruits of 'Sugarloaf' pineapple to some growth regulators, Gartenbauwiss. 43: 62.

Patrick, J. W., 1976, Hormone-directed transport of metabolites, in: "Transport and transfer processes in plants, J. F. Wardlaw, and J. G. Passioura, eds., Academic Press, New York.

Pecheur, J., et Ribaillier, D., 1975, La regulation de la maturation des fruits par l'ethephon, Coll. Inter. C.N.R.S. Paris No. 238:121.

Peynaud, E., and Ribereau-Gayon, P., 1971. The grape, in: "The biochemistry of fruits and their product," Vol. 2, A.C. Hulme, ed., Academic Press, London.
Proebsting, E. L., and Mills, H. H., 1966, Effect of gibberellic acid and growth regulators on quality of 'Early Italien' prunes (Prunus domestica L.), Proc. Amer. Soc. Hort. Sci., 89: 135.
Purvis, A. C., 1980, Sequence of chloroplast degreening in 'Calamondin' fruit as influenced by ethylene and $AgNO_3$, Plant Physiol., 66: 624.
Rhodes, M.J.C., Harkett, P. J., Wooltorton, L.S.C. and Hulme, A.C., 1969, Studies on the effect of N-dimethylaminosuccinamic acid on the ripening of apple fruits J. Fd. Technol., 4:377.
Rylski, J., 1979, Effects of temperatures and growth regulators on fruit malformation in tomato, Scientia Hortic. 10, 27.
Sacher, J. A., 1973, Senescence and postharvest physiology, Ann. Rev. Plant Physiol. 24: 197.
Sachs, R. M., and Weaver, R. J., 1968, Gibberellin and auxin - induced berry enlargement in vitis vinifera L., J. Hort. Sci. 43: 185.
Schwabe, W.W., 1978, Growth regulators and the control of development in fruit trees, in: Proc. British Plant Growth Reg. Group, Monograph No. 21: 143.
Sharples, R. O., 1973, Orchard sprays, in: "The biology of apple and pear storage," J. C. Fidler, B. G. Wilkinson, K. L. Edney and R. O. Sharples, ed., Commonwealth Agric. Bureau, Slough.
Sharples, R. O., and Johnson, D. S., 1977, The influence of calcium on senescence changes in apple, Ann. Appl. Biol., 85: 450.
Shear, C. B., 1975, Calcium related disorders of fruits and vegetables, Hort Science, 10: 361.
Stahly, E. A., and Benson, N. R., 1976, Calcium levels of 'Golden Delicious' apples as influenced by calcium sprays, 2,3,5-triiodobenzoic acid, and other plant growth regulator sprays, J. Amer. Soc. Hort. Sci., 101: 120.
Stembridge, G. E. 1973. Effect of growth regulators on the size and shape of fruits, Acta Horticulturae 34:435
Stösser, R., 1978. Untersuchungen über die Entstehung der Lakunen bei der Gummibildung des Steinobstes, Mitt. Kosterneuburg, 28: 119.
Streif, J., 1976, Einfluß der Temperatur auf verschiedene Reifemerkmale von Apfeln, Erwerbsobstbau, 18: 168.
Streif, J., and Bangerth, F., 1976, The effect of different partial pressures of oxygen and ethylene on the ripening of tomato fruits, Scientia Horticulturae 5: 227.
Taylor, B. K., 1975, Reduction of apple skin russeting by gibberellin $A_{4/7}$, J. Hort. Sci. 50: 169.
Unrath, C. R., 1974, The commercial implications of gibberellin $A_{4/7}$ plus benzyladenine for improving shape and yield of 'Delicious' apples, J. Amer. Soc. Hort. Sci., 99: 381.
Wang, C. Y., Mellenthin, W. M., and Hansen, E., 1971, Effect of temperature on development of premature ripening in 'Bartlett'

pears, J. Amer. Soc. Hort. Sci., 96: 122.
Wareing, P. F., 1978, Hormonal regulation of assimilate movement in: Proc. British Plant Growth Reg. Group, Monograph No. 21:105.
Webster, A.D., and Goldwin, G. K. 1981, The hormonal requirments for improved fruit setting of plum, Prunus domestica L. cv. Victoria, J. Hort. Sci., 56: 27.
Wertheim, S. J., 1971, The drop of flowers and fruits in apple, with special reference to the June drop of 'Cox's Orange Pippin' and its control with growth regulators, Med. Landbouwhogeschool Wageningen 71-17: 1.
Williams, M. W., and Stahly, E. A. 1970, N-Malonyl-D-tryptophan in apple fruits treated with succinic acid 2,2-dimethylhydrazide, Plant Physiol., 46: 123.
Zucconi, F., and Bukovac, M. J., 1978, Stimulation of growth of frost-injured peach fruit by gibberellin A_3, Acta Horticulturae 80-159.

EFFECT OF POST HARVEST TREATMENTS OF GROWTH AND BIOREGULATORS ON QUALITY AND LONGEVITY OF FRUITS AND VEGETABLES

Elias D. Dekazos

The Agricultural College of Athens, Votanikos
Athens, Greece

Fruits and vegetables after harvest continue to carry on most of the life processes that were predominate just before harvest. Harvested fruits and vegetables are living things, using O_2 and stored substrates while giving off CO_2, water vapor, and heat in the process of respiration. The more rapid the rate of respiration after harvest, the more quickly the fresh produce ripens. The rate of respiration varies with stage of maturity, kind of produce, temperature, chemical treatment, and composition of surrounding atmosphere.

Modern technology has greatly increased the yields of fruits and vegetables. Higher yields are worth little, however, if the harvested crops are not consumed by people. Approximately 30% of the harvested crops per year are lost as they move from the farmer to the consumer. Such losses are much greater in the developing countries where storage of food is a perennial problem. A sizable reduction also occurs in the nutritive values and general quality of much of the perishables such as fruits and vegetables that do reach the consumer. In addition, such wastage is reflected in higher consumer prices. It is of prime importance to preserve what has already been produced. By preventing or minimizing crop losses, millions more can be fed. It is highly desirable to inhibit their ripening and senescence until they are to be consumed. Ideally we would like to be able to inhibit the ripening process at will to insure maximum quality at a predetermined time. The storage life of fruits and vegetables can be prolonged by inhibiting the ripening process either with chemical treatments and/or with controlled atmosphere storage at lower temperatures.

During ripening and senescence, the fruits and vegetables pass through a series of overt changes in color, texture and flavor indicating the compositional changes are taking place. Many of the physiological and bio-chemical changes that occur in fruits and vegetables, as they ripen, have been extensively studied. The endogenous factors that control and regulate these chemical changes remain largely obscure.

The ripening and senescence processes of fruits and vegetables involve plant hormones, a group of chemical substances produced by the plants themselves. Men have synthesized compounds and have achieved some success in controlling the vital processes of ripening and senescence of fruits and vegetables. These include all categories of plant hormones, auxins, gibberellins, cytokinins, ethylene and ethylene generators, growth retardants, metabolic inhibitors and other chemicals. The achievement of specific agricultural objectives may depend upon the proper balance of natural and applied growth regulators. The balance changes throughout the growing season. Both timing and concentration are critical in achieving specific responses.

Attention will be given to growth substances and other chemical compounds that have been shown to have a pronounced effect on the storage life of fresh fruits and vegetables. At this point, it can be said that we are actually only at the beginning of the commercial exploitation of plant growth substances in the production, harvesting, and storage of foods.

Auxins

The application of a synthetic auxin such as 2,4-dichlorophenoxyacetic acid (2,4-D) to harvested lemons and limes before storage was shown to affect markedly the retention of the buttons in a green condition, thus preventing the undesirable blackening of buttons that frequently develop during storage (50,67,96). Comparable results were obtained when the fruit on the tree was sprayed with these chemicals before harvest. Some oranges and grapefruit were also found to have better keeping qualities after treatment with 2,4-D (96). In citrus, storage life was prolonged by applications of 2,4-D and 2,4,5-T (96,97,98), mainly as a result of delayed degreening (59,60). Storage qualitiies of 2,4,5-Trichlorophenoxyacetic acid (2,4,5-T) treated fruits were also improved. The Monsoon Coorg mandarin orange treated with 2,4,5-T (25 and 50 ppm) and chlorophenoxyacetic acid (CLPA) (25 ppm) 8 weeks before harvest had 34, 35 and 28% more weight, respectively, than the control. When stored for 35 days at 5.5-7.2° C and 85-90% RH, the treated fruits suffered less physiological loss in weight, had high marketable percentage, and less vitamin C loss as compared with the untreated ones (88). It was reported that 2,4,5-T was better than 2,4-D in promoting early ripening of mandarins (25). Later, using Nangpur mandarin oranges, it was found that 2,4,5-T was in fact less effective than 2,4-D in prolonging the storage life of the fruit (74b).

The keeping quality of fruits in storage is a primary concern of the citrus industry. Prior to the late 1940's Alternaria decay was one of the most serious fungus diseases of stored lemons. In both preharvest and postharvest treatments, 2,4,5-T has proven more effective in control of the disease than 2,4-D, and ester formulations have been more effective than amine salt formulations. Although 2,4-D is relatively ineffective in inhibiting the growth of Alternaria in vitro (46), treatment of fruit with this compound prior to storage reduces the fungus to a minor role in storage decay (45,62,96,98). The primary reason for success appears to be that treated lemons are more resistant to the entry of fungus. Once the fungus has gained entrance, it grows as well in treated as in non-treated fruits (35). Alternaria spores are present under the button (receptacle and calyx) of most lemons (5). Once the button dies, invasion can readily occur. Treatment with 2,4-D delays the development of the abscission layer, the button remains in a living condition, and entry of fungus is reduced. The usual practice is to apply 2,4-D as a final step in the washing procedure for lemons. Where wax is applied prior to storage, 2,4-D is added with the wax.

A reduction in black buttons, reduced Alternaria decay, and delayed aging also has resulted from the use of 2,4-D on oranges and grapefruits (96). Delayed yellowing of Persian limes (59) and improved storage life of Clementine mandarin (58) has been effected by postharvest treatments with 2,4,5-T.

Treatments with 2,4-D also delays aging and reduces external decay which is probably a result of retardation of the aging process. Delayed aging is characterized by firmer fruits and a delay in development of the lemon-yellow color. If the dosage is too high, the delay in coloring can become a problem.

The effectiveness of 2,4,5-T and 1-naphthaleneacetic acid (NAA) in markedly retarding further ripening of pineapple fruit and thus extending its marketable life as a fresh fruit has been demonstrated. As little as 1 ppm of 2,4,5-T has noticeable effect, and 100 ppm appears optimum for senescence delay (52,54,55). For NAA, 500 ppm is at optimum level for dipping fruit. A brief wetting of the fruit is adequate. The crowns remain in better condition when not treated with growth regulators. Refrigeration can supplement the effect of the chemical in retarding senescence. This process extended the shelf-life of pineapples for several days by delaying senescence.

Among leafy vegetables application of 2,4-D, especially if combined with N^6-benzyladenine retarded yellowing of cauliflower(66) and other green vegetables (115). Thus, during storage, leaf drop in cauliflower and cabbage was reduced by an application of 100 to 500 ppm 2,4-D, 1 to 7 days before harvest (98).

Delaying effect on ripening was obtained with NAA by Lampe (68) on tomatoes, by Stewart (98) on Smooth Cayenne pineapples, and by Soni et al. (93) on Basrai Dwarf banana. After 16 days of storage at 20.5 to 30.2° C, bananas treated with 100 ppm NAA and 4% wax emulsion had lower reducing, non-reducing, and total sugars acidity and total soluble solids than the controls (93).

Storage life of onion bulbs was also prolonged when 0.40 ppm NAA was used as a foliar spray (15). Cauliflower and cabbage wrapped with shredded paper containing 50 to 100 mg NAA and stored at 32° F showed a reduction in leaf abscission and weight loss (98).

Gibberellins

The primary aim of growing citrus, for fresh fruit consumption, is to produce fruit of impeccable appearance. Applied growth regulators have already been a commercial success in this field.

Rind quality: The Navel orange is prone to softening, puffy rinds, sticky rinds, and damage. The incidence of these disorders may be reduced by applying GA (10-20 ppm) when the fruit is beginning to acquire its orange color (20). The events leading to senescence in navel orange rind appear to include pigment changes, softening of the tissue, increases in carbohydrates, some changes in carbohydrate metabolism, and changes in cation concentrations (the ratio of potassium to calcium plus magnesium increases) (71). In Australia, where late harvesting is practiced, 2,4-D, which has been used for some time to prevent the pre-harvest drop of Navel oranges, is now applied at color break at 30 ppm plus GA at 10 ppm (40). Sometimes the albedo breaks down giving rise to a serious condition known as creasing. The cause is still unknown. Since ethylene has been reported to increase activity of an albedo macerating factor (89) then natural ethylene may contribute to this disorder. Applied GA reduces but does not completely prevent the incidence of creasing (43,69).

GA delays color maturity in citrus fruits (70). This can be a disadvantage when using GA to prevent rind damage but the navel in Navel oranges of GA treated fruit, remains greenish and some people would consider that this enhances the fruit's appearance.

With late Valencia oranges stored on the tree, regreening of the fruits is an ever present problem. No growth regulator has been found that will prevent this or overcome it. Where trees have been previously treated with GA regreening is increased.

It has been demonstrated that GA delays loss of green rind pigments by Navel orange, Valencia orange, grapefruit, tangerine, lemon, lime, and Satsuma mandarin (12,16,17,18,19,84,94). Lemons from treated trees with GA were greener as early as one month and for as long as seven months after treatment. Fruits treated with GA also appeared to develop a lemon-yellow color less rapidly in storage. Delayed yellowing has been induced by GA applied to lemon trees at any time of the year. The response reflects an over-all delay in maturity rather than a simple delay in rind maturity as appears to be the case with Navel orange. The primary benefits of GA treatment are a more desirable seasonal harvest pattern in relation to market demands, a large percentage of fruit with a long storage life and a decrease in the number of small yellow lemons. Postharvest immersion of 'West Indian' limes in solutions of gibberellic acid increased chlorophyll retention during subsequent storage at several combinations of temperature and relative humidity (9).

Size: Several growth regulators increase fruit size through their effects on the growth of the fruit. 2,4-D increased the size of Bearss limes (44) and enhanced the growth rate of orange fruitlets (83). GA is also known to enhance fruit size, and like kinetin, tends to give longer shaped fruit (83). Recently, a practical problem arose with lemons exported from Australia for a specialized market in Europe. In order to delay maturity and to prevent over-ripening in transit, GA was applied prior to color formation. Since GA increased fruit size, the fruit became too large for this market. This problem can be overcome by applying a combination of GA (10 ppm) and (2-chloroethyl) trimethylammonium chloride (CCC) (1,000 ppm) which gives the required color and prevents increase in size. Also winter crop fruit held on the tree after similar treatment until midsummer still exhibits excellent rind quality (103).

Storage: GA applied as a preharvest spray (5 to 20 ppm) on Navel oranges reduces the incidence of storage disorders (8). Conversely, ethylene treatment increases susceptibility to such disorders (65). Benomyl is now the preferred postharvest fungicide. It also exhibits cytokinin-like activity (92) and prevents degreening of citrus fruit (41). Thus, other cytokinins might be worth investigating for their effects on the storage qualities of citrus fruits.

Residues of plant-growth regulators

The possibility of residues of plant growth regulators remaining in fruit at harvest presents a problem that will become increasingly important as commercial applications are developed for these compounds. Preharvest sprays of the isopropyl ester of 2,4-D are registered in some countries for oranges, grapefruit, and lemons. Postharvest sprays of 2,4-D on lemons also are registered. Gibberellic acid is registered for preharvest use on fruiting lemon and

Navel orange trees. GA is employed in preharvest treatments for the purpose of delaying fruit maturity and rind senescence.

Cytokinins

The purpose of most postharvest research is the prolongation of market acceptability of fresh fruits, vegetables and cut flowers. Much attention is being devoted to the influence of cytokinins on plant respiration. The cytokinins which are N^6-substituted adenine derivatives have been used, both natural and synthetic, to affect plant growth. N^6-benzyladenine delays the senescence of many harvested crop plants. Treatment results in a decrease in respiration, retention of chlorophyll, decreased desiccation and the retention of an increased proportion of the total phosphorus in an organic form (82). These factors enhance the postharvest life of perishable commodities. N^6-benzyladenine was effective over an extremely wide range of concentrations (0.1 ppm for broccoli to 300 ppm for apples). However, the most effective concentration varies with the tissue and method of application (7,91,114). Dedolph et al. (27) observed that senescence inhibition in treated asparagus with BA and held in the dark at 21° C was concomitant with respiration inhibition. However, Lipton and Ceponis (74a) observed a retardation of senescence of head lettuce sprayed with 10 ppm BA and stimulation of O_2 consumption of treated lettuce tissue. MacLean et al. (81) found that pre- and post-harvest applications of BA at a concentration of 10 ppm reduced the respiration rates and extended the storage life of harvested broccoli at 15° C.

For the both the European and North American markets, limes are required to be green, with minimum yellowing and wilting (108). Treatment of limes with certain plant hormones has beneficial effects on the rate of degreening. Thus, postharvest immersion of lime fruits in either solutions of kinetin (6-furfurylaminopurine) or seaweed extracts of known cytokinin activity significantly increased the degreening time (12,86). Also, postharvest immersion of limes (cvs. 'Persian' and 'West Indian') in solutions of N^6-benzyladenine increased chlorophyll retention during subsequent storage at several combinations of temperature and relative humidity (9).

It was reported (100) that lower concentrations of BA and higher concentrations of GA_3 effectively reduced senescence in fruits of Trichosanth dioica Roxb which are generally consumed as vegetables. Accompanied with delay in senescence, levels of chlorophyll, skin and pulp protein were maintained at higher level than control.

It was known that cytokinins delay citrus rind coloration, in accordance, with their universal senescence-delaying effects (52, 100). Goldschmidt et al. (53) found that there is an antagonism between ethylene and the senescence delaying regulators GA_3 and BA which seems to operate mainly within the chloroplast. Treatment of mature citrus fruit (Citrus sinensis) with ethylene induced rapid chlorophyll destruction, a rise in respiration, a release of free amino acids and an accumulation of reducing sugars. BA and GA_3 opposed the effects of ethylene on chlorophyll, amino acids, and to a lesser extent, reducing sugar levels.

Dilley (36) suggested that the ripening process is controlled by the endogenous cytokinins. Fruit ripening, essentially a senescence phenomenon, has been shown to be retarded by applied cytokinins (104). Davey et al. (26) reported that ripening fruits of a normal strain (Rutgers) of tomato (Lycopersicon esculentum Mill.) were found to contain lower levels of endogenous cytokinins than fruits of a non-ripening mutant rin. The non-ripening rin contained also high levels of zeatin glucoside, a storage cytokinin.

Ethylene and ethylene-generators

Degreening: Early varieties of citrus fruit will usually meet legal maturity standards before the peel attains the characteristic varietal color. With oranges and mandarins, the practical aim is for a deep orange-red color. The characteristic orange color does not fully develop in warmer regions, and may never be attained in the tropics. At present ethylene gas is used in degreening. A concentration of 5-10 ppm in the temperature range 21-30° C over a period of 4 days is optimal (65). Although the best color is considered to develop if fruit is exposed to ethylene at 15-25° C for a longer period of time (109). However, ethylene treatment has the disadvantage that degreening rooms are required and there is a delay before color is attained.

It is possible that solutions of ethephon could one day replace ethylene gas although high concentrations (8,000 ppm) are needed (64). Preharvest applications of ethephon (CEPA) have been investigated (63) as an aid to degreening, but the results were variable due to interference from endogenous growth regulators within the tree. Postharvest degreening with ethephon appears more promising than preharvest degreening because it eliminates the problem of defoliation and its effects are not inhibited by such packing house operations as washing and waxing (49,56). Postharvest dipping in ethephon solutions has been shown to degree citrus fruits comparable to, or better than, those obtained with commercial degreening using ethylene gas (10). Recent findings indicate that preharvest degreening of 'Robinson,' 'Nova,' 'Lee' and 'Dancy' tangerines and 'Hamlin' oranges can be hastened by spraying with 25 to 500 ppm ethephon (23,81,111, 112). Higher rates of 500 to 1000 ppm were required to induce coloring of 'Satsuma,' 'Shamouti' and 'Washington Navel' oranges (39,47, 61). Postharvest dipping in ethephon solutions at 500 to 12,000 ppm for 1 to 10 minutes induced and accelerated color changes of 'Bearss, 'Eureka' and Taiwan lemons (49,63,101); 'March' grapefruit (49,56); 'Valencia,' 'Hamlin' (56,82), 'Valle Washington' (13), 'Shamouti' oranges and 'Clementine' tangerines (49). Gautam et al.(51) found that mature green fruit of Kagazi-lime dipped in solutions of 500 to 2000 ppm ethephon for 1 minute before storage at room temperature caused degreening of fruit, which develop a yellow color. Degreening of ethephon-treated citrus was more efficient at 22.8° C than at 17.2° C and was inhibited at 6.1° C.

Application of ethephon to fruits accelerated the rate of chlorophyll breakdown, carotenoid accumulation, respiratory activity, ethylene production and abscission of buttons (49,112,113). Since ethephon will not improve the internal quality care has to be taken not to degreen fruits before minimum quality standards.

It would be of considerable interest to determine the effects of commercial fungicides (e.g. benomyl, thiabendazole, sodium orthophenylphenate) prior to degreening followed by anti-transpirant spray (Vapor Guard, rate 1 to 100) on the degreening, decay and storage of citrus fruits.

Ethylene promoting compounds applied at postharvest will induce degreening. For example, cycloheximide, cyclamic acid and ethephon (all at 4,000 ppm), have affected the coloring of summer lemons (42).

Coloring of citrus fruits

Exogenous growth regulators have a major impact on coloration. The major exogenous responses are as follows. Ethylene causes the loss of chlorophyll and produces some minor changes in carotenoids. Gibberellic acid causes a substantial delay in the loss of chlorophyll and substantial reduction in the rate of accumulation of carotenoids. 2,4-D and benzyladenine cause delay in loss of chlorophyll. However, another class of compounds deserve attention. 2-(4-chlorophenylthio)-triethylamine hydrochloride (CPTA) and some related compounds cause high amounts of lycopene, and certain other carotenoids, to accumulate in flavedo and albedo tissues of orange, lemon, lime, grapefruit cultivars (21,110). Studies conducted on the mode of action of the high lycopene inducers strongly suggest that these bio-regulators act at the enzyme level by inhibition of the cyclase(s). No growth regulator, to date, has been found that specifically increasea the level of β citraurin, the main pigment responsible for the orange color (11). A compound that promotes the accumulation of the carotenoid lycopene would clearly be a bio-regulator, not a hormone. Fruit color is of great economic importance to the citrus industry because consumers favor oranges having a deep orange color over those having a yellow or yellow-orange color. Treatment of the Valencia orange, Eureka lemon and Satsuma mandarin with CPTA caused the fruit to develop a tomato-red color due to accumulation of lycopene in the pericarp (rind). Preharvest and postharvest treatments (immersion in CPTA solution) were equally effective.

Daminozide-ethephon interaction

Recent pomological studies include the introduction of growth regulators for modifying fruit quality and ripening behavior (78). The growth regulator, succinic acid-2,2-dimethylhydrazide (Daminozide, SADH) was found to delay ripening of apples (75), improve flesh color of peaches (6) and sour cherries (102). Ethephon (2-chloroethyl phosphonic acid) shows considerable promise for commercial application to accelerate color development, advance fruit maturity and promote abscission of fruits (11,37,57,87,90).

Application of SADH and/or ethephon as preharvest fruit sprays to rabbiteye blueberries (Vaccinium ashei Reade cv. T-19) at specific stages of berry development, resulted in advanced and concentrated fruit ripening, and quality improvement. Successive applications of SADH and ethephon were more effective in the promotion of highly-colored, firm fruit. Storage quality of blueberries was maintained for 40 days--a prolongation of their marketable life. The waxy bloom, freshly harvested appearance, of the treated berries was retained throughout storage (28,31). SADH-ethephon treatment drastically suppressed the rate of softening of stored berries (Table 1). Furthermore, the Hunter negative b values which measure blueness were significantly higher for all treated berries than for the control. In addition the L values which measure lightness or darkness were considerably lower for the control meaning that these berries were darker than the treated (Table 1). The greatest percentage of marketable fruit after 40 days of storage was obtained from the SADH-ethephon treated berries and the double ethephon treated (Table 1). The double applications of ethephon and the application of SADH-ethephon have imparted some important advantages to the rabbiteye blueberries. The improved fruit texture obtained from the application of SADH-ethephon will not only be an asset to immediate fresh market acceptability but will also enhance storage life of the fruit. Anatomical studies of treated blueberry tissue showed that a relationship exists betseen cell wall thickness and texture of sprayed berries.

Ripening and quality of detached fruit

The ripening of detached fruit by treatment with growth regulators has received some attention. Postharvest treatments with ethephon, ethephon followed by CPTA, and/or controlled storage conditions for ripening of green, full-sized 'Babygold' peaches (29, 30) accelerated ripening and color development and produced peaches of highest quality as measured by the various quality parameters (Table 2). Of the quality attributes considered, the most striking trend was an enhancement of fruit color by ethephon as exhibited by a decrease in hue angle for both skin and flesh color and an increase in soluble solids as compared to the control fruit. The ethephon treated peaches followed by CPTA developed an orange

Table 1. Influence of SADH and/or Ethephon on Color, Firmness, Chemical Characteristics and Storage of 'T-19' Rabbiteye Blueberries after 40 Days at 3°C and 95+% Relative Humidity

Treatment & dates of application	Hunter color values			Texture (puncture-force) gm-cm	Soluble solids (%)	Titratable acidity[y] (%)	Marketable fruit (%)
	L	a	b				
Control	18.70c[z]	+0.18b	-2.51d	1.60×10^3c	14.2c	.84b	70.35c
SADH (500 ppm May 20 & June 3)	20.67b	+0.25a	-3.01c	2.20×10^3a	14.1c	1.03a	85.00b
SADH (500 ppm May 20) followed by Ethephon (500 ppm June 3)	21.63a	+0.18b	-3.62a	2.39×10^3a	16.3a	.65d	94.25a
Ethephon (500 ppm May 20 & June 3)	20.54b	+0.20b	-3.25b	1.88×10^3b	14.7b	.76c	93.55a

[z]Means within a column followed by different letters are significantly different at the 5% level.

[y]As percent citric acid.

Table 2. Effects of Postharvest Treatments on Ripening of Immature 'Babygold 7' clingstone peaches as measured by various quality parameters

Postharvest treatments	Hunter color values: Skin, Hue angle $(\tan^{-1} b/a)$	Hunter color values: Skin, Saturation index $\sqrt{(a^2 + b^2)}$	Hunter color values: Flesh, Hue angle $(\tan^{-1} b/a)$	Hunter color values: Flesh, Saturation index $\sqrt{(a^2 + b^2)}$	Firmness (Magness-Taylor) (kg)	Soluble solids %	Citric acid %
Immature (Green)	97.52°	18.35	81.99°	30.30	7.12	9.1	0.36
Tree-ripened	65.77°	23.51	71.13°	32.40	5.44	10.9	0.37
23.9°C + 90% RH	77.65°	22.95	70.69°	32.75	5.50	10.9	0.35
23.9°C + 90% RH E 250 ppm	70.49°	24.19	67.44°	32.56	4.88	10.8	0.35
26.7°C + 90% RH	72.79°	22.95	68.13°	33.18	5.58	11.3	0.40
26.7°C + 90% RH E 250 ppm	64.68°	25.21	65.61°	35.96	5.36	11.3	0.31
26.7°C + 90% RH E 250 pp + CPTA 2500 ppm	55.89°	25.27	64.99°	33.66	5.37	11.8	0.29
40.5°C + 90% RH	65.82°	25.37	66.11°	34.91	5.57	11.0	0.29
26.7°C + 90% RH E 250 ppm							
40.5°C + 90% RH	67.19°	25.77	65.70°	34.34	5.48	11.1	0.29
26.7°C + 90% RH							

color, exhibited the greatest decrease in hue angle in both skin and flesh color.

As the ripening temperatures decreased from 26.9° to 23.9° C, the hue angle reading increased in the first quadrant going from red to yellow. With postharvest ripening of immature peaches, an intensity of saturation was exhibited. Saturation was proportional to the strength of a given color. Soluble solids were higher in fruit treated with ethephon and ethephon followed by CPTA treatments. Softening of fruit was increased and the total acid decreased by ripening with ethephon. Peaches treated with ethephon and placed at 26.7° C and 90% RH resulted in the development of better color. Fruit dipped with ethephon before being dipped in CPTA and then placed at 26.7° C and 90% RH resulted in an orange color. Peaches at 40.5° C and 90% RH prior to the ripening temperatures and ethephon dipping were not only more desirable than the tree-ripened fruit but were also practically free of decay losses.

These quality peaches were more desirable and contained higher levels of carotenoids than the tree-ripened fruit. Lycopene accumulation was stimulated and detected for the first time in this peach tissue treated with CPTA (Table 3). These peaches developed an orange color. Ethephon appeared to hasten the development of carotenoids. The value for total carotenoids reached 2.7 g/100 g for fresh peaches. Minor amounts of lycopene were found in the untreated fruit but lycopene in the treated fruit increased 4.5 times. Lycopene accumulated mainly in the skin but also in the flesh. Carotenoids were compared with the hue angle and saturation index. There seemed to be a relationship of hue angle to carotenoids in the flesh of the postharvest ripened fruit. Peaches held at 40.5° C and 90% RH for 24 hours (in hot-air rooms) before ripening at 26.7° C and 90% RH after dipping in 250 ppm ethephon for 1 minute produced one of the most efficient methods of fruit ripening. Ethephon proved most effective in advancing maturity, that is, when the immature fruit had developed sufficiently to respond.

<u>Aminoethoxyvinylglycine (AVG)</u>

Ripening inhibitors are essential to prlong shelf life and to control problems such as premature ripening. Daminozide suppresses ethylene production and delays ripening (76,77,79). Cycloheximide inhibits protein synthesis and ethylene production (48).

The discovery that rhizobitoxine (2-amino-4-(2'-amino-3'hydroxypropoxy)-trans-3-butenoic acid) can inhibit ethylene production from methionine in fruit and other plant tissue (73,85) has led to the investigation of related structural analogs. It was found that L-2-amino-4-(2-aminoethoxy)-trans-3-butenoic acid (74), equally inhibits ethylene production in fruit and other plant tissues. Now when the detailed pathway of ethylene biosynthesis was defined an

Table 3. Carotenoids and Color of Postharvest Ripened 'Babygold 7' Non-melting, Clingstone Peaches

Postharvest Treatment	Fresh fruit: Relative intensity of carotenoids: Crude carotene O.D. (450nm)	Lycopene O.D. (502nm)	Tristimulus Color: Hue angle $(\tan^{-1} b/a)$	Saturation index $\sqrt{(a^2 + b^2)}$
Immature	0.661	0.057	81.99°	30.30
Tree-ripened	0.810	0.056	71.13°	32.09
23.9°C + 90% RH E 250 ppm	0.965	0.109	67.44°	32.56
26.7°C + 90% RH	0.923	0.105	68.13°	33.18
26.7°C + 90% RH E 250 ppm	1.090	0.130	65.61°	35.96
26.7°C + 90% RH E 250 ppm + CPTA 2500 ppm	1.340	0.262	64.99°	33.66
40.5°C + 90% RH 26.7°C + 90% RH E 250 ppm			66.11°	34.91
40.5°C + 90% RH 26.7°C + 90% RH	1.200	0.125	65.70°	34.34

E = ethephon

important conclusion emerged--that ACC synthase, which converts S-adenosylmethionine (SAM) to 1-aminocyclopropane-1-carboxylic acid (ACC) is the rate-controlling enzyme in ethylene biosynthesis and is inhibited by AVG (2).

AVG inhibits ethylene production on both 'Anjou' and 'Bartlett' pears (107), green tomato slices (3) and other plant tissue and increases the longevity of various cut flowers (4,72,106,108). In the case of broccoli (105), the AVG markedly reduced ethylene production and respiration, and effectively retarded its yellowing and senescence at warm temperature (20° C). Samples of broccoli treated with 5×10^{-3}M AVG retained much of the green color and compactness and were still in salable condition on the third day.

A series of studies were conducted and designed to determine whether the ethylene inhibitor, AVG, applied after harvest would effectively delay senescence and extend the longevity of fresh berries (with or without preharvest treatments) (33).

Rabbiteye blueberries (Vaccinium ashei Reade cvs. 'T-19' and 'Tifblue') sprayed before harvest with daminozide [butanedioic acid mono(2,2-dimethylhydrazide)] and/or ethephon at specific stages of berry development and placed in storage at 3° C and 95-100% relative humidity effectively maintained storage quality for 2½ months--a prolongation of their market life. The waxy bloom, which is characteristic of freshly harvested blueberries, was retained throughout storage. The Hunter negative b values, which measure blueness, were significantly higher for the sprayed berries than for the control (Tables 4, 5). In addition, the L values, which measure lightness or darkness, indicated that the control berries were considerably darker than the sprayed berries (Table 4, 5). Firmness of berries stored for 75 days was greatly influenced by preharvest growth regulator treatments. The daminozide/ethephon treatment was the more effective, drastically suppressing the rate of softening of stored berries (Table 4,5). The percentage of marketable 'T-19' fruit after 75 days of storage was higher for the daminozide/ethephon-sprayed berries than for the berries sprayed twice with ethephon (Table 4). Daminozide, a growth retardant, decreases ethylene production (121); and this effect may have caused the increase in longevity. If so, the decrease in marketable fruit between 75 days and 7 months storage (Table 6) means that the suppression of ethylene production by daminozide declined with time.

AVG minimized weight loss (Table 7) of the berries treated with preharvest growth regulators. AVG was especially effective in lightening the color of the sprayed berries. Treating berries with AVG as a dip improved firmness and color retention during storage. The values of texture (firmness) for 'T-19' berries treated with AVG 2000 ppm and for the daminozide/ethephon followed by AVG 2000 ppm were 2.77×10^3 g-cm and 3.33×10^3 g-cm, respectively, while

Table 4. Influence of Aminoethoxyvinylglycine (AVG), Daminozide and/or Ethephon on Color, Firmness, Chemical Characteristics and Storage Life of 'T-19' Rabbiteye Blueberries after 75 Days at 3°C and 95+% Relative Humidity.

Cultivar and treatment groups	Hunter Color Values[y]			Texture (puncture-force in g-cm X 10^3)	Soluble solids (%)	Titratable acidity[x] (%)	Marketable fruit (%)
	L	a	b				
T-19							
A Control	17.49b[z]	+0.16a	-2.50c	2.47c	13.90b	0.48b	65.60c
B 2X Ethephon	19.59a	+0.16a	-3.15a	2.68b	13.93b	0.64a	86.19a
C Daminozide + ethephon	20.24a	+0.17a	-3.60a	3.07a	15.55a	0.63a	92.10a
F Control + Tween-20	16.66c	+0.19a	-2.42c	2.40c	13.40b	0.53b	58.10d
G Control + Tween-20 + 1000 ppm AVG	18.19b	+0.22a	-2.69b	2.69b	12.63c	0.52b	79.68c
H Control + Tween-20 + 2000 ppm AVG	18.69b	+0.20a	-2.82b	2.77b	13.20b	0.46b	85.80b
I_2 2 X Ethephon + Tween-20 + 2000 ppm AVG	18.93b	+0.20a	-2.94a	2.76b	13.70b	0.60a	92.20a
J Daminozide + ethephon + Tween-20 + 2000 ppm AVG	19.53a	+0.20a	-3.03a	3.37a	14.68a	0.62a	95.00a

[z]Means within a column followed by different letters are significantly different (Duncan's multiple range tests, 5% level).

[y]Hunter L, a b terms: L, defines lightness or darkness; a measures redness-greeness; and b, describes yellowness-blueness.

[x]As percent citric acid.

Table 5. Influence of Aminoethoxyvinylglycine (AVG), Daminozide & Ethephon on Color, Firmness, Chemical Characteristics and Storage of "Tifblue" Rabbiteye Blueberries after 75 Days at 3°C and 95+% Relative Humidity.

Cultivar and treatment groups	Hunter Color Values[y]			Texture (puncture-force in g-cm X 10^3)	Soluble solids (%)	Titratable acidity[x] (%)	Marketable fruit (%)
	L	a	b				
Tifblue							
A' Control	17.78b[z]	+0.26a	-2.45c	2.21b	15.23b	0.73a	57.76c
C' Daminozide + Ethephon	21.29a	+0.24a	-3.38a	3.14a	16.55a	0.70a	84.50a
F' Control + Tween-20	17.30c	+0.24a	-2.41b	2.15b	14.23a	0.67a	57.24c
G' Control + Tween-20 + 1000 ppm AVG	17.89b	+0.22a	-2.64a	3.17a	13.60b	0.68a	79.60b
H' Control + Tween-20 + 2000 ppm AVG	18.13a	+0.22a	-2.69a	3.19a	14.00a	0.65a	87.45a

[z]Means within a column followed by different letters are significantly different (Duncan's multiple range tests, 5% level).

[y]Hunter L, a, b terms: L, defines lightness or darkness; a, measures redness-greeness; and b, describes yellowness-blueness.

[x]As percent citric acid.

Table 6. Influence of Aminoethoxyvinylglycine (AVG), SADH and/or Ethephon on Storage Life of 'T-19' and 'Tifblue' Rabbiteye Blueberries after 7 Months at 3°C and 95+% Relative Humidity.

Cultivar and treatment groups		Deteriorated fruit (%)	fruit (%) Marketable
T-19			
A	Control	81.27	0
B	2X Ethephon	46.86	35.50
C	Daminozide + ethephon	50.68	37.00
F	Control + Tween-20	81.80	0
I_1	2X Ethephon + Tween-20 + 1000 ppm AVG	39.30	49.50
I_2	2X Ethephon + Tween-20 + 2000 ppm AVG	28.64	61.00
J	Daminozide + ethephon + Tween-20 + 2000 ppm AVG	30.60	60.00
Tifblue			
A'	Control + Tween-20	74.24	0
C'	Daminozide + ethephon	46.00	40.00
F'	Control + Tween-20	74.80	0
G'	Control + Tween-20 + 1000 ppm AVG	35.10	49.50
H'	Control + Tween-20 + 2000 ppm AVG	25.70	62.00

Table 7. Influence of Aminoethoxyvinylglycine (AVG), SADH and/or Ethephon, CMC, and Storage Upon Weight Loss of 'T-19' and 'Tifblue' Rabbiteye Blueberries.

Cultivar and treatment	Fruit characteristic and length of storage	
	Weight loss after 2½ mos. storage (%)	Weight loss after 7 mos. storage (%)
T-19		
A Control	6.3	19.0
B 2 X Ethephon	6.3	16.7
C Daminozide + ethephon	4.4	12.3
D 2 X Ethephon + CMC	4.4	11.1
E Control + CMC	5.8	17.5
F Control + Tween-20	6.5	18.2
G Control + Tween-20 + 1000 ppm AVG	5.0	14.0
H Control + Tween-20 + 2000 ppm AVG	4.0	11.2
I_2 2 X Ethephon + Tween-20 + 2000 ppm AVG	3.5	9.8
J Daminozide + ethephon + Tween-20 + 2000 ppm AVG	3.0	8.4
Tifblue		
A' Control	9.0	25.2
G' Control + Tween-20 + 1000 ppm AVG	6.5	15.4
H' Control + Tween-20 + 2000 ppm AVG	4.5	12.6

the Tween-20 was 2.40 x 10^3 g-cm. At 2000 ppm, AVG tended to enhance the effectiveness of the preharvest sprays in maintaining fruit firmness. The effects of AVG and the sprays on fruit color, firmness and weight were positive and should be of significance to producers, processors and consumers. Treatment with AVG did not exert the same degree of inhibition on various ripening reactions. The percentage of marketable fruit after 75 days of storage was higher for AVG treated berries than for the corresponding berries without AVG, and was highest for the berries with preharvest spray treatments plus AVG (Table 4,5). The daminozide/ethephon-treated berries (with or without AVG) had the highest soluble solids content and the highest percentage of marketable fruit. Doubling the concentration of AVG from 1000 to 2000 ppm increased but did not double the percentage of marketable fruit (Tables 4,5). AVG, which was effective in reducing ethylene production from methionine in other tissue (72), extended the postharvest longevity of 61% and 62%, respectively of 'T-19' and 'Tifblue' rabbiteye blueberries to 7 months (Table 6). An informal panel indicated that the AVG-treated berries had superior taste and retained a fresh appearance. At 2000 ppm, AVG increased the amount of marketable 'T-19' berries treated twice with ethephon by about 72% and of 'T-19' berries treated with daminozide plus ethephon by about 62%. Also, the amount of marketable 'Tifblue' treated only with 2000 ppm AVG was about 55% greater than that of 'Tifblue' treated only with daminozide plus ethephon. This marked effect of AVG in increasing storage life of berries is probably due to its potential in retarding ethylene-induced senescence. Both AVG treatments (1000 and 2000 ppm), immediately after harvest, with or without preharvest sprays of daminozide and/or ethephon effectively retarded senescence and deterioration of blueberries during storage at 3° C. A direct effect of AVG on increasing longevity was demonstrated and is of practical significance. This is the first time that longevity of rabbiteye blueberries has been appreciably increased by chemical treatments. Of the compounds tested, AVG a suppressant of ethylene, was the most effective inhibitor of senescence of rabbiteye blueberries. It maintained firmness and increased the longevity of 'T-19' and 'Tifblue.' The effect of AVG in lengthening the storage life of rabbiteye blueberries was marked.

Recently, applications of high concentrations of AVG on intact branches of 'Tifblue' and 'Woodard' rabbiteye blueberries, and of 'Loring' and "Rio Oso Gem' peaches after the completion of the rest period effectively delayed blossoming; thus, reducing the risk of spring freeze (32, 34). In addition, in both blueberry cultivars and the 'Rio Oso Gem' peaches firmness was increased in fruit from branches treated with AVG. Textural properties are important attributes and contribute to the overall quality of the fruit. Controlling C_2H_4 synthesis, as per example by AVG, is the key to controlling ripening and consequently the storage life of climacteric fruits. Thus, for the first time it is possible to control the quality of the fruit from its bud stage.

REFERENCES

1. Abeles, F. B. and B. Rubinstein. 1964. Regulation of ethylene evolution and leaf abscission by auxin. Plant Physiol. 39: 963.
2. Adams, V. O. and S. F. Yang. 1979. 1-Amimocyclopropane-1-carboxylic acid: A metabolic intermediate in the conversion of methionine to ethylene. Proc. Nat. Acad. Sci. 76:170.
3. Baker, J. E., M. Lieberman, and A. T. Kunishi. 1976. Inhibition of ethylene production in tomato and avocado fruit slices by a rhizobitoxine analogue and free radical scavengers. Plant Physiol. Suppl. 57:97 (Abstr.).
4. Baker, J. E., C. Y. Wang, M. Lieberman, and R. E. Hardenburg. 1977. Delay of senescence in carnations by a rhizobitoxine analog, sodium benzoate. HortScience 12:38.
5. Bartholomew, E. T. 1926. Alternaria rot of lemons. Univ. Calif. Agr. Expt. Sta. Bul. 408:39 pp.
6. Baumgradner, R. A., G. E. Stembridge, L. O. Van Blaricom and C. E. Gambrell, Jr. 1972. Effect of succinic acid-2,2-dimethylhydrazide on the color, firmness and uniformity of processing peaches. J. Amer. Soc. Hort. Sci. 97:485.
7. Bessey, P. M. 1964. Effects of a new senescence inhibitor on lettuce storage. Univ. Arizona Expt. Sta. Rep. 189:5.
8. Bevington, K. B. 1973. Effect of gibberellic acid on rind quality and storage of coastal navel orange. Aust. J. Exptl. Agric. Animal Hus. 13:196.
9. Blunden, G. and E. M. Jones. 1979. Increases in chlorophyll retention times of limes after post-harvest immersion in N_6-benzyladenine and gibberellic acid. Trop. Agric. (Trinidad).
10. Bondad, N. D. 1971 Postharvest ripening and degreening of banana and citrus fruits with 2-chloroethyl-phosphonic acid (Ethrel). Philip. Agr. 50:310.
11. Bukovac, M. J., E. Zucconi, R. P. Larsen and C. D. Kesner. 1969. Chemical promotion of fruit abscission in cherries and plums with special reference to 2-chloroethylphosphonic acid. J. Amer. Soc. Hort. Sci. 94:226.
12. Burns, R. M., D. O. Rosedale, J. E. Pehrson, and C. W. Coggins, Jr. 1964. Preliminary trials indicate gibberellin delays lime maturity. Calif. Citrog. 49:488.
13. Castro, R. S. 1969. The action of Ethrel (2-chloroethylphosponic acid) on the external color of the orange variety Valle Washington. Proc. Amer. Soc. Hort. Sci. Trop. Reg. 13: 100.
14. Chauhan, K. S. and C. Parmar. 1972. Degreening of 'Mosambi' orange with Ethrel (2-chloroethylphosphonic acid). Proc. 3rd Int. Symp. Trop. Subtrop. Hort. p. 82.
15. Chhongkar, U. S. and B. N. Sengupta. 1972. Effects of plant growth regulators on growth, yield and quality of some important vegetables. Proc. 3rd Int. Symp. Trop. Subtrop. Hort. 4:47.

16. Coggins, C. W., Jr. and H. Z. Hield. 1958. Gibberellin on orange fruit. Calif. Agr. 12(9):11.
17. Coggins, C. W., Jr., H. Z. Hield, and S. B. Boswell. 1960. The influence of potassium gibberellate on Lisbon lemon trees and fruit. Proc. Amer. Soc. Hort. Sci. 76:199.
18. Coggins, C. W., Jr., H. Z. Hield, and M. J. Garber. 1960. The influence of potassium gibberellate on Valencia orange trees and fruit. Proc. Amer. Soc. Hort. Sci. 76:193.
19. Coggins, C. W., Jr., H. Z. Hield, and R. M. Burns. 1962. The influence of potassium gibberellate on grapefruit trees and fruit. Proc. Amer. Soc. Hort. Sci. 81:223.
20. Coggins, C. W. 1969. Gibberellin research on citrus rind aging problems. Proc. 1st Int. Citrus Symp., Riverside, Calif. 3:1177.
21. Coggins, C.W., G. L. Henning, and H. Yokoyama. 1970. Lycopene accumulation induced by 2-(4-chlorophenyl-thiotriethylamine hydrochloride). Science 168:1589.
22. Cooper, W. C., G. K. Rasmussen, and J. J. Smoot. 1968. Induction of degreening of tangerines by preharvest applications of ascorbic acid, other ethylene-releasing chemicals. Citrus Indus. 49(10):25.
23. Cooper, W. C., W. H. Henry, P. C. Reece, G. K. Rasmussen, and B.J. Rogers. 1969. Ethylene participation in natural and chemically induced senescence and abscission of citrus fruits and tissues. Proc. Conf. Trop. Subtrop. Fruits. Trop. Prod. Inst., London. pp. 121.
24. Daito, H. and K. Hirose. 1970. Studies on acceleration of coloring or degreening of citrus fruits. II. Effects of ethrel (ethylene-releasing compound) on the acceleration of coloring and carotenoid pattern of the 'Natsudaida' (Citrus natsudaidai Hayata). Bul. Hort. Res. Sta. (Japan) Ser. B., No. 10,35.
25. Damigella, P. 1962. Experiments on the effectiveness of 2, 4-D and 2,4,5-T on mandarins. Tech. Agric. 14,430.
26. Davey, J. E. and J. Van Staden. 1978. Endogenous cytokinins in the fruits of ripening and non-ripening tomatoes. Plant Science Letters, 11:359.
27. Dedolph, R. R., S. H. Wittwer, and V. Tuli. 1961. Senescence inhibition and respiration. Science 134:1075.
28. Dekazos, E. D. 1976. Effects of preharvest applications of ethephon and SADH on ripening, firmness and storage quality of rabbiteye blueberries ('T-19'). Proc. Fla. State Hort. Soc. 89:266.
29. Dekazos, E. D. 1977. Effect of postharvest treatments on ripening, carotenoids and quality of canned 'Babygold 7' peaches. HortScience 12(4):38 (Abstract)
30. Dekazos, E. D. 1977. Effect of postharvest treatments on ripening, carotenoids and quality of 'Babygold 7' peaches (in Greek). Panhellenic Conference of Agricultural Research. Kallithea-Halkidiki. May 5-8, 1981.

31. Dekazos, E. D. 1978. Maturity and quality responses of 'Tifblue' rabbiteye blueberries to SADH and ethephon. Proc. Fla. State Hort. Soc. 91:168.
32, Dekazos, E. D. 1979. Effects of aminoethoxyvinyl-glycine (AVG) on bloom delay, fruit maturity of 'Tifblue' and 'Woodard' rabbiteye blueberries. Proc. Fla. State Hort. Soc. 92:248.
33. Dekazos, E. D. 1980. Effect of aminoethoxyvinyl-glycine, carboxymethylcellulose and growth regulators on longevity of fresh rabbiteye blueberries. Proc. Fla. State Hort. Soc. 93:145.
34. Dekazos, E. D. 1981. Effect of aminoethoxyvinyl-glycine on bloom delay, fruit maturity and quality of 'Loring' and 'Rio Oso Gem' peaches. HortScience (in press).
35. DeWolfe, T. A., L. C. Erickson, and B. L. Brannaman. 1959. Retardation of Alternaria rot in stored lemons with 2,4-D. Proc. Amer. Soc. Hort. Sci. 74:367.
36. Dilley, D. R. 1969. Hormonal control of fruit ripening HortScience 4(2):111.
37. Edgerton, L. J. and G. D. Blanpied. 1970. Interaction of succinic acid-2,2-dimethylhydrazide, 2-chloroethylphosphonic acid and auxins on maturity, quality and abscission of apples. J. Amer. Soc. Hort. Sci. 95:664.
38. Eilati, S. K., E. E. Goldschmidt, and S. P. Monselise. 1969. Hormonal control of color changes in orange peel. Experientia 25:209.
39. El-Zeftawi, B. M. 1970. Chemical induction of mature citrus abscission. J. Aust. Inst. Agr. Sci. 36:139.
40. El-Zeftawi, B. M. 1971. 2,4-D plus GA for quality of late Navels. Aust. Citrus News 47:4.
41. El-Zeftawi, B. M. 1973. Degreening of Valencia oranges. Food Tech. in Australia 25:449.
42. El-Zeftawi, B. M. 1973. Chemical coloring of lemon fruit. Food Tech. in Australia 25:151.
43. Embleton, T. W., W. W. Jones, and C. W. Coggins. 1973. Aggregate effects of nutrients and gibberellic acid on 'Valencia' orange crop value. J. Amer. Soc. Hort. Sci. 98: 281.
44. Erickson, L. C. and B. L. Brannaman. 1950. Some effects on fruit growth and quality of a 2,4-D spray applied to Bearss lime trees. Proc. Amer. Soc. Hort. Sci. 56:79.
45. Erickson, L. C. 1952. Plant growth regulators for lemons. Calif. Citrograph 37:179, 201.
46. Erickson, L. C., T. A. DeWolfe, and B. L. Brannaman. 1958. Growth of some citrus-fruit pathogens as affected by 2,4-D and 2,4,5-T. Botanical Gazzette 120:31.
47. Fishler, M. and S. P. Monselise. 1971. The use of ethephon (2-chloroethylphosphonic acid) to promote color development of 'Shamouti' orange fruits. Israel J. Agr. Res. 21:67.
48. Frenkel, C., I. Klein, and D.R. Dilley. 1968. Protein synthesis in relation to ripening of pome fruits. Plant Physiol. 43:1146.

49. Fuchs, Y. and A. Cohen. 1969. Degreening of citrus fruits with Ethrel (Amchem 66-329). J. Amer. Soc. Hort Sci. 94:617.
50. Gates, C. M. 1949. The possibility of 2,4-D for the control of post-harvest decay in Florida limes. Proc. Fla. State Hort. Soc. 62:220.
51. Gautam, D. R., R. P. Dhar, V. P. Bhutani, and H. S. Dhuria. 1977. Ethephon for post-harvest degreening of Kagazi lime. Indian J. Agric. Sci. 47(6):282.
52. Gortner, W. A. 1963. Delaying senescence of pineapple fruit. U. S. Patent No. 334697.
53. Goldschmidt, E. E., Y. Aharoni, S. K. Eilati, J. W. Riov, and P. Shaul. 1976. Differential counteraction of ethylene effects by gibberellin A_3 and N_6-benzyladenine of senescing citrus peel. Plant Physiol. 59:193.
54. Gortner, W. A. 1969. Relation of chemical structure to plant growth regulator activity in the pineapple plant: retarding senescence of pineapple fruit with application of 2,4,5-trichlorophenoxyacetic acid and α-naphthalene acetic acid. J. Food Sci. 34, 577.
55. Gortner, W. A. and R. Leeper. 1969. Studies on the relation of chemical structure to plant growth regulator activity in the pineapple fruits. V. Post-harvest delay of senescence of pineapple fruit. Bot. Gaz. 130:87.
56. Grierson, W., F. H. Ismail, and M. F. Oberbacher. 1972. Ethephon for post-harvest degreening of oranges and grapefruit. J. Amer. Soc. Hort. 97:541.
57. Griggs, W. H., B. T. Iwakiri, R. B. Fridley, and J. Mehlschau. 1970. Effects of 2-chloroethylphosphonic acid and cycloheximide on abscission and ripening of 'Bartlett' pears. HortScience. 5:264.
58. Gutter, Y. 1956. The influence of growth regulators on the keeping quality of stored Clementines. Ktavim 6:119. (Israel).
59. Hatton, T. T. 1958. Effects of waxes and 2,4,5-trichlorophenoxyacetic acid as post-harvest treatment of 'Persian' limes. Proc. Fla. State Hort. Soc. 71:312.
60. Hield, H. Z. and C. C. McCarthy. 1956. 2,4-D control of preharvest drop of 'Navel' oranges in Tulare county. Univ. Calif. Dept. Hort. Newsletter. 10:1.
61. Hirose, K., M. Yamamoto, and H. Daito. 1970. Studies on acceleration of coloring or degreening of citrus fruits. I. Effects of Ethrel (ethylene-releasing compound) on the acceleration of coloring of 'Satsuma' orange. Bul. Hort. Res. Sta. (Okitsu) 10:17.
62. Iwasaki, T., M. Nishiura, and T. Chichijo. 1956. Effects of 2,4-D and 2,4,5-T on fruit drop and abscission of stem button in some citrus varieties. Bul. Hort. Division NAT TOKAI KINKI Agr. Exp. Sta. 3:17.
63. Jahn, O. L. and R. Young. 1972. Influence of the tree on the response of citrus fruit to preharvest applications of

(2-Chloroethyl phosphonic acid).J. Amer. Soc. Hort. Sci. 97:544.
64. Jahn, O. L. 1973. Degreening citrus fruit with post-harvest applications of (2-chloroethyl phosphonic acid)(Ethephon). J. Amer. Soc. Hort. Sci. 98:230.
65. Jahn, O. L., W. G. Chase, and R. H. Cubbedge. 1973. Degreening response of 'Hamlin' oranges in relation to temperature, ethylene concentration and fruit maturity. J. Amer. Soc. Hort. Sci. 98:177.
66. Kaufman, J. and S. M. Ringel. 1961. Tests of growth regulators to retard yellowing and abscission of cauliflower. Proc. Amer. Soc. Hort. Sci. 78:349.
67. Kessler, K. L. and J. R. Allison. 1948. Use of growth regulators on lemons in storage. Calif. Citrograph 34:24-25, 28.
68. Lampe, C. H. 1971. Response of tomato fruits to certain growth regulators with emphasis on pectolytic enzymes, cellulose and ethylene. Diss. Abst. 32:1308B.
69. Leggo, D. 1968. Puffing and creasing of oranges--effect of gibberellic acid. Agri. Gaz. N.S.W. 79:112.
70. Lewis, L. N., C. W. Coggins, and M. J. Garber. 1964. Chlorophyll concentration in the Navel orange as related to potassium gibberellate, light intensity and time. Proc. Amer. Soc. Hort. Sci. 84:177.
71. Lewis, L. N., C. W. Coggings, Jr., C. K. Labanauskas, and W. M. Duggen, Jr. 1967. Biochemical changes associated with natural and gibberellin A_3 delayed senescence in the 'Navel' orange rind. Plt. Cell Physiol. (Tokyo) 8:151.
72. Lieberman, M., A. T. Kunishi, and L. D. Owens. 1975. Specific inhibitors of ethylene production as retardants of the ripening process in fruits. In: R. Ulrich, ed., Facteurs et Regulation de la Maturation des Fruits. Colloques Internationaux des C.N.R.S. No. 238. Editions du C.N.R.C. Paris. p. 161.
73. Lieberman, M. 1975. Biosynthesis and regulatory control of ethylene in fruit ripening. A review. Physiologie Vegetale. 13:489.
74a. Lipton, W. J. and M. J. Ceponis. 1969. Retardation of senescence and accumulation of oxygen consumption in head lettuce treated with N_6-benzyladenine. Proc. Amer. Soc. Hort. Sci. 81:379.
74b. Lodh, S. B., S. K. Mukherjee, and A. N. Bose. 1962. Storage of mandarin oranges. II. Effects of hormones and wax coatings. Proc. Amer. Soc. Hort. Sci. 81:519.
75. Looney, N. E. 1967. Effect of N-dimethylaminosuccinamic acid on ripening and respiration of apples. Can. J. Plant Sci. 47:549.
76. Looney, N. E. 1971. Inhibition of apple ripening by succinic acid 2,2-dimethyl hydrazide and its reversal by ethylene. Plant Physiol. 43:1133.

77. Looney, N. E. 1972. Interaction of harvest and maturity, cold storage and two growth regulators on ripening of 'Bartlett' pears. J. Amer. Soc. Hort. Sci. 97:81.
78. Looney, N. E. 1973. Control of fruit maturation and ripening with growth regulators. Acta Horticulturae 34:397.
79. Looney, N. E. 1975. Control of ripening in 'McIntosh' apples. II. Effect of growth regulators and CO_2 on fruit ripening, storage behavior and shelf life. J. Amer. Soc. Hort. Sci. 100:332.
80. MacLean, D. C. and R. R. Dedolph. 1964. Phytokinins and senescence in broccoli. Amer. Jour. Bot. 51(6):618.
81. MacLean, D. C., R. R. Dedolph, and S. H. Wittner. 1964. Respiratory responses of broccoli (Brassica olerecea var. Italia) to pre- and post-harvest treatments with N^6-benzyladenine. Proc. Amer. Soc. Hort. Sci. 83:484.
82. Majeski, E. and C. A. Brasley. 1968. New chemical aid to fight regreening passes its early test. West Fruit Grow. 22 (11): 22.
83. Moss, G. I. 1972. Promoting fruit-set and yield in sweet orange using plant growth substances. Aust. J. Exptl. Agric. Animal Hus. 114:96.
84. Nishiura, M. and Y. Iba. 1964. Effects of gibberellin spray on citrus. Bul. Hort. Res. Sta. (Japan).
85. Owens, L. D., M. Lieberman, and A. T. Kunishi. 1971. Inhibition of ethylene production by rhizobitoxine. Plant Physiol. 48: 1.
86. Pantastico, E. B., W. Grierson, and G. Soule. 1966. Peel injury and rind color of 'Persian' limes as affected by harvesting and handling methods. Proc. Fla. State Hort. Soc. 79:338.
87. Proebsting, E. L., Jr. and H. H. Mills. 1969. Effect of 2-chloroethane phosphonic acid and its interaction with gibberellic acid on quality of 'Early Italian' prunes. J. Amer. Soc. Hort. Sci. 94:443.
88. Rodriques, J. and H. Subramanyam. 1966. Effect of preharvest spray of plant growth regulators on size composition and storage behavior of 'Coorq' mandarin (citrus reticulata Blanco). J. Sci. Food Agric. 17:425.
89. Rogers, D. J. and C. Hurley. 1971. Ethylene and the appearance of an albedo macerating factor in citrus. J. Amer. Soc., Hort. Sci. 96:811.
90. Rom, R. C. and K. R. Scott. 1971. The effect of 2-chloroethylphosphonic acid (Ethephon) on maturation of a processing peach. HortScience 6:134.
91. Salunkhe, D. K., A. S. Dhaliwal, and A. A. Boe. 1962 N^6-benzyladenine as senescence inhibitor for selected horticultural crops. Nature 195:724.
92. Skene, K. G. M. 1972. Cytokinin-like properties of the systemic fungicide benomyl. J. Hort. Sci. 47:179.
93. Soni, S. L., K. S. Chauhan, and S. C. Jain. 1972. Effect of plant growth regulators, wax emulsions and their combinations

on the storage behavior and physio-chemical changes during storage of banana (Musa paradisiaca Linn). Proc. 3rd Int. Symp. Trop. Subtrop. p. 77.

94. Soost, R. K. and R. H. Burnett. 1961. Effects of gibberellin on yield and fruit characteristics of Clementine mandarin. Proc. Amer. Hort. Sci. 77:194.

95. Stewat, I. and T. A. Wheaton. 1971. Carotenoids in citrus: their accumulation induced by ethylene. Agr. and Food Chem. 20: 448.

96. Stewart, W. S. 1949. Effects of 2,4-dichlorophenoxyacetic and 2,4,5-trichlorophenoxyacetic acid on citrus fruit storage. Proc. Amer. Soc. Hort. Sci. 54:109.

97. Stewart, W. S., J. R. Palmer, and H. Z. Hield. 1952. Packing-house experiments on the use of 2,4-D and 2,4,5-T to increase storage life of lemons. Proc. Amer. Soc. Hort. Sci. 59:327.

98. Stewart, W. S. 1956. Maturity and ripening as influenced by application of plant regulators. In: Tukey, H. B. (ed.), "Plant Regulators in Agriculture," J. Wiley and Sons, Inc., New York. p. 132.

99. Stother, J. 1970. 'Markets for fresh limes,' Report. Trop. Prod. Inst., London.

100. Tosh, S., M. A. Choudhuri, and E. E. Chatterjee. 1977. Hormonal regulation of senescence in fruits of Trichosanthes dioica Roxb. Indian J. Exp. Biol. 15:1234.

101. Tasi, C., C. Huang, and M. Chiang. 1970. The effect of ethylene-producing chemicals on the degreening rate of detached lemons and Satsuma oranges (Citrus unshiu More). Mem. Coll. Agr, Nat. Taiwan Univ. 11:14.

102. Unrath, C. R., A. L. Kenworthy, and C. L. Bedford. 1969. The effect of Alar, succinic acid-2,2-dimethylhydrazide on fruit maturation, quality and vegetative growth of sour cherries, Prunus cerasus L., cv. 'Monmorency.' J. Amer. Soc. Hort. Sci. 94:387.

103. Victorian Department of Agriculture. 1973. Extension Leaflet.

104. Wade, N. L. and C. J. Brady. 1971. Effects of kinetin on respiration, ethylene production and ripening of banana slices. Aust. J. Biol. Sci. 4(1):165.

105. Wang, C. Y. 1977. Effect of aminoethoxy analog of rhizobitoxine and sodium benzoate on senescence of broccoli. HortScience 12(1):54.

106. Wang, C. Y., J. E. Baker, R. E. Hardenburg, and M. Lieberman. 1977. Effects of two analogs of rhizobitoxine and sodium benzoate on senescence of snapdragons. J. Amer. Soc. Hort. Sci. 21:517.

107. Wang, C. Y. and W. M. Mellenthin. 1977. Effect of aminoethoxy analog of rhizobitoxine on ripening of pears. Plant Physiol. 59:546.

108. Wang, C. Y. and J. E. Baker. 1979. Vase cut flowers treated with rhizobitoxine analogs, sodium benzoate, and iso-

pentenyl adenosine. HortScience 14(1):59-60.
109. Wheaton, T. A. and I. Stewart. 1973. Optimum temperature and ethylene concentrations for post-harvest development of carotenoid pigment in citrus. J. Amer. Soc. Hort. Sci. 98: 337.
110. Yokoyama, H., C. DeBenedict, C. W. Coggins, and G. L. Henning. 1972. Induced color changes in grapefruit and orange. Phytochem. 22:1721.
111. Young, R., O. Jahn, W. C. Cooper, and J. J. Smoot. 1970. Preharvest sprays with (2-chloroethyl phosphonic acid) to degreen 'Robinson' and 'Lee' tangerine fruits. HortScience 5:268.
112. Young, R. and O. Jahn. 1972. Degreening and abscission of citrus fruits with preharvest applications of (2-chloroethyl phosphonic acid)(ethephon). J. Amer. Soc. Hort. Sci. 97: 237.
113. Young, R. and O. Jahn. 1972. Ethylene-induced carotenoid accumulation in citrus fruit rinds. J. Amer. Soc. Hort. Sci. 97:256.
114. Zink, F. W. 1961. N^6-benzylademine, a senescence inhibitor for green vegetables. J. Agric. Food Chem. 9:304.

MANIPULATION OF THE POSTHARVEST ATMOSPHERE FOR PRESERVATION OF FOOD CROPS

David R. Dilley

Department of Horticulture
Michigan State University
East Lansing, Michigan USA 48824

INTRODUCTION

Numerous handling, transportation and storage systems have evolved over the years for postharvest preservation of fresh fruits and vegetables. Depending upon the commodity and the specific preservation objective, there is a wide selection of techniques and systems to choose from (1). Table 1 outlines preservation systems available for fresh perishables. They vary in simplicity from common storage involving little or no control of the postharvest environment to highly sophisticated systems such as hypobaric storage (2,3) controlling within very narrow limits the temperature and humidity, concentration of oxygen, carbon dioxide and other gases that may affect product preservation. The choice and successful application of postharvest preservation technology depends largely on understanding certain fundamental aspects of biology, engineering and economics that are important in the maintenance and distribution of perishable commodities. Biological consideration include; the physical attributes of the commodity, the physiological response to the postharvest environment and susceptibility to pathogens. Engineering aspects include heat and mass transfer in maintenance of the ideal environment of temperature and gas atmosphere and product protection. Economic principles of supply of and demand for the product in relation to cost, benefit and practicality will ultimately determine what postharvest technology will be employed. The emphasis of this paper will be the biological consideration important in storage of apple and pear fruits.

Table 1. Storage Systems and Techniques for Postharvest Preservation of Fruits and Vegetables

I Common
II Refrigerated
III Controlled Atmosphere
1. Ventilated sum: $O_2 + CO_2 = 21\%$
2. Scrubbed sum: $O_2 + CO_2 < 21\%$
   a. caustic
   b. water
   c. dry lime
   d. activated carbon
   e. molecular sieve
3. Inert gas generators
4. Nitrogen purge
5. Diffusion membrane
6. Modified atmosphere
7. Ethylene scrubber
8. Hypobaric

FACTORS ASSOCIATED WITH MAINTENANCE OF QUALITY OF HARVESTED FRUIT

Apples and pears are marketed from storage in some countries on nearly a year-around basis. Large production volume of suitable varieties in relation to the market place, and demand requires controlled atmosphere (CA) storage. Achieving high quality and marketability of the product from refrigerated and CA storage requires that the fruits be harvested at optimum maturity and stored under ideal conditions of temperature and gas composition according to the expected storage duration. Establishing the proper time to harvest some apple cultivars to ensure high quality after long-term CA storage requires special consideration. Quality will be poor if fruits are harvested too early or too late. This is particularly true for the early-coloring strains of Red Delicious which are fully red while still very immature. When harvested before they are mature they are much more apt to develop scald when stored. Chemicals like diphenylamine and ethoxyquin used as a postharvest treatment to control scald are often ineffective when applied to immature fruits. Bitter pit and shrivel are other common problems associated with storing immature apples. Furthermore, the apples are likely to be of low dessert quality from the standpoint of texture and flavor. Importantly, fruit size and tonnage is reduced by harvesting too early. The adverse consequences of storing apples and pears for extended periods after ripening has begun include; excessive softening and internal breakdown and mealiness - physiological disorders associated with senescence. Ripe fruits are also more prone to bruising and decay. Harvesting apples and pears at the proper time according to when and how they will be utilized is thus a very important decision.

GROWTH AND DEVELOPMENT OF FRUIT AND THE USE OF CHEMICALS IN CONTROLLING MATURATION AND RIPENING

The relationship between growth and development and metabolism for fruits of the climacteric type such as apples and pears is shown in Figure 1. This figure also relates how the important growth regulating chemicals, ethephon and daminozide (Alar) and controlled atmosphere storage affect fruit ripening. During the first 3 to 4 weeks following bloom, 30 to 80 million cells are formed in the apple and pear fruit. This is followed by a period of several months during which growth is by cell enlargement. The growth rate during the last few weeks is about 1% per day and this is accompanied by a significant increase in the intercellular air space between the cells. This, in large measure, is responsible for the gradual decrease observed in flesh firmness before harvest. It is important to recognize that this decrease in flesh firmness is a result of growth rather than ripening and imparts to the apple the desired textural characteristic of crispness as opposed to the tough and woody texture of immature fruits.

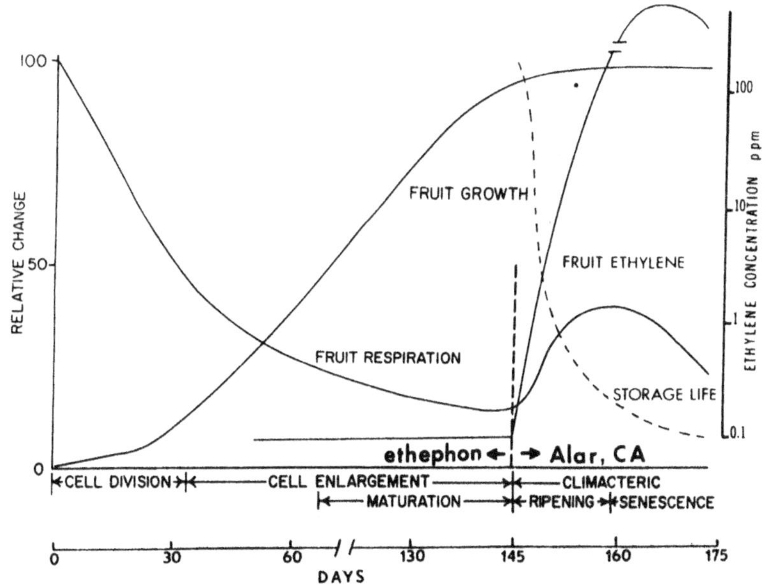

Fig. 1. Growth and development of apple fruits in relation to the effects of ethylene, ethephon, Alar and controlled atmosphere storage on ripening and storage life.

It has been known for more than 50 years that ethylene gas produced by the fruits causes them to ripen. The relation of ethylene to fruit ripening and storage life is also shown in Figure 1. Note particularly the precipitous and logarithmic increase in ethylene which marks the beginning of the ripening process. During the last few weeks of the maturation period the ethylene level in the fruit is fairly constant and mostly below 0.1 ppm (4). This low rate of production then gives way to a more rapid rate and when the concentration rises too much above 0.1 ppm it stimulates further ethylene production in an autocatalytic manner responsible for the log-linear increase. This marks the beginning of the ripening process which involves qualitative and quantitative changes in metabolism and this will not occur without ethylene action. Metabolism is redirected. New enzymes are synthesized (5) that catalyze cell wall softening (6,7) and development of flavor and aroma. New biochemical pathways are initiated and the metabolic rate, exemplified by the respiratory climacteric, increases dramatically as shown in Figure 1. When this occurs, the course of ethylene action cannot be reversed - only slowed.

Figure 1 then describes important features of apple and pear fruit growth and development that are essential to understand the processes of maturation and ripening and the role of growth regulators and storage technology. The time scale roughly approximates the situation for 'Red Delicious' but it is not fixed.

The effects of ethephon and Alar on apples are also best explained by hastening or delaying, respectively, the onset of accelerated ethylene production. Ethephon is absorbed into the tissues, releases ethylene and, if the fruits are sufficiently mature, this initiates ethylene production and subsequent action. One of these acts is to stimulate the pathway responsible for production of the anthocyanin pigments. Considerable enhancement of red color can sometimes be achieved without overly stimulating ethylene production by the fruits, particularly when ethephon is applied to trees sprayed earlier with Alar. In many instances these highly colored fruits can even be stored successfully in CA. But this requires knowledge of ethylene levels in the fruits in order to ensure safe storage. Ethephon advances and Alar delays the "time-line" of ripening as shown in Fig. 1.

Fruits become more responsive to ethylene as they mature. Thus if ethephon is applied too early, red color enhancement is minimal. And when fruits are receptive but ethephon is applied to trees without prior Alar treatment, ethylene production can be overly stimulated resulting in excessive drop.

Studies at Michigan State University on 'McIntosh' apples

established the delaying action of Alar on fruit maturity (8). Alar was applied about 60 days prior to the estimated harvest date for long-term storage, and 2,4,5-TP was applied 21 days before the same target date. These treatments were employed to ascertain if Alar could attenuate the acceleration of ripening sometimes observed with the use of 2,4,5-TP for preharvest drop control.

The effects of treatment on maturity and ripening are seen in Fig. 2 a,b, and 3. Fruits from all treatments on the harvest of Sept. 15 had identical and low respiration rates, indicating the fruits had not reached physiological maturity. At harvest, the fruits from trees sprayed with 20 ppm 2,4,5-TP entered the climacteric sooner than the control fruits which were judged to be of ideal maturity for long term CA storage on Sept. 22. Fruits which received Alar alone at 500 or 1000 ppm (1/2 to 1 lb. per 100 gal. - 400 gal. basis) showed no tendency to enter the climacteric even after 1 week at 20°C. However, 20 ppm of 2,4,5-TP treatment overcame the retarding effect of the 500 ppm Alar concentration.

The same general response was observed for fruits harvested one week later. Fruits from all treatments containing Alar were less advanced, and those receiving 20 ppm 2,4,5-TP more advanced than the controls. 2,4,5-TP stimulates ethylene production, and in this way overcomes the influence of Alar unless very high dosages are used.

That Alar delays the onset of the ripening process is clear from the data in Fig. 3 in which the loss in firmness during a holding period of 7 days is given for each harvest in relation to Alar concentration. Here it is seen that fruits receiving Alar at 500 ppm (1/2 lb. per 100 gal.) ripen like control fruits do, but about one week later. Fruits that received 1000 ppm (1 lb. per 100 gal.) Alar ripen like control fruits, but about two weeks later.

To slow down ethylene production and the events it initiates in ripening requires prompt establishment of controlled atmosphere conditions. Storage life diminishes rapidly as fruits gain capacity to produce ethylene. This is depicted as the dotted line in Figure 1. The ability of CA storage to retard ripening diminishes as the ripening process is initiated. Carbon dioxide in the CA atmosphere is a natural inhibitor of ethylene action but it becomes ineffective when ethylene accumulates to above 1 ppm in the fruit. Oxygen is required for fruits to make ethylene and low temperature slow down the rate of the chemical reactions of metabolism. These considerations serve as the basis for CA storage. They explain why CA storage is most successful, in terms of quality retention, when this technique is applied to fruits before they begin to produce much ethylene. CA storage can thereby delay the 'time-line' of ripening as indicated by the arrow in Figure 1.

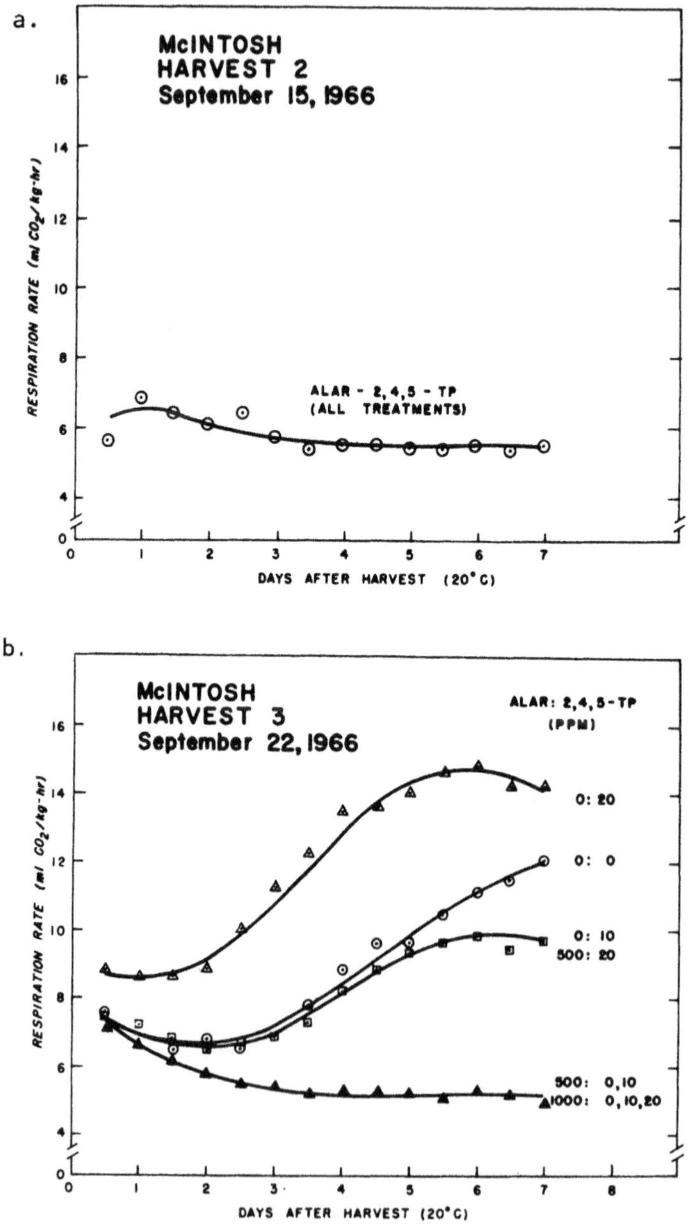

Fig. 2 a, and b, The effect of Alar and 2,4,5-TP on the initiation of the respiratory climacteric and ripening of McIntosh apples.

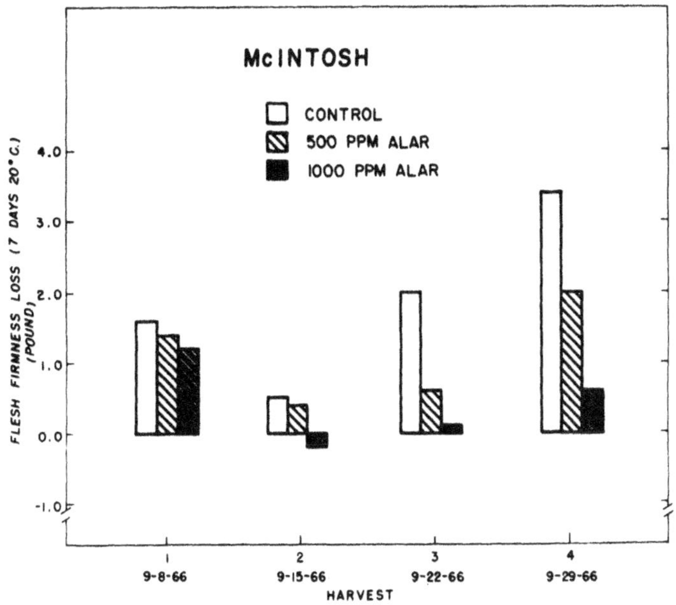

Fig. 3. The effect of Alar on delaying the loss in flesh firmness of McIntosh apples held at 20°C following harvest.

ETHYLENE AS AN INDEX OF FRUIT MATURITY, RIPENESS, AND STOREABILITY

Bartlett pears provide another example of ethylene receptivity. This is shown in Figure 4 where ethylene was unable to stimulate flesh softening of fruits harvested on August 7 while it worked one week later. In this particular season, it was not until August 21 that the fruits had gained sufficient maturity and ethylene production capacity of their own to initiate ripening. There is an important lesson to be learned from these data. Bartlett pears can gain as much as 30 percent in weight during the final three weeks before ethylene begins to accumulate sufficiently to stimulate ripening. Unfortunately many of the Bartlett pears grown are harvested while still quite immature. One of the factors contributing to this problem is placing too much reliance on the use of flesh firmness in establishing the harvest date. Studies have repeatedly shown that physiological maturity correlates poorly with flesh firmness. That is, fruits of 15 to 16 lbs. firmness (5/16" tip) from one orchard can in fact be less mature than fruits of 18 to 19 lbs. from another orchard. A good example of this can be seen in Figure 5 where ethylene concentration in fruit is plotted over a sequence of four weekly harvests. Flesh firmness at harvest is given and the value in parenthesis is the firmness after one week at ripening temperature. Note in particular that fruits from the last two harvests had initial firmness of 18 lbs. yet only the fruits from the last harvest were sufficiently mature and contained enough ethylene to cause ripening. Fruits harvested during the week beginning September 1 ripened very well after long-term storage in the air. This clearly demonstrates that internal ethylene level is

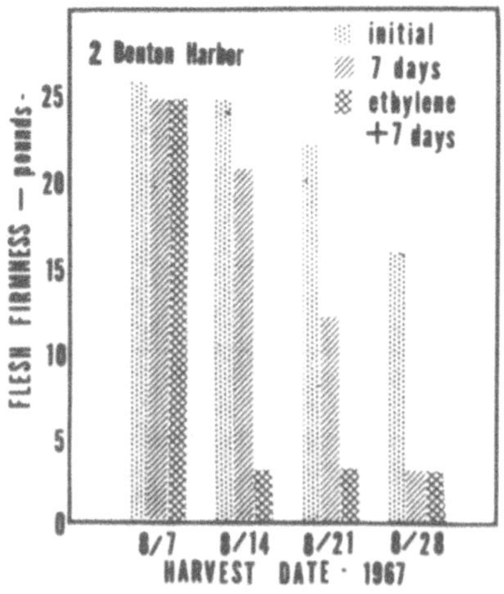

Fig. 4. Flesh firmness of Bartlett pears at harvest and after one week at ripening temperature in response to ethylene treatment at 500 ppm for 24 hours.

a better measure than flesh firmness as a criterion of harvest maturity. Moreover, these data show that Bartlett pears <u>need</u> <u>not</u> and <u>should</u> <u>not</u> be harvested for storage until at least several fruits in a sample of ten contain more than 0.1 ppm of ethylene. Be mindful that fruit size and therefore tonnage increases significantly during the last few days without sacrificing storeability. This is assuming that good storage and handling procedures are employed. When properly handled and promptly cooled in storage to

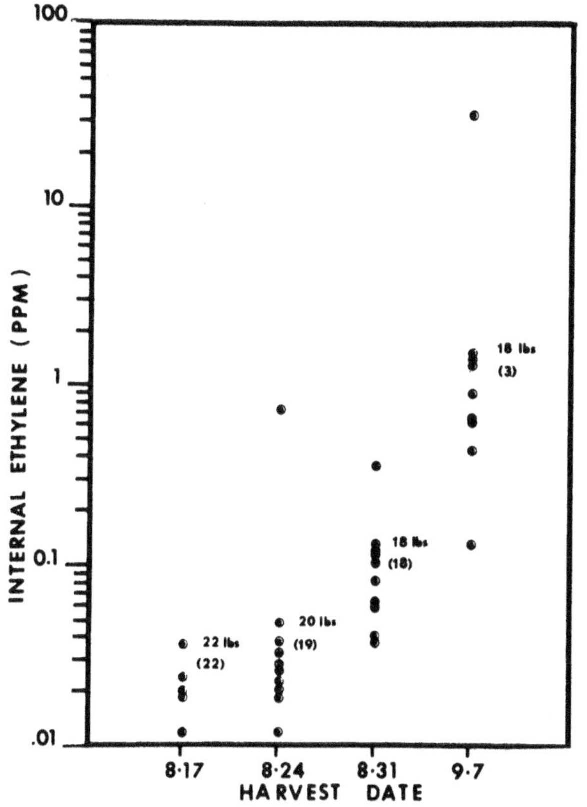

Fig. 5. Internal ethylene levels in Bartlett pears at harvest at weekly intervals. Each point is an individual fruit. Firmness is given at harvest and after one week at 20°C.

to 30°F it takes several weeks before the fruits gain capacity to produce ethylene.

Ethylene is clearly the best measure to assess maturity, ripening and storability (9). Extensive studies with apple fruit substantiate this out and in addition, considerable insight has been gained that helps explain why orchards can vary greatly from grower to grower and between years. The leaves of apple (4), and presumably pear, trees produce an inhibitor of ripening that is translocated to the fruits in the phloem. There are several lines of evidence for this. Firstly, fruits ripen more rapidly at the same temperature 'off' as compared to 'on' the tree. This suggests that the substance is metabolized. Secondly, and more definitively, fruits borne on spurs that have been defoliated and girdled ripen much earlier than fruits on normal spurs. These studies help to interpret observations of ripening behavior and reinforce the necessity to make an orchard by orchard and block by block assessment of maturity and ripening development. This is because there are many orchard management practices and environmental stresses that can hasten or delay fruit ripening. These include: biotic, insects and diseases; nutrient deficiencies; wound injury; and environmental stresses of heat, cold, drought and flooding.

Ethylene measurements can now be done in the orchard or warehouse with a new portable instrument called "Snoopy," developed in 1978 in the Postharvest Research Laboratory of the Horticulture Department at Michigan State University (10). 'Snoopy' is a highly sensitive portable gas chromatograph that detects as little as 0.1 ppm ethylene in a 1 cc gas sample taken from the fruit. The instrument separates ethylene from the other gases and measures the ethylene concentration. A 1 cc sample of intercellular atmosphere from a fruit or a composite sample taken from several fruits is injected into the instrument and the response is followed on the meter or a recorder which provides a permanent record of the analysis. Analysis takes about two minutes. This instrument is being widely used by individual growers, fieldmen, storage operators, extension agents and researchers around the world to assess fruit maturity, ripening and storability.

All fruits on a tree do not begin to ripen at the same time. The onset of ripening will vary by variety, and from orchard to orchard and year to year. When immature, all fruits in the sample will show uniformly low ethylene levels of 0.1 ppm or lower. Subsequently, some fruits in the sample will begin to show accumulation of ethylene while the majority remain low. The king-blossom fruits and those on spurs adjacent to vigorous terminal growth will generally begin to produce ethylene before others. It is important then to monitor the more advanced fruits on a tree to make the maturity and ripening assessment. Eventually, the majority of fruits in a sample will begin to accumulate a significant amount of ethylene indicating that ripening is well underway. Table 2. provides guidelines for storability of apples according to internal ethylene levels.

An example of how ethylene data for the Empire apple was used to make storage duration decisions for a Michigan orchard in 1979 is shown in Figure 6. Samples taken up until the first of October had uniformly low ethylene levels below 0.1 ppm. Although the fruits were fully red and attractive they were still immature. By October 2, some fruits began to show significant ethylene while the remainder were low. Fruits harvested during the ensuing week were judged to be excellent for long-term CA storage. By October 8, fruits were still low enough in ethylene for mid-term CA storage and had flesh firmness of 16 lbs. Not until the 16th of October were fruits from this orchard judged to be suitable for only short-term storage. Performance of fruits from this orchard harvested over a three week period beginning September 28th and stored in CA storage up to 9 months bore out the correctness of the maturity and ripeness assessment.

Table 2. Fruit Ethylene Levels As A Guide For Harvest And Storage Decisions

No. of Fruits	Fruit Ethylene	Suggested Action
10 out of 10	Less than 0.1 ppm	May delay harvest for better size, color, and quality.
3 out of 10	0.1 to 0.5 ppm	Suited for long-term CA. 'CO_2 treatment' maximally effective.
3 out of 10	0.5 to 1 ppm	Suited for mid-term CA. 'CO_2 treatment' marginally effective.
3 out of 10	1 to 5 ppm	Suited for short-term CA. 'CO_2 treatment' ineffective.
3 out of 10	5 to 10 ppm	Suited for up to 4 months of refrigerated ro CA storage.
3 out of 10	More than 10 ppm	Suited for short-term storage and processing.

It should be clear from the foregoing discussion that apples destined for long-term CA storage should be harvested and stored before they have gained much capacity to produce ethylene. Further, that fruits intended for shorter storage duration can and should be allowed to remain on the trees longer in order to gain the extra measure of quality and sales appeal required of fruits in this more competitive marketing period.

Even fruits of low ethylene content at harvest begin to make significant amounts of the ripening gas during several months in CA storage. This is because ethylene accumulates in the sealed storage atmosphere. This stimulates the less advanced fruits to initiate ethylene production and eventually levels of ethylene in excess of 100 ppm are commonly found.

Recent research (11, 12, 13) as well as our own has shown that ripening of apples, providing the fruits haven't begun autocatalytic ethylene production, can be markedly delayed during CA storage employing low oxygen levels of 1 to 1.5 percent and judiciously scrubbing ethylene from the storage atmosphere. Current devices for removing ethylene are ineffective or cost prohibitive. Some acti-

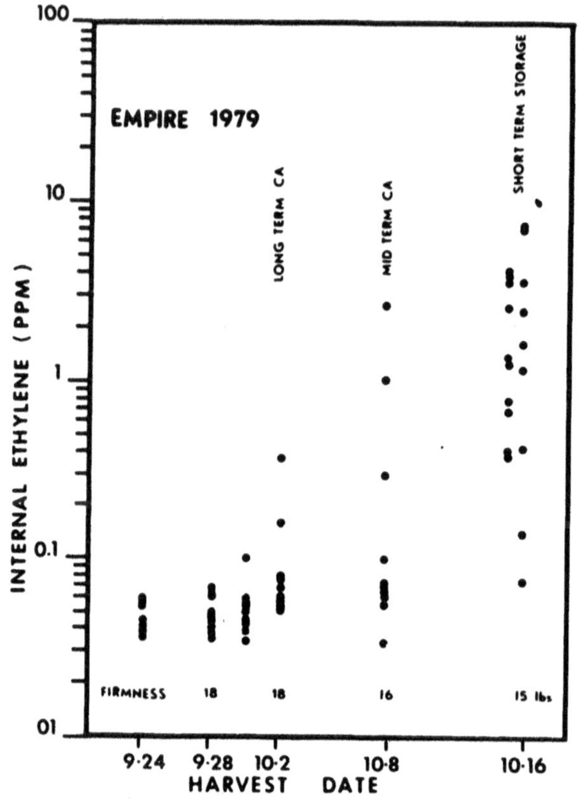

Fig. 6. Internal ethylene levels in Empire apples in relation to harvest date and assessment of storability. Flesh firmness values are given above the harvest dates.

vated carbon scrubbers used to remove carbon dioxide from CA rooms are marginally effective in reducing ethylene level in the storage atmosphere about 10-fold in comparison to the level in CA rooms scrubbed with dry lime (e.g. 50 vs. 500 ppm).

It is important to keep the ethylene level below 1 ppm in the CA storage room to effectively delay ripening. This is one of the primary benefits derived from hypobaric storage at absolute pressures of 0.1 atmos. or less (14). By 1977 it was generally concluded that other systems for removing ethylene from the storage atmosphere would be ineffective because the ethylene in the internal atmosphere of the fruit could not be significantly reduced. This concept is being reevaluated now that ultra-low oxygen levels (e.g. 1 to 1.5% O_2) are being employed for CA storage of apples and pears. And it is being recognized that fruits must enter CA storage before they enter their ethylene climacteric.

These considerations led to the development of a high capacity ethylene scrubber based upon catalytic oxidation (9). The system is hermetically sealed and consists in sequence, of a large air blower (ca. 200 cfm), a heater to raise the storage atmosphere to about 200°C, a monolithic catalyst similar to that employed for engine exhaust emission control, and a heat exchanger to cool the ethylene-free storage atmosphere as it is returned to the CA room. The ethylene scrubber was tested on a 20 thousand bushel CA storage of 'Red Delicious' apples in Michigan in 1980. Figure 7 shows that the scrubber effectively removed ethylene down to less than 1 ppm. The high initial ethylene level was due to inefficient propane combustion by a CA generator used for oxygen pull-down. Continuous operation of the scrubber kept the ethylene level in the CA storage below 1 ppm. Fruit ripening was prevented during the balance of the CA storage period. Whether or not this type of ethylene scrubber becomes a commercial reality will depend on the cost/benefit analysis.

HYPOBARIC STORAGE

Hypobaric storage deserves special consideration as new technology for postharvest preservation of perishables. Research was initiated in numerous laboratories around the world following the disclosure of this storage technology by Burg and Burg (2) in 1966 and was reviewed recently (3). Hypobaric or low pressure storage (LPS) is currently being commercially developed by the Grumman/ Dormavac Division of Grumman Allied Industries, Inc. The commodity is placed in a vacuum-tight and refrigerated container and evacuated by a vacuum pump to the desired low pressure which, depending upon the commodity may vary from 10mm Hg to 76mm Hg. The oxygen level varies in direct response to the absolute pressure. When the desired low pressure is obtained, fresh air is admitted to the chamber through a pressure regulator and then humidified. In the continuously ventilated partial vacuum, carbon dioxide, ethylene, and waste volatile by-products of metabolism rapidly diffuse out of the commodity and are flushed from the storage chamber. The oxygen content or partial pressure is set and varies within narrow limits by adjusting an absolute pressure regulator. Air, water, and electricity are the consumable items involved in the system. Hypobaric storage conditions can be achieved within minutes; so the chamber can be opened to inspect, remove or add commodity and then closed again without any deleterious effects.

Hypobaric storage is the most significant breakthrough in postharvest preservation technology since controlled atmosphere was introduced in England in the 1930's. Extensive investigations have been conducted on hypobaric storage of a wide range of fruits, vegetables and flowers. These studies have confirmed and extended Burg's

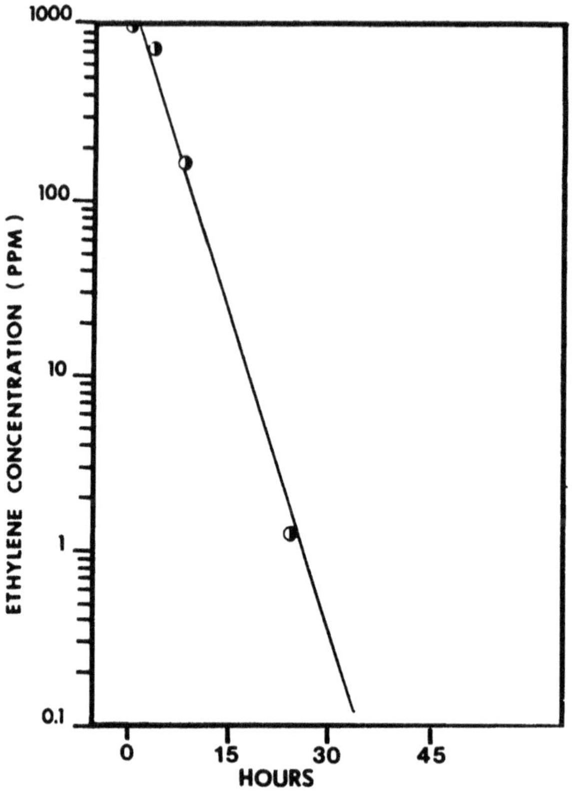

Fig. 7. Scrubbing ethylene from a 20,000 bu. CA storage of Red Delicious apples with a catalytic oxidizer at 200°C and 130 CFM. The high initial ethylene level was from incomplete oxidation of propane by an Arcat CA generator during oxygen pull-down.

hypothesis that hyponormal ethylene levels and low O_2 partial pressure in this rarefield environment are the factors responsible for delaying ripening and extending the storage life of fresh perishable commodities.

Hypobaric storage markedly extends the postharvest preservation period of a wide range of fresh fruits and vegetables beyond the useful period achieved by conventional storage methods. The principles of hypobaric storage apply to commodities which benefit solely by reducing the oxygen partial pressure and those which benefit additionally by lowering the internal atmosphere concentration of ethylene and other by-products of metabolism. Hypobaric storage maintains apples and pears in their pre-climacteric stage and markedly arrests ripening of post-climacteric fruits. Hypobaricallystored apples can become insensitive to ethylene action in ripening until they are acclimated in air at atmospheric pressure while under refrigeration. Hypobaric storage: retains sweetness in sweet corn, delays loss of ascorbic acid in asparagus, prevents internal browning of mushrooms, prevents scald of apples, maintains freshness of sweet cherries and blueberries, delays leaf abscission of leafy vegetables and retards bacterial soft-rot development.

REFERENCES

1. Dilley, D. R. Approaches to maintenance of postharvest integrity, J. Food Biochem. 2:235 (1978).
2. Burg, S. P. and E. A. Burg. Fruit storage at subatmospheric pressure, Science 153:314 (1966).
3. Lougheed, E. C., D. P. Muir and Luce Berad. Low pressure storage for horticultural crops, HortScience 13:21 (1978).
4. Sfakiotakis, E. M. and D. R. Dilley. Internal ethylene in apple fruits attached to or detached from the tree. J. Am. Soc. Hort. Sci. 98:501 (1973).
5. Frenkel, C., I. Klein and D. R. Dilley. Protein synthesis in relation to ripening of pome fruits, Pl. Physiol. 43: 1146 (1968).
6. Knee, M. Changes in structural polysaccharides of apples during storage. Coll. Inter. Centre Recherche Sci.: 238 (1975).
7. Sawamura, Masayoshi, E. Knegt and J. Bruinsma. Level of endogenous ethylene, carbon dioxide, and soluble pectin, and activities of pectin methyl esterase and polygalacturonase in ripening tomato fruits. Plant and Cell Physiol. 19: 1061 (1978).
8. Dilley, D. R. and W. W. Austin. The effect of Alar (N-dimethylamino succinamic acid) on maturation and storage quality of apples. 96th Ann. Rpt. Mich. Hort. Soc. p. 102 (1966).
9. Dilley, D. R. Assessing fruit maturity and ripening and techniques to delay ripening in storage. 110th Ann. Rpt. Mich. State Hort. Soc. p. 82 (1972).
10. Dilley, D. R., J. Lee and M. E. Saltveit, Jr. Measuring fruit ethylene concentrations for proper harvest and storage decisions. 108th Ann. Rpt. Mich. State Hort. Soc. p. 121 (1978).
11. Forsyth, F. R., C. A. Eaves and H. J. Lightfoot. Storage quality of McIntosh apples as affected by removal of ethylene from the storage atmosphere. Can. J. Plant Sci. 49:567 (1969).
12. Stoll, K., F. Hansen and D. Datwyler. The ripening of apples in CA storage as affected by reduction of the ethylene content of the atmosphere. Coll. Inter. Centre Nat'l. Recherche Sci. No. 238:81 (1975).
13. Liu, F. Interaction of daminozide; harvesting date, and ethylene in CA storage on 'McIntosh' apple quality. J. Am. Soc. Hort. Sci. 104:599 (1979).

METABOLISM, HEAT TRANSFER AND WATER LOSS UNDER HYPOBARIC CONDITIONS

Stanley P. Burg and Robert Kosson
Grumman/Dormavac Division
Grumman-Allied Industries
90 Crossways Park Drive W.
Woodbury, NY 11797

INTRODUCTION

Evaporative cooling is the dominant mode of heat transfer under hypobaric conditions. This creates a dilemma in water conservation for plant products. Can they dispel their respiratory heat and remain cold without drying excessively, or does water loss ultimately limit storage? A nodal model for heat transfer is used to analyze this problem. The present report summarizes results obtained with this model, and in addition discusses several other important differences between hypobaric and atmospheric pressure storage.

DESCRIPTION OF THE HYPOBARIC SYSTEM

The hypobaric method requires a vacuum tank to store the cargo, a vacuum pump continuously withdrawing air from the tank, a pressure regulator leaking air into the tank at a rate sufficient to maintain the set pressure, a humidifier which injects enough cool steam into the incoming rarified air to saturate the atmosphere in the vacuum tank, and a refrigerator to control the cargo temperature (Fig. 1).

The vacuum/water subsystem of Grumman/Dormavac's intermodal container is illustrated in Figure 2. The dew-point of air passing through the vacuum tank is controlled by setting the electrical energy input to the humidifier. Boiling takes place at a low temperature, between 12 to 47°C when the pressure is 10 to 80 mm Hga. The vacuum pump, a dry, lobed-blower super-charging a water sealed stage, is continuously cooled to the temperature of the container walls. During isothermal compression most of the water contained in air leaving the container is condensed and recovered in the

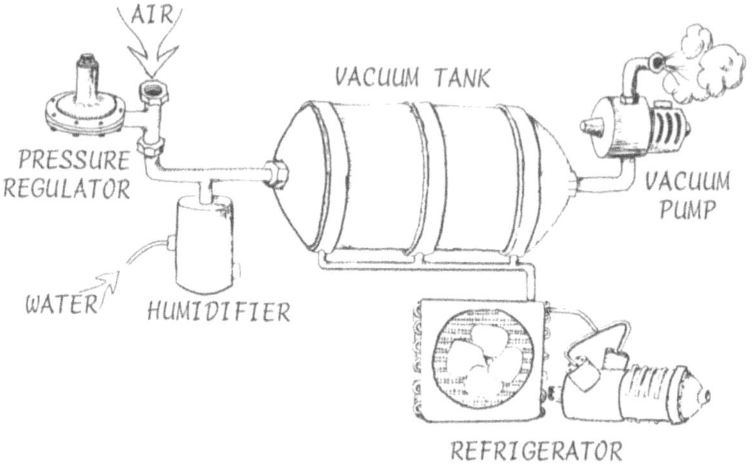

Fig. 1 - Schematic diagram of hypobaric system

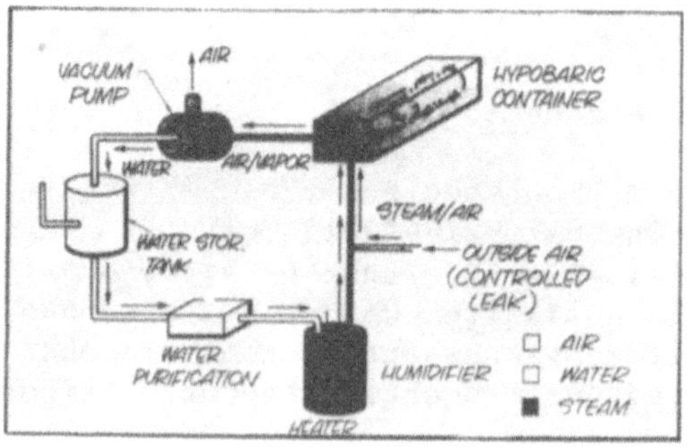

Fig. 2 - Vacuum/water subsystem

water-seal stage. This distilled water is gravity drained into a storage tank and passed through a deionizer before it is supplied to the boiler on-demand. The deionizer only needs to remove salt from the initial supply of stored water. The boiler is further protected from scale build-up by automatically draining whenever the container is shut-down. This nearly-sealed system conserves enough water to operate continuously for 6 weeks at the most demanding combination of pressure and temperature, with only 50 gallons of water initially stored.

The refrigeration subsystem is illustrated in Figure 3. The inner aluminum surface of the container is used as a cold-plate, with refrigerated glycol flowing through closely spaced integral wall tubes, entering and leaving in paired-tubes to eliminate front-back temperature gradients. Wall gradients are $\pm$ 0.6°C compared to the average wall temperature. This makes it possible to maintain the RH close to 96%. (In the hypobaric pressure range, the dew-point and wet-bulb are equivalent, and both can be converted to RH from a wet-bulb table.) There is no mechanical air-moving system, except the vacuum pump, which draws air diagonally through the container's cross section at a net velocity of 0.3 ft/min from the wall-roof juncture at one side to the floor-wall juncture at the other. The vacuum pump supplies approximately 100 cfm, sufficient to change the air in an empty container 3.3 times per hour independent of pressure and temperature.

DIRECT EFFECTS OF PRESSURE ON METABOLISM

Lowering the atmospheric pressure reduces the cellular hydrostatic pressure, thereby decreasing the cellular water potential. Although water participates in many biochemical reactions, and the rates of these reactions are controlled in part by the cellular activity of water, reaction rates are only slightly altered by a pressure reduction. The relationship between water activity (a) and water potential (Ψ_ω) is:

$$\Psi_\omega = (Rt/V_1) \ln a \qquad (1)$$

Ψ_ω = Water potential
R = Gas Constant
T = Temperature
V_1 = Liquid volume
a = Water Activity

From this equation it can be predicted that changing Ψ_ω by 20 atmospheres only changes (a) by 3 to 4%.

Changes in turgor pressure (Ψ_p) also have no direct effect on chemical and metabolic processes. The electrochemical potential

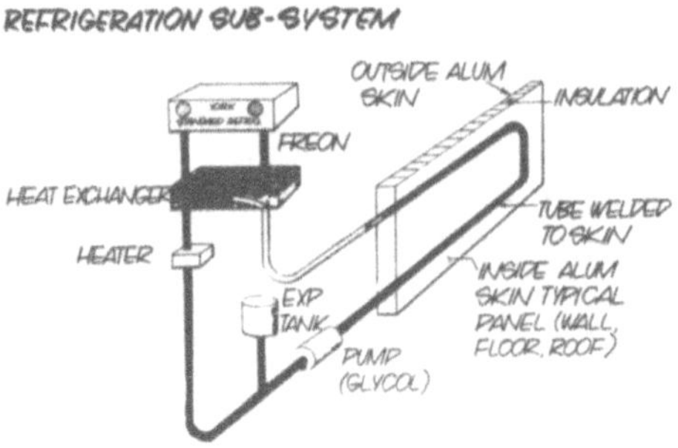

Fig. 3 - Refrigeration subsystem

of ions is virtually insensitive to physiological shifts in pressure (about 0.2 mV per atmosphere). To markedly effect the rate of chemical or metabolic reactions involving solid or liquid phases, more than 100 atmospheres of pressure are required. Therefore shifts in turgor can only influence metabolism by altering cellular structures.

Cells typically possess a pressure sensing system which responds by negative feedback to a change in hydrostatic pressure, setting in motion a series of events which restore the original hydrostatic pressure. The mechanism by which turgor is sensed is thought to be a thickening or thinning of the cellular membrane (11), in response to which the rate of passive permeability and active transport of potassium ion are altered in directions which restore the original turgor (19,21,48,55,56). This pressure sensing-negative feedback system participates in such processes as osmoregulation, opening and shutting of stomata, and the movement of leaves on sensitive plants. Another process dependent upon turgor is the growth of walled cells (43). Turgor affects cell expansion in two ways: a pressure in excess of a critical value is necessary for biochemical modification of the cell wall properties, and the subsequent physical extension of the wall also requires turgor. Growth of fungi and bacteria precedes by processes similar to those in higher plants, so far as is known, with the exception that the wall materials are different and growth seems not to be under hormonal control.

Two methods have been used experimentally to modify turgor: osmotic up- or down-shock, and changes in atmospheric pressure. Similar results are obtained in both cases. Increasing the atmospheric pressure by 1 atmosphere doubles the rate of potassium efflux and reduces potassium influx by two-thirds in Valonia cells (19). Hypotonic solutions or water enhance and hypertonic solutions inhibit solute leakage in a wide variety of higher plant products (10,14,22,31,37,44,51) and these treatments have an opposite effect on potassium influx (14). Animal cells and bacteria osmoregulate in like manner (23). Extreme turgor shock with 0.5 M glucose under appropriate conditions causes the release of many of the components of the active transport system of bacterial cells (23). During the upshock period the cells become very leaky, and can be loaded from external solutions with molecules which normally would not permeate. Cells in the log phase of growth are particularly sensitive to turgor shock.

The freezing point of mushrooms, roses, lamb and chrysanthemum cuttings is elevated nearly 0.5°C under hypobaric conditions. There is no obvious physicochemical basis for this shift, so the explanation must be biological. The first water to freeze usually is extracellular where the solute concentration is lowest (40). If the product osmoregulates in response to a decrease in atmospheric pressure, it will withdraw solute from the extracellular fluid, thereby increasing the freezing point of that water.

Products stored under hypobaric conditions typically are not growing, but microbes must grow to cause decay, and therefore they are subject to inhibition in response to a pressure-induced change in turgor. Cells of E.coli in the resting state possess no turgor and do not osmoregulate. Their contents resemble that of the media in which they are bathed. They begin to transport actively and osmoregulate during the stationary phase of growth. The osmotic content of such cells typically is about 0.15 M, mainly potassium ion. In response to a pressure reduction to $<$ 80 mm HgA, these cells must increase their osmotic content by nearly 0.045 M. Computation based on an estimate of the respiratory energy available at a low pressure and temperature, presuming one ATP per potassium taken up, suggest that even psychrophylic bacteria would require many hours to restore their original osmotic balance, a period roughly comparable to their generation time. Subsequently, they can develop the additional turgor needed for growth, and thereafter hypobaric storage should not affect their rate of multiplication except by influencing the pO_2 and pCO_2 available to support growth. If at a later time, however, the pressure is returned to atmospheric, the cell now would need to leak out nearly 0.045 M solute. Theoretically, by adjusting the frequency of a pressure cycle it should be possible to induce bacteria to pump and leak often enough to consume all of their respiratory energy, with too little left over to support the synthetic processes needed for growth. In a preliminary test of this concept during 8 weeks at 13°C, mature green tomatoes on a pressure cycle of 22 hours at 80 mm Hg - 2 hours at 760 mm HgA, developed 90% less decay than fruits continuously at 80 mm HgA. Surprisingly, the rate of ripening during storage was slightly lower in the cycled fruits.

Another direct effect of pressure reduction is to decrease the boiling point of certain volatile metabolic products. At 0°C, acetaldehyde boils when the pressure is less than 320 mm HgA. Data for ethanol is given in Table 1.

PRESSURE-INDUCED CHANGES IN GASEOUS COMPOSITION

The P_{CO_2} (atm) in the vacuum tank's air depends on the total pressure (P) and water vapor pressure (P_v):

$$P_{CO_2} = .00035 \quad (P - P_V) \qquad (2)$$

At atmospheric pressure, even when the product is ventilated adequately with fresh air, its internal CO_2 is much higher than that in the ambient air due to restricted outward diffusion of respiratory gas. In addition, CO_2 accumulates in the air of "sealed" Controlled Atmosphere (CA) and Modified Atmosphere (MA) storages, where it may be controlled at a pressure as high as .05 atm, but cannot be reduced below .005 atm with the best chemical or mechanical scrubbers. Under hypobaric conditions the internal CO_2 is decreased in proportion to the pressure reduction because of increased gaseous diffusivity.

Elevated CO_2 has the advantage of suppressing ethylene action by competitive inhibition, and in addition directly inhibits respiration. The first advantage is not beneficial under hypobaric

Table 1. Vapor Pressure of Ethanol at Various Temperatures

Temperature (°C)	Vapor Pressure (mm HgA)
-1	11
0	12
2	13
10	20
13	25

At each indicated temperature, if the total pressure is equal to or lower than the corresponding vapor pressure value in the table, ethanol boils from the product. For example, at 0°C, 10 mm Hg, neither ethanol nor aldehyde can remain in the tissue, and this conceivably could have a beneficial effect on freezing damage or low O_2 (fermentation) injury.

conditions because the outward diffusivity of ethylene from the tissue is so enhanced that the internal concentration is kept below a biologically active level (7,8). With many products, any accumulation of CO_2 is harmful, producing disorders such as scald, and in general a P_{CO_2} greater than .05 mm atm is not desirable.

Very high concentrations of CO_2, ranging from 30 to 70% at atmospheric pressure, have been used to control bacterial decay, especially in meat and fish storage. But before this practice became common it was known that many bacteria require CO_2 to grow, and that their multiplication is greatly enhanced by CO_2 in the range of .0001 to .001 atm (12,38). In fact, CO_2 incubators are commonly used to promote bacterial growth! Conceivably, the low CO_2 produced under hypobaric conditions may be beneficial for controlling bacterial decay.

The P_{O_2} in the vacuum tank's air also is related to P and P_v:

$$P_{O_2} = 0.21\ (P - P_v) \qquad (3)$$

Regulation of P $\pm$ 1 mm Hg by an altitude compensated vacuum breaker (Figs. 1 and 2), and P_v by a precise, controlled boiler wattage, is accurate enough and the flow of make-up fresh air fast enough to control the P_{O_2} of the vacuum tank's air $\pm$ .0003 atm, irrespective of the product's O_2 consumption. This contrasts favorably to the best control under CA conditions, P_{O_2} $\pm$ .05 atm. Moreover, due to the leak-tightness of the vacuum tank and associated systems, O_2 can be kept close to complete anaerobiosis, whereas CA and MA cannot bring the P_{O_2} lower than .05 atm. This feature of the hypobaric system makes it suitable for use with meat and fish, which require nearly complete anaerobiosis to prevent pigment oxidation and to retard aerobic and microaerophylic bacterial growth.

Low O_2 damage occurs under CA conditions when the P_{O_2} is less than .01 to .02 atm, but under hypobaric conditions products often store best at pressures in the 10 to 20 mm HgA range where the P_{O_2} is .001 to .003 atm. This difference in behavior probably is caused by an effect of pressure on gaseous diffusivity. At atmospheric pressure the inward diffusivity of O_2 is slow enough compared to the tissue's O_2-consumption to create an O_2 gradient of .005 to .02 atm between the product surface and center. Apparently low O_2 damage occurs when the center becomes almost anaerobic. Under hypobaric conditions the gradient is lowered in relation to the pressure reduction, so that the surface and center have a closely similar P_{O_2}. Even at 10 mm HgA this P_{O_2} is not low enough to markedly inhibit cyanide sensitive respiration mediated by cytochrome oxidase, and therefore does not usually induce substantial fermentation, but it profoundly inhibits the cyanide insensitive alternative pathway of respiration, reducing the production of peroxide without depleting the energy supply or inducing fermentation (47). A further advantage gained is that the entire product, surface and center, is brought to an optimal P_{O_2} for storage, and in addition the surface P_{O_2} is lowered enough to inhibit the growth of aerobic molds and bacteria without damaging the product's center.

HEAT LOADS UNDER HYPOBARIC CONDITIONS

The only significant heat which has to be transferred at steady-state from the product to the cold plate or changing air is the respiratory load. This is much smaller in a partial vacuum than it is at atmospheric pressure because less O_2 is available to support respiration. The amount of heat produced can be computed by comparing the O_2 or CO_2 content of the air entering the container with that leaving at the vacuum pump exhaust, assuming 673 Kg-Cal of heat produced per 6 moles of O_2 consumed or CO_2 evolved. Data obtained with carnations, corn, apples and mushrooms show a closely similar relationship between CO_2 production, O_2 consumption and pressure, and an R.Q. close to unity even at very low pressures (Figure 4). Using these data as a nomograph, values at atmospheric pressure (50) have been adjusted to estimate hypobaric heat loads at optimal pressure and temperature (Table 2).

Heat infiltration and controlled (and uncontrolled) in-leakage contribute little to the heat load. The pounds of air which have to be passed through a container under hypobaric conditions to effect an air change are much less than those needed at atmospheric pressure, in proportion to the pressure reduction. Therefore the amount of sensible heat admitted per air change is correspondingly lower. In addition, the incoming air decreases in RH when it expands because now the same amount of water vapor is present in a much larger volume. This air does not reach its dew-point even when it is cooled to 0°C, and no latent heat must be removed. Assuming ambient conditions of 80% RH at 38°C, only 79 Btu/Hr must be removed to change the air 3.3. times each hour at an interior temperature of 10°C and a pressure of 20 mm HgA. At atmospheric pressure 20,000 Btu/Hr would have to be rejected to accomplish the same rate of air change at 10°C. Uncontrolled in-leakage is so small relative to the controlled leakage even at 10 mm HgA that it can be neglected. Heat infiltrating from outside is captured at the hypobaric container's cold plate before it enters the cargo area, and therefore does not normally reach the product. At atmospheric pressure in a conventional forced air container this heat infiltrates into the cargo area and warms the product.

NODAL MODEL OF HEAT TRANSFER

A schematic model for the transfer of heat from a papaya fruit (P) contained in a box (B) to the container air (C) and container

Table 2. Respiratory Heat Loads at Optimal Storage Pressure and Temperature

Product	Pressure (mm HgA)	Temp (°F)	BTU/Ton/Day
Apple	50	32	252
Asparagus	10	32	2200
Avocado	20	50	2800
Banana	40	56	1656
Carnation (bloom)	15	36	1950
Chrysanthemum (bloom)	10	32	462
Corn	50	32	2685
Cucumber	80	55	1050
Lettuce	10	32	550
Lime	150	50	780
Mango	80	55	1500
Melon (Honeydew)	20	50	1400
Mushroom	10	32	1760
Papaya	20	50	625
Pepper (sweet)	80	50	1848
Pineapple	20	50	385
Rose (bloom)	10	32	2376
Strawberry	10	32	726

The heat load at atmospheric pressure and the indicated temperature (50) has been corrected to an optimal pressure according to Figure 2.

wall (W) is illustrated in Figure 5. The model assumes uniform temperatures for the papaya (T_P), air in the box (T_A), dew-point in the box (T_{DP}, A), box wall (T_B), container air (T_C), container dew-point ($T_{DP,C}$) and container wall (T_W). Other symbols used are for papaya respiratory heat output ($\dot{Q}_G$), convective heat transfer from the papaya to the box air ($\dot{Q}_{C,PA}$). convective heat transfer from the box air to the box wall ($\dot{Q}_{C,AB}$), convective heat transfer from the box wall to the container air ($\dot{Q}_{C,BC}$), radiation from the papaya to the box wall ($\dot{Q}_{R,PB}$), radiation from the box wall to the container wall ($\dot{Q}_{R,BW}$), evaporative heat flux ($\dot{Q}_{V,PA}$) from the papaya to the box air, and evaporative heat flux ($\dot{Q}_{V,AC}$) from the box air to the container air. All convective couplings are presumed to be natural convection.

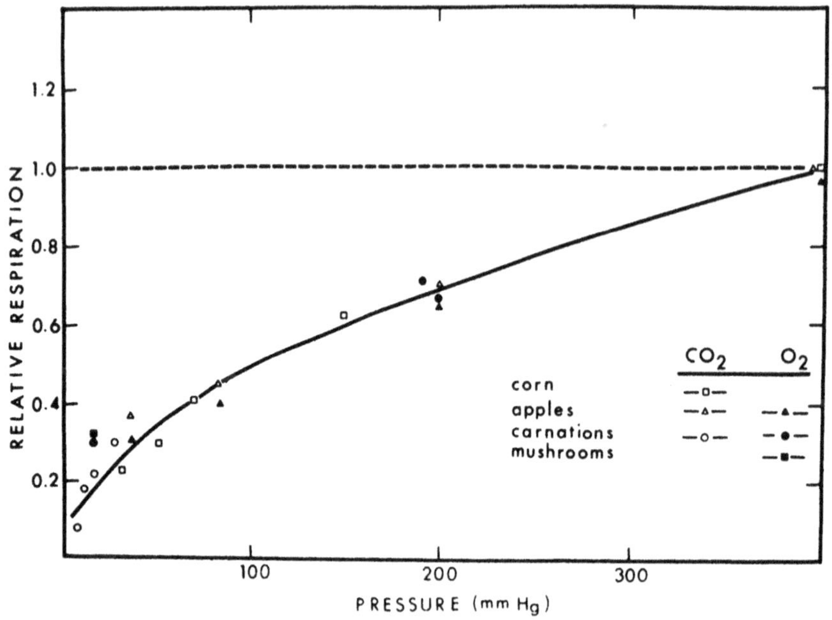

Fig. 4 - Relation between respiration rate and ambient pressure.

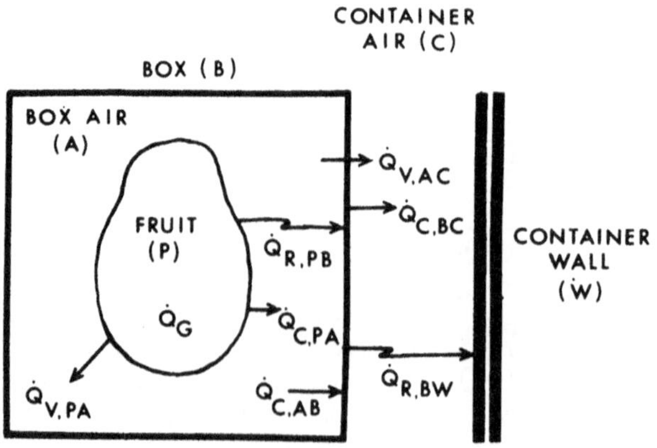

Fig. 5 - Schematic model for heat transfer

The radiative coupling from the box to the all is set = 0 for interior boxes.

The corresponding nodal model, shown in Figure 6, involves the determination of 4 nodal temperatures (T_P, T_A, $T_{DP\ A}$, T_B), given a specified respiratory heat source applied to node P, and three fixed boundary nodal temperatures (T_C, $T_{DP,C}$, T_W). There are seven temperature dependent heat transfer coupling -- three convective ($\dot{Q}_{C,PA}$, $\dot{Q}_{C,AB}$, $\dot{Q}_{C,BC}$), two radiative ($\dot{Q}_{R,PB}$, $\dot{Q}_{R,BW}$) and two evaporative ($\dot{Q}_{V,PA}$, $\dot{Q}_{V,AC}$). Equations for each of the heat transfer couplings are presented in the discussion which follows.

CONVECTIVE COUPLINGS

Convection from the papaya to the air is given by:

$$\dot{Q}_{C,PA} = K_1 \ (T_P - T_A) \quad (4)$$

with conductance

$$K_1 = h_{PA}A_P$$

when h_{PA} is the convective film coefficient and Ap is the surface area of the papaya. Values for h_{PA} are based on data correlations for natural convection from short vertical plates to air (33) in the form:

$$\frac{hL}{K} = aX^b \quad (5)$$

where

$$X = G_rP_r = \frac{L^3\rho^2 g\beta\Delta T}{\mu^2} \left(\frac{C_P\mu}{K}\right)$$

Constants a, b depend on the value of X as shown below:

X	a	b
1-10	1.44	0.120
$10-10^2$	1.37	0.141
10^2-10^3	1.20	0.170
10^3-10^4	1.04	0.190
10^4-10^9	0.59	0.250
10^9-10^{12}	0.13	0.333

The bouyancy force for natural convection arises from the changes in density, represented by the ($\rho\ \beta\Delta T$) product in the numerator, and, for a perfect gas in the absence of mass transfer, $\beta = 1/T$. With evaporation, there is an additional bouyancy term due to the lower molecular weight of water vapor compared with air. While

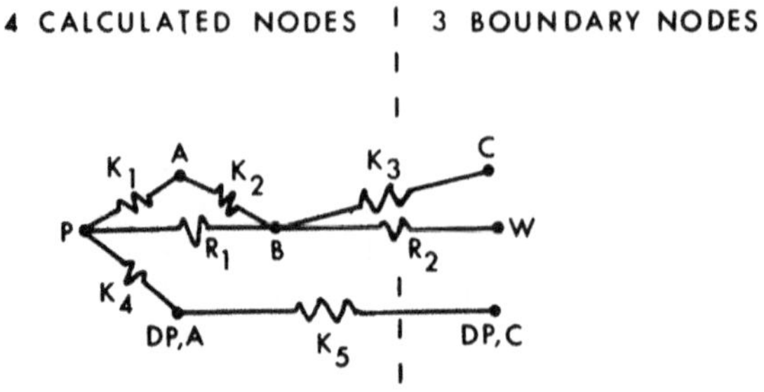

Fig. 6. Where $K_1, K_2, K_3, K_4, K_5, R_1$ and R_2 are the rate constants for convection and radiation respectively, as defined in Equations 4,17,18,21,23,25,27

small at atmospheric pressure, this can be significant at hypobaric conditions where the water vapor represents a much larger fraction of the container air.

The expression for β is then:

$$\beta = -\frac{1}{\rho}\frac{d\rho}{dT} = \frac{1}{T} - \frac{1}{M}\frac{dM}{dT} \tag{6}$$

where, for a mixture of air and water vapor:

$$M = M_a - (M_a - M_v)\frac{P_v}{P} \tag{7}$$

This expression for β can be written

$$\beta = \frac{1}{T}(1 + \alpha_1) \tag{8}$$

where the bouyance term, α_1, is given by

$$\alpha_1 = (M_a - M_v)\ \frac{CpK_{ge}}{h}\left(\frac{dP_v}{dt}\right)\left(\frac{T_P - T_{DP,A}}{T_P - T_A}\right) \tag{9}$$

In this expression, the ratio $(T_P - T_{DP,A})/(T_P - T_A)$ accounts for the fact that evaporation depends on dew point differences, while the β term is multiplied by the dry bulb temperature difference in the expression for Grashof number. K_{ge} is the effective mass transfer coefficient, defined by the relation:

$$\dot{m}_v = K_{ge}\ A_P\ (P_P - P_{DP,A})$$

For papaya, like all other fruits, the skin resistance, r_p, and the natural convective boundary layer resistance act in series, so that:

$$\frac{1}{K_{ge}} = \frac{1}{K_g} + \frac{R_\mu \, T r_P}{M_v} \tag{11}$$

where K_g is the mass transfer coefficient for the convection boundary layer. The mass transfer coefficient is related to the heat transfer coefficient through the expression (28):

$$K_g = \left(\frac{h}{C_P\rho}\right) \left(\frac{PM_v}{R_\mu T\bar{P}_a}\right) \left[\left(\frac{C_P\mu}{k}\right)\left(\frac{\rho D_v}{\mu}\right)\right]^{0.67} \tag{12}$$

where $\bar{P}_a$, the mean air partial pressure in the film, is close to the air partial pressure in the box. D_v, the binary diffusion coefficient, is inverse with mixture total pressure and proportional to the square root of temperature:

$$D_v = D_{vr}\frac{P_R}{P}\sqrt{\frac{T}{T_r}} \tag{13}$$

where the subscript R denotes the reference conditions for D_v:

Equation (12) can be rewritten in the form:

$$K_g = \frac{hM_v}{C_P M\bar{P}_a}\left[\left(\frac{C_P}{K}\right)\left(\frac{P_R D_{vR}}{R}\right)\sqrt{\frac{Tr}{T}}\right]^{0.67} \tag{14}$$

from which it can be seen that K_g is roughly inverse with $\bar{P}_a$.

Similarly the skin resistance, r_p, which is a function not only of the geometry of the skin, but also of the resistance to diffusive flow of vapor through the pores of the skin, can be expressed as:

$$r_p = r_{P,R}\frac{\ln\left[\frac{P_r - P_{v,o}}{P_r - P_{v,i}}\right]}{\ln\left[\frac{P - P_{v,o}}{P - P_{v,i}}\right]} \tag{15}$$

where $r_{P,R}$ is the skin resistance measured at a reference pressure P_R for particular vapor pressure values $P_{v,i}$ and $P_{v,o}$ inside and outside the skin, respectively. Most skin resistance measurements are made at 1 atmosphere. Using equation (15), skin resistance values under hypobaric conditions are computed to be one or two orders of magnitude lower.

Equations (9) - (15) permit evaluation of β using equation (8). Under hypobaric conditions, cases can arise in which the papaya is being heated by the air ($T_P < T_A$) and at the same time losing heat by evaporation ($T_P > T_{DP,A}$). In such cases, the sensible and evaporative bouyancy effects are in opposite directions, and it is necessary to set:

$$\beta = /\beta/ \tag{16}$$

to prevent computational difficulties in evaluating equation (5)

Convection from the air in the box to the box is given by:

$$\dot{Q}_{C,AB} = K_2(T_A - T_B) = h_{AB}(A_{BX} + A_{BE})\ (T_A - T_B) \qquad (17)$$

where A_{BX} is the portion of the surface area of a box within the stack which is exposed to a ventilating air passage (chimney), and A_{BE} is the additional box surface area for a box at the outer surface of the stack exposed to the container walls. Note that portions of the box surface which contact adjacent boxes in the stack are considered adiabatic and are excluded from the heat transfer analysis. The film coefficient h_{AB} for the inner surfaces of the box is simply taken equal to that of the external surface of the fruit, h_{PA}, in part because of the strong coupling between the two natural convective flows, but also because the internal packing and wrapping geometry is rather complicated for analysis.

Convection from external surfaces of the box to the container air is given by:

$$\dot{Q}_{C,BC} = K_3(T_B - T_C) = h_{BC}(A_{BX} + A_{BE})(T_B - T_C) \qquad (18)$$

where h_{BC} is the natural convective film coefficient for exterior surfaces of the box. Values of h_{BC} are determined using the same equations as for h_{PA}, except that the characteristic length is taken as the height of the stack and the bouyancy augmentation term is given by:

$$\alpha_2 = (M_a - M_v)\ \frac{C_P}{R_\mu r_{BOX} h}\ \left(\frac{dP_v}{dt}\right)\ \frac{(T_{DP,A} - T_{DP,C})}{(T_B - T_C)} \qquad (19)$$

where r_{BOX} represents the resistance to vapor flow of the box, expressed as a "skin" resistance.

EVAPORATIVE COUPLINGS

The heat transferred by evaporation from the papaya to the air in the box is given by:

$$\dot{Q}_{V,PA} = \dot{m}_v h_{fg} = K_4\ (T_P - T_{DP,A}) \qquad (20)$$

where the conductance:

$$K_4 = A_P\ K_{ge}\ \left(\frac{dPv}{dt}\right)\ h_{fg} \qquad (21)$$

and K_{ge} is the same mass transfer coefficient defined in equation(11).

The heat transferred by vapor flow through the box walls is given by:

$$\dot{Q}_{V,AC} = \dot{m}_v h_{fg} = K_5\ (T_{DP,A} - T_{DP,C}) \qquad (22)$$

where the conductance:

$$K_5 = A_B \frac{M_v}{R_\mu T r_{BOX}} \left(\frac{dP_v}{dt}\right) \quad (23)$$

In this case, A_B is taken as the total box surface area (including surfaces in contact with adjacent boxes). Box resistance, r_{BOX} is the same value as used in equation (19). Values of r_{BOX} were obtained directly from box tests under hypobaric conditions using simultaneous dew point measurements inside and outside the box. Because of the way the measurements were made, r_{BOX} combines both box wall resistance and the resistance of the convection films on the box surfaces, resulting in the slightly simpler form for K_5 compared with K_4.

RADIATIVE COUPLINGS

The heat transferred by radiation from the papaya to the box wall is given by:

$$\dot{Q}_{R,PB} = R_1\sigma\ (T_P^4 - T_B^4) \quad (24)$$

where σ is the Stefan-Boltzman constant (.1714 Btu/(ft^2-hr-$^oR^4$)) and the radiation coupling:

$$R_1 = F_{PB}\ A_P \quad (25)$$

F_{PB} is the grey body shape factor for radiant exchange between papaya and box. F_{PB} is set = 0.1 to account for the restrictive radiative path due to intervening layers of plastic and paper.

Radiant heat transfer from an exterior box surface to the container wall is given by:

$$\dot{Q}_{R,BW} = R_2\sigma(T_B^4 - T_W^4) \quad (26)$$

where

$$R_2 = F_{BW}A_{BE} \quad (27)$$

and F_{BW} is set = 1

SKIN RESISTANCE MEASUREMENTS

Gas exchange in plants has been studied extensively in connection with transpiration (water vapor movement), respiration and photosynthesis (CO_2 and O_2 movement), and ethylene evolution. The respective gases move through the plant product's skin in an air phase of restricted area, the stomates and lenticles, and then through a boundary air layer so the behavior of one gas might be inferred from that of others by comparing their diffusion coefficients in air. Water vapor diffuses twice as readily as ethylene and 1.7 times more rapidly than CO_2 in air.

In leaves, the resistance of the boundary air layer to water vapor ($r_{a,v}$) ranges from 0.4 to 3.5 sec^{-1} cm depending upon air velocity and the size and shape of the leaf (24,42,46); the minimal stomatal resistance ($r_{s,v}$) is 0.2 to 17.8 sec^{-1} cm at full opening (24,42,46); and the cuticular resistance to water ($r_{c,v}$) is 4 to 170 sec^{-1} cm (24). During hypobaric storage the stomata must be closed because the product has been harvested and is kept in the dark. Under these conditions air velocity has little or no influence because $r_{a,v}$ now constitutes a minor part of the total resistence (r_v), which depends mainly on $r_{c,v}$ (42).

The cuticular resistance to CO_2 is much higher than to water $r_{c,v}$ (24), and that to ethylene must also be high since the overall resistance to ethylene in various types of leaves, cotyledons, stems and buds under conditions where the stomata must be closed is large, ranging from 6000 to 18,000 sec^{-1} cm (1,2,5,9,15,18,26, 34,35,54). This is not due to a mesophyll resistance, as defined in ref. 24 for that term is intended to account for the additional resistance afforded by liquid water when gases penetrate through the cellular fluid prior to being metabolized. The ethylene values do not include this step in permeation since they are based on the rate of ethylene loss from the leaf surface and the ethylene gradient between the intercellular air spaces and atmosphere. Ethylene and CO_2 data clearly are not useful for estimating $r_{c,v}$ in leaves, so only values for water vapor have been used in the heat transfer computations.

The surface to mass ratio of leafy tissue is so large that even the highest reported value of $r_{c,v}$ (170 sec^{-1} cm) constitutes an insignificant constraint on water evaporation when account is taken of the influence of a low pressure on water mass movement. However, the surface to mass ratio of fruits is several orders of magnitude lower, and skin resistance might in this case significantly influence evaporative cooling and the temperature gradients which arise. Only a few measurements of overall resistance, r_v, have been made with fruits, but there are many measurements of the resistance to ethylene and CO_2 (3,4,6,7,12,27,29,30,32,39,41,49,52). Can these be used to compute r_v?

The skin resistance of apples to water vapor varies from 116 to 398 sec^{-1} cm depending upon variety and RH (16,25), and that of oranges from 56 to 77 sec^{-1} cm (36). These values are far lower than resistances to ethylene and CO_2, e.g. about 8000 sec^{-1} cm for apples (4,6,7,9) and 2000 sec^{-1} cm in oranges (3,7). Two dozen fruits tested all have resistances to ethylene in the range 1600 to

50,000 sec^{1-} cm and in ten cases where the resistances to CO_2 and ethylene have been compared, they are nearly identical.

Why is the resistance to water vapor far lower than that to other gases in fruits? The resistance of the isolated apple peel to water vapor (25) and to CO_2 and ethylene (7) is equal to and accounts for the entire resistance of the intact apple to these vapors. Isolated pulp slices have very little resistance to CO_2 (7), and when the peel is removed from the fruit the normal CO_2 and ethylene gradients in the pulp are dissipated (7,49). Therefore, an explanation must be sought in the properties of the peel.

Ethylene and CO_2 pass through the apple peel in an air phase of limited area. This is proven by the fact that the rate of passage of these gases out of an apple is inversely proportional to the atmospheric pressure, and also dependent upon the density of the constituents of the ambient atmosphere (7). If passage were limited by a liquid phase, the resistances to CO_2 and ethylene would be inversely proportional to their solubility in that phase; e.g. much higher to ethylene if the phase were water, much higher to CO_2 if it were lipid, and in no event would the resistance be similar to both gases, as is the case. Differential permeability is manifest even when water vapor and the other gases are applied at one side of an isolated apple peel and measured at the other side, so it is independent of "pore length." The only plausible explanation is that the lenticular and sublenticular chambers are partially occluded with water. This would decrease the pore area available for mass transfer of all gases except water vapor. Small amounts of water occluding intercellular spaces markedly increase the skin resistance to various gases (4,7,13,53); and when fruits ripen the intercellular spaces progressively occlude with water and the resistance to gas exchange increases (49). However, the resistance to water vapor movement in bananas does not increase, but rather the transpiration at all times is proportional to the respiratory heat generated, even when the internal O_2 falls to very low values, and the CO_2 accumulates to high levels during ripening (45).

In summary, the large amount of data available on the skin resistance of various fruits to ethylene and CO_2 cannot be used to compute a skin resistance to water vapor. For analytic computations values ranging from 400 to 1200 sec^{-1} cm at atmospheric pressure have been used in the papaya computation.

HEAT TRANSFER AND WATER LOSS: ANALYTIC RESULTS

The nodal model described previously was run for papaya fruit for conditions and properties as given below:

Container conditions:

Pressure	P_C	=	20 mm HgA
Air Temperature	T_C	=	52°F
Dew Point Temperature	$T_{DP,C}$	=	46 - 52°F
Wall Temperature	T_W	=	46 - 52°F

Papaya Properties:

Respiration at 20 mm HgA, $\dot{Q}_p$ = 625 Btu/ton/day
Density ρ = 69.3 pcf
Surface to Volume Ratio, A_p/V_p = 20.5 ft^2/ft^3
Skin Resistance at atmospheric pressure,
r_p = 400 - 1200 sec^{-1} cm
Characteristic Length for natural convection,
x_p = 0.5 ft

The papaya are assumed to be packed 10 lbs. per box. To determine surface areas, the boxes are assumed to have internal dimensions of 14" x 10" x 6", providing a total inside surface area of 2.94 ft^2. With the chimney stack arrangement, 10 high, 6 wide, and 32 long, there would be 1920 boxes, of which half would be interior boxes, surrounded by other boxes or interior gaps. The others, called exterior boxes, would see one or more walls.

The interior boxes would have the equivalent of 19" of side length and a 2" x 10" strip on top and bottom surfaces (total area = 0.86 ft^2) exposed to the container air. The remaining surfaces (2.08 ft^2) would be in contact with other boxes. The exterior boxes, on average, have an additional 0.78 ft^2 of surface area facing a wall.

The effective resistance to vapor transport of the box surfaces was determined from cool down tests with simultaneous internal and external dew point measurements. A value for r_{BOX} = 10.9 sec^{-1} cm was determined at 20 mm HgA, equivalent to a value of 660 sec^{-1} cm at 760 mm HgA.

Some representative results for an assumed skin resistance at one atmosphere of 400 sec^{-1} cm are shown in Table 3. The results for 100% relative humidity ($T_{DP,C}$ = 52°F) show that the papaya temperature is very close to air temperature, and the weight loss is only 0.51 to 0.61% for exterior and interior boxes, respectively, in 3 weeks. The results for 86% relative humidity ($T_{DP,C}$= 48°F) show a markedly higher weight loss, with the exterior box rising to 4.0%, well above the value of 1.2% calculated for the interior box in 3 weeks.

The high weight loss at 86% relative humidity is due to the low resistance to vapor loss of both the skin and box under hypobaric conditions (11 and 7 sec^{-1} cm respectively). The papaya tends to cool by evaporation to a temperature close to dew point temperature and well below the container temperature. In this condition the papaya receives heat by convection from the air in the box and by radiation from the box itself. This heat, in addition to the respiratory load, leads to a high evaporation rate. The exterior box receives more heat from its surroundings than does the interior box, leading to a higher weight loss.

Figure 7 (left and middle) shows the heat loss by mode for interior and exterior boxes as a function of dew point temperature. At 100% relative humidity ($T_{DP,C}$ = 52°F), the papaya is marginally warmer than its surroundings, with all three modes of heat transfer in the same direction, and evaporation the dominant mode of heat loss. At lower dew point temperatures, radiation and convection provide heat gains, and evaporation increases substantially. The results shown in figure 7 (middle) are for the wall temperature equal to the air temperature of 52°F. The low relative humidity in this case would be due to air being supplied to the container in a relatively dry state. The results in figure 7 (middle) indicate the importance of proper humidification to keep evaporation rates low.

If saturated air was supplied to the container hotter than the temperature of the walls, the container dew point would be kept very close to wall temperature by condensation on the walls. This case is illustrated in figure 7 (right). Exterior boxes, in this case, can radiate some heat to the container walls and hence are colder than interior boxes. The colder box provides less heat to the papaya, reducing the evaporation heat loss. The evaporation rate for interior boxes is shown in figure 7 (right) for comparison.

The effect of 3-fold increase in skin resistance is indicated by the dashed curves in figure 7 (left and middle). The effect is modest, because the 1200 sec^{-1} cm value at atmospheric pressure drops to only 21 sec^{-1} cm at 20 mm HgA. A somewhat stronger effect should be exhibited at higher storage pressures.

While not directly shown, the papaya weight loss during storage is directly proportional to the evaporative heat loss. Expressed as a percentage, for a 21 day storage period, the weight loss is simply the evaporative heat loss in Btu/hr multiplied by 4.73.

The weight loss problem with leafy products such as cut flowers is even more severe than that with fruits because the surface/mass ratio of the flowers is so much higher. To prevent excessive weight loss it is essential that the flowers not be allowed to decrease in temperature below that of the container air and wall. This can be accomplished by adding additional resistances such as a plastic wrap in series with the papaya skin and box resistances. The effect is

Table 3. Analytic Results, Papaya at 20 mm HgA

Computed Values	$T_{DP,C}$ = 52°F Interior Box	Exterior Box	$T_{DP,C}$ = 48°F Interior Box	Exterior Box
Papaya Temp., T_p	52.13	52.10	48.31	48.99
Box Air Temp., T_A	52.13	52.08	48.40	49.73
Box Temp., T_B	52.12	52.03	48.71	51.08
Box Dew Point, $T_{DP,A}$	52.03	52.02	48.06	48.16
21 Day Wgt. Loss, %	0.61	0.51	1.19	4.00

Container Air (T_C) = 52°F (11.11 C); Walls (T_W) = 52°F (11.11 C); Container Dew Point ($T_{DP,C}$) = 52°F (100% RH) or 48°F (8.89 C, 86% RH); Papaya Skin Resistance = 400 sec^{-1} cm @ 1 Atmosphere. 6.9 sec^{-1} cm @ 20 mm HgA.

similar to that shown in figure 7 (left and middle) for the two values of skin resistance. By selecting an appropriate total resistance of the box plus plastic wraps it is always possible to prevent weight loss in excess of that required to dispel respiratory heat without causing the product to overheat.

The respiratory process generates only about 9% of the water needed to remove respiratory heat by evaporative cooling. The remainder must be drawn from the cellular reserve, but there is a limit to how much can be lost without adverse consequences. Weight loss in excess of about 5% not only causes wilting, but also hormonal changes such as enhanced ethylene production and increased ABA levels, with concomitant loss of storage and shelf-life. Due to the marked reduction in heat generation under hypobaric conditions, water loss need not limit storage life provided it does not exceed that needed to dispel the product respiratory heat load.

If the only air movement in the container was the flow induced by the vacuum pump from the roof to floor, the dew-point of the air passing through a stack of papaya boxes would progressively increase in that direction. One might expect that the upper layers of the stack would be cooler and experience noticeably more weight loss than the lower layers. Observations made during a 21-day papaya test indiate that this does not occur. Since the net flow must be vertically downward, it would appear that there is significant recirculation within the stack, with downward flow in the narrow gap regions and near the box surfaces, and upward flow in the center of the chimneys, due to the bouyancy effect.

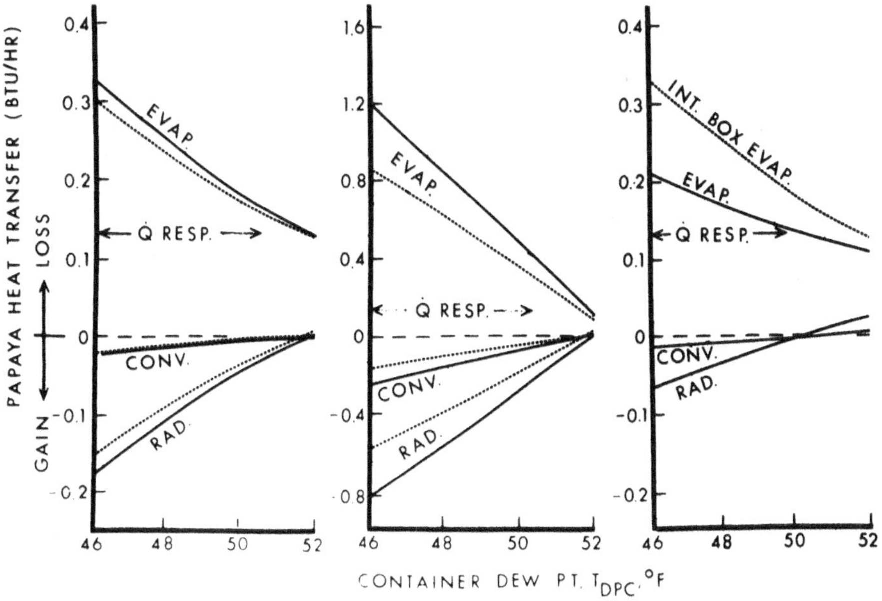

Figure 7. (LEFT) Papaya heat transfer by mode (Interior box). 10 lbs/box; r_{box} = 10.9 sec^{-1} cm at 20 mm HgA; $T_C = T_W =$ 52°F; 20 mm HgA pressure; r_{skin} at 1 atmosphere = 400 sec^{-1} cm (———) or 1200 sec^{-1} cm (-----).

(MIDDLE) Papaya heat transfer by mode (Exterior box). Same conditions as in Fig. 7 (left).

(RIGHT) Papaya heat transfer by mode. 10 lbs/box; r_{BOX}; r_{BOX} = 10.9 sec^{-1} cm at 20 mm HgA; $T_{wall} = T_{DPC}$; T_C = 52°F. Exterior box (———); Interior box (-----).

COOLDOWN

The mechanism discussed for steady-state condition also apply with minor reservations to the cooldown process. Theoretically, during evaporative cooling, the load loses about 1% of its weight per 5.6°C decrease in temperature to remove sensible heat. During a 6.7°C cooldown at 15 mm Hg, mushrooms lost the expected 1.0 - 1.2% of their weight in a container-load test.

In a cooldown of carnation flowers from 15.6° to 1.7°C with the pressure regulator set to open at 10 mm HgA, product water flashed at 12 mm HgA. During the next five hours the water vapor pressure in the container (computed from dew-point measurements), total container pressure, and vapor pressure of water at the product temperature all were equal and slowly declined as the product cooled. During this period the cooldown rate was limited only by refrigeration capacity. When the product's temperature decreased below 11.7°C, all pressure reached 10 mm HgA and the vacuum breaker opened. Thereafter, in the absence of product "boiling", the cooldown was much slower, limited by the rate of heat transfer from the product to the cold plate and air.

Vacuum cooling is used routinely and successfully with many products, but we do not recommend cooldowns in Dormavac which cause the product's water to "flash" and "boil." The duration of time that a product withstands anaerobic conditions depends in part on the temperature. Carnations are not damaged by 0.4 to 0.5% O_2 at atmospheric pressure during 6 to 9 weeks at 0.6°C, or 3 weeks at 7.2 to 10°C, but they are damaged in a matter of days above 10°C (20). Experimentally we have determined that under hypobaric conditions carnations withstand 2 to 3 hours at their boiling point when the temperature is 11.1 to 15.6°C, but in 5 hours they are damaged. With full loads it may take longer than 5 hours to bring the temperature below 11.1°C. We have also observed damage with chrysanthemum cuttings during hypobaric cooldown, but roses are more tolerant. To avoid this problem either the product should be precooled, or else it must be cooled in Dormavac at an initial set-pressure substantially higher than the vapor pressure of water at the initial product temperature, followed by a reduction in pressure after the product has cooled. The latter procedure provides a relatively slow cooldown, however, and, in our experience, precooling is clearly preferable.

REFERENCES

1. Aharoni, N., M. Lieberman and H. D. Sisler, Patterns of ethylene production in senescing leaves. Plant Physiol. 64: 796 (1979).

2. Baur, J. R. and P. W. Morgan, Effects of Picloram and ethylene on leaf movement in Huisache and Mesquite Seedlings, Plant Physiol. 44:831 (1969).
3. Ben-Yehoshua, S. and I. Eaks, Ethylene production and abscission of fruit and leaves of oranges, Bot. Gaz. 131:144 (1970).
4. Ben-Yehoshua, S., R. N. Robertson and J. B. Biale, Respiration and internal atmosphere of avocado fruit, Plant Physiol. 38: 194 (1963).
5. Beyer, E. M., Jr., and P. W. Morgan, Abscission: The role of ethylene modification of auxin transport, Plant Physiol. 48: 208 (1971).
6. Burg, S. P. and E. A. Burg, Role of ethylene in fruit ripening, Plant Physiol. 37: 179 (1962)
7. Burg, S. P. and E. A. Burg, Gas exchange in fruits, Physiol. Plantarum 18:870 (1965).
8. Burg, S. P. and E. A. Burg, Fruit storage at subatmospheric pressure, Science 153: 314 (1966).
9. Burg, S. P. and E. A. Burg, The interaction between auxin and ethylene and its role in plant growth, Proc. Natl. Acad. Sci. USA 55: 262 (1966).
10. Burg, S. P., E. A. Burg and R. Marks, Relationship of solute leakage to solution tonicity in fruits and other plant tissues, Plant Physiol. 39:185 (1964).
11. Coster, H. G. L., E. Steudle and U. Zimmerman, Trugor pressure sensing in plant cell membranes, Plant Physiol. 58: 636 (1976).
12. Dagley, S. and C. N. Hinshelwood, Physicochemical aspects of growth. I. Quantitative dependence of the growth rate of Bact. lactic aerogenes on the carbon dioxide content of the gas atmosphere, J. Chem. Soc. 1938: 936 (1938).
13. DeVaux, H., Etude experimentale sur l'aeration des tissues massifs, Ann. Sci. Nat. Bot. XIV: 297 (1891)
14. Enoch, S. and Z. Glinka, The effect of cell's turgidity on influx and efflux of potassium ion, Plant Physiol. 65:S-61 (1980).
15. Erez, A., The effect of different portions of the sunlight spectrum on ethylene evolution in peach (Prunus persica) apices, Physiol. Plantarum 39: 285 (1977).
16. Fockens, F. H. and H. F. Meffert, Biophysical properties of horticultural products as related to loss of moisture during cooling down. J. Sci. Food Agric. 23: 285 (1972).
17. Forsyth, F. R., I. V. Hall and H. J. Lightfoot, Diffusion of CO_2, O_2 and ethylene in cranberry fruit, HortScience 8:45. (1973).
18. Goeschl, J. D., L. Rappaport and H. K. Pratt, Ethylene as a factor regulating the growth of pea epicotyls subjected to physical stress, Plant Physiol. 41:877 (1966).
19. Gutknecht, J., Salt transport in Valonia: inhibition of potassium uptake by small hydrostatic pressures, Science 160:68 (1968).

20. Hanan, J. J., Experiments with CA storage of carnations, Proc Amer. Soc. Hort. Sci. 90:370 (1967).
21. Hastings, D. F. and J. Gutknecht, Turgor pressure regulation: modulation of active potassium transport by hydrostatic pressure gradients, in: U. Zimmerman and J. Dainty, Eds., "Membrane Transport in Plants," Springer-Verlag, Berlin, pp 79 (1974).
22. Helder, R. J., The loss of substances by cells and tissues (salt glands), Handb. d. Pflanzenphysiol. 2:468 (1956).
23. Hellebust, J. A., Osmoregulation. Annu. Rev. Plant Physiol. 27: 485 (1976).
24. Holmgren, P., P. G. Jarvis and M. S. Jarvis, Resistance to carbon dioxide and water vapor transfer in leaves of different plant species, Physiol Plantarum 18:557 (1965).
25. Horrocks, R. L., Wax and the water vapor permeability of apple cuticle, Nature 203: 547 (1964).
26. Kang, B. G. and P. M. Ray, Ethylene and carbon dioxide as mediators in the response of the bean hypocotyl hook to light and auxins, Planta 87:206 (1969).
27. Kosiyachinda, S. and R. E. Young, Ethylene production in relation to the initiation of respiratory climacteric in fruits, Plant and Cell Physiol. 16:595 (1975).
28. Kreith, F., "Principles of Heat Transfer," International Textbook Co., Scranton, PA, p. 500 (1958).
29. Lipe, J. A. and P. W. Morgan, Ethylene as a regulator of young fruit abscission, Plant Physiol. 50:S-18 (1972).
30. Lipe, J. A. and P. W. Morgan, Ethylene a regulator of young fruit abscission, Plant Physiol. 51:949 (1973).
31. Luttge, U., E. Ball and H. Greenway, Effects of water and turgor potential on malate efflux from leaf slices of Kalanchoe diagremontiana, Plant Physiol. 60:521 (1977).
32. Lyons, J. M., W. B. McGlasson and H. K. Pratt, Ethylene production, respiration and internal gas concentrations in cantaloupe fruits at various stages of maturity, Plant Physiol. 37: 31 (1962).
33. McAdams, W. A., "Heat Transmission," Third Edition, McGraw-Hill Book Co., Inc., New York, pp 172 (1954).
34. Morgan, P. W. and J. R. Baur, Involvement of ethylene in picloram-induced leaf movement response, Plant Physiol. 46: 655 (1970).
35. Morgan, P. W. and J. A. McAfee, A comparison of internal ethylene levels with production rates in various tissues of the vegetative cotton plant, Plant Physiol. 45:S-32 (1970).
36. Moreshet, S. and G. C. Green, Photosynthesis and diffusion conductance of the Valencia orange fruit under field conditions, J. Exptl. Bot. 31:15 (1980).
37. Munch, E., Die stoffbewegungen in der pflanze, Verlag Gustav, Fischer, Jena, p. 23 (1930).
38. Pappenheimer, A. M. and G. A. Hottle, Effect of certain purines and CO_2 on growth of strain of group A haemolytic Streptococ-

cus, Proc. Soc. Expt. Biol. Med. 44:645 (1940).

39. Pharr, D. M. and A. A. Kattan, Effect of air flow rate, storage temperature and harvest maturity on respiration and ripening of tomato fruits, Plant Physiol. 48:53 (1971).

40. Potter, N. N., "Food Science," Avi Publishing Company, Westport, Conn., pps. 204 (1973).

41. Pratt, H. K. and J. D. Goeschl, The role of ethylene in fruit ripening, in: F. W. Wightman and G. Setterfield, eds., "Biochemistry and Physiology of Plant Growth Substances," Runge Press, Ltd., Ottawa, Canada, pp. 1295 (1968).

42. Raschke, K., Heat transfer between the plant and environment, Annu. Rev. Plant Physiol. 11:111 (1960).

43. Ray, P. M., P. B. Green and R. Cleland, Role of turgor in plant cell growth. Nature 239: 163 (1972).

44. Simon, E. W., Leakage from fruit cells in water, J. Exptl. Bot. 28:1147 (1977).

45. Simmonds, N. W., "Bananas," 2nd Ed., Longmans, Green and Co., Ltd., London, p. 222 (1966).

46. Slaytor, R. O. and J. F. Bierhuizen, Transpiration from cotton leaves under a range of environmental conditions in relation to internal and external resistances, Austr J. Biol. Sci. 17:115 (1974).

47. Solomos, T., in press, Respiratory and energy metabolism in relation to senescence, M. Lieberman ed., "Post Harvest Physiology and Crop Preservation," Plenum Press, New York (1982).

48. Steudle, E. and U. Zimmerman, Turgor pressure regulation in algal cells: pressure dependence of electrical parameters of the membrane in large pressure ranges, in: U. Zimmerman and J. Dainty, eds., "Membrane Transport in Plants," Springer-Verlag, Berlin, pp.72 (1974).

49. Trout, S. A., E. G. Hall, R. N. Robertson, F. M. V. Hackney and S. M. Sykes, Studies in the metabolism of apples. I. Preliminary investigations on internal gas composition and its relation to changes in stored Granny Smith apples Austr. J. Expt. Biol. Med. 20-21: 219 (1942).

50. USDA, The commercial storage of fruits, vegetables and florist and nursery stocks, Agr. Handbook No. 66 (1968).

51. Wachter, W., Untersuchungen uber den austritt von zucheraus den zellen der speicherorgane von Allium cepa und Beta vulgaris, Prings. Jahr. fur wiss. Bot. 41:165 (1905).

52. Wardlaw, C. W. and E. R. Leonard, Studies in tropical fruits. IX. The respiration of bananas during ripening at tropical tempratures, Ann. Rev. Botany 4:264 (1940).

53. Woolley, J. T., Potato tuber tissue respiration and ventilation, Plant Physiol. 37:793 (1962)

54. Zeroni, M., P. H. Jerie and M. A. Hall, Studies on the movement and distribution of ethylene in Vicia faba L., Planta 134: 119 (1977).

55. Zimmerman, U. and Steudle, The pressure dependence of the hydraulic conductivity, the membrane resistance and membrane potential during turgor pressure regulation in Valonia ultricularis, J. Membrane Biol. 16:331 (1974).
56. Zimmerman, U., E. Steudle and P. I. Lelkes, Turgor pressure regulation in Valonia utriculares. Effects of cell wall plasticity and auxin, Plant Physiol. 58:605 (1976).

MAINTAINING NUTRITIONAL AND PROCESSING QUALITY IN GRAIN CROPS DURING HANDLING, STORAGE, AND TRANSPORTATION

Philip C. Williams

Canadian Grain Commission
Winnipeg, Canada

INTRODUCTION

Production of grain in excess quantities sufficient for export and distribution to other areas where it is in demand for human and animal food is only practiced in a few areas of the World at present. North America, including Canada, Australia, and Argentina account for over 95% of the World's grain exports. The development of other areas with the potential of growing grain surpluses and the distribution of this grain to food-poor regions of high population density is essential to lasting World peace. Significant steps are presently in progress to achieve this aim. The initiation and continuity of the "Green Revolution" has been largely responsible for dramatic improvements in the grain production of countries such as Mexico and Turkey, and in the establishment of International Centers for Agricultural Research, such as C.I.M.M.Y.T., I.C.R.I.S.A.T., I.R.R.I., I.I.T.A., and I.C.A.R.D.A.,[1] which are aimed at the improvement of agricultural production in the regions covered by their respective mandates. The successful completion of such projects depends not only upon the production of continually increasing quantities of grains, pulses, root crops, and other food raw materials, but also (a) on their proliferation at

[1]C.I.M.M.Y.T. - Centro Internacionale por Majorifiacion de Mais y Trigo.
I.C.R.I.S.A.T. - The International Centre for Research in the Semi-arid Tropics.
I.R.R.I.- The International Rice Research Institute.
I.I.T.A.- The International Institute for Tropical Agriculture.
I.C.A.R.D.A. - The International Centre for Agricultural Research in the Dry Areas.

quality levels suitable for processing and acceptability in socio-economic patterns, (b) on the provision of incentives to growers of the increased quantities of food by developing an efficient and equitable system for marketing, distribution, and processing of the food materials which is both economically and politically acceptable, and (c) on the development and maintenance of practicable methods for transportation of the food materials from their source of origin to the points of distribution.

GRAIN TRANSPORTATION SYSTEMS

An efficient system for transportation and distribution must consist of several components, which are summarized in Table 1. This paper will limit its content to matters concerning the quality and handling of grains and seeds of all types. The most formidable task facing the overall improvement of agricultural production and distribution of food materials unquestionably features post-harvest technology, rather than the relatively more simple aspect of growing two grains where one grew before. The biggest reasons for this lie (a) with the complexity of the concept, outlined briefly in the first paragraph, and also (b) with the cost of improving post-harvest technology. We are not asking for ploughs, drills, tractors, combine harvesters, or even the extra staff to operate them. These are relatively inexpensive. Instead post-harvest technology calls for building grain silos of many different sizes and degrees of complexity, also building roads and railroads,

Table 1. Components of a grain transportation system

1. Collection of grains from farms into consolidated deposits.
2. Facilities for storage--short- and long-term.
3. Loading, unloading, and conveying systems.
4. Methods of packaging or bulk handling.
5. Roads, railways, and waterways.
6. Systems for grading the commodities into categories of different visual quality and establishing equitable price scales.
7. Systems for servicing and maintaining equipment and facilities for all aspects of the system.
8. Systems for recruiting and training personnel for operation and administration.
9. Systems for education and extension of information to farmers, grain merchants, and other personnel involved with the overall handling operation.

improving wharf facilities, and establishing marketing and pricing systems and concepts which do not conflict with the individual viewpoints and aspirations of the people called upon to formulate them. In other words, post-harvest technology is expensive and complicated, a very tough job. At present, losses of grain annually amount to about 10-20% of World production. This proportion will certainly increase if post-harvest technology does not keep pace with increased production. In order to derive full benefits from the wealth of expertise and equipment presently devoted to increased crop production, post-harvest technology must improve at a much greater rate than crop production itself.

GRAIN QUALITY

Quality in grains is a term which may be subdivided into at least three categories; namely, nutritional quality, processing quality, and marketing quality. These in turn are also divisible into different aspects of each quality category, and the overall picture is summarized in Table 2. These aspects of quality are not listed in order of importance since all are of equal importance. Nutritional quality includes actual eating quality and is affected by chemical composition, texture, color, flavor, and general appearance of the grain. It can also be seriously affected as a result of infestation with fungi, insects, or rodents, which themselves may produce toxic alkaloids and other substances. Processing quality means the usefulness of the grains for industrial or domestic processing into other substances which may be the end products themselves, such as edible oils, or intermediate products such as flour, semolina, or malt. Finally, the marketing of grains can probably be more seriously affected by storage and transportation than either nutritional or processing quality, since marketing involves vendor-customer relationships which may depend on rather subjective evaluation of the potential quality of the commodity before negotiations are completed.

Nutritional Quality

Changes in nutritional quality of a grain as a result of storage and transportation are easier to define than those in processing or marketing quality. The chemical composition of the grain has usually been established before purchase or assignment of the grain to a particular purpose, so after storage and delivery and prior to use, the grain must be inspected to verify that it meets the specifications defined for its particular end use. If it does not meet the specifications, either it can be used only by modifications of processing technique or diet formulation, or it cannot be used for the purpose originally specified, and must be assigned to other uses. The moisture content of grains is a chemical constituent, and often changes during storage and trans-

Table 2. Aspects of grain quality

A. Nutritional	B. Processing	C. Marketing
1. Chemical composition	1. Physical condition	1. Appearance
2. Flavor	2. Chemical composition	2. Physical condition
3. Appearance	3. Physico-chemical properties	3. Chemical composition
4. Toxicities	4. Inclusions (foreign material)	4. Infestation
5. Infestation	5. Infestation	5. Inclusions (foreign material)
		6. Financial considerations

portation. Changes in moisture proportionally introduce changes in all other constituents so that changes may become necessary in the formulation of diets. Also, the amounts of minor but important constituents such as vitamins can change significantly as a result of adverse conditions between harvest and utilization. These changes also call for modifications in diet or feed formulation. Overall appearance and infestations can be readily detected, and if deterioration has occurred, either in appearance or due to infestation, further testing for possible toxic or otherwise harmful contaminants must be carried out. Of recent years, considerable emphasis has been laid on the detection and determination of aflatoxin and other mycotoxins. The flavor of a grain, or the products derived therefrom can be affected by the development of taints caused by packaging, mold development, bacterial development, or by the application of protection chemicals such as bag preservatives, pesticides, fungicides, and fumigants. Nutritional quality is affected by insect infestation in two ways: first, the repugnant effect the sight of crawling insects or larvae exerts on the consumer and second, the actual composition of the grain may be changed due, for example, to preferential consumption of the germ areas of grains by many types of insects and by contamination by excreta, secretions, and insect parts. The accidental incorporation of insects into foods during processing also affects the nutritive value of the food, since insect parts are often heavily chitinized, which makes them difficult to digest by monogastrics.

Processing

Processing quality is affected to a marked degree by changes induced during storage and transportation. Breakage and abrasion of grains affect such procedures as tempering, which is a prerequisite of flour and semolina milling, and also forms sites for preferential attack by fungi, bacteria, and insects. This last aspect is of particular importance in the case of crops which are harvested and stored at moisture contents in excess of 14-15%, since molds can establish themselves very rapidly in such material. Chemical composition can change in the same way as described above. Changes in moisture content affect such procedures as tempering of wheat prior to milling and steeping of barley prior to malting. Milling of other cereals such as sorghums is also affected by the moisture content of the grain. Heating of grains during storage usually causes discoloration and usually affects the state of proteins and fats causing varying degrees of denaturation and polymerization. This in turn affects the physicochemical behaviour of the end products, and flours that have been subjected to these conditions often will not perform satisfactorily in the baking of breads or sweet goods. Foreign material such as straw, chaff, weed seeds, sticks, stones, soil, and bits of metal affect the physical makeup of end products, for example, by changing particle size distribution and maybe also causing serious damage to equipment such as mill rolls. Pieces of metal or stones may also generate sparks during processing, which can cause fires or explosions in dust-charged atmospheres. The removal of foreign material from grains is inherent to most processing plants that handle grain as a raw material. Some of the foreign materials, such as straw, chaff, and weed seeds, can be recycled into other products, but inorganic inclusions have to be disposed of as detritus, and their presence is deleterious to processing quality. Finally, infestations by insects or rodents seriously impair processing quality in most cases. Both spoil rather than consume the commodity, but the processing of infested grain calls for removal of the causes of infestation by such processes as fumigation and extra cleaning, and the extra costs and losses involved represents a direct depreciation in processing quality.

Marketing

Marketing of grains can probably be affected by storage and transportation more seriously than either processing or nutritional quality. Grains in many countries are priced according to their physical condition in terms of size, soundness, and general appearance. Grain size is not likely to be affected by either storage or transportation, but both soundness (in terms of density) and physical appearance can undergo drastic changes. Storage for even short periods under conditions of excessive moisture and temperatures above 10°C can result in heating and mold development.

Heating may cause discoloration which may extend through the grain, and under extreme cases, the grain can ignite. The physical condition of grains can deteriorate due to breakage, abrasion, and the introduction of foreign material during both storage and transportation. The introduction of 1% foreign material such as dust or soil and stones means that the customer has purchased 1% of rubbish for the same price as the grain, and furthermore, has to pay more to remove the extra 1%. Although 1% may not appear to be a significant increase, 1% of a 10,000 ton shipment is 100 tons of rubbish which has to be removed both from the grain and from the cleaning area. Some grain exporting countries impose upon themselves rigid specifications for foreign material; for example, the Government of Canada stipulates maximum total content of foreign material of 0.75% including cereal grains other than wheat, in their top milling grade of hard red spring wheat. A concise and well-monitored system for separating different grains into grades based on visual appearance and soundness is fundamental in marketing of grains.

Moisture Content

Changes in moisture content affect the physical amount of grain present in a parcel. Moisture may change by up to $\pm 5\%$ or more during storage and transportation. Most grain marketing agencies recognize this and negotiate compensatory clauses based on moisture tests at the time of purchase and delivery. The development of taints and odors as a result of molds increases in free fatty acids, and rancidities impair marketing quality and usualy develop in storage rather than during transportation. Infestation by insects or rodents both reduces the marketable quantities of grains and reduces the market value. Though fumigation may destroy all insects, larvae, and even eggs, it will not disguise holes bored in grains by weevils and boring-type beatles. Finally, financial considerations can affect marketing quality indirectly. Inequitable pricing and generally inefficient systems for buying from producers or merchants and distribution to customers and processors can prolong the necessity for storage, and increase the likelihood of deterioration by any of the causal agencies to be outlined below. This is particularly true of grain marketing in hot climates.

The remainder of this paper deals with the principal causes of damage to grains during storage and transportation, and the steps to be taken to eliminate or minimize these factors. The agencies are not discussed in order of importance since under individual circumstances practically any of the agencies may become the most important. Space and time limit the scope of the paper to make no more than a brief introduction to the many aspects which have to be considered in order to maintain quality. Several excellent monographs exist, notably references 1-12, which between them list a comprehensive reference library.

CAUSAL AGENCIES OF DAMAGE

Storage

Table 3 summarizes the principal sources of damage to grains during both storage and transportation. Bacterial infection is generally not a serious hazard, and is not discussed further in this paper. In alphabetical order, the first factor to be discussed is complacency and ignorance. Other than genuine charitable organizations, most grain-handling involves some sort of agreement for payment in money or concessions, which becomes the focal point of the transaction. The assumptions are usually made that grain can and will be collected from place A and delivered to place B in accordance with the terms of the agreement. People who make the agreements are often remote from the grain itself, and the possibility of damage or spoilage of the grain frequently falls under the "it can't happen to us," or "it's going to__________ (country) and they'll never notice" category. The ignorance component usually means simply a lack of knowledge either of the hazards of grain storage, or of the state of either the grain or the storage container prior to storage. A 1963 estimate put grain storage losses in the U.S. at between 13 and 23 million tons (13) which occurred in spite of the wealth of knowledge and sophisticated equipment available. Complacency also includes the attitude prevalent among some organizations to increase productivity of

Table 3. Sources of damage to grain during storage and transportation

Storage	Transportation
Complacency and ignorance	Complacency and ignorance
Containers	Fungi
Conveying	Handling (loading and unloading)
Dockage	Insects
Fungi	Moisture
Insects	Rodents
Material itself	Temperature
Moisture	Transport (railraods, roads, ships)
Rodents	

grains without consideration of what will happen to the grain after it has been grown. Actually, if all of the resources presently being devoted to increased productivity in the next 5 years could be directed instead to improvements in storage and damage prevention, the increase in grain available for food would probably be far greater, and the on-going advantages of the improved conditions and technology would be beneficial to further researches on improved productivity.

Containers include storage receptacles from sacks to bulk storage bins. Bins, pits, and cribs can vary in size from a few quintals at the farm level to hugh flat storage bins which are really covered grain piles, the biggest of which accommodates up to 100,000 tons. The shape, age, and the construction material are the most important features of a container affecting the possibility of damage to grain during storage. Covered pits and mud walled stores are satisfactory for short-term storage, especially where moisture levels are low. Several types of such storage are described in (4). Bags made of hessian or other material allow free passage of air to and from the contents and are highly susceptible to attack by insects and rodents. On the other hand, plastic bags restrict moisture-movements, and although they offer more protection, they can also constitute a hazard when temperatures are liable to fluctuate, due to condensation inside the bags. Insect eggs and larvae and mold spores persist well in the seams and interstices of bags, and the bags themselves can constitute sources of infestation. Grain bins made of wood are usually square and by virtue of their construction possess a multitude of cracks, crevices, and angles which are havens for insect eggs and larvae. Furthermore, certain types of insects bore into the wooden fabric itself and emerge from the actual walls to infest new charges of grain. Concrete bins are usually round, star-shaped, or hexagonal. Star-shaped bins possess crevices which can harbor grain residues constituting a source of infestation. Furthermore, concrete itself possesses certain sorptive powers and chemical reactivity, and unless coated with an impervious material such as paint, the walls of concrete bins can interfere with normal fumigation procedures. Nevertheless concrete bins, particularly the more modern hexagonal type, are probably the safest form of long-term storage for grain. Metal bins are usually round and again may be very large. They carry few crevices and do not react significantly with protective chemicals and are also a very safe method of storage. Airtight storage has been in use for storage of high moisture grain in temperate climates. Success depends on reduction of oxygen and inhibition of respiration which restricts all infestations. This type of storage has not been widely used in hot climates to date. Neither concrete nor metal allows interchanges between stored grain and the atmosphere, and moisture movement due to temperature fluctuations, convection, and condensation can result in extensive deterioration and even internal combustion of grains. Due to the superior conductive properties of steel, this is particularly true of metal bins.

Conveying of grain into and out of storage areas may be achieved by hand, bucket elevators, augers, belts, box conveyors, gravity, or pneumatics. Of these systems, hand carrying, bucket elevators, and belts are least likely to cause physical damage to the grain, since the grain itself is static when in transit. The other four methods involve movement of the grain, and this can cause abrasion and breakage of kernels. There are about 30 million kernels in a ton of hard red spring wheat, so a conveyor, such as an auger or box conveyor which moves grain at about 100 metres a minute, can generate considerable kernel to kernel abrasion. Bucket conveyors and augers possess several areas where grain residue may accumulate and cause a source of infestation. Belts pneumatics, air slides (gravity flow), and box conveyors are more or less self-clearning

Dockage is foreign material other than the grain itself. It consists of grains of other cereals, weed seeds, bits of straw, chaff, wood, and metal, stones and soil, string, and many other minor items. Weed seeds and fibrous dockage increase the severity of infestations by insects (14,15) and fungi. Fine particles of dockage, such as grain dust, blocks the interstices between grains, inhibits movement of air, and changes convection patterns. Fibrous dockage such as chaff and straw are often contaminated with fungal and bacterial spores. Stones and bits of metal increase the degree of abrasion of grains during movement.

Fungi

The fungi which affect grains can be subdivided into field fungi and storage fungi. Field fungi attack crops in the field before harvest and in the swathe or stook. They exist in spore form on the grains either in the crease or in the interstices of the outer layers of the seed coats, and cause discoloration of the seed coat (and eventually of flours and other derived products), affect germination capacity, and may produce mycotoxins which affect the subsequent nutritive quality of the grain. Principal field fungi include the general Alternaria, Fusarium, Cladosporium, and Helminthosporium. Storage fungi include several species of Aspergillus and Penicillium. Storage fungi vary widely in their conditions for optimum growth, and Table 4 summarizes the lower limit of grain moisture levels for growth for several species, together with principal damage caused. The data are taken from reference 1

Insects

Insects have been estimated to cause annual losses in grain of over half a billion dollars (16). Insect pests cause proportion-

Table 4. Principal species of storage fungi responsible for damage to wheat

Genus and species	Minimum grain moisture content for fungal development	Principal types of damage
Aspergillus restrictus	13.5-14.5%	"Mustiness," discolors and kills germs
A. glaucus	14.0-14.5%	As for A. restrictus. Also creates favorable conditions for other species
A. candidus	15.0-15.5%	As above, rapid action, causes heating and decay
A. ochraceus	15.0-15.5%	Discolors & kills germs
A. flavus	18.0-18.5%	Discolors & kills germs, causes heating
Penicillum sp.	16.5-19.0%	Kills & discolors germs, causes mustiness, may cause some heating

ally greater losses in tropical and subtropical climates. The most prominent insect pests are tabulated in Table 5. Of these, five species develop inside the kernels. Weavils actually deposit their eggs inside the grains; the grain borer and Angoumois moth lay eggs on the surface of the grains and their larvae bore into the kernels. The larvae of all these organisms feed on the endosperm of the grains so grain which appears unharmed may, in fact, be heavily infested with organisms hidden beneath the surface. The beetle species, such as the rusty and flat grain beetles, and the red and confused flour beetles develop outside the kernels so infestations are more readily apparent. All insects contaminate grain with excretion, cast skins, and body parts. In addition, they may produce odoriferous secretions which can taint the entire parcel of grain. Most of the insect pests attack the germ preferentially since this is both the softest part of the grain and contains the richest concentrations of nutrients.

The material itself; i.e., the type of grain, affects the type and extent of damage which can occur to it. The size and shape of grains affects the area of interstitial space between the grains. There are about 30 million grains in 1 ton of wheat, but over 120 million in a ton of mustard seed, with consequent increase in surface area. This affects the movement of air and moisture in the grain mass. In general, softer grains such as soft wheat and corn, sorghums, and the softer lines of chickpeas and lentils are more readily attacked by insects. Oilseeds such as flax, rapeseed, and sunflowers are rather less susceptible to insect damage. Since these crops are hydrophobic, due to their high oil content, their moisture content at time of storage can be very important since moisture is lost rapidly and can cause serious accumulation of moisture at different areas of bins due to convection and condensation. As a result, heating can develop quite rapidly in bins containing oilseeds, particularly if the dockage content is high.

Moisture

Moisture is probably the most important single factor affecting damage of all types to grain during storage. Grains consist of starch and protein, with smaller proportions of oil and cellulose-type substances. Starch and protein can absorb, respectively, 75% and 200% of their own weight of water, and the pentosan/cellulose-type materials which make up much of the external layers of the grain can absorb relatively even greater quantities. Consequently, the seeds of most cereals and pulses are fairly hygroscopic. Oilseeds are generally hydrophobic and do not absorb water as readily. The interchange of water between grain and the atmosphere is zero when the partial pressures of the water in the grain and the air immediately surrounding it are equal. This state of equilibrium is not static and will depend on temperature changes within the grain mass, relative moisture contents of different areas of grain within the mass, movements of the grain, and secondary influences set up by infestations with insects or fungi. Generally, the equilibrium

Table 5. Principal insect pests of stored grain

Species	Common name
Acarus siro (L)	Grain mite
Crypolestes ferrugineus (Stephens)	Rusty grain beetle
C. pusillus (Schönherr)	Flat grain beetle
Oryzaephilus surinamensis (L)	Saw-toothed grain beetle
Plodia interpunctella (Hübner)	Indian meal moth
Rhizopertha dominica (F)	Lesser grain borer
Sitophilus granarius (L)	Granary weevil
S. oryzae (L)	Rice weevil
S. zeamais (Motschulsky)	Maize weevil
Sitotroga cerealella (Olivier)	Angoumois moth
Tenebroides mauritanicus (L)	Cadelle
Tribolium castaneum (Herbst)	Red flour beetle
T confusum (Duval)	Confused flour beetle
Trogodema granarium (Everts)	Khapra beetle

within a large bulk storage bin will change fairly extensively during the initial period of storage, then will settle down and undergo little further change provided no such infestations develop.

The matter of whether the environment within a storage area is safe for storage or likely to lead to heating, mold developments, and spoilage depends mainly on the initial moisture content of the grain at the time of storage, together with the relative humidity of the atmosphere within the storage area (the environmental relative humidity or ERH), with a rather lesser but still important input from the temperature of both the grain and the storage area. Due to their individual composition and structures, different grains have different critical moisture levels above which prolonged storage is unsafe. Prolonged storage in this paper means storage for more than 2 weeks, but under conditions of high moisture content and high temperature, the period after which spoilage can occur may be only a few hours. The "safe" moisture levels are generally recognized to be the moisture levels of the grains which are in equilibrium with an ERH of 70% for temperate climates and 60% for tropical climates. Table 6 gives the approximate levels of moisture at which several commodities are in equilibrium with ERH of 60 and 70% at 20° and 30°C. Two features are apparent: first, the significant difference between oilseeds, and cereals and pulses; second, the relatively small influence of temperature. These figures have been derived as a result of painstaking determinations of absorption isotherms for the various cereals. Practical use is made of these during grain inspections in North America for establishing legal levels of moisture above which grain cannot be accepted for storage unless it is predried, the expense of which has to be borne by the producer or the vendor of the grain.

Temperature

Temperature affects the movement of air in grain stored in bins and other storage areas. Water tends to move from warmer areas to cooler areas of the grain mass. The movement of air affects the movement of water and eventually the accumulation of areas of higher moisture in the grain mass. The temperature of a grain mass is affected by the ambient temperature of the storage area at the time of storage, by the temperature of the grain itself, by the size of the storage container, and by the material of which it is made. Sheet metal, for example, conducts heat well, so that the effects of wide fluctuations in air temperature outside the storage container are likely to be transmitted into the grain inside, in contact with the walls, with consequent secondary effects on the movement of air currents and water within the grain mass. Concrete, mud, and wood are less effective conductors. Grain itself, partly due to the air-filled interstices within the grain mass, is an effective insulator and is not a good

Table 6. "Safe" moisture levels for storage of grains at 25°C

Grain	70% ERH		60% ERH	
	20°C	30°C	20°C	30°C
Wheat	14.5[a]	13.5	13.0	12.5
Barley	15.0	14.5	14.0	13.5
Oats	14.5	14.0	13.0	12.5
Corn(maize)	14.5	13.5	13.0	12.5
Sorghum	14.5	13.5	13.0	12.5
Millet	16.5	16.0	15.5	15.0
Rice	13.5	13.0	12.5	12.0
Soybean	12.0	11.5	10.0	9.0
Beans	15.5	15.0	14.5	14.0
Cowpeas	15.5	15.0	14.5	14.0
Flax	9.0	8.5	8.0	7.5
Sunflower	8.0	7.5	7.0	6.5
Mustard	8.5	8.0	7.5	7.0

[a]To nearest 0.5% moisture

conductor of heat. As a result, grain entering a storage bin immediately after harvest at a temperature of possibly 30° or 40°C will likely retain its heat for a considerable period after entering storage. Due to differential temperatures near the top and sides of grain in storage, whether in bins or piles, moisture movements by convection tend to result in increased moisture contents in these areas. When conditions reach a critical stage, molds develop. This may occur at other areas randomly distributed within the grain mass, and the increased temperature caused by respiration and other metabolic activities of the developing fungi lead to the development of "hot spots" in the grain. Hot spots, or localized areas of higher temperature within the grain mass, may develop in two ways. The first occurs at relatively low moisture levels, for example 12-14%, and is usually caused by insect metabolism. Temperatures of about 40°C may be reached. This is sometimes referred to as "dry grain" heating. At higher relative humidity levels, fungal infestations develop. These produce higher temperatures of up to and over 60°C, and the condition is known as "damp grain" heating. In association with this, and due to the favorable conditions of temperature and moisture, germination of grain may take place, and the increased metabolism of the grain may cause even greater heating.

Rodents

Rodents are the final causal factor of damage to grain in

storage and transportation. Rats and mice of the genera Rattus and Mus are the rodents which most frequently infest grains. Damage is caused by direct consumption of grain; by fouling the grain with faeces, urine, hairs, and dead bodies; by transmission of diseases in body excreta; and by destruction of containers and bags. Indirect damage may be brought about by fires caused by rats gnawing through electrical cables. Losses in individual stores can amount to 70% of the total grain, due to direct spoilage and contamination. Mouse faeces are particularly difficult to remove from grain because they are approximately the same size as grains of wheat, rye, and barley. In 1953, it was estimated that rats alone caused an annual loss of approximately 1 billion dollars in terms of consumption and spoilage of grain and other foods and damage to property. The most predominant rodent pests are the Norwegian rat, *R. norvegicus*; the black rat, *R. rattus rattus*; the grey-bellied rat, *R. rattus alexandricus*; the Polynesian rat, *R. exulans*; and the house mouse, *Mus musculus*. All of these rodents are capable of climbing and can easily enter storage sheds, flat storage bins, and warehouses, and infestation of terminal elevators is possible through basements. The interstices of galleries, elevator boots, and other areas can house colonies of rodents, and contamination of grain moving along the belts or conveyors can become continuous.

Transportation

Turning to transportation, damage to grains arising as a result of transportation follows basically the same pattern as for storage. Of the agencies associated with storage, most of them are equally applicable to transportation conditions. This applied to complacency, fungi, insects, moisture, rodents, and temperature. Conveying in the transportation of grains usually involves loading and unloading of ships, trains, and trucks, and lateral movement of grains from collection points into and out of elevators. Physical damage can occur to the grains at any stage of conveying. Also, sources of infestation can be introduced. Ships and railcars can serve as sources of infestation due to small amounts of residual grain in holds and railcars. Also, certain types of insect larvae bore into the woodwork of wooden railcars. On long trips or when cars are at rest in sidings awaiting unload, the larvae can infest the new grain in the car.

PREVENTION OF DAMAGE

Storage of grain can be short-term (up to 2-4 weeks) or long-term (up to several years). In general, long-term storage is of more concern in temperate climates and in countries producing large surpluses of grain for export. In developing countries, storage periods are generally shorter, parcels of grain are smaller, and rather more use is made of pit- and mud-walled silos and bags for storage and transportation. In many cases, grain marketing in developing countries is carried out on a large scale but with small scale facilities--horse and oxen-drawn carts and smaller trucks--with bag storage as an interim between producer and consumer or processor. Longer term storage occurs in private homes and village communes where grains may be stored up to 1 year between crops. Another reason for fairly long-term storage in developing countries concerns those countries which are obliged to import significant quantities of grain for subsequent distribution within the country. Damage to grains over short-term storage is usually limited to physical damage, or rapid fungal degeneration of the material due to excessive moisture content. Insect generations usually take several weeks to develop and are sensitive to physical activity and changes in environment. As a result, movement of the grain out of temporary storage may decrease damage due to reduction of the insect population at the larval and pupal stage. Temporary storage facilities if already infested or contaminated may cause new infestation to any grain passing through the facilities. The most effective measures for prevention of damage to grains are listed in Table 7.

Aeration

Aeration is an effective method of reduction both of moisture content and temperature. It is particularly applicable to smaller storage facilities and finds its forte in the open corn-crib type of storage. Turning of grain in bins is practiced in farm storage and in elevator bins. In some areas, grain elevators have built-in aeration facilities. Grain stored in piles packs itself fairly closely and movement of air inside the pile is restricted. Turning such piles can effectively reduce moisture content by 1-4%, depending upon the initial moisture level and may reduce insect and rodent infestation.

Design of storage facilities can assist in prevention of damage. Avoidance of sharp angles and unnecessary recesses and crevices, choice of building materials, use of conveying equipment such as pneumatics, which causes less damage to grain and provides less scope for infestation all helps in prevention of damage. This extends from the largest, most sophisticated storage facilities down to the farm storage bin, grain pit, or bagged grain warehouse, where well-planned stacking is important. New designs of railcars of the hopper-type have reduced time for loading and unloading and made the hoppers self-cleaning, so that the risk of infestation is minimized. The use of the tanker-type ship for moving grain introduced the concept of many small holds (tanks) which does not seriously impair loading and unloading and reduces infestation risks since one or two small infested tanks are easier to restore than a hold with a capacity of 2,000 tons. The extra work involved in clean-up and fumigation of the many smaller tanks is offset by the increased effectiveness and flexibility.

Table 7. Measures for prevention of damage to grains during storage and transportation

Measure	Area of Application
Aeration	Storage bins, short-term and long-term
Moisture control	Storage bins and containers
Temperature control	Storage bins and warehouses
Fumigation	Storage bins, warehouses, ships' holds, railcars, trucks, sacks
Fungicides	Storage bins, warehouses
Insecticides	Storage bins, warehouses, ships' holds, railcars, trucks, sacks
Design	Storage areas
Hygiene	All areas
Legislation	All areas

Fumigation

Fumigation and treatment with protective chemicals are widely used and are effective methods for prevention of infestation. Fumigants act in the gaseous or vapor states. They can be applied by several methods, and storage areas such as collections of bins can be fitted permanently with equipment for periodical fumigation at the same time as aeration. Fumigants penetrate to all parts of the storage area, including tiny crevices and ideally are effective against all stages of insect life including eggs, larvae, pupae, and adults. The most widely used fumigants include methyl bromide, which is effective against eggs, larvae, and adult insects of many species (but rather less effective at the pupal stage). Phosphine in the form of aluminum phosphide is another excellent fumigant which can be applied in tablet form. Hydrogen cyanide and dichlorvos, an organic phosphate, are two other commonly used fumigants. Fumigants can be applied to piles of grain stored under tarpaulins or in bags by covering the pile with a gasproof sheet such as polyethylene or polyvinyl chloride. The fumigant is applied under the sheet, and the ends sealed down with tubes of fabric filled with sand or any suitable weight. Strict precautions must be observed to protect workers from exposure to the fumigant. Gas masks and other protective clothing must be worn by workers responsible for removing tarpaulins, etc. after fumigation.

The efficiency of fumigation is affected by temperature, moisture, the presence of dockage, the type of container, the dosage rate, the types of insects, and the characteristics of the grain itself. Temperature affects efficiency at both ends of the scale. At high temperatures, the fumigant tends to evaporate too quickly without realizing maximum effect. At low temperatures, permeation is affected and the dosage reaching the insects may be sublethal. However, for practical purposes these drawbacks are of relatively little moment, since most insects cannot survive at temperatures much above 40°C, and at temperatures below 10°C (50°F) grains do not require fumigation. High moisture reduces the effectiveness of fumigants by increasing the sorptive capacity of the grain. The presence of excessive dockage reduces aeration space between kernels and therefore the ease of permeability by fumigants, and also absorbs significant amounts of the fumigant. The efficiency of fumigants and the overall ease of sanitation is facilitated by minimizing crevices and rough surfaces of containers.

Fungicides are not generally recommended as being amoung the most satisfactory methods for controlling fungal infestation of stored grains, mainly due to residues toxic to animals, difficulty of application and most of all, limited efficiency in combating the fungi. Insecticides on the other hand are widely used to control a variety of insects. The most effective of these are malathion, pyrethrins, lindane (although use of this is decreasing due to legislation against residues) and dichlorvos. Malathion may be applied as sprays, aerosols, or dusts directly to the grain, or during the filling of bins, etc. Dosages depend upon the amount and type of grain, storage area, type of insect, and environmental conditions. Pyrethrins are one of the oldest forms of insecticide and have been in use for more than 100 years. Recent research has increased their effectiveness by the use of synergists, such as piperonyl butoxide, which significantly increases toxicity. Pyrethrins are often applied as space treatment; i.e., to empty bins, ships, or railcars in advance of the grain, as well as directly to the grain itself. Dichlorvos is an organophosphate which has great potential as a conventional insecticide, particularly for use in areas where the insect populations are exhibiting resistance to malathion. It is also effective when applied directly to bagged grain. Resistance to malathion and some other pesticides is increasing and necessitates continuous research into the development of new insecticides. Legislation against infestations of all types provides a negative incentive to grain-handling organizations and companies to install, improve, and maintain protective measures. Inspections, heavy fines, and condemnation of grains has the effect of forcing intensified measures to combat infestation by insects or fungi. Hygiene means simply keeping all areas and equipment free of infested and otherwise contaiminated grain and other residues. This is achieved by sweeping, vacuum cleaning, aspiration, and washing. Grain-handling areas are very dusty, and dust settles over all equipment and floor, wall, and ceiling space providing ideal locations for insect multiplication. Exhausting of dusts, frequent sweeping of floors, etc. can go a long way to reduce the incidence of infestation. Hygiene can also be interpreted to include removal of dockage from grain, which is beneficial to the grain itself by improvement of the value and purity of the commodity and by improvement in pest control. In addition, the cleanings (screenings) may be sold or processed as animal feed.

Temperature

Grain stored at low temperatures (up to 10°C) and low moisture content (up to 12-13° for cereals and pulses and 9% for oilseeds) will maintain its quality for very long periods. Hard red spring wheat tested for baking quality after storage for up to 13 years in Western Canadian wooden grain elevators showed no signs of deterioration in quality, in terms of loaf volume or such parameters as fat acidity (18). By far the most effective methods of prevention of infestation or spoilage of grains involve the control of moisture and temperature, and of these, moisture is the most important single variable. Temperature control of large masses of grain is difficult without excessive and expensive bin-turning (moving grain from one bin to another) due to the insulating properties of grain. When grain enters bulk storage, even on a relatively small scale, the temperature on the inside of the bulk remains fairly stable, and in the absence of infestations, appreciable changes in temperature may take several months. If grain harvested at 40°C is immediately placed in storage, the temperature of the bulk grain or bags will remain close to 40°C for a considerable time. Similarly, if grain enters storage, for example, from railcars or ships at low temperatures, these temperatures will prevail with only very slow change. For grain to be stored safely at elevated temperatures (30°C and higher), the moisture content must be reduced to at least 1% below the values given for the equilibrium moisture content of the grains at given ERH. High ambient temperatures are advantageous to moisture reduction, and under field conditions in warm climates, moisture reduction to quite low levels can be attained simply by allowing the grain to lie in the sun for a day or two before or after harvest.

Moisture

Moisture control is overall the most effective method of prevention of many types of damage and infestation to grain. High moisture grain is more readily damaged physically during elevation and conveying then lower moisture grain (provided that the moisture content is not sufficiently low as to make the grain brittle). Neither insects nor fungi can survive sufficiently well at low moisture levels (below 12-13% for cereals and 9% for oilseeds) to cause a serious infestation. Moisture content can be reduced by direct exposure to radiant heat, such as the sun, by aeration, and by forced draught hot air, as in grain driers. It can also be reduced to a certain extent by blending high moisture with low moisture grain, provided that the equilibrium moisture content is likely to be sufficiently low (4). Grain drying by means of various types of crib or open storage container is well suited to areas where sunshine and high temperatures are abundant, and to crops which are harvested in cobs or in sheaves when the heads can be exposed to the action of hot sun and wind. The drying of bulk or bagged grain cannot be achieved in this manner. Bagged grain can be dried by aeration, either naturally (e.g. by stacking in narrow rows in open-sided covered sheds) or in batch driers. Batch hot-air driers are efficient, but the bags must be moved periodically to avoid overheating and heat damage to bags that are closest to the heat source. The most efficient method of drying bulk grain is the grain drier, of which many versions exist (1,17). These are available for use at farm or commune level, as well as for use in large grain terminals, and all employ some form of hot-air forced-draught aeration.

It is important to obtain an accurate test result for the moisture content of the grain before it is placed in storage and during its period in storage. The methods for testing for moisture are legion, but most accurate, modern moisture testers utilize some form of electrical conductivity, with capacitance measurement foremost among the inexpensive instrumental methods. Sampling for moisture testing is rather more important than the actual test method employed, since the error of sampling large bulks of grain is far greater than the error of the moisture test itself. Sampling of large bins and grain piles can be carried out by vacuum probes, which can withdraw samples from any part of a bin or pile. It is also important to record the actual points and depths from which the samples have been taken, in order to map the moisture/temperature relationships of the grain mass as a whole. Grain which has been dried should not be tested for moisture by electric meters too soon after drying, since moisture may migrate outwards from the center of the kernel, resulting in significant increases in moisture until about 16 hours after drying (19). During the period in storage for the greatest protection, grains should be tested periodically for moisture to verify that significant buildup of moisture in the upper levels of bins and piles is not taking place. Monitoring of temperature throughout the grain bulk using remote sensing is another effective preventive precaution, since the development of "hot spots" can then be detected and the corrective precautions taken.

In summary, this paper has presented some of the hazards inherent in storage and transportation of grain, the causal agencies, and methods for remedy. To reach for the ultimate objectives of no spoilage and losses is practically and economically unrealistic, but the establishment of and strict adherence to sound preventive maintenance and hygiene practices will achieve highly significant annual savings in grain and food production and utilization.

REFERENCES

1. C. M. Christensen, ed., Storage of Cereal Grains and Their Products, The American Assoc. of Cereal Chemists, St. Paul, Minn. (1974).
2. C. M. Christensen and H. H. Kauffman, Grain Storage, Univ. of Minnesota Press, Minneapolis, Minn. (1969).
3. S. S. Easter, ed., Preservation of Grains in Storage, F.A.O. Agricultural Series No. 2, Washington, D.C. (1948).
4. D. W. Hall, Handling and Storage of Food Grains in Tropical and Subtropical Areas, F.A.O. Agricultural Development Paper No. 90, Rome, Italy (1970).
5. L. A. Trisviatsky, Storage of Grain, Nat. Lending Library for Science and Technology, Boston Spa, England (1969).
6. T. A. Oxley, The Scientific Principles of Grain Storage, The Northern Publishing Co. Ltd., Liverpool, England (1948).
7. A. Wexler, ed., Humidity and Moisture Measurement and Control in Science and Industry, Vol. 4, Van Nostrand, New York, N.Y. (1965).
8. C. M. Christensen and H. H. Kauffman, Grain Storage - The Role of Fungi in Quality Loss, Univ. of Minn. Press, Minneapolis, Minn. (1969).
9. G. Wogan, ed., Mycotoxins in Foodstuffs, Mass. Inst. of Technol. Press, Cambridge, Mass. (1965),
10. R. T. Colton, Pests of Stored Grain and Grain Products, Burgess Pub. Co., Minneapolis, Minn. (1963).
11. H. E. Hinton and A. S. Corbet, Common Insect Pests of Stored Products - A guide to Their Identification, Brit. Museum Economics Ser. S., London, England (1955).
12. E. A. Parkin, Insects and Stored Food: World Losses and Control Measures Surveyed, Food Manuf. 34:164 (1959).
13. W. H. Pawley, Possibilities of Increasing World Food Production, F.A.O. Basic Study No. 10, Rome (1963).
14. R. T. Colton, H. H. Walkden, G. D. White, and D. A. Wilbur, Causes of Outbreasks of Stored Grain Insects, Kansas Agr. Exp. Stn. Bull. No. 416 (1960).
15. H.E. MacGregor, Preference of Tribolium Castaneum for Wheat Containing Various Percentages of Dockage, J. Econ. Entomol. 57:
16. USDA ARS Handbook No. 291, Losses in Agriculture (1965).
17. A. P. Gerzhoi and V. F. Samochetor, Grain Drying and Grain Driers, translated by B. Shapiro, Office of Tech. Services, U.S. Dept. of Commerce, Washington, D.C. (1960).
18. K. H. Tipples, Offsite Storage Study, Grain Research Lab. Report, unpublished (1972).
19. P. C. Williams and J. T. Sigurdson, Grain-drying Study, Grain Research Lab. Report, unpublished (1974).

NEW POST-HARVEST TREATMENTS OF HORTICULTURAL PRODUCE AND DEVELOPMENTS TO MAINTAIN QUALITY AND TO PREVENT DAMAGE IN WESTERN EUROPE WITH SPECIAL REFERENCE TO THE NETHERLANDS

W. S. Duvekot

Sprenger Instituut
Wageningen, The Netherlands

INTRODUCTION

The purpose of this paper is to describe in short the present situation of post-harvest techniques in Western Europe, as to storage, pre-cooling, sorting, grading, packing and transport.

Competition on the European markets compels producers and the whole trade to offer high quality produce, attractively and safely packed and in such a stage of maturity, that the proper taste and flavor will develop fully.

Existing practices are going to be extended and new ones are introduced and/or developed. Restrictions made by public health and environmental authorities force in some cases to discontinue established practices and to develop new ones in order to maintain the required high quality of the produce.

STORAGE

In the field of storage of perishables attention is paid more and more to a better control of the storage climate, so the decrease of quality can be reduced and produce of high quality, with a reasonable shelf life can be offered to the consumer. Attention has been and still is focused at relative humidity, temperature, C.A.-conditions, and the rate of air-circulation with respect to quality, waterloss and energy consumption.

Traditionally the range of relative humidities recommended for fruit, vegetables and cut flowers is 90-95%. In the last seven years however research workers and commercial experiences have shown, that very high relative humidities of 98-100%, so near saturation, may reduce weight loss and the rate of senescence of some vegetables. Even in several cases rot and decay were less, compared to produce that was stored under lower relative humidities. For cut flowers it is found, that storage and transport at relative humidities, near saturation, prolong the flowering period at the vase and increase the ornamental value and pleasure. However, for some varieties of apples relative humidities, near saturation seem to stimulate the occurrence of core flush. Due to these results a diversification in the installation of refrigeration plants can be observed. For maintaining very high relative humidities, the so-called wet-cooler systems, sometimes based on the ice-bank system, are being introduced. Particularly for long-term storage of some winter vegetables as cabbage, beets, carrots, turnips, celeriac and chicory roots, wet-cooler systems are going to be installed more and more. The same is true for storage of cut flowers, some flower bulbs, and ornamental trees and shrubs. The wet-cooler systems are preferred above the jacketed store, because of the uniform climate which is produced and can be maintained.

For cut flowers the wet-cooler system also came into use as a pre-cooling system. Practical experience has made clear however, that in that case the flowers that have been pre-cooled previously, should be kept in a separate cold room. The changes in temperature occurring during loading of the pre-cooler do harm the quality of already cold flowers, mostly due to condensation.

C.A.-storage is still practically reserved for the long-term storage of fruit, apples and pears. Commercially vegetables are nearly not stored under CA conditions. In the last years it has become evident, that for certain varieties of apples the oxygen concentration in the CA atmosphere could be decreased below 3% to 2% and even 1.5%. Commercial practice in the United Kingdom, the USA, and in the Netherlands, has proved that the storage time of Cox's Orange Pippin can be prolonged for several months in an atmosphere of 2% oxygen and 0.1% carbon dioxide. Oxygen levels lower than 1.8% do effect taste and flavor in an unacceptable way and increase the danger of the forming of alcohol and low oxygen injury. The variety Boskoop appears to keep under the same conditions also well. The occurrence of core-flush is then retarded considerably.

In case this method is going to be introduced on a large scale this will make high demands upon measuring instruments and regulations of the atmosphere. In the United Kingdom a large cold store with many CA rooms is the first enterprise in Europe that has been equipped with computer registration and computer regulation of temperature, carbon dioxide and oxygen. The whole complex is completely automatically controlled by a computer. The very high investment necessary will be a handicap for a rapid growth of this type of operation. A further impediment of introducing this way of working will be the gas tightness of CA rooms and the capacity of existing scrubbers. Though the gas tightness of CA rooms has been improved considerably, it can be stated, that about 40% is not operating as it should be.

The scrubbers of the activated carbon type, are mostly calculated for atmospheres containing 3 - 4% carbon dioxide, so they have too small a capacity if a level of 0-1% carbon dioxide has to be maintained. In commercial practice the insufficient capacity of the activated carbon scrubbers is compensated by placing dry lime sacks in the rooms. In some cases the scrubbing is done exclusively by means of dry lime.

As to the construction of refrigerated- and CA rooms the development goes into the direction of ready insulated panels covered with metal or plastic. In the case of CA rooms here the problem of leaking seams exists, which demands utmost attention. In cases where rural or city regulations do forbid the application of panels and require brick buildings, gas tightness is performed by spraying inner walls and ceilings with a special elastic p.v.c. compound, resulting in a seamless cover, which lasts for years without problems. As to the capacity of new cold- and CA rooms it must be mentioned, that there is a tendency towards rooms of 80 to 150 tons, the average being around 100 tons. Reasons for this practice are that special pre-cooling rooms for fruit nearly do not exist. Filling the rooms in 5 to 7 days, thus bringing in a warm lot every day, has consequences for the already cold fruit and the same is true for the unloading period, particularly after long-term storage. For cooling down the produce as soon as possible, a rather high refrigeration capacity is necessary, but when the fruit is at the required storage temperature a small part of the compressor capacity is needed. Therefore central cooling plants are installed with capacity regulations instead of installations for each room separately. So during the storage period only one or two large compressors are operating. This saves a considerable amount of energy. In the last few years owners of cold stores have experimented with the running time of fans in order to save on energy. Part of the fans were stopped when no cold was asked for, so maintaining a slow air circulation. According to the last results of some experiments it must be possible to maintain an air circulation of 15 times/ hour the capacity of the room during the whole storage period. This could result in a saving of energy of about 40%, and at the same time give a decrease of weight loss because of slighter difference in temperature and relative humidity between the micro- and macro climate. This practice of a decrease in air circulation will undoubtedly be continued. A further point in maintaining a uniform storage climate is that a development is going on towards continuous operating installations, instead of the traditional on-off installations. Though continuous operating refrigeration plants are installed already on modern refrigerated containers for trans-atlantic transport of flower bulbs, the same investments in land plants are going very slowly. Research of the last year has shown, that in fruit stores ethylene is produced at low temperatures, even at 1°C. In The Netherlands it has been measured, that in CA stores filled with Golden Delicious apples and scrubbed with dry lime and in ventilated CA stores the ethylene concentration was between 1800 and 2700 ppm. In CA stores scrubbed with activated carbon concentrations have been found varying from 350 to 500 ppm. So from a view point of quality, it could be worthwhile to consider scrubbing of ethylene in the future. In The Netherlands an experimental scrubber, oxidizing ethylene by means of ozone has been developed, but the apparatus has not been applied due to its very high energy consumption. New storage techniques, such as the Low Pressure Storage System and the French PRAC system (pre-cooling under vacuum and CA storage) and scrubbing of ethylene are in Europe still not applied commercially and it is not to be expected that the introduction will proceed quickly. The low pressure storage system requires high investments and as long as good results are obtained with scrubbed CA storage and storage with a very high relative humidity it is doubtful whether this system will come into existence in Europe.

Regarding the introduction of the PRAC system it is not clear whether the required working method gives a better result than existing systems and whether it is to be realized commercially. The produce has to be in a vacuum cooler at the latest six hours after harvest. The produce has to be packed on pallets. The pallets are enveloped with a large plastic bag which has an opening which after vacuum cooling is closed with a special filter of silicone treated polythene (according to the Marcellin patent). This filter, which lets pass in a selective way oxygen and carbon dioxide has a surface which has to be in proper relation to the product concerned (rapid and slow respiration). It will keep the atmosphere in the bag at a maximum of 5% carbon dioxide. The bag itself takes care of a very high relative humidity inside it. As said the system works on the condition that the produce is brought to a vacuum cooler maximum six hours, but preferable two to four hours, after harvest. It should be cooled down in the vacuum cooler to the desired storage- or transport temperature. The feature of the system is that the intercellular spaces are sucked empty and that after the vacuum is abolished and air of normal composition has entered, very soon a CA condition comes into being with a maximum of 5% carbon dioxide, a low partial pressure of oxygen and a very high relative humidity. The results according to French research are similar to those of the Low Pressure system. Whether this PRAC system, which has been applied up till now on a very small commercial scale will grow in importance, will depend completely on the practical possibility of getting the produce into the vacuum cooler in a very short time.

PRE-COOLING

In Europe pre-cooling of horticultural produce is applied to a limited extent. For lettuce and other leafy vegetables vacuum cooling is applied, but mainly in exporting countries. Tomatoes, cucumbers and other vegetables are pre-cooled by means of forced air circulation and cut flowers to a very limited extent by means of pressure cooling. This again is practiced only in exporting countries and often for produce that has to be transported over long distances only. Since, however, the customer is asking for produce which is as fresh as possible, the interest in, pre-cooling is growing rapidly and it may be expected, that pre-cooling in the not too distant future will be an established practice for all produce that has to be transported over middle and long distances. In some countries pre-cooling is also applied on a limited scale for uncooked food, cut vegetables in consumer units, and for this purpose too this activity will be growing. For the pre-cooling of large hauls cooling by evaporation is now under experimentation.

SORTING AND GRADING

Sorting and grading on quality and size are progressively demanding special attention. The trade and consumers do ask for uniform produce of high quality, well graded and sized and without bruises. In many cases the existing equipment does not meet these needs, particularly not in small growers enterprises. An electronic ball, measuring impacts in 3 dimensions has been of great help in finding the hazardous spots on the machinery. For larger packing houses great efforts have already been made in improving the accuracy of sizing and avoiding of bruising. Protection against bruising has been obtained by applying better cushioning materials and by replacing mechanical dumpers and conveyor belts by water dumpers and water gutters and by using weight sizers releasing the fruit in water. The accuracy of the sizing and grading is improved tremendously by combining electronic weight sorting and colour sorting. Fruit is categorized both by its size and colour and the sorting can be programmed to release a given category of fruit at any cross conveyor or gutter. Besides the programming feature facilitates the layout and positioning of down-line packaging or bulk-filling machinery. This type of equipment is also used for pre-grading prior to storage. Despite these improvements the sizing is still not as satisfying as it should be. At present sizing requirements are such that fruit has to be measured by its largest diameter. The correlation between weight and largest diameter is not so precise, that packing in trays and cell pack can be done right away from the machine without partly resizing. In this respect the development of an electronic shadow sizer, measuring the largest diameter, the smallest diameter and the height of a fruit may be mentioned as the possible proper solution. Experiments with this electronic shadow sizer have given very hopeful results. As to economics it has to be stated, that the large electronic, computer-steered graders and sizers require a high investment. Consequently these machines can be operated economically by large central packing-houses only. For individual growers, this type of equipment is as yet not within reach. In our opinion this does not have to be a disadvantage. There is a tendency towards improvement of quality of all horticultural produce. The consumer is demanding a constant high and reliable quality and uniform products as to size and color and these demands require at its turn efficient operating packing houses and an alert trade apparatus. In such a concept, small enterprises of individual growers, with their own interpretation of uniformity and quality, do not fit very well. They will undoubtedly have to join cooperatives or private operated large enterprises or find their own market, a market for lower quality products.

TREATMENTS AFTER HARVEST

In using water dumpers and graders, it is very well possible to add chemicals to the water to control rot and possible storage diseases, as for instance scald on apples. This system is practiced at several places in the world. In the European Economic Community (Common Market) however there is a strong drive to exclude all chemicals for treatment of fruits and vegetables after harvest. In several European Countries application of chemicals after harvest for controlling rot and storage diseases are forbidden for reasons of public health or pollution of the environment. This may have far reaching consequences for storage life, shelf-life and final quality of produce. As an example it may be mentioned, that in the Common Market the residue tolerance for the scald controlling chemical Ethoxyquin has been dropped per January 1, 1981 from 3 to 0 ppm. Only when the manufacturers of these types of chemicals are able to prove, that there is nothing to fear at all for human health nor for pollution of the environment and that after the treatment the remaining chemical can be collected safely from the water and destroyed, the authorities concerned may consider admission. Since this type of research is very expensive and time consuming, and the quantities of the chemicals concerned that can be sold relatively small, most of the manufacturers have little or no interest to undertake such research. A similar attitude of public health and environmental authorities can be observed as to the application of artificial waxes.

For cut flowers a post-harvest treatment with bactericides and nutritional substances is going to be common practice. A treatment against the injurious influence of ethylene with silverthiosulfate is up till now only applied for carnations.

Also here environmental authorities are ready to issue

restrictions. It is necessary and advisable that research is carried out to find substances that are friendly to the environment. Maybe growth regulators could provide an acceptable solution.

PACKING

The developments in the distribution channels, the wishes of the consumers to be able to buy good looking, good tasting undamaged products, the demands of the produce as to packages and the necessity to reach an optimal degree of loading of transport means and warehouses, forces the packaging industry to appear with new designs, better materials and better constructions. This is particularly true for cardboard packages. The manufacturers of paper and cardboard do have already reasonable wet-strength material, but it needs improvement. The manufacturers have research underway to increase the stacking strength of corrugated cardboard by using thicker flutings, by applying different material for the fluting, as semi-chemical and by laminated flutings. It is not to be expected, that the height of the flutes, a very important factor as to the strength, will deviate very much from the current ones. The packaging industry has already developed better liners, wet-strength liners, with resin impregnated ones, liners coated with plastics and laminated liners. Experiments with polythene laminated liners have shown, that a high stacking-strength can be achieved. Combinations of different packing materials such as cardboard, wood and plastic have been tried, but did not prove satisfactory as a whole. It may be expected, that cardboard, solid and corrugated will be used in the near future to a greater extent than nowadays. Cardboard is preferred to wood and plastic by the majority of the trade, especially the retail trade. Wooden and plastic packages provide at the points of sale problems as to storage and transport of the empties. Cardboard can be pressed in bulk and recycled. Many supermarkets and chain-store enterprises do have paper-presses at their points of sale. The old-paper trade collects regularly, sometimes daily the pressed paperbulk, so the retailer gets easily rid of all packaging material. In several branches of industry the manufacturers of cardboard packages and their clients have stimulated the system-approach, by which cardboard and machines for setting up the packages will be offered as a deal. The packages are delivered flat and set up at the point of dispatch according to need. This system saves on costs of transport, warehousing and administration. In horticulture this system-approach is hardly known. It is applied on a limited scale for packing of flower-bed plants, but it is expected to gain much interest, specially when products as tomatoes are going to be packed in cardboard packages. The majority of this product still moves in the international trade in wooden trays.

In recent years a certain standardization of bottom dimensions of packages has been accomplished. The Economic Commission for

Europe of the United States at Geneva has, after long years of discussions and commercial trials, recommended to standardize the bottom dimensions of fruit and vegetable packages in the dimensions 60 x 40 cm, 50 x 30 cm and 40 x 30 cm. In many countries these recommendations have been followed and the majority of the packages moving in the international trade do have more or less the recommended bottom dimensions. These packages fit the international standard of I.S.O. fixed pallet dimensions of 100 x 120 cm and 80 x 120 cm. However problems arise with refrigerated loads over long distances. Even when the loads are properly pre-cooled, there must be air space available between the packages in order to remove the heat generated by the produce. These air spaces cannot be created when bottom dimensions of the packages, loaded on standard pallets, are up to the recommended ones. Therefore in practice bottom dimensions are chosen mostly close to the recommended dimensions, and well in this respect, that they are slightly, one or two cm smaller. Only wooden packages with extending corner posts do as a rule conform to the recommendation, since the extended corner posts allow for sufficient air movement over and through the produce.

TRANSPORT

The majority of fresh horticultural produce is transported by road. At present only a minor part move in open or not completely closed vehicles. Depending on the distances that have to be covered and the demand of the receivers, closed, insulated or refrigerated vehicles are used. For the sake of efficiency the use of pallet loads is increasing rapidly. To achieve a stable load, pallet loads are usually shrink wrapped in large polythene bags which are put over the pallet load. This method has the disadvantage, that ventilation and air circulation in the shrunk bag are very much limited. In many cases condensation will occur, which can harm the produce. This is avoided by pinching several holes in the shrunk bag, particularly in the top. Since recently a low-priced machine became available, pallet loads are wrapped also in stretch nets, so no problems of air circulation and condensation exist any more. In general there is a tendency towards insulated and refrigerated trucks and trailers, especially thin-walled ones, equipped with 7 cm high floor racks, so sufficient air circulation under the load is guaranteed. As to standardization of pallets it has to be mentioned, that at present there is no unanimity as to which pallet dimensions are to be preferred. For many reasons the 100 x 200 cm pallet, which is used world wide, is to be preferred, compared to the 80 x 120 cm pallet. On the long-run it is cheaper as to handling and energy consumption. Loading or unloading a certain quantity of goods on a 100 x 120 cm pallet demands less moving and handling of fork-lift trucks than when the goods are stacked on 80 x 120 cm pallets. Besides the 100 x 120 cm pallet has more possibilities for cross stacking and consequently, that pallet loads will be more stable and safe. Though both pallet sizes are fixed by ISO standards the European Union of Railway Companies has established a palletpool with the 80 x 120 cm pallet. And following the railway policy for not very clear reasons, the largest importing country on the continent has also chosen for the 80 by 120 cm pallet. This decision has given ground for heavy discussions between producers, trade circles, receivers, transporters and even international organizations. Undoubtedly these discussions will continue for years before a reasonable, and for all parties concerned, acceptable solution will be achieved. Due to increasing costs, the movement of piece-goods is increasing in the roll on - roll off system by railroad. Whether horticultural produce is going to be transported this way, will depend on the development of costs of energy and the speed the service and the guarantee the railway companies can give. In Europe an international drive exists to come to standard procedures for transport conditions. Proposals are worked out to come to an obligation for registration of the transit temperature, which could be of importance for the receiver in connection with further handling and distribution. At present the use of generators on road vehicles for maintaining certain conditions in the load is not a common practice. They are used to a limited extent on containers for overseas transport. A rather new development is the application of ethylene absorbing devices, like impregnated blankets and small towers with potassium permanganate. Particularly for mixed loads of vegetables, fruit and cut flowers purafill and exten-O-life blankets have proved to have some favorable effects. According to our latest research results they are most effective at about 15°C.

QUALITY AND CONSUMER

The applications of quality and grading regulations shall, through a better and above all an uniform inspection in all countries, guarantee that trade and consumer can buy a reliable product. For this purpose a central training course is established at Saragossa, Spain. The demand for a top quality product shall induce the further development of test methods for the internal quality. Disappointments with regard to internal quality are growing, particularly of products that have been stored too long or of imported produce from many countries that have been harvested far too early. The consumer is getting more and more quality conscious and is also asking more information of the produce. As a result of this the Common Market has made the decision that after 1982 all foodstuffs sold in shops have to be provided with eight basic data, namely: name, list of ingredients, net weight, names of manufacturer or producer, packer or importer, origin and directions for use. For a number of products exceptions are made as to the mentioning of certain data on the label or the package. For fresh fruit and vegetables it is not obliged to mention an outmost date of sale and a list of ingredients. Consumer organizations however have announced already, that they have the strong wish to know at least in the not too distant future what the most important ingredients are of fresh fruit and vegetables with regard to the nutritional value. Whether this will materialize in the future has to be waited for.

SUMMARY

- -In the field of storage of perishables, attention is focused on improvement of the climate in refrigerated stores and CA rooms. Special attention is paid to the relative humidity and CA storage under low partial pressures of oxygen under absence of carbon dioxide.
- -It is not to be expected that new storage techniques as storage under low pressure and the French PRAC system will be introduced rapidly on a large commercial scale.
- -Pre-cooling will grow in importance, particular for middle- and long-distance transport and for long-term storage of fruit.
- -In sorting and grading plants the accuracy has been improved by the introduction of computer controlled electronic machinery.
- -It is rather likely that chemical treatments after harvest will have to be replaced by different techniques, due to requirements of public health and environmental authorities.
- -Packing materials, in particular cardboard will be improved as to quality and stacking strength.
- -Standardization of packages and pallets still form a subject of discussion as to dimensions.
- -In transport the development goes towards closed, insulated and refrigerated thin-walled vechicles.
- -Labeling of packages with the aim to inform the customers will put a heavier responsibility on producers and trade.

POSTHARVEST QUALITY MAINTENANCE OF FRUITS AND VEGETABLES IN DEVELOPING COUNTRIES

Adel A. Kader

Department of Pomology
University of California
Davis, CA 95616

QUALITY COMPONENTS

Quality of fruits and vegetables is a combination of attributes or properties that give them value in terms of human food. Components of quality include appearance, texture, flavor, and nutritive value (Table 1). Growers and shippers are concerned that their commodities have good appearance and few visual defects. But to them a useful cultivar of a given commodity must score high on yield, disease resistance, ease of harvest, and shipping quality. Plant breeders have given these characteristics higher priority over flavor and nutritional quality. To receivers and market distributors, quality of appearance is most important; they are also keenly interested in firmness and long storage life. Traditionally, postharvest biology and technology research has concentrated on using appearance and texture as parameters for quality evaluation (28). Consumers see quality fruits as ones that look good, are firm, and offer good flavor and nutritive value. Although they buy on the basis of appearance and feel, their satisfaction is dependent upon good eating quality.

Fresh fruits, nuts, and vegetables play a very significant role in human nutrition, especially as sources of vitamins, minerals, and dietary fiber (4, 12, 25, 30). Postharvest losses in vitamin content, particularly vitamin C, can be substantial. These losses are enhanced by extended storage, higher temperature, low relative humidity, physical damage, and chilling injury (8). A large volume of data on composition and compositional changes of fruits and vegetables is available (14, 15, 29). But many gaps exist and further research is needed. As more becomes known about human

Table 1. Quality Components of Fresh Fruits and Vegetables.

Main Factors	Components
A. Appearance (visual)	1. Size: dimensions, weight, volume 2. Shape & form: diameter/depth ratio, smoothness, compactness 3. Color: uniformity, intensity 4. Gloss: wax 5. Defects: external, internal a. Morphological b. Physical & mechanical c. Physiological d. Pathological e. Entomological
B. Texture (feel)	1. Firmness, hardness, softness 2. Crispness 3. Succulence, juiciness 4. Mealiness, grittiness 5. Toughness, fibrousness
C. Flavor (taste and smell)	1. Sweetness 2. Sourness (acidity) 3. Astringency 4. Bitterness 5. Aroma (volatile compounds) 6. Off-flavors and off-odors
D. Nutritive value	1. Carbohydrates (including dietary fiber) 2. Proteins 3. Lipids 4. Vitamins 5. Minerals
E. Safety	1. Naturally-occurring toxicants 2. Contaminants (chemical residues, heavy metals, etc.) 3. Mycotoxins 4. Microbial contamination

nutrition, additional compositional data will be needed. For example, it is not adequate to know how much total sugars are contained in a certain fruit. Information about individual sugars is important from the human nutrition as well as the sweetness standpoints.

Flavor is a complex sensation that involves perception of the tastes and aromas of many compounds. It is difficult to deal effectively with flavor in research programs because meaningful widespread taste testing is virtually impossible. An effective approach is to define the crucial components of flavor and then look at how these are affected by genotypes, cultural practices, and postharvest handling procedures (18, 26). Objective analytical evaluation of critical components coupled with subjective evaluations by a taste panel can result in meaningful and useful information about flavor.

Numerous methods are available for evaluation of color, texture, and other quality attributes (1, 3, 6, 9, 10, 11, 20). However, there is a need for developing new objective and non-destructive methods for quality evaluation. It is also important to better define the interrelationships among various components of quality (appearance including color, texture, flavor, nutritive value) in various fruits and vegetables. In each case, an attempt should be made to correlate subjective and objective methods of quality evaluation. Such information is essential for selection of new cultivars by plant breeders, choice of optimum production practices by production physiologists, and redefinition of optimum postharvest handling procedures by postharvest biologists. This total effort will no doubt result in the best quality fruits and vegetables possible for the consumers.

MATURITY AND QUALITY INDICES

Maturity indices are important for deciding when a given commodity should be harvested to provide some marketing flexibility and to insure the attainment of acceptable eating quality to the consumer. These two goals are not always compatible. The frequent need for shipping fruits and vegetables long distances has necessitated harvesting them at less than ideal maturity. This, in turn, has resulted in less than optimum quality to the consumer. Indices used for determining the legal maturity of fruits in most cases coincide with their minimum palatability.

Maturity Indices Currently Used and Their Limitations

For decades substantial effort has been directed by horticulturists towards the evaluation of maturity indices. Extensive data are available on morphological, physiological, and biochemical changes in fruits and vegetables during development, maturation, and ripening (14, 15). However, only a small portion of these data has been used in the establishment of maturity standards. In the

U.S. standards for grades, maturity is considered as one parameter of quality for many fruits and vegetables. It is defined as "that stage which will ensure proper completion of the ripening process."

Table 2 includes a listing of maturity indices used for selected fruits and vegetables. It is necessary for some commodities to define maturity indices for specific cultivars, production areas, and seasons. Although numerous objective indices for maturity are available, very few are actually used in practice because they are in most cases destructive and difficult to do in the field or orchard. Emphasis is on appearance factors, i.e., harvesting stage is determined by experience and judged largely by the visual appearance of the commodity.

Maturity vs. Quality

Most maturity indices are also factors of quality, but there are many important quality indices which are not used in determining optimum harvesting stage. The eating quality of fruits and vegetables cannot be accurately determined by appearance factors alone.

Timing of harvest (based on maturity indices) is complicated by the great differences which occur in the rate of development and maturation of individual plants, or organs on the same plant, bush or tree. This variability in maturation and ripening is especially important when once-over mechanical harvesting is used. Variability is related to preharvest cultural practices and environmental factors.

With a few exceptions (e.g., pears, avocados bananas), all fruits reach peak eating quality when fully ripened on the plant. Because of the constraints of the postharvest distribution system, fully-ripe fruits cannot be successfully delivered to the consumer except for roadside or pick-your-own type marketing situations. So compromises between optimum maturity and optimum quality have to be made.

For many vegetables, the optimum eating quality is reached before full maturity, e.g., leafy vegetables, immature fruits (cucumbers, sweet corn, green beans, peas, etc.). In this case, the problem frequently is delayed harvest which results in lower quality.

FACTORS INFLUENCING QUALITY AND ITS MAINTENANCE AFTER HARVEST

Many pre- and postharvest factors influence the composition and quality of fresh fruits and vegetables; these are:

1. Genetic factors: selection of cultivars, rootstocks.

Table 2. Maturity Indices for Selected Fruits and Vegetables.

Index	Examples
Elapsed days from full bloom to harvest	Apples, pears
Mean heat units during fruit development	Peas
Development of abscission layer	Cantaloupe
Surface morphology and structure	Cuticle formation on grapes, tomatoes Netting of cantaloupes Gloss of some fruits (development of wax)
Size	All fruits and many vegetables
Specific gravity	Cherries, watermelons, potatoes
Shape	Angularity of banana fingers Full cheeks of mangoes Compactness of broccoli and cauliflower
Solidity	Lettuce, cabbage, brussels sprouts
Textural properties	
Firmness	Apples, pears, stone fruits
Tenderness	Peas
Toughness	Asparagus
Color, external	All fruits and most vegetables
Internal color and structure	Formation of jelly-like material in tomato fruits Flesh color of some fruits
Compositional factors	
Starch content	Apples, pears
Sugar content Acid content sugar/acid ratio	Apples, pears, stone fruits, grapes, pomegranates, citrus, papaya, melons
Juice content	Citrus fruits
Oil content	Avocados
Astringency (tannin content)	Persimmons, dates
Internal ethylene concentration	Apples

2. Preharvest environmental factors
    a. Climatic: temperature, light, wind, rainfall, pollutants, etc.
    b. Cultural conditions: soil type, nutrient and water supply, mulching, pruning, thinning, control of pests and diseases, time and method of harvest, etc.

3. Harvesting stage: maturity, ripeness, physiological age.

4. Postharvest treatments: environmental factors, handling methods, duration between harvesting and consumption, etc.

5. Interactions among various factors.

Both quantitative and qualitative losses take place in horticultural crops between harvest and consumption. Our aim is to minimize these losses and to do so we must: 1) understand the biological and environmental factors involved in deterioration, and 2) use those postharvest technology procedures which will slow down senescence and maintain the best possible quality (13, 24, 27). Qualitative losses include loss in edibility, in nutritional quality, in caloric value, and in consumer acceptability of the products. Qualitative losses are much more difficult to assess than quantitative losses, especially since standards for quality and consumers' purchase power in developing countries are different than those in developed countries. For example, elimination of defects for a given commodity before marketing is much less rigorous in developing than in developed countries. This, however, is not necessarily bad, since appearance quality is somewhat over-emphasized in developed countries. A fruit or vegetable which is misshaped or has some blemishes may be as tasty and nutritious as one that is perfect in appearance. Any produce that is not spoiled (rotten) or totally unusable will have a market, if the price is right, in developing countries (16). Miles (21) proposed that an investigation should be conducted to determine the degree to which current trends in the consumption of fresh fruits and vegetables in the U.S.A. are indirectly the result of retailer demand for cosmetic qualities that are unrelated to flavor and nutritional quality; the degree to which there would be consumer acceptance of less blemish-free fruits and vegetables; the effect that lowered quality standards would have on the prices of fresh fruits and vegetables; and the effect that lower prices would have on consumption.

Biological and Environmental Factors Involved in Deterioration

Fresh horticultural crops are diverse in morphological structure (roots, stems, leaves, flowers, fruits, etc.), in composition, and in general physiology. Thus, commodity requirements and recommendations for maximum postharvest life vary among various groups of commodities. However, they are all subject to the following biological (internal) causes of quality deterioration:

1. Metabolic changes associated with development, maturation, and senescence (respiration, compositional changes);

2. Mechanical injuries (cuts, bruises, abrasions, etc.);

3. Transpiration or water loss (shrivelling, wilting, desiccation);

4. Growth and development (sprouting and rooting in some commodities, elongation, etc.);

5. Incidence of physiological disorders (sunburn, freezing injury, chilling injury, etc.); and

6. Pathological breakdown (decay caused by bacteria and fungi).

The rate of biological deterioration depends on various environmental (external) factors such as temperature, relative humidity, atmospheric composition, pressure, etc. (7, 13, 24, 27). The relative importance of the deterioration factors depends upon the commodity (Table 3). Losses in quality and quantity in fresh fruits and vegetables can occur throughout the harvesting and postharvest handling systems and are cumulative (Table 4). Losses closer to the consumer level are more costly because they include not only direct losses of the commodity but also losses of energy and natural resources used in postharvest handling (16).

There is a need for an accurate and specific identification of the causes and extent of losses in quantity and quality for each commodity at each stage between harvest and consumption. Such detailed information is essential to pinpoint problem areas in the handling system and to set priorities for loss prevention efforts in every developing country. The principal objectives of postharvest technology are: 1) to maintain quality of the commodity between harvest and consumption, and 2) to reduce losses during harvesting, preparation for market, transport, storage, and marketing operations. In general, the level of technology currently used in postharvest handling of fruits and vegetables in developing countries is not adequate for realizing the above-stated objectives (2, 5, 7, 16, 22, 23). Adoption of new technological procedures is badly needed provided they fit local conditions. Transfer of "advanced" technology such as that used in developed countries without some adaptation and modification to suit specific local conditions can be counterproductive.

Table 3. Principal Causes of Postharvest Losses and Poor Quality for Various Groups of Fruits and Vegetables.

Group	Examples	Principal Causes of Postharvest Losses and Poor Quality (in order of importance)
Root vegetables	Carrots Beets Onions Garlic Potato Sweet Potato	1-Mechanical injuries 2-Improper curing 3-Sprouting and rooting 4-Water loss (shrivelling) 5-Decay 6-Chilling injury (subtropical and tropical root crops)
Leafy vegetables	Lettuce Chard Spinach Cabbage Green Onions	1-Water loss (wilting) 2-Loss of green color 3-Mechanical injuries 4-Relatively high respiration rates 5-Decay
Flower vegetables	Artichokes Broccoli Cauliflower	1-Mechanical injuries 2-Yellowing and other discolorations 3-Abscission of florets 4-Decay
Immature-fruit vegetables	Cucumbers Squash Eggplant Peppers Okra Snap beans	1-Overmaturity at harvest 2-Water loss (shrivelling) 3-Bruising and other mechanical injuries 4-Chilling injury 5-Decay
Mature-fruit vegetables	Tomato Melons Citrus Bananas Mangoes Apples Grapes Stone fruits	1-Bruising 2-Over-ripeness and excessive softening at harvest 3-Water loss 4-Chilling injury (chilling sensitive fruits) 5-Compositional changes 6-Decay

Table 4. Most Common Causes of Postharvest Losses in Fresh Fruits and Vegetables in Some Developing Countries.

Postharvest Handling Operation	Causes of Losses
Harvesting	1-Immaturity or overmaturity of the commodity 2-Inadequate field containers 3-Mechanical damage due to improper harvesting methods 4-Failure to protect the commodity from the sun 5-Delays before delivery to packinghouse or transporting to market
Preparation for market (in the field or at the packinghouse)	1-Failure to sort-out produce with serious defects and decay; inadequate cleaning 2-Inappropriate packaging resulting in mechanical damage, inadeuate ventilation and cooling, and increased decay 3-Failure to remove field heat (lack of cooling prior to shipment) 4-Lack of sanitation
Transport	1-Rough handling causing increased mechanical injuries 2-Lack of proper management of temperature, relative humidity, and ventilation during transit 3-Mixing of non-compatible commodities in the transport vehicle (different types of containers which are not easily stackable together, different temperature requirements, ethylene-producing and non-producing commodities) 4-Delays during transport
Handling at destination	1-Rough handling during loading and unloading 2-Exposure to undesirable enviornmental conditions 3-Delays in getting the commodity to the consumer 4-Improper ripening and storage practices 5-Lack of sanitation
Handling at home	1-Delays before consumption 2-Improper storage (lack of home refrigerators or other means of storage)

Non-biological (Socio-economic) Factors Involved in Deterioration

Several indirect and non-biological, but very important, factors contribute to postharvest qualitative and quantitative losses of fresh fruits and vegetables (16). These include:

1. Inadequate marketing systems. Growers can produce large quantities of good-quality fruits and vegetables, but if they do not have a dependable, fast, and equitable means of getting such commodities to the consumer, losses will be extensive. This problem exists in many locations within developing countries. It is accentuated by lack of communication between producers and receivers, and lack of market information.

Marketing cooperatives should be encouraged among producers of major commodities in important production areas. Such organizations are especially needed in developing countries because of the relatively small farm size. Advantages of marketing cooperatives include: 1) providing central accumulation points for the harvested commodity, 2) purchasing harvesting and packing supplies and materials in quantity, 3) providing facilities for proper preparation for market and storage when needed, 4) facilitating transportation to the markets, and 5) acting as a common selling unit for the members, coordinating the marketing program and distributing profits equitably.

Alternative distribution systems such as direct selling to the consumer (roadside stands, produce markets in cities, local farmers' markets in the countryside, etc.) should be encouraged. Production should be maintained as close to the major population centers as possible to minimize transportation costs.

Wholesale markets in most of the developing countries are in desperate need of improvement in terms of facilities and sanitation. These are overcrowded, unsanitary, and lack adequate facilities for loading, unloading, ripening, consumer packaging, and temporary storage. In several countries, there are plans to build better wholesale marketing facilities, but their implementation has been delayed more because of social and political than because of financial considerations.

2. Inadequate transportation facilities. In most developing countries, roads are not adequate for proper transport of horticultural crops. Also, transport vehicles and other modes, especially those suited for fruits and vegetables, are in short supply. This is true whether for local marketing or export to other countries. The majority of producers have small holdings and cannot afford owning their own transport vehicles. In a few cases, marketing organizations and cooperatives have been able to

acquire transport vehicles, but they cannot do much about poor road conditions.

3. Governmental regulations and legislations. The degree of governmental controls especially on wholesale and retail prices of fresh fruits and vegetables varies from one country to another. In many cases, price controls are counter-productive. Although intended for consumer protection, such regulations encourage fraud and provide no incentive for producing high quality produce or for postharvest quality maintenance. On the other hand, regulations covering proper handling procedures and public health aspects during marketing are, if enforced properly, very important to the consumer.

4. Unavailability of needed tools and equipment. Even if the growers and handlers of fresh fruits and vegetables were convinced of the merits of using some special tools and/or equipment in harvesting and postharvest handling, they most likely will not be able to find them on the domestic market. This is true of harvesting aids, containers, equipment for cleaning, waxing and packing, and cooling facilities. Most of these tools are neither manufactured locally nor imported in sufficient quantity to meet demand. Various governmental regulations in some countries do not permit direct importation by producers of their needs. It is imperative that the tools that will enable handlers to use recommended technology for a given situation be available for them to use. In many cases such tools can be manufactured locally at much lower cost than the imported ones.

5. Lack of information. The human element in postharvest handling of fruits and vegetables is extremely important. Most handlers directly involved in harvesting, packaging, transporting, and marketing in developing countries have limited or no appreciation for the need for or how to maintain quality. An effective and far-reaching educational (extension) program on these aspects is critically needed now and will continue to be essential in the future.

6. Poor maintenance. In many developing countries some good facilities which were built a few years ago are currently "out-of-order" or not functioning properly because of lack of maintenance and unavailability of spare parts. This problem is especially true of public sector facilities. Any new project should include in its plan adequate funds for maintenance to ensure its success and extended usefulness.

Kriesberg and Steele (19) stated that at each stage of the marketing system in developing countries there are forces exogenous to it which influence its development. Among political and economic factors are public policies, the general stage of technology, and income levels and their distribution. Among social and cultural factors are urbanization, education and population growth and its characteristics. Other factors working more directly on the marketing systems include the kinds and quantities of commodities available for market, and consumer demand and commodity preferences.

All non-biological factors influencing quality deterioration are much more difficult to change than the environmental factors which control biological deterioration. However, to be successful, any improvement program must take all factors into consideration.

QUALITY STANDARDIZATION AND INSPECTION

Grade standards are developed to identify the degrees of quality in the various commodities which aid in establishing their useability and value. They are important tools in the marketing of fresh fruits and vegetables because of the following factors: a) they provide a common language for trading between growers, handlers, processors, and receivers at terminal markets; b) they assist producers and handlers in doing a better job of preparing fresh horticultural commodities for market and appropriate labeling; c) they provide a basis for making incentive payments for better quality; d) they serve as the basis for marketing reporting; and e) they help settle damage claims and disputes between buyers and sellers.

The first U.S. Grade Standards were developed for potatoes in 1917. Currently there are more than 150 standards covering 80 different commodities. In addition, several states also have mandatory minimum quality standards. The International Standards for Fruits and Vegetables, which have been defined by the Economic Commission for Europe (since 1954) and the Organization for Economic Cooperation and Development, have provided the basis for EEC standards currently in effect for 38 commodities. Developing countries must use these quality standards for their commodities which are exported to EEC countries. With very few exceptions, no quality standards are used for fresh fruits and vegetables destined for local marketing. The establishment of simple grade standards and their use for local distribution in developing countries can be major steps which would help reduce fraud and deception in packaging and encourage high and uniform quality. Enforcement of these standards will require a well-trained group of inspectors in each country and an effective extension program to inform all producers and handlers about the standards. They need to be convinced that proper standardization and inspection would promote trust and encourage commerce.

POSTHARVEST TECHNOLOGY PROCEDURES IN RELATION TO QUALITY

Commodity requirements and recommended conditions for optimum quality maintenance and postharvest life are the same regardless of the distribution system (direct marketing, local marketing, export, etc.). However, the type of appropriate technology needed to provide such conditions will depend upon the distance and time between production and consumption areas as well as intended use (fresh vs. processing). In selecting the proper postharvest technology procedures, one should remember the following:

A. The technology used elsewhere is not necessarily the best for use under conditions of a given developing country. Many of the recent modifications in postharvest technology in developed countries have been in response to the need to economize in labor, materials, and energy use, and to protect the environment. It is useful to study the currently used practices in other countries and to select those which are appropriate for local conditions.

B. Expensive equipment and facilities without proper management are useless. People who operate such facilities are more important than their level of sophistication.

C. Commodity requirements can be provided using simple and inexpensive methods in many cases. For example, proper temperature management procedures include:

   1. Protection from exposure to the sun,
   2. Harvesting during cooler parts of the day or even at night,
   3. Adequate ventilation in containers and non-refrigerated transport vehicles,
   4. Possibly simple and inexpensive cooling procedures such as evaporative cooling and use of cool-night ambient air, and
   5. Expedited handling.

D. Mechanical injuries are major causes of losses in quality and quantity of fresh horticultural commodities in all handling systems. Their incidence and severity can be greatly reduced by simple modifications in harvesting and handling procedures and by informing all personnel involved about the need for careful handling.

Solving the postharvest technology problems in a given country will require cooperation and effective communication among all the research and extension personnel involved. Postharvest horticulturists need to coordinate their efforts and to cooperate with production horticulturists, agricultural marketing economists, engineers, food technologists, and others who may be involved in various aspects of the marketing systems. In most cases, solutions to existing problems in the postharvest handling system require use of available information (17) rather than any new research. Following is a proposed program for improving the postharvest handling system in a developing country:

1. Survey the magnitude and causes of losses in quality and quantity during harvesting and postharvest handling of major commodities.

2. Survey available tools and facilities for harvesting, packing, transport, storage, etc., for each commodity in its important production seasons and areas.

3. Evaluate the impact of simple modifications in the handling system (picking stage and method, type of containers, quality sorting, etc.) on quality maintenance and losses.

4. Extend information about recommended harvesting and handling procedures to all those who can use it. All appropriate extension methods for the intended audiences should be used.

5. Identify problems which need further research, carry out research and extend any new information when completed.

FUTURE RESEARCH NEEDS

Additional research in both developed and developing countries is needed to improve our understanding of quality and its postharvest maintenance in fresh fruits and vegetables. The objectives of this research should be to:

A. Better define components of quality (appearance, texture, flavor, nutritive value) and their interrelationships for various fruits and vegetables destined for the fresh market or for processing.

B. Develop objective and non-destructive methods for determination of appearance and textural quality and optimum maturity which are related to their flavor and nutritional quality.

C. Evaluate the effects of preharvest factors (genetic, environmental, and cultural) on flavor and nutritional quality of fruits and vegetables.

D. Develop multiple maturity indices related to distance to market and intended use of the commodity (fresh market, processing, etc.).

E. Relate maturity indices at harvest to the final organoleptic acceptability by the consumer and to nutritional quality.

F. Study the effects of any proposed changes in harvesting and postharvest handling practices on quality and safety attributes.

REFERENCES

1. Arthey, V. D. 1975. Quality of horticultural products. Halstead Press, John Wiley and Sons, NY, 228 p.
2. Bourne, M. C. 1977. Post-harvest food losses -- the neglected dimension in increasing the world food supply. Cornell University International Agriculture Mimeograph No. 53.
3. Bourne, M. C. 1980. Texture evaluation of horticultural crops. HortScience 15:51.
4. Bradley, G. A. 1972. Fruits and vegetables as world sources of vitamins A and C. HortScience 7:141.
5. Claypool, L. 1976. Consultancy report on postharvest physiology and technological problems in some countries of the Near East and North Africa. FAO Proj. REM/74/015. 31 p.
6. Dull, G. G., G. S. Birth, and J. B. Magee. 1980. Nondestructive evaluation of internal quality. HortScience 15:60.
7. FAO. 1981. Food loss prevention in perishable crops. FAO Agr. Serv. Bul. 43, United States Food and Agriculture Organization, Rome, Italy, 72 p.
8. Fennema, O. 1977. Loss of vitamins in fresh and frozen foods. Food Technol. 31:32.
9. Finney, E. E., Jr. 1970. Measurement techniques for quality control of agricultural products. Amer. Soc. Agric. Eng. Spec. Publ., ASAE, St. Joseph, MI, 53 p.
10. Francis, F. J. 1980. Color quality evaluation of horticultural crops. HortScience 15:58.
11. Gaffney, J. J. (compiler). 1976. Quality detection in foods. ASAE Publ. 1-76, Amer. Soc. Agric. Eng., St. Joseph, MI, 240 p.
12. Goddard, M. S. and R. H. Matthews. 1979. Contribution of fruits and vegetables to human nutrition. HortScience 14: 245.
13. Harvey, J. M. 1978. Reduction of losses in fresh market fruits and vegetables. Annu. Rev. Phytopathol. 16:321.
14. Hulme, A. C. (ed.). 1970. The biochemistry of fruits and their products. Vol. 1. Academic Press, NY, 620 p.
15. Hulme, A. C. (ed.). 1971. The biochemistry of fruits and their products. Vol. 2. Academic Press, NY, 788 p.
16. Kader, A. 1979. FAO consultancy report on postharvest handling of fruits and vegetables in the Middle East. 133p.
17. Kader, A. A., L. L. Morris, and M. Cantwell. 1979. Postharvest handling and physiology of horticultural crops: a list of selected references. Vegetable Crops Series 169, University of California, Davis.

18. Kader, A. A., L. L. Morris, M. A. Stevens, and M. Albright-Holton. 1978. Composition and flavor quality of fresh market tomatoes as influenced by some postharvest handling procedure. J. Amer. Soc. Hort. Sci. 103:6-13.
19. Kriesberg, M. and H. Steele. 1972. Improving marketing systems in developing countries -- an approach to identifying problems and strengthening technical assistance. U.S. Dept. Agr., Econ. Res. Serv., Foreign Agr. Econ. Rept. No. 93, 85 p.
20. Lipton, W. J. 1980. Interpretation of quality evaluations of horticultural crops. HortScience 15:64.
21. Miles, G. H. 1977. Alternative food delivery systems: an exploratory assessment. National Science Foundation, RANN Grant No. AER 77-07184 Report, 299 p.
22. National Academy of Sciences. 1978. Postharvest Food Losses in Developing Countries. National Academy of Sciences, Washington, D.C., 202 p.
23. National Academy of Sciences. 1978. Postharvest Food Losses in Developing Countries: A bibliography. National Academy of Sciences, Washington, D. C., 356 p.
24. Rhodes, M. J. C. 1980. The physiological basis for the conservation of food crops. Prog. Food Nutr. Sci. 4(3-4): 11.
25. Senti, F. R. and R. L. Rizek. 1975. Nutrient levels in horticultural crops. HortScience 10:242.
26. Stevens, M. A. and M. Albright. 1980. An approach to sensory evaluation of horticultural commodities. HortScience 15: 48.
27. Tindall, H. D. and F. J. Proctor. 1980. Loss prevention of horticultural crops in the tropics. Prog. Food Nutr. Sci. 4(3-4):25.
28. Watada, A. E. 1980. Quality evaluation of horticultural crops: the problem. HortScience 15:47.
29. Watt, B. K. and A. L. Merrill. 1963. Composition of foods, raw, processed. USDA, Agr. Hb. No. 8, 190 p.
30. White, P. L. and N. Selvey (eds.). 1974. Nutritional qualities of fresh fruits and vegetables. Futura Publ. Co., Mount Kisco, NY. 186 p.

INSTRUMENTAL TECHNIQUES FOR MEASURING QUALITY OF AGRICULTURAL CROPS

Karl H. Norris

Instrumentation Research Laboratory
BARC, ARS, USDA
Beltsville, MD 20705

INTRODUCTION

Many instrumental techniques have been developed for measuring the quality of agricultural products. However, instrumentation specialists often feel frustrated in this effort because quality as related to agricultural products is difficult to define in terms of physical parameters which can be readily quantified. In general, it is the chemical composition of the product which really determines the quality, but most often, we define the quality by our senses of sight, touch, or smell, rather than by the composition. It has been our experience that it is much easier to develop techniques to measure composition than it is to measure quality. Therefore, this paper will concentrate on the measurement of composition as a means of indicating quality. Spectrophotometry by transmission, reflectance, and fluorescence offers the best technology for instrumental measurement of composition, so we will examine the visible and near-infrared region of the spectrum for this purpose.

NEAR-INFRARED REFLECTANCE (NIR)

Diffuse reflectance spectrophotometry is now widely used for rapid prediction of composition of grains and oilseeds. This technology is based on the fact that each of the major constituents of grains and oilseeds has definitive absorption characteristics in the 1000 to 2600 nm region, and these absorption characteristics can be sensed by diffuse reflectance. The spectra for the major constituents are shown in Fig. 1. The data are plotted as log (1/R) where R is the diffuse reflectance. The spectra plotted

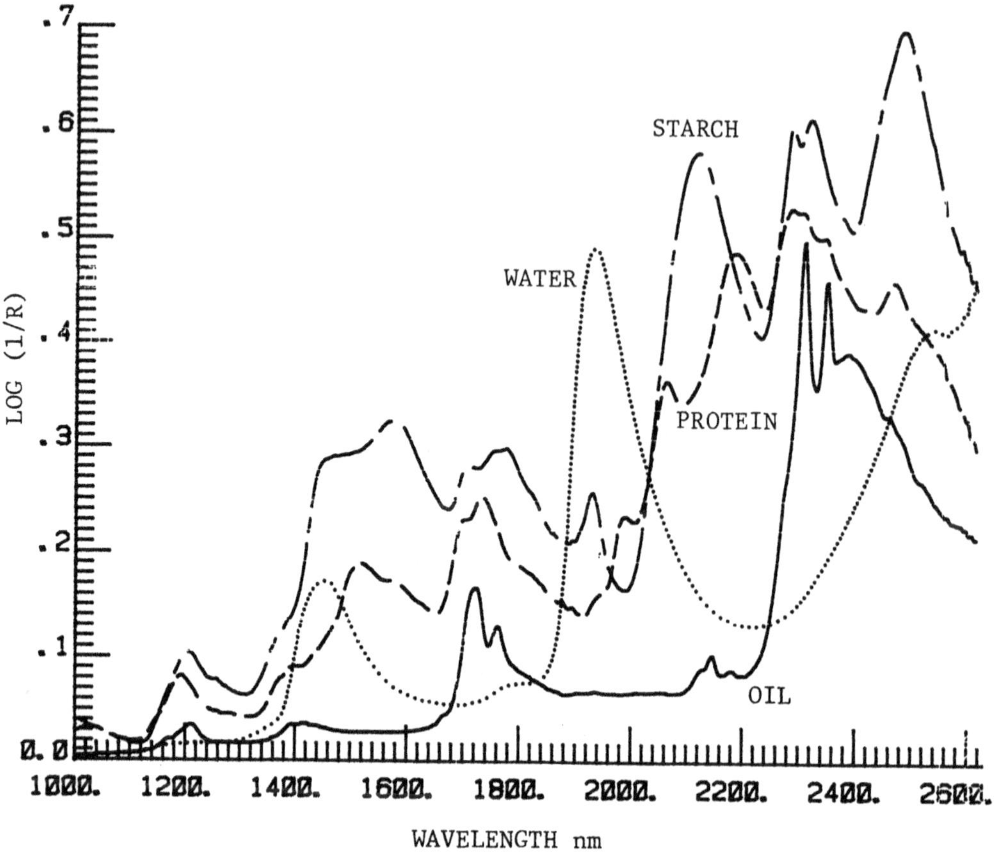

Fig. 1. Spectra of major components of grains and oilseeds.

as log (1/R) are very similar to absorption spectra, and it has been found that log (1/R) varies linearly with concentration of a specific àbsorber in a mixture with other materials.

These spectra were recorded with an inhouse-designed computerized spectrophotometer. This instrument is built around a Cary Model 14 prism-grating monochromator with optics optimized for the infrared. The monochromator is coupled to a digital computer for collection and analysis of data. The instrument is operated in a single-beam mode with a reference spectrum stored in the computer. The sample, packed into a cell and covered with a quartz window, is illuminated through the window and diffusely reflected radiation is collected with four lead sulfide cells surrounding the sample. Several different materials have been used as reflectance reference standards, but we now use a ceramic material because its reflectance is the same for all wavelengths. Solid teflon, powdered telfon, calcium carbonate, magnesium carbonate, and barium sulfate, previously used as standards, all exhibit absorption bands within the 1000 to 2600 nm region. The signal

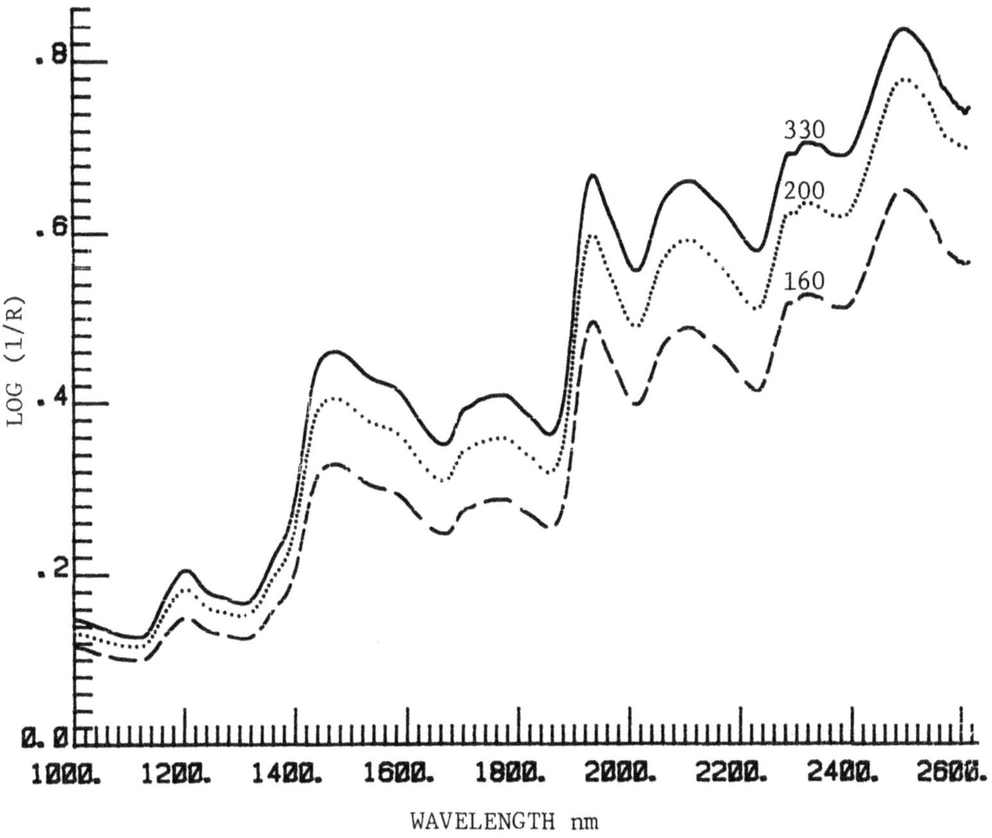

Fig. 2. Spectra of a sample of wheat ground with three different grinders to produce mean particle sizes of 330, 200, and 160 μm respectively.

from the detectors is amplified, digitized, and fed to the computer. The computer corrects for the system response, including the reflectance of the ceramic standard, and processes the data to log (1/R) for storage on disk or magnetic tape. The scanning speed is 10 nm per second with data collected at 0.2 nm intervals. The data processing includes smoothing by averaging of 10 adjacent points on either side of each data point and shrinking of points to a maximum of 1024. The resulting spectrum at 1.6 nm between points has sufficient resolution to resolve the sharpest absorption bands which have been observed in grains and oilseeds.

The near-infrared diffuse reflectance spectra of grains and oilseeds are affected by many factors, including: particle size, particle shape, packing density, sample temperature, and the composition of the sample. The effect of particle size is shown in Fig. 2 for a wheat sample ground with three different grinders to produce three samples with the same composition having mean

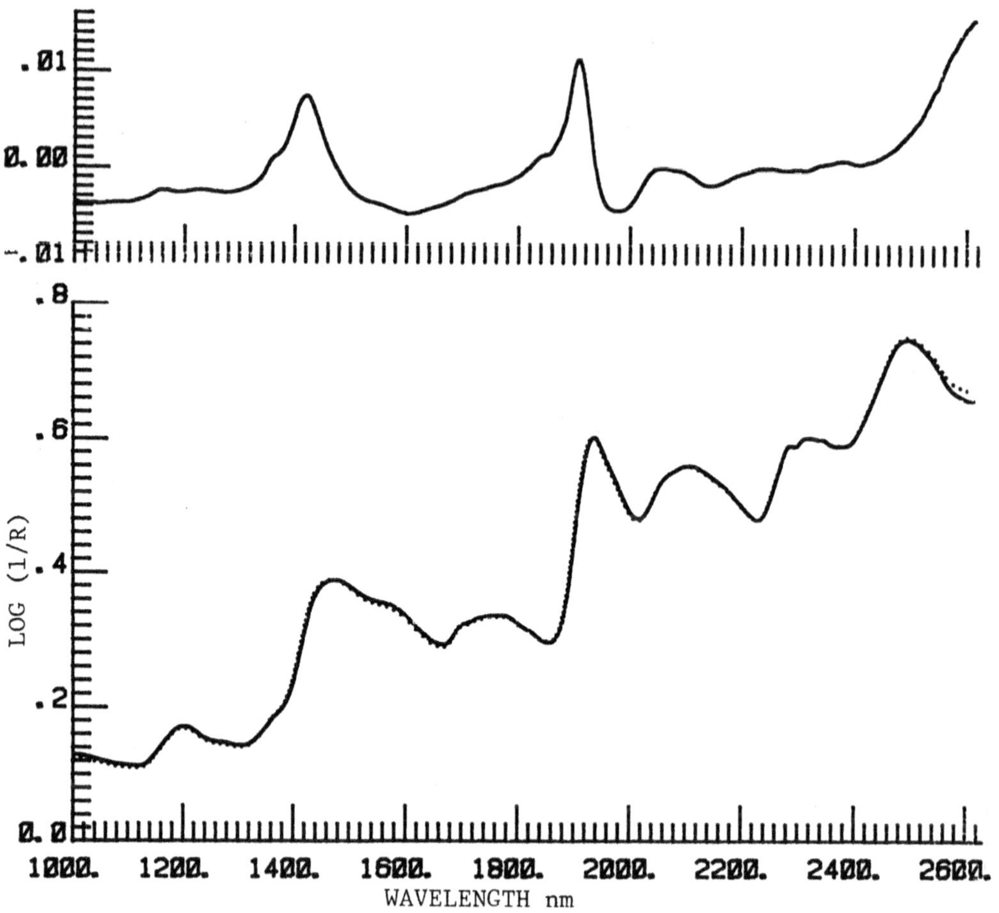

Fig. 3. Spectra of a wheat sample at 15°C ——— and 25°C ·····. Upper curve is the difference curve (25°-15°).

particle sizes of 160, 200, and 330 μm, respectively. As the particle size increases, the apparent absorption increases causing an increase in log (1/R). The effect of particle size is greater for strong absorbers than for weak absorbers, so the effect is greater at the longer wavelengths. A similar effect is observed for packing density with more tightly packed samples having lower log (1/R) values.

Sample temperature affects the near-infrared spectra because the water absorption band is temperature sensitive. This effect is difficult to observe in Fig. 3 for a wheat sample at two different temperatures. However, a plot of the difference curve, subtracting one curve from the other, clearly shows the temperature effect. The greatest effect is in the region of the water absorption bands at 1450 nm and 1940 nm. The dip in the curve at 2100 nm

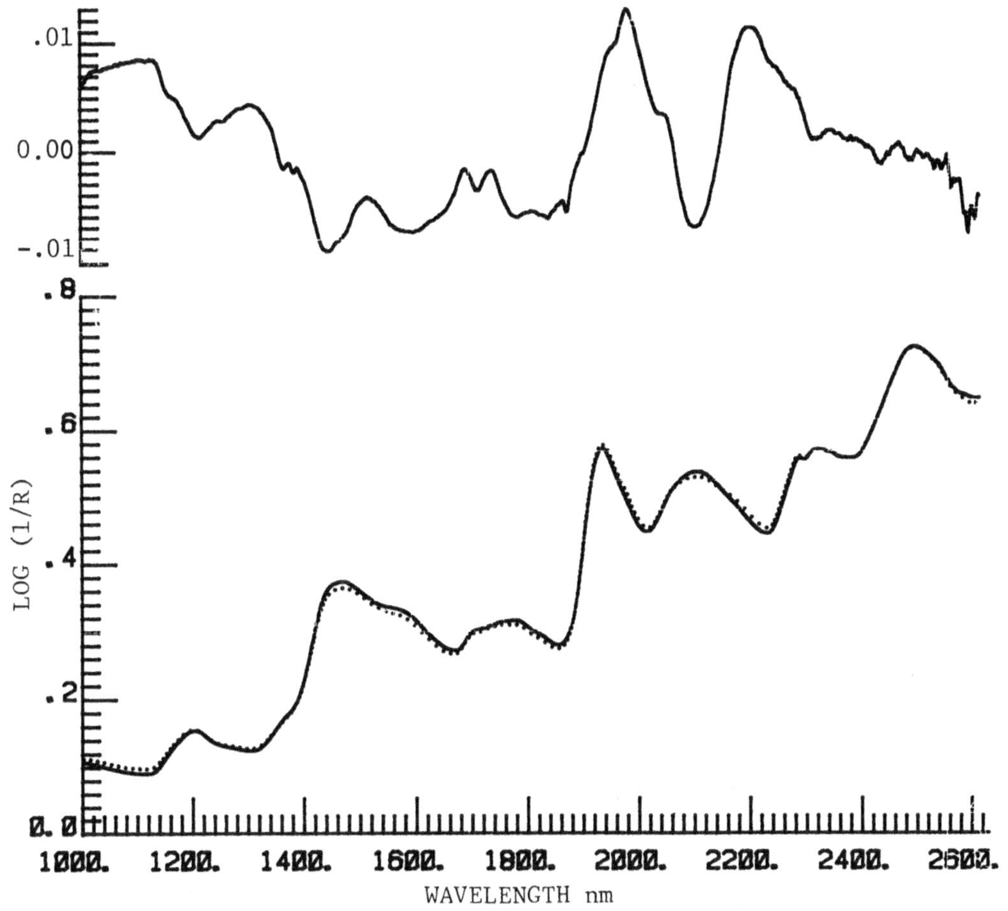

Fig. 4. Spectra of two wheat samples. ····· 20.0% protein, ——— 11.8% protein. Upper curve is the difference curve (20.0%-11.8%).

suggests an interaction between carbohydrates and water with respect to temperature, but this has not been explored.

Relatively large differences in protein cause only a small change in the log (1/R) curve for wheat. Fig. 4 shows the curves for two samples with 20.0 and 11.8% protein, respectively. The difference curve, subtracting the curve for 11.8% protein from the curve for 20.0% protein, more clearly shows the effect of the protein difference with major effect in the carbohydrate band at 2100 nm and the protein bands at 1980, 2060, and 2180 nm. It should be noted that an increase in protein in wheat is accompanied by a decrease in carbohydrates, and the difference curve shows this relationship.

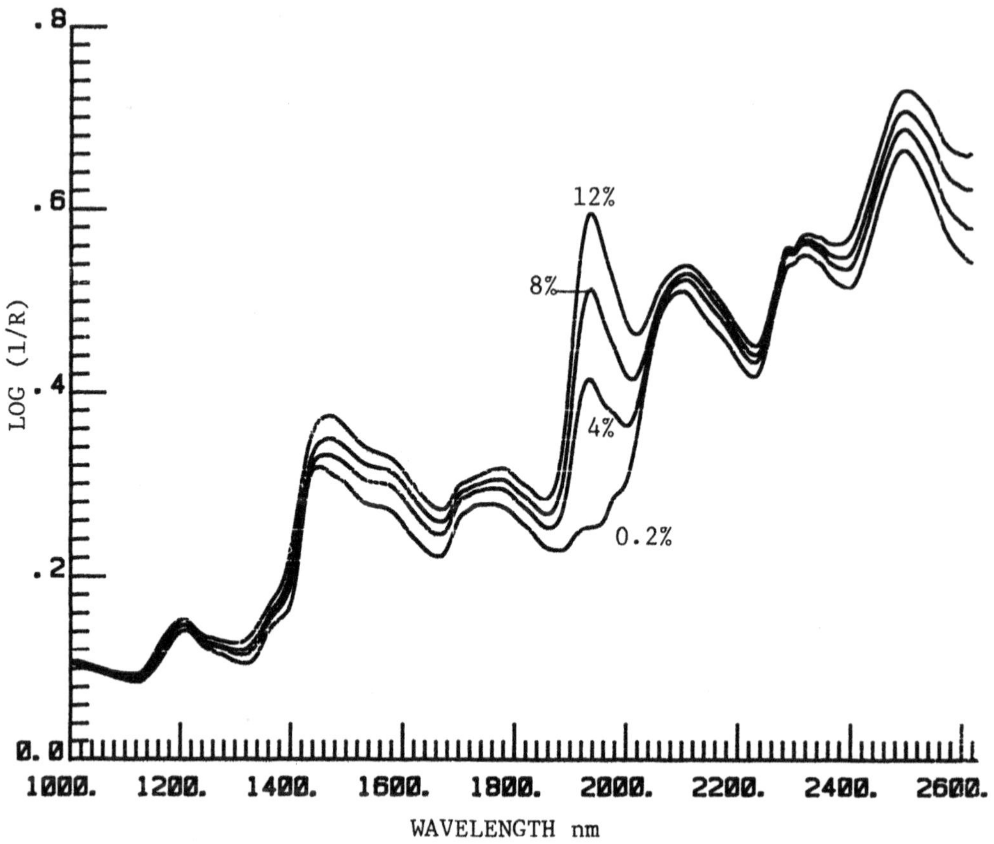

Fig. 5. Wheat spectra at four moisture levels.

Differences in moisture content are readily apparent as shown in Fig. 5 for wheat at different moisture levels. The greatest effect is in the regions where water absorbs at 1940 and 1450 nm. Another factor which affects NIR spectra is the form of the constituents. The spectra of two breakfast cereals (Fig. 6) are quite different, although they are very similar in composition. The difference between the two spectra results from the fact that the sugar is in the amorphous form for one sample and in the crystalline form in the other. The spectrum of crystalline sucrose shows all the sharp spectral features which distinguish the spectrum of one cereal from the other. The sharp spectral features of crystalline sucrose have not been observed with natural products, even those with very high sucrose content, such as dates. However, this type of effect could occur with processed products.

The problem of analyzing the diffuse reflectance curve of a sample to predict its composition is obviously not a simple problem because of the many factors which affect the curve. Several methods of data treatment have been used to analyze NIR spectra. The

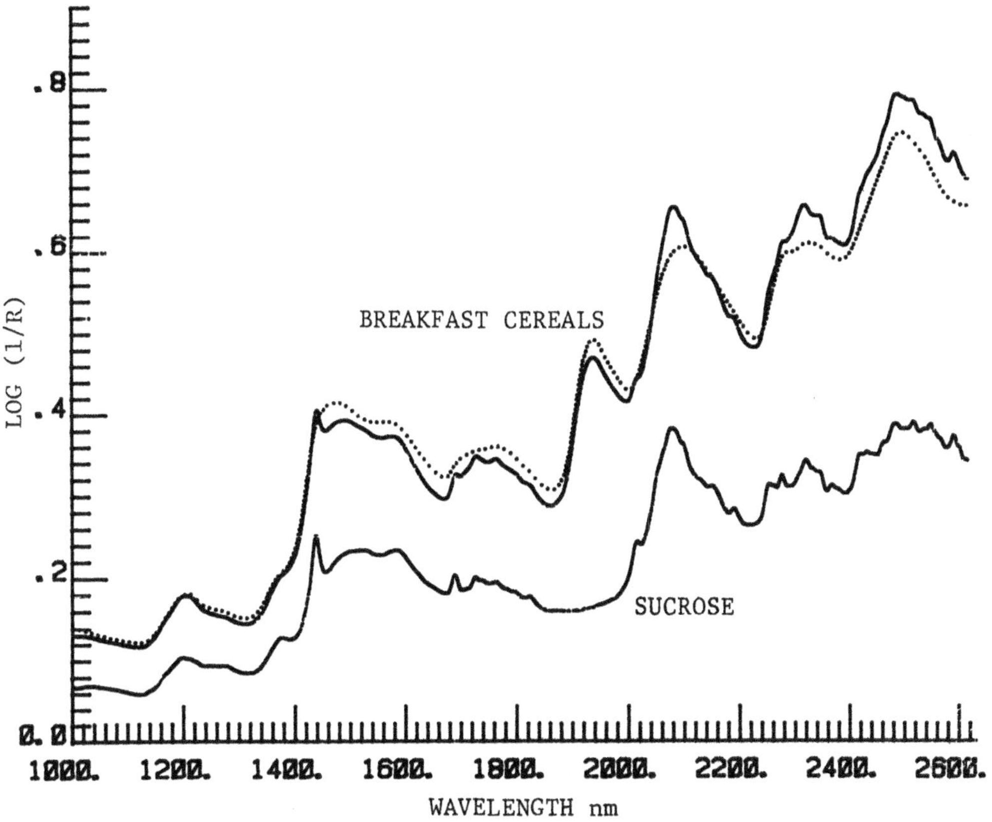

Fig. 6. Spectra of two breakfast cereals with the same sucrose content (36%), and the spectrum of sucrose.

simplest is a measurement of log (1/R) at each of the wavelengths for the maximum absorption of the constituents. These data are then combined in a multiple regression program to produce a calibration for each constituent. For example, the prediction of oil, moisture, and protein is made with measurements at six wavelengths. These are: 2308 nm for oil, 2180 nm for protein, 1940 nm for water, 2100 nm for starch, and then two reference wavelengths of 1680 nm and 2230 nm at which the absorption is a minimum. Varying combinations from three to six of these wavelengths are used for each of the constituents of oil, moisture, and protein in grain and oilseeds.

A second data treatment uses the difference signal at two wavelengths for each constituent, where one wavelength is at a peak absorption and the other is at a minimum absorption. Measurements at 2308 nm and 2230 nm are used for oil, 2180 nm and 2230 nm for protein, and 1940 and 1800 nm for water. These three sets of differences are then combined in a multiple regression program to produce a calibration for oil, protein, and moisture.

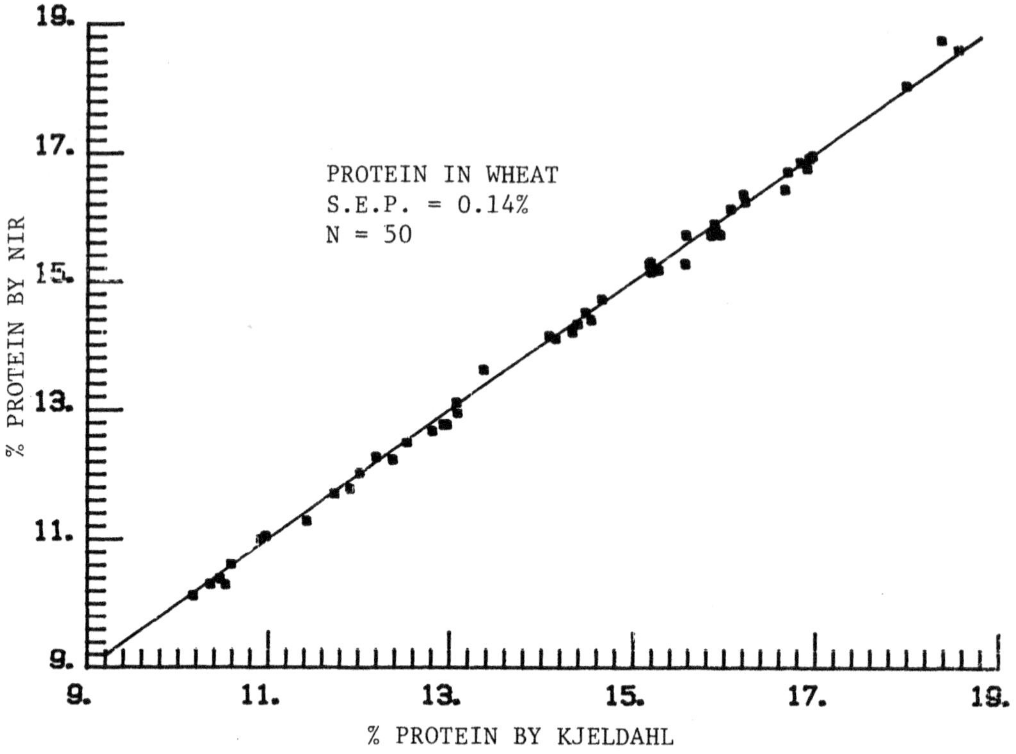

Fig. 7. Scatter plot of protein predicted from NIR spectra of ground wheat.

A more complex data treatment uses the first or second derivative of the log (1/R) data, with two to four first or second derivative terms combined in a multiple regression equation. The second derivative at 2308 nm is most sensitive to oil, 2180 nm is most sensitive to protein, and 1940 nm is most sensitive to water. More than four derivative terms can be used for calibration, but overfitting of the data may occur if the number of samples is inadequate. It is recommended that the number of samples should be greater than fifteen times the number of regression terms.

A single-term regression using a ratio of first derivatives at two wavelengths has been very successful for predicting the protein content of wheat samples with wide variation in particle size, temperature, and moisture content. The wavelengths used are 2160 nm for the numerator and 2270 nm for the denominator. This data treatment gave a standard error in predicting the protein content of Hard Red Spring wheat of 0.14% protein for samples varying from 10.1% to 18.7% protein (Fig. 7).

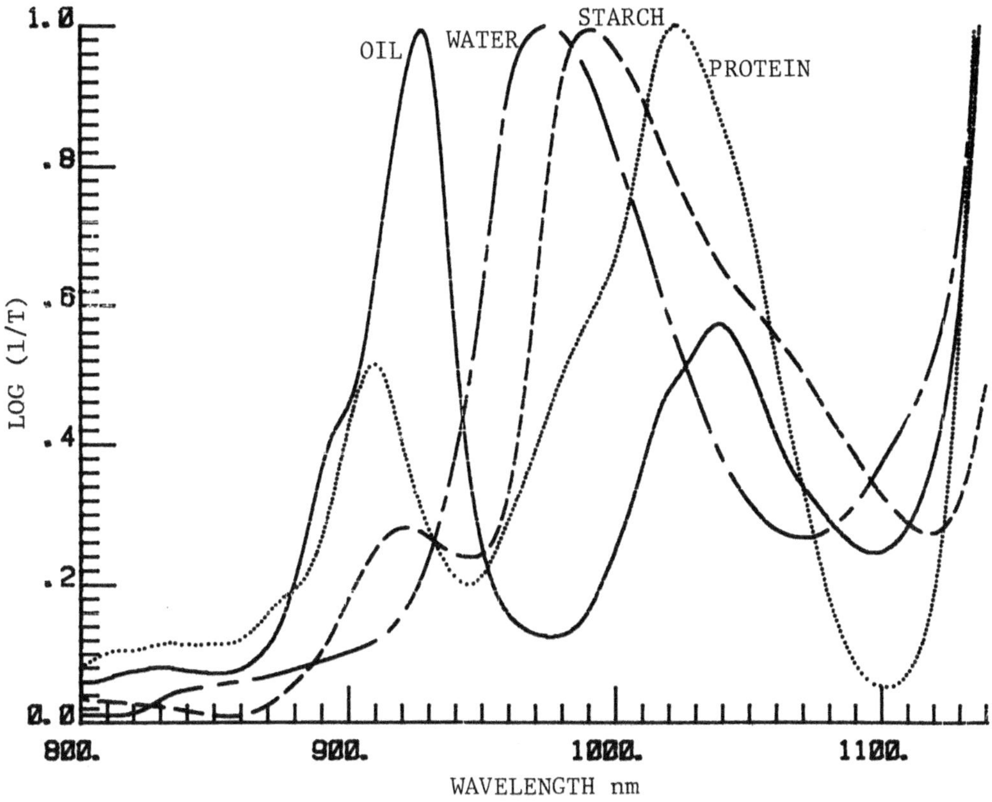

Fig. 8. Short-wavelength absorption spectra of oil, water, starch, and protein.

Other data treatments include multiterm first or second derivative ratios, curve fitting, and factor analysis. The optimum data treatment appears to be dependent on a number of sample-related parameters, and on the available computer capability.

Three manufacturers are now marketing NIR instruments. They are: Neotec Instruments Division of Pacific Scientific, Silver Spring, MD; Dickey-john Corp., Auburn, IL.; and Industrial Systems Division of Technicon Corp., Tarrytown, NY. These three firms market filter instruments designed for analyses of grains and oilseeds but their designs are applicable to other products as well. In addition, Neotec and Technicon market computerized spectrophotometers for research applications.

NEAR-INFRARED TRANSMITTANCE

The initial work in rapid infrared analysis used transmittance of a thin slurry of sample and carbon tetrachloride to measure moisture content (Norris and Hart, 1965; Ben Gera and Norris, 1968). The magnitude of the absorption coefficients over the 1400 to 2700 nm region limits the sample thickness to less than 1 mm for dry powders and the problems of working with a hazardous chemical has limited the application of the transmittance technique. However, the major constituents of grains and oilseeds also have definitive absorptions in the 800 to 1100 μm region. In this region, the absorption coefficients are weak enough to permit transmittance measurements on relatively thick samples. The relative absorption characteristics of starch, protein, oil, and water are compared in Fig. 8.

Large area, high sensitivity, silicon detectors make it possible to measure the transmittance of 10 cm of clear water or oil, 2 cm of potato tissue, 2 cm of whole grain, 1.5 cm of homogenized meat, and 0.3 cm of finely ground grain or other dry powder in the 800 to 1100 μm region. Therefore, this region offers the possibility for a rapid analysis on a wide range of agricultural products. This region has been explored for the measurement of moisture and protein of samples of intact wheat. Spectral transmittance measurements were made on 2.1 cm thick samples of whole-grain wheat of 8 different classes. A second derivative ratio calibration for each class of wheat was developed using log (1/T) data where T is the transmittance relative to air. These calibrations were used to predict the protein content of another set of samples with the results as shown in Table 1. The mean error of 0.29% protein is more than adequate for many applications. An instrument based on this development is now marketed by Trebor Industries, Gaithersburg, MD.

Table 1. Protein Errors for Eight Classes of Whole-Grain Wheat.

Class	HRS	HRW	SRW	DURUM	Semi-Dwarf	Western White	White Club	Soft White
S.D.[a]	.93	1.11	.62	1.09	1.89	1.59	1.79	1.31
S.E.P.[b]	.207	.291	.232	.305	.216	.288	.316	.446
BIAS	-.022	-.065	.053	-.04	-.02	.003	-.010	.026

[a]S.D. is the standard deviation for the protein content within each class.

[b]S.E.P. is the standard error in predicting the protein compared to the Kjeldahl results.

Transmittance measurements have been used on individual seeds to measure moisture content of peanuts (Norris and Hart, 1965) and of corn kernels (Finney and Norris, 1978). This type of measurement is now being explored in the author's laboratory to measure the oil, protein, and moisture content of individual seeds.

Reflectance measurements can be useful in this region although the reflectance signals are relatively weak. The fat content of ground beef has been measured by reflectance in this region using a gallium arsenide infrared-emitting diode as the radiation source (Massie, 1976). In this work, the temperature of the diode was changed to scan a limited spectral region so that no filters or other type of spectral-discriminating elements were needed. Infrared-emitting diodes, being developed for this region, may greatly enhance the possibilities for low-cost instruments to measure the composition of agricultural commodities.

VISIBLE TRANSMITTANCE AND REFLECTANCE

Reflectance measurements are widely used to evaluate color and to detect defects. Many products are sorted automatically based on reflectance measurements in the visible region. Because of the wide availability of easy-to-use color-measuring instruments, they are often used to predict composition of agricultural products. However, it should be noted that such instruments were not designed to measure composition, and for this reason, they do not represent the optimum design for composition analysis. The computerized spectrophotometric techniques developed for NIR can be applied in the visible to analyze for anthocyanins, chlorophyll, carotenoids, and other pigments.

Much of the effort for transmittance requirements in the visible has concentrated on intact commodities. Multiplier-type phototubes provide the sensitivity needed to measure the light transmitted through an object such as an apple, peach, or tomato. A computerized spectophotometer has been developed for such measurements (Massie and Norris, 1975) and a two-filter instrument has been described for specific measurements (Birth and Norris, 1965). A four-filter instrument, using a silicon detector in place of the phototube, has been marketed by Neotec. The silicon detector limits the range of the instrument both in wavelength and density of sample which can be measured. The Neotec instrument is most useful in the 600 to 1000 μm region, and for samples with an optical density less than 6.0.

The visible absorption spectra of intact tomatoes (Fig. 9) show strong chlorophyll absorption in the 600 to 690 nm region for the mature green stage. The chlorophyll absorption which

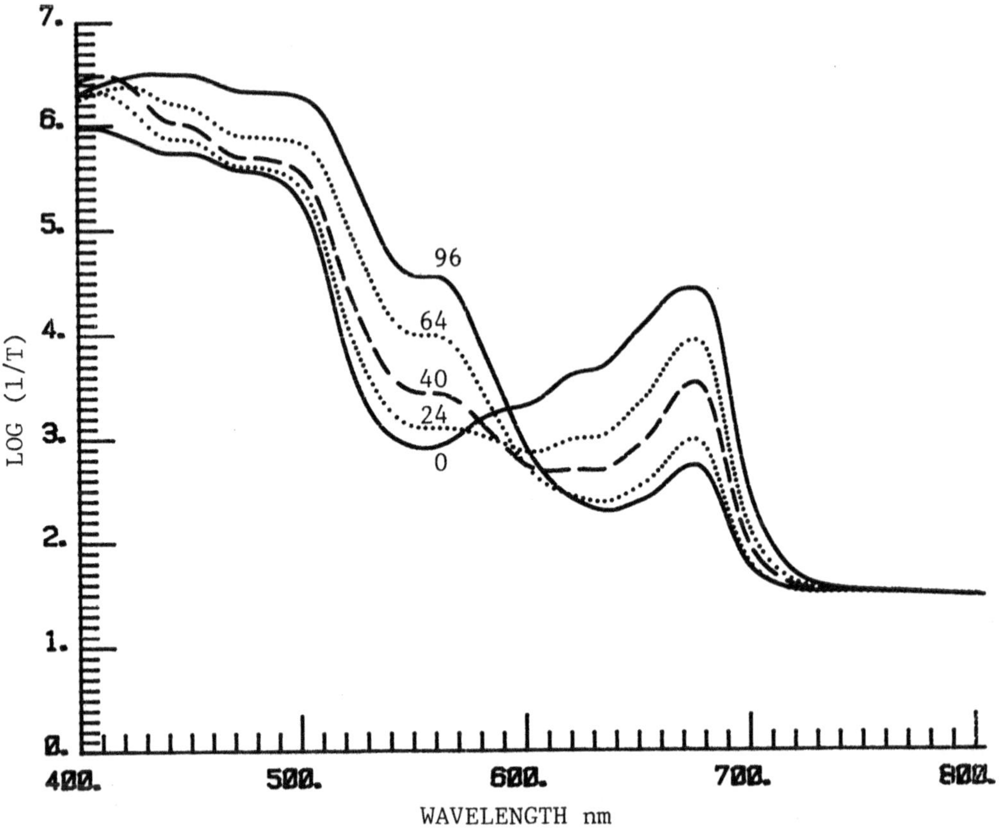

Fig. 9. Absorption spectra of a ripening miniature tomato. The numbers are the hours of storage at 22°C after harvesting the fruit in the mature green stage.

peaks at 675 nm, gradually declines as the tomato ripens. The increase in absorption in the 500 to 600 nm region is mainly from the lycopene which shows its peak absorption at 570 nm. Another pigment, probably a carotenoid, having a maximum absorption at 526 nm also increased with time as the tomato ripened. These spectra of ripening tomatoes illustrates a problem which is common to all spectrophotometric measurements of biological tissue. The spectra observed on biological tissues do not match the spectra anticipated for the pigments known to be present. This results from the fact that our knowledge of absorption spectra for pigments are based on spectra of the pigments in a solvent and solvent interactions change the absorption spectra. Thus, we find lycopene in tomatoes with maximum absorption at 570 nm rather than at 510 nm as we observe in a solvent. The pigment absorbing at 526 nm may be β-Carotene although the solvent spectra for β-Carotene is at a much shorter wavelength.

These spectra were measured with a computerized spectrophotometer built around a Cary Model 14 monochromator optimized for the visible region. This spectrophotometer uses a large area multiplier-type phototube to measure the light transmitted by the sample. Dense light-scattering samples such as tomatoes are measured using the principles described by Butler and Norris, 1960; and Norris, 1965. Data handling for this visible spectrophotometer is the same as for the NIR instrument discussed earlier.

Light-transmittance techniques have been used to measure the chlorophyll and carotenoid content of intact tomatoes (Watada, et al., 1976); the chlorophyll content of intact peaches (Sidwell, et al., 1961); and chlorophyll content of apples (Yeatman and Norris, 1965). These techniques have also been used to detect defects such as water core in apples (Birth and Olsen, 1964); hollow heart in potatoes (Birth, 1960); and scald damage in cherries (Yeatman, et al., 1961). In general, any internal defect which causes a general change in the internal flesh color can be detected by visible transmittance measurement.

SUMMARY

Computerized spectrophotometric measurements provide a technology for rapid prediction of the major constituents of agricultural commodities. Protein, oil, starch, and moisture content have been measured on grains and oilseeds using NIR techniques on ground samples. Protein and moisture content have been measured on wheat samples without any sample preparation using near-infrared transmittance. Chlorophyll, anthocyanin, and carotenoid content have been measured on intact fruits and vegetables using visible transmittance. These developments make it possible to measure the composition of a product and this compositional information can then be used to indicate quality.

REFERENCES

Ben Gera, I. and Norris, K. H., 1968, Determination of moisture content in soybeans by direct spectrophotometry, Israel J. Agric. Res. 18:125.

Birth, G. S., 1960, A nondestructive technique for detecting internal discolorations in potatoes, Amer. Potato J. 37:53.

Birth, G. S. and Olsen, K. L., 1964, Nondestructive detection of water core in Delicious apples, Proc. Amer. Soc. Hort. Sci. 85:74.

Birth G. S. and Norris, K. H., 1965, The difference meter for measuring interior quality of foods and pigments in biological tissues, USDA Tech. Bull. No. 1341.

Butler, W. L. and Norris, K. H., 1960, The spectrophotometry of dense light-scattering material, Arch. Biochem. and Biophys. 87:31.

Finney, E. E. and Norris, K. H., 1978, Determination of moisture in corn kernels by near-infrared-transmittance measurements, Trans. ASAE 21:581.

Massie, D. R., 1976, Fat measurement of ground beef with a gallium arsenide infrared emitter, in: "Quality Detection in Foods," ASAE Publication.

Massie, D. R. and Norris, K. H., 1975, A high intensity spectrophotometer interfaced with a computer for food quality measurements, Trans. ASAE 18:173.

Norris, K. H., 1965, Measuring and using light transmittance properties of plant materials, in: "Electromagnetic Radiation in Agriculture," Illuminating Engineering Soc. and ASAE, New York.

Norris, K. H. and Hart, J. R., 1965, Direct spectrophotometric determination of moisture content of grain and seeds, in: "Humidity and Moisture," Vol. 4, "Principles and Methods of Measuring Moisture in Liquids and Solids," Reinhold Publishing Corp., New York.

Sidwell, A. P , Birth, G. S., Ernest, J. V., and Golumbic, C., 1961, The use of light transmittance techniques to estimate the chlorophyll content and stage of maturation of Elberta peaches, Food Tech. 15:75.

Watada, A. E., Norris, K. H., Worthington, J. T., and Massie, D. R., 1976, Estimation of chlorophyll and carotenoid content of whole tomato by absorbance techniques, J. of Food Sci. 41:329.

Yeatman, J. N., Birth, G. S., Ernest, J. V., Bender, R. W., and Sidwell, A. P., 1961, Spectrophotometric evaluation of anthocyanin pigment development and scald damage in intact red tart cherries, Food Tech. 15:521.

Yeatman, J. N. and Norris, K. H., 1965, Evaluating internal quality of apples with new automatic fruit sorter, Food Tech. 19:123.

POST HARVEST LOSSES IN PERISHABLE FOODS OF THE DEVELOPING WORLD

D. G. Coursey

Tropical Products Institute
56/62 Gray's Inn Road
London WC1X 8LU England

INTRODUCTION

The majority of the papers presented to this Advanced Study Institute have been related primarily to situations relevant to the technologically advanced countries of Europe, North America and Australasia. This paper will do something to redress the balance, for it must be remembered that about a third of the world's population lives within the Less Developed Countries (L.D.C.'s); that it is generally within the L.D.C.'s that living standards and nutritional standards are the lowest; while rates of population increase tend to be the greatest. Thus, it is within these countries, mostly within the tropics, that there is the most urgent need to increase food availability. This is not solely an ethical issue; whatever an individual's, or a government's, political or moral stance may be it must be accepted that a situation where such a large proportion of the world's population are living under severely sub-optimal conditions is highly unsatisfactory, and is a destabilising force within the established world economic order.

The world's population is expected to increase by around 50% between 1980 and the end of the century with a disproportionately large increase in the tropical world: a similar increase in food availability is necessary merely to maintain even the current position, which is itself seriously inadequate. It is a matter of speculation as to how far this increased food availability will be achieved, but it is generally accepted that only in a few limited areas is there much scope for bringing new lands into cultivation. Increased food production will need to be derived to a greater extent from the intensification of agricultural production, both extension and intensification of agriculture may have severe environmental impacts. A large part of tropical food production is derived from small scale farmers operating at or near the subsistence level, with only modest excess production to supply the rapidly growing urban centres. These farmers usually have no, or severely limited, access to such inputs as credit, mechanical power or fertilizers. These constraints will tend to increase with increasing energy costs, rendering intensification difficult (71).

A partial alternative to increased agricultural production as a means of increasing food availability is provided by improved storage and conservation, leading to reduced post harvest loss. Food loss reduction is normally cheaper than equivalent increases in food production, whether the cost is expressed in terms of economics, energetics, or adverse environmental impact. Nevertheless, only recently has serious scientific attention been given to post harvest problems of the tropical world, compared with those of agricultural production. Good conservation of fresh produce is especially important in the tropics, where the proportion of food which is processed is much lower than in technologically advanced countries. Post harvest studies initially developed from the need to protect the large bulk grain reserves held in Europe during the 2nd World War, and subsequent work has concentrated largely on the problems of "durable" or "dry" produce, such as grains, grain legumes, and oilseeds, which have essentially similar storage problems, mainly attack by external factors such as insects, rodents and spoilage microorganisms (28,29).

Attention to post harvest food loss reduction as a significant means to increase food availability was first given prominence at the international level at the World Food Conference in Rome in 1974, following which the 7th Special Session of the U.N. General Assembly in 1975 passed a Resolution calling for a 50% reduction of post harvest losses by 1985. This recognition of the potential value of post harvest loss reduction has found practical expression in the continuing debate among a number of international and national institutions, and several practical initiatives have been taken, aimed at making an effort to reduce losses at all the post harvest stages of the food production process. In the Food and Agriculture Organization (F.A.O.), food loss prevention became a priority area in 1978, but initially concentrated on grain storage. Perishables received some attention at a series of meetings held in 1977 by the U.S. National Academy of Sciences (41) but the first formal, international recognition to the importance of post harvest losses in perishable foods came only when an Expert Consultation on the Reduction of Food Losses in Perishables of Plant Origin was held jointly by F.A.O. and the United Nations Environmental Programme (U.N.E.P.) in 1980 (26). Several individual studies on the subject had nevertheless been made earlier, notably at the Tropical Products Institute on a limited scale, before this, (11,14,16,17) but it attracted only limited interest among a relatively small number of workers.

THE MAGNITUDE AND SCOPE OF THE PROBLEM

After grains and similar durables, perishable plant foods are the next in importance, especially staple foods such as cassava and cooking bananas followed by fruits and vegetables. According to recent F.A.O. statistics production of perishable plant foods in L.D.C.'s represents 85% of cereal production. The comparable global figure is only 70%, indicating the higher relative importance of perishables in the tropics. These figures, expressed as simple tonnages, somewhat overstate the case for perishables, as they have higher water contents than the grains. If this is allowed for, their contribution to L.D.C. diets in calorific terms is of the order of 29% of that of the grains. Total production in L.D.C.'s of the principal types of perishable plant foods are indicated in Table 1, the grand total being 372,300,000 tonnes (24), and, allowing for the substantial production of unrecorded minor "backyard" vegetable crops, the actual production is probably substantially higher. A classification based on utilization may be convenient, although a commodity may sometimes fall into more than one group.

Staple foods. In many tropical regions, perishable staples take the place of grains in providing the calorific base of the diet. Cassava is the most important in spite of the problems associated with its toxicity (42), followed by cooking bananas, yams, sweet potatoes, potatoes, aroid root crops, breadfruit and some minor staples. Root crops alone were estimated a decade ago to provide staple food for around 400 million people (16): the figure today would probably be more than 500 million. Under humid tropical ecosystems, many of these crops are the most efficient producers of carbohydrate (72), being far more productive than grains.

Secondary foods. This includes the culinary category of 'vegetables': e.g. green or salad vegetables, onions, tomatoes, brassicas, legumes and a great variety of others, many unfamiliar to European diets; and also fresh dessert fruit. Although consumed in smaller quantities than staples, they are of importance in balancing the diet by providing minerals and vitamins; in adding flavour and variety to make it more attractive; and making minor contributions to calorific and protein nutrition.

Export crop. Substantial quantities of fruit, e.g. bananas, pineapples, citrus, mangoes, avocados and out-of-season vegetables are exported from the tropics to the developed

TABLE 1

Production of Vegetable and Fruit Crops in Less Developed Countries
(after F.A.O., 1979)

Commodity	Production ('000 metric tons)	Commodity	Production ('000 metric tons)
Cassava	110,617	Tomatoes	15,161
Potato	32,053	Melons	13,257
Sweet potato	16,324	Onion (dry)	7,612
Yams and miscellaneous root crops[1]	29,931	Cabbage	3,980
Root and tubers total	188,925	Pumpkins, squashes and gourds	2,685
Grapes	13,078	Green chillies and peppers	2,691
Oranges	22,139	Eggplant	1,601
Other citrus	5,602	Cucumbers and gherkins	1,735
Mango	13,771	Cauliflower	1,046
Pineapples	5,884	Carrot	865
Bananas	37,383	Green peas	566
Plantains	20,584	Green beans	598
Apples	5,038	Garlic	1,355
Dates	2,525	Artichokes	178
Peaches and nectarines	1,319	Vegetables and melons total	53,330
Avocado	1,196		
Papaya	1,526		
Fruits (excluding melons) total	130,045		

[1] Yams (*Dioscorea* spp.) constitute about 19 million tons, and taros, tannias, etc. most of the remainder.

countries. Although these contribute to the nutrition of the developed world, they are valuable earners of foreign exchange for the exporter countries, many of which are net food importers.

Within the general category of "perishables," the post harvest characteristics of the different commodities vary enormously. At one extreme, some leafy vegetables and soft fruit keep only a day or two under tropical ambient conditions, while at the other extreme, yams and potatoes will store under the right conditions for many months.

DEFINITIONS OF LOSS

The definition of loss has been a subject of considerable debate, and the situation is especially complex with perishable food products, owing to factors as the high water content, which permits substantial transpiratory weight loss with only minor loss of actual food; senescence, or sprouting, which can reduce the quality of the food product without necessarily rendering it a total loss, and the varying levels of quality deterioration at which a food product becomes unacceptable in different societies, or even at different social levels within a society. Some studies on the subject (20) concentrated on the economic aspects of loss, which is difficult to relate to actual food loss as the product acquires added value at every stage of the complex chain from producer through purchaser, transporter, wholesaler, retailer(s) to final consumer at a greater rate than it suffers deterioration or loss. The economic loss is thus inflated in relation to the loss of food.

At the recent F.A.O./U.N.E.P. Expert Consultation (26), considerable effort was devoted to establishing a manageable definition of "post harvest food loss," and the form ultimately determined follows:-

> 'Post Harvest' begins at the moment of separation of the edible commodity from the plant that produced it by a deliberate human act with the intention of starting it on its way to the table. The post harvest period ends when the food comes into the possession of the final consumer.
>
> 'Food' means weight of wholesome edible material that would normally be consumed by humans, measured on a moisture free basis. Inedible portions such as skins, stalks, leaves and seeds are not food. Potential foods (e.g. leaf protein) are not foods; they do not become food until they are accepted and consumed by large populations. Feed (intended for consumption by animals) is not food.

> The method of measuring the quantity of food in the post harvest chain should be on the basis of weight expressed on a moisture free basis. There will be times when information on losses in nutritional units and economic losses will also be needed but these should not be the prime means of measuring post harvest food losses.
>
> '<u>Loss</u>' means any change in the availability, edibility, wholesomeness or quality of the food that prevents it from being consumed by people.

Some comment is needed: although it is difficult to estimate loss once the product is in the hands of the final consumer, losses occurring at this stage are of importance, even in market economies. Within L.D.C.'s, substantial proportions of the populations live within the subsistence agriculture sector, where most food produced on the farm is consumed within the extended family, and most food eaten is home-grown, so that there is virtually no distinction between producer and consumer. That food loss must be calculated on a moisture free basis is of the greatest importance, especially with the perishables, which are usually of high moisture content: nevertheless, gross changes in water content, although not necessarily destroying the nutritional value of the produce, could render it unacceptable, and so a loss.

A briefer definition (5,30) seems to be adequate for practical purposes, although perhaps a trifle simplistic.

> "that weight of wholesome edible product (exclusive of moisture content) that is normally consumed by humans and that has been separated from the medium and the site of its immediate growth or production by deliberate human action with the intention of using it for human feeding but which, for any reason, fails to be consumed by humans"

CAUSATIVE FACTORS IN POST HARVEST LOSS: TECHNICAL

Perishable produce differs in physical and biochemical characteristics from durable produce, as summarized in Table 2, and the mechanisms and causative factors of loss are thus quite different within the two groups. Loss in durable produce is very largely brought about by external factors, mainly moulds, insects and rodents: endogenous factors are of minimal importance, and if the produce is of adequate quality and is protected from external biodeteriogens it may be kept for very long periods: certainly longer than from one harvest to the next, which is all that is usually necessary, and under optimal conditions for many years. Perishable produce is, however, still metabolically active, and

Table 2. Characteristics of Durable and Perishable Crop Products (adapted from F.A.O./U.N.E.P., 1981)

Durables	Perishables
Low moisture content, usually 10% to 15% or less.	High moisture content, typically between 50% to 90%.
Small unit size, typically less than 1 gram.	Large unit size, typically 5 g to 5 kg., occasionally even larger.
Very low respiration rate, with very small generation of heat.	High to very high respiration rate, heat production is therefore high.
Hard texture, not easily damaged.	Soft texture, easily damaged.
Stable, natural shelf life of several years	Perishable, natural shelf life a few days to at best several months, according to type of produce.
Losses mainly caused by external agents, such as moulds, insects and rodents.	Losses caused partly by external agents, e.g. rotting by bacteria and fungi and partly by endogenous factors, respiration, senescence and sprouting.

the very metabolic processes necessary for its survival are themselves a cause of loss, although external factors can also be of importance. The very term "perishable" implies that such produce cannot be kept for any very great length of time, hardly ever as long as from one harvest to the next, and in many cases only for a few days or weeks.

This leads to the concept of inherent storage life, which is of fundamental importance in the understanding of the post harvest behaviour of perishables. Although much can be done to reduce post harvest losses in these commodities, and extend their storage life, there is a limit beyond which they cannot effectively be kept, when the produce is either shrivelled or rotted to destruction, or has been so changed by its own metabolism as to have become unacceptable as food. This inherent life is related to the natural biological function of the plant organ, and may be only a few days, or up to several months, but in every case the life of the produce as an acceptable food item will eventually be terminated.

Thus, to minimize losses in perishables, the first essential is to maintain the physical and physiological integrity of the detached but still living plant organ, as losses arise from assaults on its integrity - a principle first clearly enunciated by the writer and his colleagues (13,17). Secondly, the natural life may be prolonged within certain limits, by the provision of optimal environments, or by manipulation of the physiological state of the material. Thirdly, there should be selection of material for storage that is entirely sound, and also at the ideal state of maturity. Thus, losses are minimized by choosing healthy material and keeping it healthy, but at the same time it must always be remembered that all living material will eventually die, and some detached organs inherently live longer than others. The factors that affect the health and length of life of perishable produce, and so post harvest loss, can be considered under three main headings:

Physical Considerations

The most obvious, but frequently ignored, factors in deterioration are purely physical. Mechanical injury takes many forms and arises at all stages in the history of the produce from pre-harvest operations, through harvesting, handling, grading, packaging, transporting and storage to exposure in the market and finally at the hands of the consumer. Losses due to mechanical injury are frequently overlooked and although difficult to estimate, are no doubt serious. Mechanically injured produce will normally deteriorate much more rapidly than sound produce and should never be considered for long-term storage. An example of the possible magnitude of loss through direct mechanical injury was revealed in recent surveys in Britain (54) which showed that up to the stage of grading on the farm, 33% of the potato crop is seriously damaged.

Pre-harvest attack by predators must be mentioned, but as already indicated, damaged produce should normally be diverted to immediate use. The harvesting process is liable to inflict wounds on the produce, except in those relatively rare cases where abscission leads to a situation where the organ can be removed from the parent plant without forming a wound. Usually, removal of the edible organ at harvest is a traumatic process, necessitating the formation of a wound where it is detached from the plant. After harvest, perishable produce is often as thoughtlessly and roughly handled as are bags of grain and extensive bruising or breakage of the individual organs frequently takes place: the damage caused greatly enhances further deterioration from both physiological and phytopathological causes. When unsuitable containers are used to remove the produce from the field, severe cuts may be inflicted on hitherto sound produce. During transportation, especially to a remote market, package design and construction needs to be optimized, if further damage by bruising or cutting is not to be incurred (40, 43). Apart from damage caused by the harvest trauma, transportation is probably a major cause of mechanical damage. As an example, the effect of transportation on mechanical damage to bananas in Jamaica is shown in Table 3. This example is typical of the kind of fundamental information on loss which is often needed, but seldom available. Different types of mechanical damage differ in the seriousness of their effects: with tubers and starchy storage organs, bruising or surface abrasions may initially be less obvious than breakage or cutting, but can be far more damaging during long-term storage. Decay will often spread from a bruise through the entire tissue, but a clean cut may be healed (cured) under favourable conditions (2,51).

Impairment of the produce's life processes from extremes of heat and cold must be considered. Produce exposed to the tropical sun can rise far above the ambient, to as high as 40° or even 50°C, which can have major deleterious effects (58,59). Damagingly low temperatures seldom occur in the tropics, but the phenomenon of chilling injury provides a major constraint on the use of refrigeration for the storage of tropical produce (21,22). Extremes of relative humidity can also adversely affect produce. Excessively low humidity can bring about rapid wilting of leafy vegetables, and shrivelling of the softer fruit, while at the other extreme, although high relative humidity may help to maintain turgor and reduce water loss, it can also favour the growth of pathogens.

Physiological Considerations

A distinction must be made between the unavoidable losses during storage of perishable produce that arise from the endogen-

Table 3. Damage to Bananas Caused During Transportation (after Thompson, 1972)

Type of transportation	% Damage	
	Scarring of fruit	Damage to 'neck'
Head-loading	1.1	0.4
9-mile lorry journey	7.1	15.1
45-mile lorry journey	25.1	12.2

ous metabolic processes of the detached plant organs and avoidable losses which are brought about by exposure to adverse environmental factors. All perishable produce consists of organs, which, although detached from the parent plant, are living, and whose metabolic processes must continue. The essential metabolic processes are energy-consuming, and depend on the energy supplied by respiration. This being a process which requires the availability of oxygen, adequate ventilation of stored perishable produce is obviously a fundamental requirement. Tropical ambient temperatures generally lie between 25° and 35°C, compared with the 10° to 20°C of temperate summer temperatures, and near or below zero of the temperate winters through which much long term storage of temperate perishable crop products takes place. The endogenous metabolic processes of tropical produce therefore proceed at substantially higher rates than those of temperate produce stored under its own native ecological conditions. The requirment for good ventilation is thus higher and the losses of weight also higher, than in temperate crops. This is illustrated by studies on losses in yams (9, 52) where respiratory losses, representing actual destruction of carbohydrate, were between 0.01 and 0.1% per day, representing from around 7% to as much as 35% of the total weight loss, according to the temperature and state of dormancy (Table 4): these total weight losses are about an order of magnitude higher than those reported for potatoes, stored under temperate winter conditions (8). Associated with destruction of carbohydrate matter by respiration, is water loss by transpiration: in the same example of the yams, this constitutes the other main factor in weight loss. However, water is of no nutritional value, and so, using the definition of loss already given, this factor should not be considered, except in the situation where the water loss is sufficiently great to affect the acceptability of the produce as food. Losses of weight associated with transpiratory loss of water may, of course, be of substantial economic importance in a trading context, where they are often referred to as shrinkage. Thus, perishable produce, even when kept under ideal conditions, will always suffer some loss of weight during storage from the combined effects of respiration and transpiration: this is in marked contrast to durable produce, where losses are miniscule under optimum conditions.

Reference has been made to inherent storage life, which is normally determined by physiological factors. All fruits mature, then ripen and finally senesce, and when the process of senescence is sufficiently far advanced, the flavour and textural characteristics will have so changed that the produce is inedible, and so must be regarded as a total loss. The onset of ripening, and senescence can often be delayed, for example in bananas by techniques for the removal of ethylene (62) and more recently by the use of gibberellin treatment (35) but even though senescence may be delayed it will eventually occur. In organs of dormancy, such as bulbs and tubers, the endogenous period of dormancy is terminated

Table 4. Relative Contribution of Respiration to Total Weight Loss of Yam Tubers During Storage at 25°C and 35°C (after Passam et al, 1978)

	Total daily weight loss (%)		Daily weight loss due to respiration (%)	
Condition of tubers	25°C	35°C	25°C	35°C
Immediately after harvest	0.22±0.02	0.36±0.02	0.058	0.108
Dormant	0.15±0.03	0.28±0.06	0.011	0.028
Sprouting	0.21±0.02	0.34±0.07	0.074	0.068

by regrowth and sprouting: sprouting leads to rapid transfer of both dry matter and water from the edible organ into the sprouts, and so to massive loss. A number of sprout suppressants are effective in delaying the appearance of sprouts in potatoes (8) but in the yam have little or no effect (49) probably because the bud primordia are formed beneath the skin only shortly before the end of dormancy. The dormancy of onion bulbs can be extended to some extent by the careful selection of temperature (66) but sprouting will eventually occur. Fruit which has started to senesce, or bulbs or tubers that have passed the natural breakage of dormancy, increase greatly in their susceptibility to phytopathological attack; become softer and so more liable to mechanical damage; and develop undesirable flavour characteristics, which make them unacceptable as food. Thus some loss will always occur in perishable produce as a result of its natural physiological activity, and its storage life ultimately terminated for physiological reasons.

These physiological processes are normal, and inevitable. Additionally, consideration must be given to abnormal physiological losses, which arise from the stresses of adverse environmental factors already outlined. Mechanical damage usually has only fairly moderate physiological effects, wounding or especially bruising leads to increased respiration rate, and enhanced loss of dry matter: evaporation of water through the damaged surface will also be increased. Wounds can be cured, but considerable physiological losses will occur before the curing process is complete. This effect has been well demonstrated in the yam (51). An important exception is that of cassava. When the roots are damaged, when they are cut from the plant, or otherwise, the primary phase of deterioration is purely physiological. Known as vascular discolouration, it consists of darkening of the vascular bundles, and the formation of occlusions within them, involving migration of material from the xylem parenchyma. Only after this physiological response has occurred is there any involvement of microorganisms, which eventually lead to total decay (3,45,60).

More generally however, abnormal physiological deterioration is caused by extremes of heat and cold or of relative humidity. Unlike temperate perishable produce, which can be stored at temperatures only marginally above zero, and is usually damaged only by actual freezing of the tissue, most tropical produce is liable to "chilling injury," at temperatures well above zero. The temperature below which chilling injury occurs is often around 10°C, but in some crops it is lower, but still well above zero, and occasionally it is as high as 14° or 15°C. There is much variation in chilling sensitivity between cultivars, with the state of ripeness, maturity or dormancy, within a species or cultivar. Visible symptoms of chilling injury vary greatly from one type of produce to another. In bananas, the first signs are discolouration of the under-peel vascular bundles, followed by a failure to ripen normally, and a grey-green peel colour when ripe (39); avocados also show vascular discolouration; in yams, there is greyish discolouration of the tissue followed by water-logging arising from breakdown of the Na^+ and K^+ pump mechanisms, leading to massive phytopathological invasion (10): rather similar symptoms occur in sweet potatoes and electrolyte leakage is also involved in chilling in tomatoes. The ultimate cause is believed to arise from phase changes in the lipid structure of the cell and mitochondrial membranes, which although reversible, can lead to irreversible changes in enzyme activity, and thus to energy supply within the cell, and relative metabolite levels (21,22,36,37).

At the other end of the temperature scale, attention has already been drawn to problems of high temperatures developed in produce when it is exposed to insolation (58, 59). The effects are greatest in dark coloured produce, which absorbs radiant heat more readily than light coloured: in some of the darker items, such as aubergines, potatoes and yams, internal temperatures during the afternoon can closely approach the "black bulb" temperature. Little information is available as to the mechanism of the deteriorative processes involved, but, especially in the larger organs it is likely that oxygen deficiency in the tissues may result in a change to anaerobic respiration.

Phytopathological Consideration

These probably constitute the largest single factor in bringing about avoidable post harvest loss in perishable produce, although endogenous physiological processes will always be the cause of some unavoidable losses. It is particularly important to distinguish between quantitative and qualitative losses, as these will be of greatly differing relative importance in different situations.

Quantitative phytopathological losses result in rapid, extensive and eventually complete breakdown of host tissues by microorganisms. The pattern of attack is typically an initial infection by one or a few specific pathogens, which establish the initial lesions in the previously healthy tissue: there follows a massive infection by a broad spectrum of non-specific saprophytes which are only weakly, if at all, pathogenic but are capable of surviving on the dead or moribund tissue remaining from the primary infection and enhancing the damage done (14,69). These secondary invaders thus play a major role in post harvest pathology by multiplying and extending the damage initiated by the primary pathogens. Qualitative pathogenic losses result from blemish or surface disease which render the produce less attractive and so less marketable, though little actual destruction of food matter has occurred. Such diseases are particularly important in fruit and vegetable export industries where increasing emphasis is being placed on visual quality. In the developed countries, much research in post harvest phytopathology has been devoted to these purely "cosmetic" factors and similar efforts in L.D.C.'s are needed if exports to sophisticated markets are to be successful.

Bacteria are often the most important casual factor in the spoilage of vegetables, the bacterial soft rot group (Erwinia spp.) being the commonest, but in the case of fruit and root crop spoilage, fungi pathogens are more frequently involved. The taxonomy can be very complicated: over 20 different fungi have been associated with post harvest decay in tomatoes or in crown rot of bananas. Virus diseases, while seldom of post harvest importance, may render produce less marketable, for example, "internal brown spot" of yams (14). Nematodes have been shown to affect the quality and storage life of yams (65). Pathogens differ in their response to the host, to environment, to various control measures, and in their time and mode of infection. Successful disease control thus depends on a thorough knowledge of these factors in each individual situation.

Post harvest diseases may also be classified, according to whether initial infection occurs before harvest; or during or after harvest. The latter are often caused by wound pathogens, associated with mechanical damage or physiological injuries. Control of the former may be effected by the pre-harvest use of protectant or eradicant sprays in addition to post harvest treatments after infection has occurred. Control measures against the latter are restricted to post harvest treatments, either physical or chemical. Physical treatments include the use of high and low temperatures to reduce the growth of the pathogens, and curing. Low temperature storage, widely used for many commodities, slows down the metabolism of both host and pathogens and so frequently arrests rotting temporarily, but when the produce is returned to ambient temperature, rotting may recommence rapidly. Heat treatment suffers from the drawback that certain pathogens are heat tolerant and the high temperatures required to achieve control can damage the produce. The pathogens and disease symptoms of the major tropical perishables, and chemical control measures have been summarized (14).

Insects, rodents and birds are usually of minor concern in the storage of perishable produce, compared with their important role in the storage of durables (14).

CAUSATIVE FACTORS IN POST HARVEST LOSS: NON-TECHNICAL

The whole problem of post harvest loss in perishables is greatly affected by factors of an entirely non-technical nature, which are of even greater importance in L.D.C.'s than in developed countries. These problems arise ultimately from the shortage of managerial and administrative skills that still exist in many L.D.C.'s; inadequate extension services; lack of appropriate educational facilities; and, in many cases, shortages of capital or of foreign exchange facilities. These are sometimes compounded by the advice or technical/managerial inputs supplied by "experts" from the technologically advanced countries, which are not always relevant, properly motivated or based on appropriate experience. A recent F.A.O. publication (25) has drawn attention to the need for improved co-operative facilities, transport systems and especially training in post harvest loss reduction, as well as to various technical considerations: the importance of training, and of publications for the dissemination of information, specifically in connection with losses in perishables was re-emphasized in another (26). Shortage of administrative and managerial expertise tends to lead to situations in which staff are often over-graded and through lack of experience and even of basic knowledge mistakes and errors of judgement occur more frequently than they should: this can apply from national administrators to packhouse operatives, and every level in between. Farmers are often unaware of the basic requirements for produce for sophisticated storage systems, or high-quality markets, although their traditional post harvest technology may be well adapted to the conditions of subsistence economies (12). The produce presented to buying organizations is often merely the excess production of the subsistence agriculture sector: supplies thus tend to be irregular, and of variable quality.

The operation of these factors can be seen clearly in many cold storage operations set up in L.D.C.'s. Firstly, at national planning level, the installation of cold stores is often seen as a universal panacea for all situations, even when it is not - an attitude often encouraged by the less scrupulous representatives of manufacturing organizations in the developed world. Secondly, it is seldom realized that a cold store, even where it is appropriate, is often useful only as a link in the "cold chain" - including cold stores in the production areas, refrigerated transport to urban centres and different cold stores in those centres. Thirdly, at store manager level, there may be lack of knowledge of the various technical requirements for produce storage. Even the concept of chilling damage may be unknown, and highly chill-sensitive produce packed into stores maintained near or even below zero. Incompatible produce can be mixed together in the same store. Fourthly, even the best machinery eventually breaks down. Local mechanics are unfamiliar with it: the operating manual (if any) may be written in a language which no one locally understands; the firm that sold the entire plant in the first instance has no agent in the country who stocks spare parts; and no allocation of foreign currency can be made available to obtain the parts from abroad, even though they may cost only $100 or so. Even when the plant is working, farmers may receive no advice of the type, quantity or quality of produce required or to supply them or advise them on suitable seed, planting regimes or harvesting techniques, and often there will be little financial incentive for them to produce what is required. When all these factors are taken into account, it is not surprising that the tropical world is littered with cold stores that are either shut down; broken down; grossly under-utilized or maintained at a loss by official subsidies. The worst possible situation is described here in order to bring out the essentials of the case most effectively. Many highly efficient cold storage and cold chain systems do exist and function in many developing countries, but a combination of any or all of the factors mentioned above do, only too frequently, apply.

Reference has been made to training and information supply needs. Although most of the L.D.C.'s now have excellent universities and institutions of higher learning, the courses available there are usually academically oriented. There is a serious need for relatively short term, academically rigorous, but practically oriented and relevant training schemes in the essentials of the post harvest technology of perishable produce, geared to various levels of seniority and responsibility, and conducted by those who have genuine and appropriate experience of L.D.C. conditions. An important aspect of such training needs to be guidance on national and international sources of up-to-date information, and on its interpretation to suit local conditions (26).

THE MAGNITUDE OF LOSSES

Although this topic has been the subject of very considerable debate over the last decade, remarkably little reliable information is yet available on the magnitude of post harvest losses of perishable produce. It has been conservatively estimated from the very limited data scattered in the literature that the global loss is of the order of 25% (11,17). However, these early papers are based very largely on what might be described as anecdotal information, that being all that was available at the time. An independent estimate (48) for Africa and India is 30%.

Even in the world's most highly developed and non-tropical country, the United States of America, perishable produce has been described (6) as "the victim of phenomenally high waste because of incredibly poor handling practices," and "the loss rate as a result of multiple handling ..... is frightful." The annual economic loss in that country arising from deterioration during shipment, storage and marketing was estimated 15 years ago at $200 million (20). Estimates of material loss in fruit and vegetables as percent of total production were made by the U.S. Department of Agriculture as 11% for the period 1942-1951 and more recently for 1957-1960 at 8% which may indicate the benefits that resulted from a decade of research and application (17). However, another analysis of the situation in the U.S.A. over the past 30 years (53) concluded that although there had been some reduction in post harvest loss with some commodities, with others losses had actually become more serious. A recent review (30) quotes a survey of produce in New York markets, selected data from which are given in Table 5. Total losses range from as little as 1.7% for apples to 22.9% for strawberries: most surprisingly the commodity which exhibited the second highest loss was sweet potatoes (15.1%), which would be considered by most to be relatively rugged.

In the tropical world, the situation is doubtless worse than in the developed world, but even less hard information is available. A number of specific cases have been reviewed (17). The meeting concerning perishables in the series on post harvest losses held by the U.S. National Academy of Sciences (41) operated in the absence of any hard factual information, on the Delphi Principle of summarizing the estimates (or guesses) of a number of professionals who have some standing or authority in the field. Some of the figures produced by this process are given in Table 6, but few, if any, of these are based on really reliable original data. Perhaps the most useful outcome was to establish that losses in perishables are virtually impossible to quantify except with reference to a particular commodity and local situation. This concept was more fully developed at the F.A.O/U.N.E.P. Expert Consultation (26) where it became evident that under different conditions and length of storage and with different commodities, almost any loss figure between 0% and 100% may genuinely be found. Overall, therefore, losses of perishable plant foods are extremely serious, and losses in the developing world would appear most frequently to be between 10% and 30%, according to commodity and location-specific storage conditions. Each local situation needs investigation and analysis

Table 5. Post Harvest Losses in Selected Fruits and Vegetables Sampled at Wholesale, Retail and Consumer Markets in the New York Market Area (selected from Harvey,1978)

Commodity	Loss in indicated markets Wholesale (%)	Retail (%)	Consumer (%)	Total loss (%)
Apples, Red Delicious	0.2	0.2	1.5	1.7
Cucumbers	-	5.0	2.9	7.9
Lettuce, Iceberg	4.1	4.6	7.1	11.7
Oranges, Navel	1.9	1.9	2.3	4.2
Peaches	2.3	4.5	8.1	12.6
Pears, d'Anjou	-	2.5	1.6	4.1
Peppers, Bell	7.1	9.2	1.4	10.6
Potatoes, Katahdin	1.3	-	3.6	4.9
Strawberries	5.9	4.9	18.0	22.9
Sweet potatoes	-	5.7	9.4	15.1
Tomatoes, packaged	-	6.3	7.9	14.2

on a particular individual basis, and broad global figures of loss such as have been quoted are of value mainly in indicating the magnitude of the problem that exists, which in turn implies the need for urgent action.

REDUCTION OF POST HARVEST LOSS

Traditional subsistence agriculture systems account for a very large proportion of agricultural production in L.D.C.'s (23) and within these systems exist many simple techniques for the avoidance of post harvest loss in traditional crops, which are often highly effective within the situations in which they were developed (12), but are nevertheless subject to severe limitations. Where multinational companies are involved in the plantation scale production of perishable crop products for export, relatively few problems arise. Production of bananas in Central America is, perhaps, the best example. The multinationals have, through their own research departments, access to the latest scientific information. The most serious problems arise at the interface between subsistence and market economy sectors. The crops being grown are often unfamiliar to the farmers, as may be the concept of growing crops, especially perishable crops, for sale, rather than subsistence. The tropical cash crops that have been developed over the last century, primarily yield durable products, directly or after simple farm-level processing. The concept of exacting quality requirements may well also be absent. At this interface, and

Table 6. Non-Grain Staples, Vegetables and Fruits. Losses Reported by Commodity (adapted from N.A.S., 1978)

Commodity	Estimated Loss (Percent)
ROOTS/TUBERS	
Potatoes	5 - 40
Sweet Potatoes	35 - 95
Yams	10 - 60
Cassava	10
VEGETABLES	
Onion	16 - 35
Tomatoes	20 - 50
Plantain	35 -100
Cabbage	37
Cauliflower	49
Lettuce	62
FRUITS	
Banana	20 - 80
Papaya	40 -100
Avocado	43
Peaches, apricots and nectarines	28
Citrus	23 - 33
Raisins	20 - 95
Apples	14

within the marketing economies that are developing within the L.D.C.'s, numerous technical problems arise in connection with perishable food crops (68). These problems, and some solutions that do not involve highly sophisticated technologies, will now be discussed.

Pre-Harvest Consideration

Cultivars of individual crops have differing, genetically controlled post harvest behaviour due to differing inherent physical, biochemical and physiological characteristics. Plant breeders have seldom exploited these potential differences, although work has recently been carried out on the genetic factors controlling the storage life of tomatoes (67). Evaluation of post harvest characteristics is a factor which should be, but seldom is, considered in the early stages of any breeding programme. A masssive banana breeding programme carried out to develop tetraploid bananas in order to reduce dependence on triploid clones did not take such factors into consideration until a late stage, when the new clones were found unsuitable for trans-Atlantic transportation (44).

There is also scope within screening trials for the evaluation of difference cultivars of a crop for significant differences in their storage potential. Red onions, for example, usually store better than white (61, 66). However, a major problem in many tropical areas is the availability of regular supplies of viable seed of cultivar which are adapted to local conditions in any respect, including good post harvest characteristics. Few tropical countries have adequate facilities for breeding, selection and multiplication and commercial sources of seed are limited and often of unsuitable cultivars. Many growers rely on their own resources for seed production, which is often responsible for the low yields obtained and the variability in market and storage life of produce from variable seed.

Appropriate timing of sowing is important for annual vegetable crops, in order to optimize quality; crops which are sown late may not mature adequately before the required harvest period. Many annual vegetable crops are grown during the dry season and heavy rainfall occurring before the crops have fully matured can promote post harvest fungal infection. Although little is known of the effects of irrigation on post harvest quality, water stress during growth is generally deleterious to produce quality, as well as to yield. Rates of fertilizer application can also affect post harvest quality and especially crops subjected to high levels of nitrogen are likely to yield produce of diminished life.

Crops which have become infected with diseases in the field may subsequently deteriorate rapidly due to the development of these initially pre-harvest infections during storage. Field hygiene, the removal and destruction of diseased plant material and dead haulms, and the adequate spacing of crops to promote air circulation all have positive effects in reducing disease levels in the field. Preventative crop protection techniques are therefore a major requirement for the production of crops of good quality with good storage potential.

Harvesting and Field Handling

Produce of optimum quality is obtained only when harvested at the optimal stage of maturity. Fruits, when harvested immature, will be of poor quality and ripen erratically: a delay in harvesting may increase susceptibility to disease. Where produce is to be stored for protracted periods it is often preferable to harvest mature, but unripe. Maturity indices may be indicated by visual or physical means, chemical analysis, computation of time from flowering and physiological factors (47) although at the subsistence level, only visual or physical methods of fruit selection are practicable. Bananas may be harvested on the subjective basis of the shape of the fruit fingers, which are angular when immature but fill out while they are still green to a rounded or full shape, this latter stage is frequently too mature for shipping. The condition of the stem can be a basis for determining maturity in melons, while a direct relationship between skin colour and soluble solids content is an index of maturity in papaya (1). Maturity in root crops, such as yam, sweet potato and potato is usually indicated by senescence of the annual leafy growth (vines or haulms). These crops are often harvested immature for immediate use, but immature tubers are not suitable for storage. Cassava is often left in the ground until required, as it has an extremely short storage life, although the crop then occupies land which could be used for alternative cropping (31). Alternative storage systems are being investigated. Mechanized harvesting techniques for tropical root crops have not been developed far, and the manual techniques still widely used generally inflict less damage to the produce. Yams, culturally a highly esteemed crop, are traditionally harvested and handled with the greatest care, but other root crops are often as roughly handled in the field as are durables, negating the benefits of manual harvesting.

It is customary and desirable for harvesting operations to be carried out early in the day, especially for leafy green vegetables fruits and fruit-type vegetables, which would otherwise acquire excessive "field heat" leading to rapid deterioration. The treatment which produce receives immediately after harvesting is a major factor contributing to its deterioration in quality. Produce should be collected from the field in suitable containers and not allowed to be exposed to sun, rain or wind. Exceptions are onions, which may be sun-dried or "cured" on the ground after harvesting, and root crops, especially yams, which benefit from being left in the sun for a few days (but not longer) to harden the skin, and to dry adhering soil so that it may be brushed off. A collecting point convenient to the field for the reception of produce is desirable. This should be shaded; some leafy vegetable crops also need to be sprinkled with water to maintain leaf turgidity, and produce may be washed at this stage, to remove adhering soil. Rudimentary grading, selection and trimming of produce may be undertaken here, to avoid unmarketable material being transported further.

Curing

With starchy root crops and onions, a "curing" process immediately after harvesting can substantially reduce post harvest loss and prolong storage life. The term is rather loosely used: in the case of root crops, it implies a wound healing process brought about by a few days' exposure to high relative humidites and temperature (2), while with bulb crops it involves simply a drying out of the skin and outer layers. The curing of Irish and sweet potatoes and the drying of onions are well established practices in developed countries but they are often not effectively undertaken in L.D.C.'s. Curing involves the suberisation of the skin, followed by the development of a wound periderm which effectively retards water loss and acts as a barrier against infection. The drying of onions, although a different process, has a somewhat similar effect (66). Methods of curing potatoes under tropical conditions have been reviewed (4) while it has been shown that curing yam tubers can significantly reduce decay (51). Curing is essential to avoid the initial physiological phase of the deterioration of cassava. The process can often be carried out under tropical conditions at little cost by stacking and covering the produce so that the temperature and relative humidity rise naturally.

Transportation and Packaging

Many types of containers, quite unsuitable for the transport of perishable produce, are nevertheless commonly used, and few attempts are made to standardize the design or capacity of those which are satisfactory. The use of containers of unsuitable dimensions or design or which are in poor condition leads to excessive mechanical damage to produce in transit. Alternative designs for wooden containers for produce, which can be manufactured locally, have been published, but these designs and modifications on them have not been effectively evaluated in developing countries (46.70). Even where only a limited range of local materials is available, modifications to design can produce a saving in product quality. Very few studies appear to have been conducted on this subject. Investigations in India have evaluated existing and improved designs of container for the transport of both mango (34) and grapes (38). In Ghana (57), when tomatoes were packed and shipped to urban centres in conventional "inverted cone" baskets, transportation losses were around 15%. When "upright cone" baskets, i.e. flat-bottomed with sides converging towards the top were used, and the tomatoes packed on a thin layer of dried grass, losses in the same handling chain were reduced to about 3%. Chilli peppers, which have very high rates of respiration, are packed for marketing in South East Asia in baskets provided with a central open chimney thus allowing effective removal of the heat of respiration.

The quality of transport facilities in L.D.C.'s varies greatly and it is difficult to identify the factors that contribute to damage during transport. Containers are over-filled as transport charges are often related to the total number of packages of a consignment. The poor condition of road surfaces, and that lorries are often driven much too fast, are two factors which result in much mechanical damage being caused to produce consigned by road. Road conditions in the tropical world are slowly improving, but little can be done to educate drivers and middle-men in better practices, as immediate monetary gain is involved. More appropriate containers, and their proper maintenance and handling, could doubtless lead to substantial reduction in losses. Proper packaging is, of course, especially important in connection with air-freight exports, where the farm-gate price of the produce represents only a small fraction (~10%) of the final selling price (55). Investment in appropriately light, but strong packaging is therefore worthwhile (43). Many items of produce, particularly those of large individual size, low unit cost and high bulk density such as water melon, plantains and pineapple, are often transported without containers. Systematic stacking within the lorries reduces the transit damage, as also does lining a lorry with protective materials such as grass or leaves, and covering the produce during transit to protect it from the weather.

Storage Structures

However simple the materials from which a structure is designed to be made, it must achieve certain requirements. It must afford protection from sun and rain, allow adequate ventilation, permit inspection of the produce to detect early any incidence of disease, or infestation by pests, and achieve the lowest practical ambient temperature. Many traditional storage structures have been developed for medium and long-term holding of commodities such as yam, Irish potato, onions and pumpkin, under tropical conditions (56). Huts and underground pits and clamps, adapted particularly to potato storage in tropical highland conditions are also used for short term holding of root crops in the lowland tropics. In upland production areas, low night temperatures offer the opportunity of further development of ventilated stores, using either natural convective ventilation or by controlling intakes of the cool night air under a forced air ventilation system. Overall, the use of appropriate designs for storage structures for perishables can make a major contribution towards loss reduction.

Chemical Treatments

These fall into two main categories: biocides that control or limit the attack of bacterial or fungal pathogens, and those which affect the physiology of the produce, with the aim of extending storage life. The use of crop protection chemicals prior to harvest has been referred to as a means of reducing the incidence of pre-harvest infections that lead to post harvest deterioration; chemical and physical treatments applied to the post harvest phase to prevent losses due to pests and diseases are also important: they may be applied as dips, dusts, sprays and fumigants. They must only, however, be used in conformity with recognized legal standards. Post harvest chemical applications need not be excessively costly. A simple fungicide applicator (7), originally designed for the control of crown rot on green bananas, could be readily adapted for other commodities for the application of systemic fungicides such as benomyl and thiabendazole, or other water-dispersible biocides. Similarly, work in Colombia on simple application systems for the volatile fungicide 2-aminobutane, has been developed as an effective low cost disease control system for citrus (27). Waxes and plastic coatings are useful for extending shelf life and retarding respiratory activity in fruits and vegetables (33), although these techniques are generally only suitable for application on a large commercial scale. The application of wax to citrus is usually carried out in combination with a post harvest fungicidal treatment (18). Their use could probably be extended to small scale production, if the value of the crop justified the additional expenditure on materials and the labour involved.

The second group of chemicals whose use may improve post harvest storage of perishables are the growth-regulatory substances. Sprout suppressants are in common use in the temperate world to control sprouting of Irish potatoes, and sprouting and root development of onions, but their possible application under tropical conditions has yet to be fully explored. The breakage of dormancy in many tropical root crops is a major constraint on storage life, and compounds effective on potatoes are not generally effective with tropical root crops (49). Post harvest applications of gibberellins A_4 to A_7 to bananas in order to delay ripening, improve the control of crown rot and thus improve post harvest life, have been successful (35), and they have also been used to extend the green life of limes (50) and the dormancy of yams. Their use in the preservation of tropical perishable produce seems to merit further attention, especially as they are relatively cheap. Extension of shelf life of pineapples by the immersion of fruits in growth regulatory chemicals, mainly alpha-naphthalene acetic, have recently been reported in Australia (63), and it may be that this technique could be extended to other tropical fruits.

REDUCTION OF LOSS BY ENVIRONMENTAL CONTROL

In developed countries, a major factor in loss reduction in perishable produce has been the use of refrigerated storage, and more recently of even more sophisticated techniques, such as controlled atmosphere and hypobaric storage, in which an optimal or near-optimal environment is provided for the produce have been used. Reduction of temperature reduces the level of metabolic activity of the produce, and so extends inherent storage life. The activity of any pathogens is diminished, although once the produce is returned to ambient re-growth will usually commence. The effective use of cold storage usually implies the existence of a complete "cold chain" of linked storage and transportation facilities, and unless such a "cold chain" system can be evolved, the use of isolated, individual cold stores may well be contraindicated, as the return of cold produce to ambient may often lead to condensation problems; a greater degree of damage than if it had never been cooled, and enhanced liability to pathogenic attack. Within the L.D.C.'s, although the same basic considerations apply, there are major limitations on the use of cold storage, or its more sophisticated variants. The administrative and infrastructural problems or organizing and managing cold stores under L.D.C. conditions have already been discussed and the importance of these problems cannot be over-emphasized. The maintenance of a high relative humidity is also important for the preservation of produce quality; many fruits and leafy vegetables are particularly sensitive to low humidity conditions that can arise in cold stores. However, high relative humidities may pre-dispose produce to infection by micro-organisms, so a compromise has to be reached between adequate air movement by ventilation and a moderately high relative humidity.

The application of crushed ice, by the 'top-icing' process, is used for maintaining the quality of many leafy vegetable and other crops during transportation, particularly in South East Asia. Produce which would be damaged by direct contact with ice can be packed in containers so that the ice is adjacent to, but not in contact with, the produce and low temperatures still maintained.

There are some co-operatives and progressive market retailers in the tropics who do provide cool storage facilities for horticultural produce, but it is only on a relatively large commercial scale that the unit value of fruits and vegetables justifies the investment and for the relatively low-cost perishable staple foods it is rarely, if ever, justified. Most traditional market systems are adapted to a rapid throughput of perishable produce; in particular, many markets in the tropics are active during the very early, relatively cool hours of the morning with produce being rapidly sold. Cool storage facilities for produce intended for local distribution tend therefore to be limited to the larger markets. However, for export crops the justification for cool storage facilities becomes more realistic in view of the high prices anticipated, and the correspondingly high quality standards that need to be maintained (40, 55, 68).

Additionally, a fundamental constraint of the use of cold storage facilities for tropical perishable produce arises from the fact that many types of tropical produce are liable to chilling damage at temperatures, usually around 10°C. This not only means that maximum benefit in reduction of metabolic activity to low levels cannot be achieved, but also that delicate control of temperature is needed in order to secure the maximum benefit possible without risk of chilling damage.

Refrigeration nevertheless has an important role to play in the reduction of post harvest losses of some tropical perishables, so long as sound management is available, and proper attention is given to the temperature requirements of different types of produce. Optimum holding temperatures for various tropical crops produce have been given. Similar general considerations apply to the more sophisticated systems, such as controlled atmosphere and hypobaric storage. The value of increased levels of carbon dioxide or nitrogen, reduced levels of oxygen and of low pressure for extending the storage life of temperate fruits and vegetables has been extensively investigated and a number of systems are in commercial use in the more technologically advanced countries. It is felt, however, that the application of such techniques to developing countries is likely to be slight for some considerable time, with the possible exception of sea freight export industries where some of these systems may be applicable. Recently, much attention has been directed to the use of the high degree of insolation typical of the tropical world for solar-powered refrigeration systems: although still at the experimental stage, the fact that most tropical perishables require holding temperatures well above zero may well facilitate this development (21,22,32,36).

Some simpler applications of controlled atmosphere principles may be of value, especially in connection with ethylene. Its removal from a system can therefore extend the pre-climacteric life of the fruit. The removal of ethylene from bananas during sea transportation, through the use of high levels of ventilation, is an important factor in ensuring that the produce reaches its destination in the required green pre-climacteric condition. Investigations (62) on the transportation of bananas at ambient temperatures by packing them in sealed polythene bags containing an ethylene absorbant has shown that transportation can be extended by up to 18 days under tropical conditions. The limiting factor then is disease and the best results were obtained by combining ethylene absorption with fungicide to control crown rot.

CONCLUSION

Perishable plant foods - fruits, vegetables, root crops - are of great importance in the nutrition of the L.D.C.'s of the tropical world, to a rather larger extent than in temperate countries. Increased availability is essential to the improvement of diets, or because of population growth, to maintaining existing standards. Reduction of post harvest loss provides an attractive alternative or supplement to increased food production. These are currently very severe: in individual situations almost any loss between 0%

and 100% may take place, losses commonly lie between 10% and 30%. The concensus of opinion suggests that around a quarter of perishable produce grown in the tropics is lost before it can be consumed. These losses arise from a number of inter-related technical factors, most of which can be classified as physical; physiological; or phytopathological. The degree of loss from any or all of these technical factors can be greatly compounded by the lack of adequate administrative infrastructures, technical expertise and educational facilities that often prevail in L.D.C.'s.

Various precautions, many of a simple nature, can be taken which can very substantially reduce post harvest losses. A largely unexplored field, is the inclusion of desirable post harvest qualities in breeding or selection programmes. This is a long-term approach; in the short-term, good pre harvest field hygiene; sound and appropriate harvesting practices; the use of curing techniques and/or of appropriate chemical treatments; the avoidance of mechanical damage and thermal stress; the use of suitable containers and storage structures can all greatly reduce loss: these measures aim at minimizing any attack on the physical or physiological integrity of the detached plant organs that constitute the produce. Refrigeration, especially when used throughout an entire cold chain, and even more sophisticated technologies such as controlled atmosphere and hypobaric storage, have a role to play, especially with high quality produce destined for high-cost markets, but their role may be only a somewhat limited one, for both technical and infrastructural reasons. In many areas, further research could improve understanding of the processes involved in loss, but more immediately there is probably a greater need for training and extension activities, so that the information and expertise that already exists can be put more widely into practical application.

REFERENCES

1. E. K. Akamine and T. Goo. Relationship between surface colour development and total soluble solids in papaya. Horticultural Sci., 6:567 (1971).
2. R. H. Booth. Post harvest deterioration of tropical root crops: losses and their control. Trop. Sci., 16:49 (1974).
3. R. H. Booth and D. G. Coursey. Storage of cassava roots and related post harvest problems. In Cassava Processing and Storage, I.D.R.C. Monograph, I.D.R.C. -031e, Int. Develop. Res. Centre, Ottawa (1974).
4. R. H. Booth and F. J. Proctor. Considerations relevant to the storage of ware potatoes in the tropics. PANS, 18:409 (1972).
5. M. C. Bourne. Proposed definition of post harvest loss. Proc. Natl. Food Loss Conf., Boise, Idaho, 129 (1976).
6. A. L. Brody and S. Sacharow. Flexible packaging of foods. Crit. Rev. Fd. Technol., 1:71 (1970).

7. O. J. Burden and P. J. Griffee. A simple machine for the application of fungicide to harvested green bananas. PANS, 20: 358 (1974).
8. W. G. Burton. The Potato. Veenman and Zonen, Wageningen (1966).
9. D. G. Coursey. The magnitude and origins of storage losses in Nigerian yams. J. Sci. Fd. Agric., 12:574 (1961).
10. D. G. Coursey. Los temperature injury in yams. J. Fd. Technol., 3:143 (1968).
11. D. G. Coursey. Biodeteriorative losses in tropical horticultural produce. In Biodeterioration of Materials - 2, A.H. Walters and E.H. Hueck-van-derPlas, edits. Academic Press, London (1971).
12. D. G. Coursey. Traditional post harvest technology of tropical perishable staples. U.N.E.P. Industry and Environment Newsl., 4:10 (1981).
13. D. G. Coursey and R. H. Booth. Post harvest losses in perishable tropical produce. Proc. 6th Br. Insectic. Fungic. Conf., 3:673 (1971).
14. D. G. Coursey and R. H. Booth. The post harvest pathology of perishable tropical produce. Rev. Pl. Pathol., 51:751 (1972).
15. D. G. Coursey and R. H. Booth. Post harvest problems of non-grain staples. Acta Horticulturae, 53:23 (1977).
16. D. G. Coursey and P. H. Haynes. Root crops and their potential as food in the tropics. Wld. Crops, 22:261 (1970)
17. D. G. Coursey and F. J. Proctor. Towards the quantification of post harvest loss in horticultural produce. Acta Horticulturae, 49:55 (1975).
18. J. W. Eckert. Post harvest disease of citrus fruits. Outlook Agric., 9:225 (1978).
19. J. W. Eckert. (This volume.)
20. J. W. Eckert and N. F. Sommer. Control of diseases of fruits and vegetables by post harvest treatment. Ann. Rev. Phytopathol., 5:391 (1967).
21. J. C. Fidler. Low temperature injury to fruits and vegetables. In Low Temperature Biology of Foods, J. Hawthorne and E. J. Rolfe edits. Pergamon, Oxford (1968).
22. J. C. Fidler and D. G. Coursey. Low temperature injury in tropical fruit. Proc. Conf. Trop. Subtrop. Fruits, London 1969. (1980).
23. E. H. Fisk edit. The Adaptation of Traditional Agriculture. Australian National University, Development Studies Centre, Monograph No. 11, A.N.U., Canberra (1978).
24. Food and Agriculture Organization of the United States. 1978 F.A.O. Production Yearbook, Vol. 32, F.A.O./U.N., Rome (1979).
25. Food and Agriculture Organization of the United States. Analysis of an F.A.O. Survey of Post Harvest Crop Losses in Developing Countries. AGPP: Misc./27, F.A.O., Rome (1977).
26. Food and Agriculture Organization of the United Nations/United Nations Environmental Programme. Food Loss Prevention in

Perishable Crops. Report of the F.A.O./U.N.E.P. Expert Consultation on the Reduction of Food Losses in Perishables of Plant Origin, F.A.O., Rome, 1980 (1981).

27. W. Grierson and F. W. Hayward. Decay control for citrus fruits using 2-aminobutane vapour. Proc. Trop. Reg. Amer. Soc. Hort. Sci., 13:124 (1969).

28. D. W. Hall. Food storage in the developing countries. Trop. Sci., 11:298 (1969).

29. D. W. Hall. Too much waste. Rural Life, 15:16 (1970)

30. J. M. Harvey. Reduction of losses in fresh market fruits and vegetables. Ann. Rev. Phytopathol., 16:321 (1978).

31. J. S. Ingram and J.R.O. Humphries. Cassava storage - a review Trop. Sci., 14:131 (1972).

32. International Institute of Refrigeration. The carriage of refrigerated cargoes. Annex to Bull. I.I.F., Paris(1973).

33. A. A. Kader, L. L. Morris and M. Cantwell. Post harvest handling and physiology of horticultural crops - a list of selected references. Department of Vegetable Crops, University of California. Vegetable Crops Series, 169, 2nd revision (1979).

34. S. Lakshminarayana, A.R. Vijayendra Rao, N.V.N. Moorthy, B. Anandaswamy, V.B. Dalal, P. Narasimham and H. Subramanyam. Studies on the rail shipment of mango. J. Fd. Sci. Tech., 8:123 (1971).

35. F. S. Leonce. The use of gibberellins A_4-$A_7$for the improvement of crown rot control in bananas. Proc. ACORBAT Conf., Martinique (1974).

36. J. M. Lutz and R. E. Hardenburg. The commercial storage of fruits, vegetables and florist and nursery stocks. United States Department of Agriculture, Agricultural Handbook No. 66, U.S. Government Printing Office, Washington, D.C.(1968).

37. J. M. Lyons. Chilling injury in plants. Ann. Rev. Plant Physiol., 24:445 (1973)

38. M. Madalgatti Rao. Problems and prospects of post harvest handling of grapes in India. Punjab Horticultural J., 9:1 (1969).

39. J. Marriott. Bananas - physiology and biochemistry of storage and ripening for optimum quality. Crit. Rev. Fd. Sci. Nutr., 13:41 (1980).

40. J. Marriott and F.J. Proctor. Transportation and conservation of tropical fruits. Outlook Agric., 9:233 (1978).

41. National Academy of Sciences. Post harvest food losses in developing countries. National Academy of Sciences, Washington, D.C. (1978).

42. B. Nestel and R. MacIntyre edits. Chronic cassava toxicity, I.D.R.C. -010e, Int. Develop. Res. Centre, Ottawa (1973).

43. J. H. New, F. J. Proctor and V. J. Hewitt. Packaging of horticultural produce for export. Trop. Sci., 20: 21 (1978).

44. S. New, J. Baldry, J. Marriott and E. A. Dixon. Fruit quality factors affecting selection of banana clones. Acta Horticulturae, 57: 205 (1976).
45. R. A. Noon and R. H. Booth. Nature of post harvest deterioration of cassava roots. Trans. Br. Mycol. Soc., 69: 287(1977).
46. Organisation for Economic Co-operation and Development. Catalogue of types and sizes of wooden packaging for fruits and vegetables used in Europe. Report of the Organisation for Economic Co-operation and Development No. 372. Organisation for Economic Co-operation and Development, Paris (1958)
47. E. B. Pantastico. Post Harvest Physiology, Handling and Utilisation of Tropical and Sub-Tropical Fruits and Vegetables. AVI Publishing Co. Inc., Westport, Connecticut (1975).
48. H.A.B. Parpia. Post harvest losses - impact of their prevention on food supplies, nutrition and development. Proc. Symp. Nutr. Agric. Develop. Tropics, Guatamala, 1974, N.S. Scrimshaw and M. Behar edits. Plenum Press, New York (1976).
49. H.C. Passam. Dormancy in yams in relation to storage. In I.F.S. Provisional Report No. 3, Yams, Buea, Cameroun, 1978. International Foundation for Science, Stockholm (1978).
50. H.C. Passam and G. Blunden (in press). Experiments on the storage of limes at tropical ambient temperatures. Trop. Agri., Trin.
51. H. C. Passam, S. J. Read and J. E. Rickard. Wound repair in yam tubers: the dependence of storage procedures on the nature of the wound and its repair. Trop. Sci., 18:1 (1976)
52. H. C. Passam, S. J. Read and J. E. Rickard. The respiration of yam tubers and its contribution to storage losses. Trop. Agric., Trin., 55:207 (1978).
53. W. T. Pentzer. Historical perspective of food losses. Proc Natl. Food Loss Conf., Boise, Idaho (1976).
54. Potato Marketing Board. Unpublished report (n.d.).
55. F. J. Proctor. Exporting horticultural produce from the tropics. Wld. Crops, 28:198 (1976)
56.. F. J. Proctor, J. P. Goodliffe and D. G. Coursey. Post harvest losses of vegetables and their control in the tropics. In Vegetable Productivity, C.R.W. Spelding edit. Macmillan, London (1981).
57. J. Rawnsley. Crop Storage: Technical Report No. 2. Food Research and Development Unit, Ministry of Agriculture, Accra, Ghana. Food and Agriculture Organisation of the United Nations, Rome (1969).
58. J. E. Rickard and D. G. Coursey. The value of shading perishable produce after harvest. Approp. Technol., 6: 18 (1979).
59. J. E. Rickard, O. J. Burden and D. G. Coursey. Studies on the insolation of tropical horticultural produce. Acta. Horticulturae, 84:115 (1978).

60. J. E. Rickard, J. Marriott and P. B. Gahan. Occlusions in cassava xylem vessels associated with vascular discolouration. Ann. Bot., 43:523 (1979)
61. G. K. Saxena, L. H. Halsey, D. D. Gull and N. Persuad. Evaluation of carrot and onion cultivars for commercial production in Guyana. Scientia Horticulturae, 2:257 (1974)
62. K. J. Scott, W. B. McGlasson and E. A. Roberts. Potassium permanganate as an ethylene absorbent in polythene bags to delay ripening of bananas during storage. Aust. J. Expt. Agric. Anim. Husb., 10:237 (1970).
63. L. G. Smith. Alpha-naphthalene acetic acid and the shelf-life of pineapples. Paper presented at the XXth International Horticultural Congress, Sydney, Australia (1978).
64. A. K. Thompson. Transport, handling and storage of fruit and vegetables in the West Indies. Proc Seminar/Workshop on Hort. Dev. in the Caribbean, Maturin, Venezuela, 11:170 (1972)
65. A. K. Thompson, B. O. Been and C. Perkins. Nematodes in stored yams. Expl. Agric., 9:281 (1973)
66. A. K. Thompson, R. H. Booth and F. J. Proctor. Onion storage in the tropics. Trop. Sci., 14:19 (1972)
67. E. C. Tigchelaar, W.B. McGlasson and R.W. Buescher. Genetic regulation of tomato fruit ripening. Horticultural Sci., 13: 508 (1978).
68. H. D. Tindall and F. J. Proctor. Loss prevention in horticultural crops in the tropics. Prog. Fd. Nutr. Sci., 4:25 (1980).
69. R. G. Tomkins. The microbiological problems in the preservation of fresh fruit and vegetables. J. Sci. Fd. Agric., 2: 381 (1951)
70. United Nations Industrial Development Organisation. Wood as a Packaging Material in the Developing Countries. ID/72. United Nations, New York (1972)
71. A. J. Vlitos (this volume).
72. C. A. de Vries, J.D. Ferweda and M. Flach. Choice of food crops in relation to actual and potential production in the tropics. Neth. J. Agric. Sci., 15:241 (1967).

SOLVING THIRD WORLD FOOD PROBLEMS: THE ROLE OF POST-HARVEST PLANNING

Martin Greeley

Research Fellow
Institute of Development Studies
University of Sussex, Brighton, Sussex

INTRODUCTION

There is a substantial body of literature which argues that developing countries' food shortages can be reduced, even removed, by preventing post-harvest food losses. It has spawned major research and development and extension initiatives, often funded through aid, that seek to improve farm-level post-harvest systems. Attention has focused on the major cereal staples which are the subject of this paper. Section I examines why the role of post-harvest planning has been characterized in this way. Section II challenges the conventional assertion of high farm-level food losses using evidence from recent field studies. Section III provides a reassessment of the opportunities for and consequences of farm-level post-harvest technical change.

SECTION I

The literature on post-harvest losses dates back at least as far as Pliny (1) but until very recently it has enjoyed only a narrow readership. With a few exceptions (2) the post-1945 effort to increase food availability in developing countries concentrated on increased production and it was not until after the Kissinger initiatives of 1974 and 1975 (3) that post-harvest loss prevention became a widely acknowledged alternative. Kissinger's speeches were essentially opportunist, and, whilst they were undoubtedly influential in encouraging politicians' support, there were at least four more substantive causes of the heightened interest in loss prevention.

First, the repeated assertions (4) of high food losses in traditional post-harvest systems suggested an easy option for solving the growing problems of hunger. There were few systematic studies of the size and causes of losses at farm level, and in their absence, genuine but untypical 'horror' stories of high losses were influential in channelling funds to solve post-harvest problems of food losses.

The science of loss-assessment was primitive and reports that did provide "estimates" of losses were often methodologically weak, omitting definitions of loss and details of measurement and even failing to specify which operations were included.

These problems led to some odd results - losses of 105 percent! (5) - and even the most widely quoted set of figures had a bias in presentation which gave a total loss figure seventeen percent greater than the study actually found (6). These deficiencies, even if recognized, were rarely acknowledged - and official reports that did so, were not always published (7) - for the problems of hunger were real, and growing, and post-harvest research offered a new dynamic in the search for higher food output.

In many countries, neither Ministries of Agriculture concerned primarily with increased food production nor Ministries of Food concerned with better food distribution, had been responsible for farm-level post-harvest problems; knowledge of this neglect further strengthened the case for intervention.

Secondly, the introduction of HYVs(high yielding varieties), notably with rice, but also maize and wheat, had sometimes resulted in higher food losses, for both biological and physical reasons. Frequently, the new varieties were softer or thinner husked than the traditional ones and were more susceptible to insect attack (8). Dormancy, a primitive genetic characteristic, was often reduced in the breeding process and the risk of premature germination increased, particularly when, as often was the case, a second rice crop was grown under irrigated conditions during the dry season and harvested at the beginning of the monsoons. These biological changes increased the risks of deterioration during post-production operations and they were accompanied by an increased risk of physical losses as output per hectare increased. These problems were felt at farm-level where space, labour and bullock constraints caused delays in harvesting and threshing but they were most acute in market centers, for the proportionate increase in marketed surplus was very much larger than the increase in output per hectare. The marketing and purchasing operations in the North Indian wheat markets were a particularly extreme case, (9) and the documentation of these losses, suffered due to inadequate market facilities and storage space, further strengthened the case for a higher priority to post-harvest research and development.

Thirdly, despite the regional problems of surplus continuing national food shortages kept food aid, notably of course PL 480, at high levels and the deterioration of these stocks prior to distribution was becoming a cause for grave concern. The US Senate Foreign Relations Sub-Committee on Foreign Assistance was receiving mission reports (10) of substantial losses in storage and was questioning the value of continued food assistance. This highly sensitive issue - and the only influence in the debate that encouraged LDCs (less developed countires) to show low food losses - was exacerbated by repeated quality problems with grain before discharge at ports that resulted in expensive litigation.

Fourthly, multinational companies manufacturing post-harvest equipment and anxious to promote sales in LDCs were using the loss-prevention argument as a sales strategy. Both chemicals (11) and machinery (12) were marketed in this way. The products were often unsuited for the job - early Japanese rice milling equipment in countries growing indica rice varieties being a notorious and widely quoted example - but sales pressure, tied aid and the sense of urgency generated through food deficits and suggestions of high losses frequently resulted in their purchase. These commercial interests were instrumental in propagating the view of high food losses in traditional post-harvest systems and were the strongest proponents of mechanization of these systems.

The response in some LCDs was the development of local manufacturing capabilities for post-harvest equipment (13) and more generally there was a concern with 'appropriate' technology (14) but the high loss figures were accepted and it was an attempt at alternative solutions not reformulation of the problem.

Thus, through a variety of mutually reinforcing developments in the 1970s, post-harvest planning became a widely debated issue. The overwhelming view was that the allocation of funds by donors and LDCs national programmes could result in post-harvest innovation that would reduce food loss and even remove the food deficit. The FAO had taken an early initiative in the late 1960s with its War on Waste programme and during the 1970s expanded its efforts through the Prevention of Food Loss programme; the World Bank has developed its own programme, a major part of the United Nations University World Hunger Programme is concerned with post-harvest losses, several large bi-lateral agencies - including the American, British, Canadian and German - have increased funding to post-harvest R & D and many LDCs (15) have developed their own programmes.

The few earlier post-harvest programmes had been concerned largely with the handling of imports and the marketed surplus; improved urban food distribution rather than loss prevention was the objective. These new programmes concentrated on farm-level postharvest systems and within that upon storage. Between sixty and and ninety percent of cereals output was retained at farm-level and if substantial increases in food availability were to be achieved by loss-prevention, then they would necessarily be through innovation at farm level. Traditional farm-level methods were regarded as inefficient not because of any very specific evidence of high losses but because they were characterized by "ancient" often rudimentary practices, uninfluenced by modern food science and technology. Several agencies directed funds to development of more systematic and reliable loss assessment methods (16), but in the meantime the need for modernizing technical change had become a conventional wisdom.

This 'wisdom' for some was rooted in the supposed neglect or carelessness of peasant farmers; for others it was their lack of basic knowledge of food science; for others it was their lack of access to improved methods and for others it was their shortage of resources to invest in improved methods. Implicit in all was the view that a latent demand for improved technology to prevent losses existed and that by meeting it and thereby improving food availability there would be fewer hungry people. A further attraction of farm-level post-harvest innovations was the apparent income distribution advantages, through the concentration of benefits on small producer-consumers who formed the majority of the world's poorest. Several reports (17) argued that small farmers suffered higher losses than large ones and would therefore benefit more. Furthermore, farm-level storage, and more generally post-harvest improvement, would reduce risk - removing uncertainty over lean season stocks and limiting the need for expensive lean season food purchases - and for tenants, was more equitable than production innovation since the benefits accrued to them rather than the landlord.

SECTION II

The opportunity to increase food availability was the core of the developing interest in LDC farm-level post-harvest technology. This genuine concern - notwithstanding less altruistic commercial, professional and bureaucratic influences - was based on four assumptions:

1. That traditional farm-level post-harvest technology was the cause of substantial food losses .... And therefore
2. That techniques were available or could be developed that would prevent these losses;
3. That it was profitable for farmers to adopt these new techniques;
4. That food consumption of the hungry would increase once new techniques were adopted.

Recent evidence suggests that these assumptions are more often false than true, and that policy-oriented R & D would be better served by an alternative set of assumptions:

1. That the most serious food losses occur off-farm in the public sector in times of large surplus;
2. That farmer motivation to change post-harvest techniques is cost-reduction which may not involve loss-reduction; techniques concerned with loss-prevention would have to be simple, low-cost improvements to existing practices;
3. That with improvement in land productivity the development of market relations and the improvement of the rural infrastructure will determine the rate of adoption and the distribution of costs and benefits of adopting new techniques;

(And for South and South East Asia particularly)

4. That technical change is invariably labour displacing and hence often worsens income (and food) distribution; the labour displaced is sometimes from the poorest rural households and threatens the livelihood and survival of these families.

These assumptions do not denigrate the role of post-harvest systems in influencing rural development. They do suggest though that their village-level impact on food consumption will be influenced by broader questions of employment and income-distribution rather than food loss prevention. These we discuss in Section III and the remainder of this section discusses recent research which has established beyond reasonable doubt the low average farm-level food losses in post-harvest operations.

These studies provide results on all operations for two countries, on all operations except drying and milling for one country, on storage only for eleven countries and for harvesting operations only for three countries. They have been selected from the numerous figures available because they provide measured (except Malaysia, based on a farmers' questionnaire) estimates of physical loss based on field surveys of operations under normal farmer practice, and provide details of the methods of assessment used. The results are summarized in Table 1 - and references in the text to country studies are from the sources given there.

Losses during farm-level storage have received the most attention in the literature; a loss of ten percent is commonly quoted as a conservative estimate and some 'experts' consider they run as high as 40 percent (18). Why then do the results from careful measurement in thirteen countries show that losses are below five percent and often very much lower (see Table 1)? The chief explanation is the acceptance of convenient but unfounded guesses, but another important set of reasons is that the methodological difficulties in accurate loss assessment tend to lead to upward rather than downward biases in estimation.

Table 1. Measured Physical Food Losses in Traditional Post-Harvest Practices at Farm and Village Level

Country	Commodity	Operation	Per Cent Loss (rounded to one decimal place)	Note
1.Africa	Maize	Storage	1.6	Insect loss only; an average of 8 countries
2.Bangladesh	Rice	Cutting to consumption	6.9	Storage losses 2.6%
	Wheat	Cutting and threshing	1.7	
3.India	Rice	Storage	4.3	Preventable loss 3.2%
	Wheat	Storage	1.5 - 4.3	
4.Indonesia	Rice	Standing crop to threshing	3.8	A further 4.5% remains uncut much of which is recovered during gleaning
5.Korea	Rice	Cutting to consumption	11	Preventable loss 7%, storage loss 2-3% (2% preventable)
6.Malaysia	Rice	Standing crop to consumption	7.3	All operations except drying and milling. Storage loss 1.2% - data based on farmer questionnaire.
7.Nepal	Rice	Harvesting	1.2	Figures reported as incomplete
		Threshing	1.9	
8.Nigeria	Sorghum & Millet	Storage	4	
9.Philippines	Rice	Shattering before and during cutting	below 1	
10. Zambia	Maize	Storage	2 - 5	

Sources:

1. M.P. Miracle, Maize in Tropical Africa, University of Wisconsin Press 1966, p. 243.
2. M. Greeley, Rural Technology, Rural Institutions and the Rural Poorest: The case of rice processing in Bangladesh. Paper presented at the Post-Harvest Technology Workshop, IARI, Delhi, Jan. 1981, p. 10.
3. R. A. Boxall et al. The Prevention of Farm-Level Food Grain Storage Losses in India: A Social Cost-Benefit Analysis, IDS Research Reports, 1978, p. 58, and B. P. Khare, Insect Pests of Stored Grain and Their Control in Uttar Pradesh, G. B. Pant University of Agriculture and Technology Pantnagar, Research Bulletin No. 5, 1972, p. 1
4. D. Gaiser, A Brief Summary of Paddy Loss in Indonesian Rice Post-Harvest Systems. Paper presented at the Regional Grains Post-Harvest Workshop, Philippines Jan. 1981, p. 6.
5. C. J. Chung, Post-Production Rice Systems in Korea, Final Report of Phase II, College of Agriculture, Suweon, 1980, pp. 4-8.
6. N. H. Mohamed, Rice Grain Losses from Kada Area (Malaysia): An Overview vis-a-vis Traditional Storage System. Paper presented at the Regional Grains Post-harvest Workshop Philippines Jan 1981, p. 11.
7. S. K. Bhalla and T. B. Basnyat, Post-Harvest Technology and Its Impact on the Rural Poor in Nepal. Paper presented at seminar on Rural Technology, Rural Institutions and Rural Poorest, Comilla Feb. 1981, p. 7.
8. P. H. Giles, The storage of cereals by Farmers in Northern Nigeria, Trop. Agric. (Trin) 41(3):197-212, 1964.
9. R. B. Calpatura, Variety-Maturity and Length of Straw Cutting Interaction in the Grain Losses of Rice During Harvesting. Paper Presented at the Regional Grains Post-Harvest Workshop, Philippines Jan. 1981, pp 1-2.
10. J. M. Adams and G. W. Harman. The Evaluation of Losses in Maize Stored in a Selection of small farms in Zambia with particular reference to the Development of Methodology, Tropical Products Institute, 1977, p. 4.

First, all the figures quoted are averages based on a number of experiments, whereas high losses, if they are based on any sort of measurement, are often selected examples of extreme cases that should be used to illustrate the worst, but often get quoted as the usual. [Ideally estimates should specify the standard deviation and sample size to allow calculation of a confidence interval, which would indicate the reliability of the estimate. This is only available for LDCs (India (1.33) and Bangladesh (0.63)) of the cases in the tables, but if these are typical then in ninety-five percent of cases further estimates of losses will be very close to the obtained mean.]

Secondly, when grain is periodically removed there is often confusion over the summation of different sample estimates made over the period of storage (19). When no allowance is made for periodic removal and the final sample result is taken as the total loss figure there is an upward bias in the estimate. Grain stored at farm level for consumption is usually removed gradually over the storage period; an even removal and a smooth incremental loss rate would give an average true loss of only half the final sample loss. Where losses increase exponentially the bias will be even greater.

Thirdly, insect damaged grains are sometimes regarded as lost grains when the damage is only partial (20). For most cereals estimates of damage are a poor guide to actual weight loss.

Fourthly, the poor supervision or organization of fieldwork will frequently lead to overestimates. Farmers are unlikely to add grain to a store long after harvest but may very well have to remove some in the absence of field staff; unless staff are well trained and well motivated, these removals will not be accounted for and estimates of loss will be larger by the quantity removed. Similarly if samples, once collected from a farmer's store are not handled carefully, they can suffer further deterioration between the time of sampling and time of analysis. There are several problems like these which make accuracy difficult and expensive to achieve. Avoiding them often results in other biases, notably the selection of sample stores primarily from large farmers who can cooperate with less inconvenience but consequently may remove grains less frequently, store for a longer period and take less care than the average farmer, therefore suffering untypically high losses.

These methodological problems are only examples and there are several more areas - selection of stores, grain sampling techniques, simulated trails, observer bias etc. - where errors in design tend to bias results upwards.

When studies are carefully designed, like those quoted in Table 1, they involve extended fieldwork and the one characteristic most commonly distinguishing high from low estimates is the time spent in the field. Field experience suggests that even though farmers' stores are often crude in construction the critical considerations of store hygiene, temperature and humidity control are competently, often ingeniously, catered for. Anthropologists (21) have been a useful source of knowledge here though their papers are rarely accessible to - or their approach acceptable to - the more technocratic post-harvest sciences.

The cases reported here confirm studies (22) on other aspects of peasant economy in showing that poor farmers are efficient. The levels of storage investment that would be justified on the basis of losses alone are small and in many cases current practices are optimal in that the costs of loss prevention are greater than the value of losses.

However, they do not always preclude carefully selected low cost innovation, and two of the studies quoted (India and Zambia) recommended specific extension efforts to encourage small modifications of existing practices; (none of them suggested priority to the introduction of new types of store such as metal bins). The prohibiting costs in these programmes of modification are in the training and motivation of field staff and village artisans to extend the concepts (of modified cribs, rodent-proof platforms etc.) rather than the actual cost of construction/manufacture. In practice as with much 'appropriate' technology (23) the marginality of benefits and the organizational strains imposed on government bureaucracies as well as peasant economy result in little progression beyond pilot programmes.

The adoption of metal bins where total capacity is being increased or where large and untypical farmers are the purchasers, often for a variety of non-economic reasons, in parts of India and in some African countries does not contradict these conclusions; apart from the subsidies they often receive, there are cost considerations - labour, materials, loading/unloading time, durability, and space/size criteria - which are independent of loss levels; there are also conditions of storage - long periods for soft grains such as wheat in the Indian Punjab - where losses may be on average higher than the results in Table 1 and high enough to make moderate investments profitable. (Although the changes in farmer practices required may lead to unanticipated results as happened in Andhra Pradesh, India; farmers using metal bins did not always dry their paddy down to the required 12 percent as opposed to their usual 14 percent moisture content and the lack of aeration led to premature germination of seed paddy.) These programmes are exceptions however and the evidence overwhelmingly refutes the stereotyped characterization of high food losses in farm-level cereal storage.

Of all the post-harvest operations, storage has received the most attention in the literature and in R & D programmes - and this is why we have singled it out - but it is threshing, and milling (grinding) where technical change is being most rapidly promoted and therefore where the assumptions regarding its effects are most critical. [Crop drying has received much attention from engineers but there is little evidence on drying losses which is a topic we take up in the next section.]

Evidence from empirical research suggests that in these cases also the high losses hypothesis is inappropriate. To illustrate Threshing: The evidence on threshing losses (Bangladesh 1.79%, Indonesia 2.38%, Malaysia 2.15% and Nepal 1.87% - the Korean data is not sufficiently disaggregated) in traditional beating and/or bullock trampling systems is restricted to rice, excepting the Bangladesh study which gives a figure of 1.13% for wheat. Variation around these averages is due to differences in straw length, variety, maturity, condition of the threshing floor/tub and care of the farmer which affect the amount lost in the mud and cracks of the floor and the amount remaining unthreshed on the straw. Table 2 (part A) shows the variability in loss in Bangladesh and also that the introduction of an 'improved' technique, the pedal thresher, actually increases food losses. The reduction in labour from 110 to 20 man-hours per ton (and the speeding up of the operation when supervisory time and space are scarce) outweighs the costs of these additional losses though and the thresher is spreading into areas where short stiff-strawed HYVs are grown. Clearly, the level of losses was not the determining factor in the adoption decision. The chief cause of loss is grain remaining unthreshed, because typically the care taken in drying out, covering, compacting and sweeping the threshing floor minimize, to immeasurable proportions, the levels of scattering loss. These unthreshed grains are in principle easily recoverable simply by taking more time in the threshing operation or by hand-stripping the straw - which poor farmers sometimes do when they purchase straw for livestock feed. The constraint is the time involved not the absence of a technique; the loss is preventable but it is not economic to do so.

Milling For non-rice crops where the whole grain is ground and where - in village-level systems - no separations are made, physical losses are negligible, but the estimation of rice milling losses is complicated by the difficulty in establishing appropriate criteria for measuring outturn. The percent total rice recovered and the percent unbroken rice are the two criteria commonly used; some authors (24) have combined them to estimate a composite index of milling efficiency. These are arbitrary judgements though for the degree of polish and the level of brokens have price effects which reflect consumer preferences and which are unrelated to nutritional (food-value) losses. The Korean study for example shows a total loss in milling of 4-6% but of this 1.5% is due to poor pre-milling quality control which should properly be included elsewhere, 2% is due to consumer preference for overpolished rice and only 1.5% of the loss is due to the poor performance of the milling equipment.

Table 2. Technical Change and Food Losses in Bangladesh

		Per cent loss	Sample size	Labor hrs per ton	Standard error
A. Threshing					
Bullock treading (Broadcast)	Long straw	2.54	40		0.29
Hand beating followed by bullock treading (Transplanted)	Short straw	0.60	38	110	0.08
	Long straw	1.45	20		0.28
Pedal Thresher (Transplanted)[1]	Short straw	1.82	10	20	0.39
	Long straw	3.49	9		0.27

	Per cent milling yield	Per cent broken grains	Labor hrs per ton	Sample size
B. Husking and Polishing[2]				
Dheki	72.02 (0.46)	28.28 (4.50)	240	20
Engleberg-type huller rice mill	69.94 (0.33)	30.54 (4.52)	2.8	20
Laboratory model modern milling	71.87 (0.28)	8.89 (1.91)		20

[1]There was only one case of a broadcast crop being pedal threshed and the measured loss was 7.3 per cent.

[2]Figures in brackets are standard errors.

Source: p. 26, M. Greeley, Rural Technology, Rural Institutions and the Rural Poorest: The case of rice processing in Bangladesh, 1981, plus previously unpublished data on labor use from the Institute of Development Studies Post-Harvest Project.

The most satisfactory criterion, reflecting economic loss to the producer, would be the effect that poor milling has on price except that in the majority of cases subsistence production prevails and the rice is consumed on the farm. There are no satisfactory criteria for measuring absolute loss in this situation but comparisons between methods can still be usefully made. In Bangladesh (see Table 2, part B) the change to Engelberg hullers from manual methods increases losses by 2% measured by the reduction in milling yield and has a negligible effect on brokens - increasing them from 28 to 30 percent. As with threshing, the reduction in labour costs more than compensates for the reduction in yield. The traditional system also compares favourably to modern milling methods (rubber roll huller and truncated stone cone polisher) in total outturn because of consumer preference for undermilled rice. The high level of brokens has no economic implications where rice is consumed on the farm - and indeed in the market there is little premium for whole rice because of these consumer habits. At present most of Bangladesh's off-farm milling capacity is in Engelberg-type hullers which continue to increase in number and the transition to rubber-roll hullers with separators and polishers is in an early stage compared to most Asian countries. The rate of transition is determined by the commercial opportunities to invest in larger scale processing units which are limited by the size and regularity of paddy supplies. There is no evidence that the change from manual methods can increase food availability or that the direction and the rate of technical change are influenced by differences in the level of food-losses which are a marginal influence on the economic viability of different milling systems.

SECTION III

Most of the reports cited in Table 1 are of very recent origin; preliminary reports from continuing research programmes support their findings and a new concensus is developing; the myth of high average food losses in farm-level post-harvest systems is being slowly but firmly dispelled. A retrenchment because of this would be wrong for at least two reasons. First, however, inaccurate, it was the almost universal belief of high losses that attracted the funding for research and which has in turn helped in assessing more carefully what needs to be done; prior to this exposure the progress on improving loss-assessment methods had been slow and systems improvement was based on educated guesswork. The research results are still providing essential information and if anything their significance is enhanced because they are so different to the preliminary hypothesis. Secondly, the evidence does not show that farm-level post-harvest technical change is unimportant only that food loss as a motivating force is. Post-harvest operations are still the major source of rural employment outside of direct crop production and collectively the ancillary activities associated with them are commonly the largest rural industry. With the quantity and quality of field data now becoming available it is possible to identify what changes in the system are most important for development and to reassess future priorities. Three issues are of obvious relevance to this reassessment: the opportunities for low-cost innovation, the effects of cost-reducing technical change and the implications of increased food output.

Low Cost Innovation

Even if food losses are low, the gains from prevention of small losses will be large if they are widespread. This has anyway been the position adopted by many (25) who have been unwilling to rely upon old statistics of dubious value and have taken a 'conservative' view of losses; the argument has been (26) that if losses are ten percent and we can save only half of them there would be an extra 40-45 million tons of food available annually worth over $7 billion... and therefore investment of $1 billion at least to achieve this is a priority.

The evidence of field studies shows that even a five percent reduction is optimistic; when low losses are spread across many operations, wastage in any one operation is small and the effective margin for improvement is often negligible. Moreover, the statistic that matters is not total loss but preventable loss. Even with a very sophisticated post-harvest system it is virtually impossible to restrict losses to zero.

Assuming a technique is available, the decision to adopt will depend upon the costs of and benefits to the farmer; with small grain savings only a low cost improvement will yield a positive return. In storage this will normally mean modification to existing structures or use of insecticides; the difficulties posed for extension and training by this type of programme have been discussed briefly above in the context of the Indian and Zambian results and similar difficulties have been identified in Bangladesh, Ghana, Kenya, Mexico, Senegal and Tanzania (27). The Tanzanian case study was unique amongst these in trying to promote a self-help programme based on local farmers' understanding of storage problems developed through group discussions. This was very much a pilot programme - and has remained so since the response was limited though similar programmes may be developed elsewhere - but a further binding constraint was the staff training requirements to operate Frierian or similar methods on a bigger scale. This staffing and management problem is a recurrent theme in self-help programmes and in practice prohibits the approach for national strategies of extension. With well-trained and motivated extension staff the Indians have demonstrated through their Save Grain Campaign (28) that more traditional extension methods can have some effect but their experiences also show that it is easier to sell finished products, such as metal bins, to less poor and prestige-conscious farmers than it is to motivate small farmers with scarce resources - including their own labour time - to invest in marginal improvements. Cash constraints - and economic rationality - prevent these farmers from costly innovation but with minimal losses any form of innovation in storage is low on their priorities. Questionnaires on farmers' attitudes sometimes give different results to the actual outcomes when extension staff begin work but there is now sufficient experience with storage extension to know that it suffers from vapidity more than rapidity.

The discussion of these storage programmes has assumed, reasonably, that low cost improvements are available but in other operations, such as crop drying, this is not always the case. Led by international research institutes, there can be few departments of agricultural engineering that have not tried to develop crop dryers for use at farm-level. Wet grain - rice in particular - is commonly cited in their papers as a major cause of deterioration and food loss in the tropics yet 'farm-level' driers are extremely uncommon off campus. The engineers have been successful technically - the principles of safe crop drying are well established - the failure has been to identify economic applications.

The frequency with which the same farmer suffers sufficient post-harvest damage due to rain to warrant investment in a new technique performing a function now provided free by the sun is the critical and very complex question that engineers need to answer. Experience from Bangladesh (29) in measuring wet season losses has shown that this condition is very infrequent and even co-operative ownership of driers has proved uneconomic there (30). For many farmers the quantity produced is small allowing discretion over harvesting dates, use of rooms for spreading the threshed crop and use of short periods of sunshine. The economic risk is further reduced because most production is retained for self-consumption where qualitative deterioration has no cash implications and, on small farms, grain for sale is usually marketed at the earliest opportunity, (i.e. when it is still wet) to meet cash requirements. These practices all reduce the potential benefits from owning or using driers and most traditional designs, requiring blowers, heaters and purchased fuel, are hopelessly unprofitable at farm-level. With the high loss-high cost solution discredited, engineers have been examining more carefully the opportunities for very low cost improvements, through solar drying (31) or natural draught drying (32) using crop residues for fuel; these designs have created a new generation of problems relating to technical efficiency, capacity, space requirements and operating skills and even in situations where losses would justify a low-cost innovation these technical difficulties have prevented any widespread application.

The experience with innovation in storage and drying leaves

little room for optimism about low-cost solutions for low-loss situations. There is still a requirement for some carefully located research to explore areas where these general conditions may not apply but the efficiency of labour-intensive subsistence production greatly limits these opportunities.

Cost Reduction

Unlike crop production, where innovation increases total output, innovation in post-harvest operations cannot increase output except by reducing losses; with low food losses the profitability of technical change depends on cost reduction. This has two implications of general significance in planning technical change. First, since traditional post-harvest practices are extremely labour-intensive, cost-reducing change will necessarily be labour replacing. Apart from storage structures and hand pounding equipment there are no major items of equipment and even these two often only require labour time rather than capital investment. Obtaining increased labour use would depend upon innovations such as the use of locally-produced crop driers which, with low food losses, are not economic.

In situations of labour shortage or where family labour is being released for other productive tasks displacement will have no costs. Frequently however, post-harvest labour is a substantial source of earnings for the landless - especially in South Asia; the Bangladesh study identified reduced earning opportunities for landless women as the chief cost of the change from traditional to mechanical milling. We have seen (Table 2, part B) that the change increases food losses but that it also reduces the labour hours per ton from 240 to 2.8. With the constraints of purdah only a small number (about 8%) of rural women work for wages but in their household they contribute over twenty-five percent of family income, half of which was traditionally earned through rice processing (33). The larger farms that employed these women are the first to start using rice mills - which are increasing in number as the rural electrification programme develops - and the lost jobs threaten the survival of these anyway very poor families. Alternative work places are difficult to develop because there are social restrictions on female mobility and lack of assets prohibits productive employment off the farm or homestead.

Contrary to the usual assertions, the implications of technical change are a reduction in food availability overall and a reduction in the earnings and therefore food-purchasing power of the poorest. This is a particularly extreme example in a very poor country but labour-displacing technical change is a general feature of post-harvest innovation. Changes in storage structure from units locally produced by village artisans to factory-made units, typical of the thrust of current extension activity, also reduce labour earnings for a minority and poor group.

Whilst some technical change, such as pedal threshing in Bangladesh, can solve genuine seasonal labour scarcities even in surplus labour countries the implications for labour use need careful evaluation in each case. The problem of hunger in developing countries is shortage of purchasing power as well as production and labour-displacing post-harvest innovation will reduce purchasing power.

Secondly, with a few exceptions such as the hand-held maize sheller, cost-reducing change will also change the ownership pattern of post-harvest equipment. The introduction of capital equipment - threshers, mills, perhaps driers eventually - allows and sometimes requires them to be operated on a custom hire basis; in other words, the potential throughput, or with lumpier investments such as mills, the minimum profitable throughput, is larger than the output of most individual farms. The significance of the change in ownership is the opening up of opportunities it provides for entrepreneurial activities that did not exist so long as post-harvest operations were without larger units of capital equipment.

The rate of adoption will depend upon the cost reductions obtained; in addition to labour saving this will be influenced by the effects on farm management and transportation. The easing of farm management by reducing the time per operation is an additional benefit when space (for threshing particularly) and supervisory time are binding constraints; but the critical adoption issue is transportation. Costs per unit of output in threshers and mills <u>including</u> the additional transport costs (for the machine or the grain) have to be lower than traditional practices to allow adoption. Therefore the development of the rural infrastructure particularly roads (and canals/railways) and seasonal and size variation in transport costs will jointly determine the rate of adoption. Where these developments have occurred post-harvest mechanization has been a major - and in some cases <u>the</u> major - source of rural agro-industrial growth. This has been particularly true for the poorest LDC economies where the shortages of purchasing power have precluded entrepreneurial initiatives in manufacturing; crop processing does not require the development of a market for once value added in the new techniques is greater than in traditional practices, a captive market exists. Moreover, value added in post-harvest operations varies between twenty-five and forty-five percent of total value added from seed to consumption; therefore, decisions on support (subsidies/credit/fiscal benefits) and control (manufacturing and operating licences/zoning/import restrictions) of post-harvest technology can have a sweeping influence on the pace and structure of rural growth.

The separation of processing from production also allows the development of programmes for the landless by supporting their, individual and collective, ownership of equipment like threshers and mills. In land-scarce economies, recent analyses (34) argue that these sorts of initiative are necessary constituents of rural growth because of their demand linkages. This type of involvement can turn the cost-reducing (labour-displacing) nature of technical change to the advantage of labour and actually increase their share of income from post-harvest activities. Programmes of this sort are now operating (35) and their further development may well be the most important dynamic of post-harvest planning.

Increased Food Output

The efficiency of traditional post-harvest systems depends to a large extent upon the ordinary (careful) management of small quantities of output by farmers with adequate labour resources and few fixed assets. As land productivity increases it becomes increasingly difficult to maintain these levels of efficiency. The fixed assets, - particularly farm-yard space, supervisory time and draught power for transport and threshing - are insufficient to maintain throughput; at the same time labour bottlenecks appear bidding wages up and making it more expensive to get the same job done. These higher marginal costs lead to larger marginal food losses - such as inadequate threshing, because it is now too expensive to avoid them, and thereby increase the return from cost-reducing innovation. Higher costs and delays also encourage producers to sell grain early - before drying and at low harvest time prices - and in so doing they pass the risk of deterioration to traders and the public sectors procurement agencies.

The severity of these effects depends upon the magnitude of production increases and relative scarcity of the fixed assets and labour. Evidence from the Indian Punjab (36) and from the MUDA scheme in Malaysia (37) shows that they can result in high losses; threshing losses measured for rice in the MUDA scheme were over nine percent. Harvesting more susceptible HYVs in the rainy season with labour shortages is characteristic of a substantial part of the increased land productivity in Asia and in some African countries, and the experience of MUDA is being repeated now even in parts of Bangladesh where small farm size and surplus labour reduce the risk of such losses. In these situations the adoption of new techniques such as mechanical threshers or even combine harvesters to reduce labour costs can, but may not always, also reduce losses. However, farmers can and do relieve both their labour and loss problems by selling the crop. The marketed surplus will anyway normally increase by more than the increase in production and as early experience has shown the most severe risks of deterioration occur with this surplus, particularly when it is wet.

Commercial traders respond to price incentives and their investment in better drying and milling complexes will depend on the price premiums for good quality grain/flour. This is not the case with the public sector food corporations who are often operating support prices and whose commercial incentives are remote from daily transaction. Faced with sudden sharp rises in marketed surplus, they become involved in market operations lacking the storage space, the equipment, technical staff and management capacities to maintain good quality stocks. The public sector is usually a residual element in the system and provides the backstop in periods of glut. At these times, the level of losses in stocks surplus to their carrying capacity can be very high because of the difficulties in obtaining reasonable temporary storage. The evidence on losses is very thin because these agencies are sensitive about their performance and frequently they only quote selected accounting records of loss - i.e. losses not within the corporation's allowable losses. However, these problems are well known and already the subject of intensive development assistance, especially in the construction of public sector storage capacity.

The low levels of farm-level storage loss provide an alternative solution to the problems of surplus. By the use of seasonally stepped incentive prices that provide a margin for storage costs, risk and a small profit farmers could be persuaded to hold stocks until after the immediate post-harvest period. Alternatively the grain could be purchased by the government but held as a lien (a prior claim) by the farmer (38) until storage space was available. Such schemes would take advantage of the low levels of farm-level food losses and save the investment costs in large-scale capital intensive central storage facilities. These are suggestions that would only operate satisfactorily in times of surplus but that is when they are needed. Decentralization offers more effective food security at lower cost and with the continuing need for large public sector food stocks offers a viable long-term policy strategy for future post-harvest planning.

REFERENCES

1. K.D. White, "Roman Farming," Thames Hudson (1970), pp. 196-97.
2. "Reducing Post-Harvest Food Losses in Developing Countries," FAO (1975), pp. 2-4 and Appendix II. The Tropical Products Institute, London, is one of the few other organizations that has been regularly active in loss prevention programmes.
3. M.C. Bourne "Post-Harvest Food Losses - The Neglected Dimension in Increasing the World Food Supply," Cornell International Agriculture Mimeograph (1977), pp. 2-3.
4. Examples are numerous since they preface most papers on the subject and a short list of examples from government, voluntary agencies, international donors and private companies is given on pp. 4-5, "Rural Technology, Rural Institutions and The Rural Poorest: The case of rice processing in Bangladesh," (Martin Greeley 1981).

5. James Boulware, USDA attache in New Delhi drew attention to this anomaly of Indian reports, cited in Lester Brown, "Seeds of Change," Pall Mall London (1970).
6. These are figures originally published internationally in "Rice Post-Harvest Problems in Southeast Asia," International Food Technologists Meeting, Philadelphia, June 1977, and elsewhere they are cited as IRRI and FAO figures. They give total maximum losses of 31.43%, but several papers, e.g. H.A. Parpia, More than Food Would be Saved, Ceres, December 1977 report 37% losses because they fail to take account of reductions in the quantity available due to loss at earlier stages.
7. For example, Final Report of the Expert Committee on Storage Losses of Foodgrains During Post-Harvest Handling, Government of India (1971), unpublished.
8. See M. Sriramalu, Studies on the Varietal Resistance of Paddy to Lesser Grain Borer, Rhizopertha Dominica (Fab.), MSc. Thesis, Dept. of Entomology, Agricultural College, Bapatla, 1973 (unpub.); S. Srinivasan, Studies on Degree of Infestation by Rice Weevil Sitophilus Oryzae L. in Different Varieties of Paddy and Rice, MSc Thesis, Dept. of Entomology Agricultural College, Bapatla, 1972 (unpub). Both demonstrate the need for plant breeders to take account of resistance to store insect pests by showing how some modern thin-husked rice varieties are more susceptible to the rice weevil (Sitophilus Oryzae L.) and the lesser grain borer (Rhizopertha Dominica (Fab.)). This is supported by field observations of the effects of this factor on the timing of sales of, e.g. RP4-14, by paddy cultivators in West Godavari, A.P. Though see also, "Resistance to Storage Insects in Wheat Grain," IARI Research Bulletin No. 28 (1980).
9. K. S. Gill, "Wheat market behaviour: emerging problems of wheat marketing in Punjab and Haryana post-harvest period," Punjab Agricultural University, Ludhiana (undated).
10. For example, "Commodity Storage Conditions in Bangladesh," a staff report to the Subcommittee on Foreign Assistance of the Committee on Foreign Relations United States Senate, (1976).
11. For example, "Protecting the World's Crops," Shell Briefing Service (1977).
12. For example, Robert Satake, "Status of the Rice Milling Sector," AMA IX, No. 2, Spring 1979.
13. For example The Comilla Co-operative Karkhana Ltd., Ranirbazar, Comilla, Bangladesh manufacturing pedal and mechanical threshers.
14. 'Appropriate' is really shorthand here for low-cost, simple to use, local skill and raw material intensive techniques. A good example of the approach is "Small Farm Grain Storage," Appropriate Technologies for Development, Volunteers in Technical Assistance, Vita (1976).

15. India has been very active both in research with the Indian Grain Storage Institute and the All India Co-ordinated Post-Harvest Technology Scheme and in extension with the Save Grain Campaign. In Africa, Zambia, Kenya and Swaziland all developed extension programmes quite early on and in Nigeria the West African Stored Products Research Unit has made important contributions to research from an early date.
16. Of special importance was the preparation of K.L. Harris and C.J. Lindblad, "Postharvest Grain Loss Assessment Methods," USAID. (1978).
17. For example, Cost-Benefit Analysis of Crop Storage Improvements: A South Indian Pilot Study, IDS Discussion Paper No. 56, University of Sussex (1974), by M. Lipton.
18. S.K. Majumdar and H.A.P. Parpia, Possible Losses of Foodgrains in India, 1966 reprint from Vijnan Karmee - The Journal of the Association of Scientific Workers in India, Vol. 18, No. 4.
19. Bourne, op. cit. pp. 17-18.
20. A.M.A. Karim and H. Rashid, Extent of Loss of Boro Paddy during Post-harvest Operation: A Study Conducted in Bahadurpur Village of Mymensingh District, Agri-Varsity Extension Project, Bangladesh Agricultural University, Publication No. 10. (1979).
21. H. Guggenheim, Who is the Loser in Post-Harvest Losses? The Wunderman Foundation, (undated)
22. T.W. Schultz, "Transforming Traditional Agriculture," New Haven, Yale UP (1964).
23. Appropriate Rural Technology: Recent Indian Experience with farm-level Foodgrain Storage Research, Martin Greeley, Food Policy, (February 1978).
24. The Technical and Economic Characteristics of Rice Post-production systems in the Bicol River Basin, Bicol River Basin Development Program (1978).
25. "Postharvest Food Losses in Developing Countries," National Academy of Sciences, Washington (1978).
26. Most recently presented in South, No. 3 Dec. 1980, p. 27.
27. A. K. Fazlul Huq, Rice in Bangladesh: Estimation of Food Losses in Farm-level Storage at a Workshop sponsored by the Food Science and Technology Division, Bangladesh Council of Scientific and Industrial Research, Dacca December 1980. p.12-13; Post-harvest Food Losses in Developing Countries, Development Digest Vol VIII No. 4 Oct 1980, p. 107; W.D. Rolston, The Post-Harvest Food Grain System in East Africa (A Kenyan Case Study), International Development Research Centre, Edmonton (1975), p. 10; W.D. Rolston, The Post-Harvest Food Grains System in West Africa (A Senegalese Cast Study), IDRC, Edmonton (1975), pp 11-14; Mexico, the Post-Harvest Maize System in Two Pider Micro-Regions, The World Bank Rural Development Division, (1978), p. 14 and Appendix 3; Appropriate Technology for Grain Storage, Report of a Pilot Project, Community Development Trust Fund of Tanzania (1977), Section 2.

28. K. Krishnamurthy, Save Grain Campaign - Objectives and Plan of Work, Post-Harvest Technology Workshop, New Delhi, Jan 1981.
29. M. Greeley and S. Rahman, Wet Season Post-Harvest Food Losses, Paper Presented at the Post-Production Workshop on Food Grains, December 1980.
30. Danida Drying Project in Comilla, Project Report 1974.
31. R.H.B. Exell, Basic Design Theory for a Simple Solar Rice Dryer, Renewable Energy Review Journal, Vol 1. No. 2, Jan. 1980.
32. Rice Post Harvest Technology Project, 1978 and 1979, Report, TPI/BRRI, Dacca Bangladesh (1980).
33. S. Begum and M. Greeley, Women, Employment and Agriculture: Notes from a Bangladesh Case Study, IDS, University of Sussex (1980).
34. G.D. Wood, How the Interests of the Rural Poor Can be Included in the Second Five-Year Plan, Dacca (1980).
35. Through the Bangladesh Academy for Rural Development, Comilla.
36. K.S. Gill, Post-Harvest Market Technology for Cereals (Paddy and Wheat) - Needed Improvements (The Punjab Case), paper presented at Rural Technology, Rural Institutions, and the Rural Poorest, Comilla, Bangladesh, Feb. 1981.
37. D.J.B. Calverley, P.R. Street, T.J. Cree, D.A.V. Dendy, Post-Harvest Losses of Rice in Malaysia, Conference on Food and Agriculture, Malaysia 2000 (undated).
38. M. Lipton, Post-Harvest Technology in the Context of the Reduction of Hunger, Workshop on Farm-Level Post-Harvest Technology for Prevention of Food Losses, New Delhi, Jan. 1981.

UTILIZATION OF AGRICULTURAL WASTES - SOME GLOBAL CONSIDERATIONS

A. J. Vlitos

Group Scientific & Agricultural Director
Tate & Lyle, Limited
Reading, Berkshire, England

INTRODUCTION

We are living in a World where the future comes sooner than it used to, and in a World where change is as inevitable as death and taxation. My lecture this evening therefore will <u>not</u> be devoted to post-harvest physiology or for that matter to the specific techniques for converting agricultural wastes to useful products. Instead I would like to make some guesses about the place of agriculture, worldwide, say in 20 years' time - by the year 2001. No one really knows what sort of place the World will be in 20 years time. All we can do is guess. And the guesses have to be based on what is known today and on trends - and on the likely innovations which will occur as the result of the research which is underway today. So, what sort of world will it be in 20 years' time?

It will be more crowded, more polluted, shorter of food, a more expensive place to live in than it is today. Every long-term global study has come up with the same conclusion. Only the degree of overpopulation, the amount of pollution and the exact figures of inflation are at issue between one study and another. The "Global 2000 Strategic Study" which was prepared for President Carter can be summarized as in Tables 1 to 10.

GENERAL

If the economic gap between the Third World and the developed world is to become even greater than it is today, one can predict that certain carbohydrate crops, such as sugar cane, will assume

Table 1. Global-2000

1. More people (i.e. 6.35 billion)
2. Poorer
3. More pollution
4. More vulnerable to disruption
5. Outlook for food and other necessities of life no better than now

(Reference: Global 2000 Report to President of U.S.A.)

Table 2. Population Growth

			Added each year
1975	4	Billion------	75 million
2000	6.35	Billion------	100 million
2030	10	Billion------	120 million

(90% of growth in population in the poorest countries).

Table 3. World Food Production

		% Increase in Production
1.	1870 to 2000	90% (i.e. 15% global *per capita*).

2. Most of this increase goes to countries which have high *per capita* food consumption (i.e. 'richer' ones).
3. *Per capita* consumption in LDC's scarcely improves or actually declines.

Table 4. World Food Production

1. Arable land increases by 4% by 2000
2. Most increases in food has to come from increased yields on existing acreages.
3. Almost all prices to consumer will be higher.

Table 5. Fisheries

1. 1970 to 1975 → 70 million metric tons *per annum*.
   (i.e. 60 million tons marine:
   10 million tons freshwater).
2. Required to keep pace with projected increases in population - 115 million tons *per annum*.
3. Assuming catch of marine and freshwater fish rises to unlikely level of 100 million tons and that yields from aquaculture double (to 12 million tons) - the 112 million tons is not enough to provide *per capita* protein from fish provided today.

Table 6. Estimates of World Forest Resources, 1978 and 2000.

	Closed Forest[a] (millions of hectares)		Growing Stock (billions cu m overbark)	
	1978	2000	1978	2000
U.S.S.R.	785	775	79	77
Europe	140	150	15	13
North America	470	464	58	55
Japan, Australia, New Zealand	69	68	4	4
Subtotal	1464	1457	156	149
Latin America	550	329	94	54
Africa	188	150	39	31
Asia and Pacific LDCs	361	181	38	19
Subtotal (LDCs)	1099	660	171	104
Total (world)	2563	2117	327	253
			Growing Stock per Capita (cu m biomass)	
Industrial countries			142	114
LDCs			57	21
Global			76	40

[a]Closed forests are relatively dense and productive forests. They are defined variously in different parts of the world. see Global 2000 Technical Report, footnote, p. 117.

Table 7. Energy

1. By 1990 - energy demand to increase by 58%.
2. Nuclear and hydrosources increase most rapidly 226%.
3. Oil by 58%.
4. Natural gas by 43%.
5. Coal by 13%.

Oil remains leading energy source through 1990's (46-47%). Real price of oil increases 65% over 1975 - 1990.

Table 8. Environmental Consequences

Impacts on Agriculture : Acceleration of -

1. Soil Erosion
2. Loss of nutrients
3. Compaction of soils
4.* Increasing salinisation of both irrigated land and water used for irrigation
5. Loss of high-quality cropland to urban development
6. Crop damage due to increasing air and water pollution
7. Extinction of local and wild crop strains needed by plant breeders
8. More frequent and more severe regional water shortages - especially where energy and industrial developments compete for water.

* 6 million hectares per annum (area size of Maine) lost through desertification. This rate likely to increase by 2000.

Table 9. Impacts of Forest Losses

1. If present trends continue forest of South Asia, Amazon basin and Central Africa will be reduced by one-half by year 2000.
2. About one billion people live today in heavily farmed alluvial basins and valleys depending on forested mountain watersheds for their water.
3. Diversity of species lost after extensive cutting.

Table 10. Impact on Atmosphere & Climate

1. Further pollution virtually certain: (SO_2 particulates, NO_2, CO, etc.).
2. pH value of rainfall dropping from 5.7 to 4.5 due to S and Nitrogen oxides combining with water vapours.
3. Combustion of fossil fuels increasing CO_2 concentration ("greenhouse" effect). Alters temperatures and precipitation patterns?
4. Impacts of nuclear waste disposal?

even a more important place in tropical agriculture than they enjoy today. There are two reasons for this prediction. Firstly, the major increases in population are likely to occur in the poorer, tropical nations. The per capita consumption of sugar in these nations is presently amongst the lowest in the World. The major increases in consumption are likely therefore to occur here. There will be more people eating more sugar in the Third World and more of that sugar will be grown there. Secondly, most of the tropical, Third World nations are deficient in fossil fuels. There will be compelling reasons to increase the acreages devoted to sugar cane for the production of power alcohol. And by the year 2001 a large proportion of the alcohol produced in Third World countries by the fermentation of cane juice will go into ethylene production employing small-scale converters. The ethylene thus produced will serve for the establishment of relatively small-scale, almost "cottage-type" 'downstream' chemical industries - turning out some of the speciality chemicals (high added-value products) such as plastics, detergents, lubricants, etc. presently imported into Third World Countries from Europe and North America. By the year 2001 the cost of energy and of the 'feedstocks' used in industrial chemical plants of Europe and North America will be so high that a shift to the tropics will be attractive. A simple diagram illustrating the key position of ethanol and ethylene by 2001 in this chain of events is shown in Table 11.

There are of course several pre-requisites for this to happen. The first of these is financial investment. Over the past eight years there has been a global shift of wealth from the nations which manufacture finished products to those nations which supply the energy needed to manufacture such products. By the year 2001 it is hoped that the OPEC nations, which by then, shall have most of the World's money invested, will have placed a major portion of the investment in Third World agriculture. But the major funding agencies today encourage most of the investments in Third World agriculture to go into local food crop schemes rather than into major sugar projects. Although one can understand why a few years ago it did not consume the amounts being produced, this situation is unlikely to be repeated between now and the year 2001 - in view of the attraction and potential for sugarcane as a raw material for fermentation. I believe a very strong case could be made to the World Bank, and to other funding agencies, for the encouragement of "carbohydrate agriculture" in nations which lack fossil fuels and which also lack any hope of establishing small-scale agricultural industries unless an energy source is readily available. Sugar cane and alcohol and ethylene from alcohol offers hope of lessening the reliance on imported fuels. Encouragement to utilize fermentable fruit and vegetable residues is another area offering considerable scope.

A second pre-requisite is, of course, a commitment by the Third World nations themselves which will encourage carbohydrate crop cultivation for domestic consumption as much as for export. Today, of course, there is a heavy dependence by sugar cane producers on

Table 11. Third World Scenario for Sugar cane by year 2001

Sugar cane ↗ Food (and)

Sugar cane → Fermentation → Power Alcohol

Ethanol → Ethylene → Plastics + Lubricants + Other Chemicals

(Liquid Fuel)

exports and on guaranteed prices for exported cane sugar. Indeed without export markets it is unlikely that the cane industry in most ACP (African/Caribbean/Pacific) Nations could survive. But is that likely to be the case in 20 years' time? It might well be that the traditional markets for raw can sugar in North America, in Europe and in Japan may not be as attractive then as they are now for a variety of reasons which I will be discussing momentarily. The likely result of these changes in external markets is that in 20 years' time more cane sugar will be refined in the countries of origin and most of the cane will be consumed either as food or used industrially in those countries in which it is produced. The change is likely to be gradual rather than precipitous - but it will depend a great deal on the position in 20 years' time of sugar beet in Europe and of maize in N. America and Europe, and of new, non-nutritive sweeteners in the industralized nations.

Although the predictions for the growth in population by the year 2001 in Europe and in North America are not as frightening as those for the Third World, there will nevertheless be increases in population and with the accompanying problems, high energy costs and consequent inflation and increasing pollution. There are good reasons for expecting sugar beet to remain a major crop in Europe, but here again there are some pre-requisites to be met. Firstly, the cost of energy in processing beet sugar is already quite high and in 20 years' time it could be <u>the</u> most serious obstacle to the crop's survival - unless of course a major research thrust begins soon which can offer significant savings in the costs of processing. Another unknown is whether new European varieties of maize, in 20 years' time, can begin to become attractive alternatives to beet - and whether high fructose syrups will make inroads into European beet. Consumer pressures could be more significant in Europe in 20 years' time and we should not underestimate them, especially if new varieties of European maize offer cheap starch, and if European farmers find growing maize more attractive than growing beet, or if a non-nutritive sweetener becomes attractive. So the variables in the future of beet are more difficult to assess than those for cane. These are the qualitative scenarios and thus far I have avoided trying to guess the quantitative impact on total world production of sugar by 2001. The FAO figures of cane and beet production for 1979 are shown in Table 12.

By 2001 the chances are that total world production for cane could exceed the 1979 figure by a minimum of 25% which would mean 737 + 184 = 921 thousand tons but with yields <u>per hectare</u> increased by at least 15% giving 56 x .15 + 64.5 t/ha. which would still be well below the Australian figure of 75 tons per hectare. Although improvements in average beet yields especially in Europe are possible - say from 38 tons/hectare to 45 or 50 tons per hectare - the total acreage devoted to beet by 2001 could be about the same as now or slightly decreased.

Table 12. Total production (10^6 t) and yields ($t.ha^{-1}$) of the major sugar and starch crops.

These figures are on a crop basis excluding the considerable amounts of crop wastes, straw etc. The cereal figures exclude such crops grown for forage silage.

a) Sugar

	Cane		Beet	
	Total	Yield	Total	Yield
World	737	56	290	32
Developed countries	70	79	263	32
Developing countries	667	54	26	30
U.S.A.	25	82	23	46
Canada	-	-	1	39
Europe	0.3	63	143	38
U.S.S.R.	-	-	93	25
Asia	305	52	25	31
S. America	187	57	-	-
Africa	60	64	16	32
Oceania	26	75	-	-

Another aspect of the energy crop concept is as follows: By the year 2001 we may see a competition for land to be used either to grow food or to grow crops for energy. In the past when crops were grown chiefly for food we measured efficiency by the number of people employed on the land to feed so many people in the cities. Thus, as we see in Table 13 a very small percentage of the population in the U.S.A. and Australia feed a much larger percentage of people in cities. This is possible because a very high input of energy is put into agricultural production - fertilizers, machines, etc. But when crops are to be grown for energy we won't be able to put more energy into the crop than we can retrieve once the crop is harvested. Thus, I have described in Table 14 four levels of agricultural practice. You will note that the natural ecosystem requires the lowest inputs of energy. Energy crops of greatest potential interest will be those requiring the lowest inputs of energy. That's something else which we shall have to deal with by the year 2001.

SUMMARY

Now, the attractions of utilizing fruit and vegetable wastes as fermentable substrates are many. First of all, the energy inputs in producing the crop have already been paid for in most cases. However, there remains the problem of collecting the wastes and

Table 13. People Employed in Agriculture on a Global Basis

	1965 Percent of Population	1980 Percent of Population
Worldwide	54	45
Africa	70	67
S. America	45	38
U,S,A,	4.0	2.5
Australia	3.5	2.0

Table 14. Levels of Agricultural Practice and Energy Inputs

Type	Characteristics
Natural ecosystem - no human influence.	High yields; forest.
Labour intensive; no inputs from external sources.	Subsistence farming - (Third World Countries absence of cash inputs, low yields, shifting cultivation, destruction of forests.
Present-day 'average' agriculture, some mechanization reasonable soil and water management.	Yields increased by increasing inputs of energy *via* more fertilizers, more mechanization, etc. Yields high in relation to people employed.
Optimal production based on theoretical use of incident solar energy.	similar to above

transporting them to the fermenter. In Brazil an energy-efficient farming system is being devised in the Mato Grosso which utilizes trucks and tractors running on 100% ethanol produced from sugar cane - the distillation being fueled by the cane by-product, bagasse. One can envisage a collection system for other agricultural wastes using vehicles running on alcohol. Secondly, most agricultural wastes are of negative value. The only major cost is in collecting and transporting them. Thirdly, in most tropical countries the wastes are rich in fermentable sugars. In the temperate zones many of the wastes are cellulosic, but the technologies for converting cellulose to fermentable sugars are being developed rapidly and in 20 years' time will be economical. In Table 15 I have listed some of the common substrates which are being experimented today. To these may be added the numerous fruit residues, ranging from pine-apple to figs, which in future will be sought as sources of alcohol.

Finally, a word about the conversion of carbohydrate wastes to microbial proteins. As you know the major producers of microbial proteins use methanol as the primary substrate. The chances are that in future methanol will prove too valuable as a liquid fuel to convert to proteins and it may prove difficult to compete with vegetable proteins such as soya.

Therefore to leave you with the final prediction - negative value agricultural wastes - especially those right in carbohydrates will become increasingly attractive as substrates for fermentation to produce microbial proteins as non-ruminant animal feeds. Oscar Wilde once remarked that "it is dangerous to prophesy - especially about the future" - and with that sentiment I humbly agree!

Table 15. Biomass of Interest

Sugar	Starch	Crop residues
Molasses	Cellulose	Rice husk Coffee husk
Bagasse	Straw Shell-fish wastes	Spoiled grains Vegetable wastes

PARTICIPANTS

Adams, D. O.	Postharvest Physiology Laboratory, U.S. Department of Agriculture, BARC-W, Beltsville, Maryland, 20705, U.S.A.
Apeland, J.	Department of Vegetable Crops, Agricultural University, Norway, P.O. Box 22, 1432 AAS-NLH, Norway
Atwa, A. A.	21 Murad Street, Giza, Cairo, Egypt
Baker, J. E.	Postharvest Physiology Laboratory, U.S. Department of Agriculture, BARC-W, Beltsville, Maryland, 20705, U.S.A.
Bangerth, F.	Institute für Obstbau, University of Hohenheim Stuttgart 70, Federal Republic of Germany
Barreiro, G.	Departmento de Fisiologia Vegetal, Estacao Agronomica Nacional, Quinta Do Marques, 2780 Oeiras, Portugal
Ben Arie, R.	Division of Fruit & Vegetable Storage, Ministry of Agriculture, P.O.B. 6, Bet Dagan 20-500, Israel
Bohling, H.	Federal Research Center for Nutrition Engesserstr 20, Karlsruhe D 7500, Federal Republic of Germany
Bruinsma, J.	Agricultural University, Arboretumlaan 4, Wageningen, The Netherlands
Bufler, G.	Institut Für Obstbau, University of Hohenheim 7000 Stuttgart 70, Federal Republic of Germany
Burg, S. P.	3770 Kent Court, Miami, Florida, 33133, U.S.A.
Cabral, M. L.	Departmento de Fisiologia Vegetal, Estacao Agronómica Nacional, Quinto Do Marques, 2780 Oeiras, Portugal
Carmona, M.	Quinto Do Marques, 2780 Oeiras, Portugal
Chalutz, E.	Institute for Technology & Storage, P.O.B. 6 Bet Dagan (20-500), Israel
Côme, D.	Laboratoire de Physiologie des Organes Végétaux Après Récolte, 4ter Route des Gardes, 92190 Meudon, France
Coursey, D. G.	Plant Food Commodities Department, Overseas Development Administration, Tropical Products Institute, 56-62 Gray's Inn Road, London, WC2X8LV, England
Cowan, A. M.	National Program Staff, PHST, U.S. Department of Agriculture, BARC-W, Beltsville, Md. 20705, U.S.A.
Dekazos, E.	Department of Pomology, College of Agricultural Sciences, Votanikos, Athens 301, Greece

De Leo, P.	Centro Richerche Bonomo, Via Sparano, 141, 70126 Bari, Italy
Dilley, D. R.	Department of Horticulture, Michigan State University, East Lansing, Mich. 48824, U.S.A.
Duvekot, W. S.	Sprenger Institute, P.O. Box 17, 6700 AA Wageningen, The Netherlands
Eckert, J. W.	Department of Plant Pathology, College of Agriculture Sciences, University of California, Riverside, CA 92521, U.S.A.
Efthimiades, P.	Agronomy Department, Agricultural College of Athens, Votanikos 75, Athens 301, Greece
Esquerré-Tugayé, M-T	Universite Paul Sabatier, Centre de Physiologie Vegetale, Laboratoire Associe au C.N.R.S. No. 241, 118 Route de Narbonne, 31077 Toulouse Cedex, France
Fallot, J.	Ecole Nationale Superieure, 145 Avenue De Muret, 31076 Toulouse, Cedex, France
Feys, M. D.	Laboratory of Food Preservation de Croylaan 42, Catholic University of Leuven, B 3030 Heverlee, Belgium
Fountas, D.	2 Sarantaporou, Kallithea, Athens, Greece
Fuchs, Y.	Division of Fruit & Vegetable Storage, The Volcani Center, Agricultural Research Organization, P.O.B. 6, Bet Dagan, Israel
Goodenough, P. W.	Department of Agriculture & Horticulture, University of Bristol, Long Ashton Research Station, Long Ashton Bristol BS 189AF England
Gorini, D. F.	Technologica Dei Prodotti Agricoli, Via G Venezian, 26 - 20133, Milano, Italy
Greeley, M.	The Institute of Developmental Studies, University of Sussex, Brighton, Sussex BN1 9RE, England
Grierson, D.	Department of Physiology & Environment Studies, School of Agriculture, Sutton Bonington, Loughborough, LE12 5RD, University of Nottingham, England
Haard, N.	Department of Biochemistry, Memorial University, St. Johns, Newfoundland, Canada
Halder-Doll, H.	University Hohenheim, Institute fur Obst-Gemuse and Weinbau, 370, Post Fach 700562 7000 Stuttgart, Federal Republic of Germany
Hartmann, C.	Laboratoire De Physiologie De La Maturation, Universite Orleans, 45017 Orleans Cedex France

Herregods, M.	Tiense Vest 136, B01300 Leuven, Belgium
Hoftun, H.	Department of Vegetable Crops, The Agriculture University of Norway, P.O.B.22 N1432 Aas-NLH, Norway
Hussein, A.	Department of Horticulture, College of Agriculture, Alexandria University, Shatbi Alexandria, Egypt
Kader, A.	Department of Pomology, University of California, Davis, CA 95616, U.S.A.
Kaminiaris, D.	34 Nikiforou Ouranou Street, Athens, Greece
Karacali, I.	Ege University, Bornova, Izmir, Turkey
Karaoulanis, G.	Food Technology Institute, Lykourissi Amaroussion, Athens, Greece
Krogh, P.	Department of Microbiology, Royal Dental College, Juliane Maries Vej 30, DK-2100, Copenhagen, Denmark
Kuru, A.	Department of General Botany, University of Istanbul, Suleymaniye, Istanbul, Turkey
Kyriakidis, I.	Agiou Meletiou 100, Patisia, Athens, Greece
Larsen, F.	Department of Horticulture, Rolighedsvej 23 The Royal Veterinary & Agricultural University, DK-1958 Copenhagen V, Denmark
LeNard, M. D.	Station D'Amelioration de la Pomme de terre et des Plantes a Bulbes BP 5, 29207 Landerneau Cedex, France
Lieberman, M.	Postharvest Physiology Laboratory, U. S. Department of Agriculture, BARC-W, Beltsville, Maryland, 20705, U.S.A.
Lurie, S.	Fruit & Vegetable Storage Division, The Volcani Institute, P.O.B. 6, Bet-Dagen 20-500, Israel
Marcellin, P.	Laboratoire de Physiologie, Des Organes Vegetaux, CNRS, 4 ter Route Des Gardes, 92190 Meudon, France
Mazliak, P.	Universite Pierre et Marie Curie (Paris VI), Physiologie Celluaire VER 59, 4 Place Jussieu, 75230 Paris Cedex 05, France
Meghir, H.	Pennwalt France, 1 Rue Freres Lumiers, 78370 Plaiser, France
Monning, A.	Institut für Obstbau und Gemüsebau, der University Bonn, 5300 Bonn, Federal Republic of Germany
Norris, Karl	Instrumentation Research Laboratory, U. S. Department of Agriculture, BARC-W, Beltsville, MD, 20705, U.S.A.
Papadatou, P.	Department of Pomology, Agricultural College of Athens, Votanikos 75, Athens 301, Greece

Pappelis, A.	Department of Botany, Southern Illinois University, Carbondale, Illinois 62901, U.S.A.
Passan, H. C.	Tropical Products Institute, 56-62 Gray's Inn Road, London WC2X8LV, England
Pech, J. C.	Ecole National Superiure Agronomique, 145, Avenue De Muret, 31076 Toulouse, Cedex, France
Petropoulos, A.	15 Ithakis, kypseli, Athens 801, Greece
Petropoulou, S.	Eratosthenous 31, Hiliodorou 2-4, Sq. Plastira Pograti, Athens, Greece
Pontikis, K. A.	Department of Pomology, Agricultural College of Athens, Votanikos 75, Athens 301, Greece
Pratt, H.	Department of Vegetable Crops, University of California, Davis, CA 95616, U.S.A.
Rhodes, M.J.C.	Food Research Institute, Coloney Lane, Norwich, NOR 26G, United Kingdom
Richardson, D. G.	Department of Horticulture, Oregon State University, Corvallis, Oregon, 97331, U.S.A.
Rokkas, D.	Argous 6, Nafplion Argolidas, Greece
Saltveit, M. E.	Department of Horticultural Science, School of Agriculture & Life Science, North Carolina State University, Box 5216, Raleigh, NC 27650, U.S.A.
Sapoutzaki, E.	Sevastoupolis 150-B, Erythros Stavros, Athens, Greece
Shou-chun, Q.	Shansi Fruit Institute, Taigu Shanxi Province, People's Republic of China
Simon, P. J.	Research Station of Gorsen, Brede Akker 3, B 3800 Sint-Truiden, Belgium
Skytt-Anderson, A.	Royal Veterinary & Agricultural University Dept. of Plant Physiology & Anatomy, Thorvaldsensvej 40 DK 1871, Copenhagen V, Denmark
Solomos, T.	Department of Horticulture, University of Maryland, College Park, Md. 20742, U.S.A.
Sovatzoglou, G.	Palados 9, Kato Petralona, Athens, Greece
Stoll, K.	Station Fédérale de Recherches, 8820 Waedenswid, Switzerland
Streif, J.	Verchusstation Fur Intensivkulturen, 7980 Ravensburg, Federal Republic of Germany
Tobback, P.	Laboratory of Food Preservation, Catholic University of Leuven, De Croylaan 42, B 303, Heverlee, Belgium
Vlitos, A. J.	Tate and Lyle Research Laboratory, P. O. Box 68, Reading RG 5 2BX, United Kingdom
Weichmann, J.	Lehrstuhl Für Gemüsebau, Technische Universität München, 8050 Freising, Weihenstephan, Federal Republic of Germany

Williams, P.	Grain Research Laboratory, Canadian Grain Commission, 303 Main Street, Winnipeg, Manitoba, Canada REC 3G9
Woolhouse, H.	Director, John Innes Institute, Colney Lane Norwich NR4 7UH, England
Xirouhakis, S.	Department of Pomology, Agricultural College of Athens, Votanikos 75, Athens 301, Greece

INDEX

GPSR Compliance

The European Union's (EU) General Product Safety Regulation (GPSR) is a set of rules that requires consumer products to be safe and our obligations to ensure this.

If you have any concerns about our products, you can contact us on ProductSafety@springernature.com

In case Publisher is established outside the EU, the EU authorized representative is:

Springer Nature Customer Service Center GmbH
Europaplatz 3
69115 Heidelberg, Germany

Batch number: 10762209

Printed by Printforce, the Netherlands